DIGITAL S

Principles and

DIGITAL SYSTEMS

Principles and Design

Raj Kamal
Senior Professor
School of Computer Sciences
Institute of Computer Sciences and Electronics
Devi Ahilya University
Indore

PEARSON

ISBN 978-81-775-8570-4

First Impression, 2007

Published by Pearson India Education Services Pvt.Ltd,CIN:U72200TN2005PTC057128.
Formerly known as TutorVista Global Pvt Ltd, licensees of Pearson Education in South Asia

Head Office: 7th Floor, knowledge Boulevard, A-8(A) Sector-62, Noida (U.P) 201309, India

Registered Office: Module G4, Ground Floor, Elnet Software City, TS -140,Block 2 & 9
Rajiv Gandhi Salai, Taramani, Chennai, Tamil Nadu 600113.,Fax: 080-30461003,
Phone: 080-30461060, www.pearson.co.in email id: companysecretary.india@pearson.com

Digitally Printed in India by Repro India Ltd. in the year of 2015.

Dedicated to my students at Devi Ahilya University, Indore,
and at Arulmigu Kalasalingam College of Engineering, Krishnankoil

CONTENTS

PREFACE

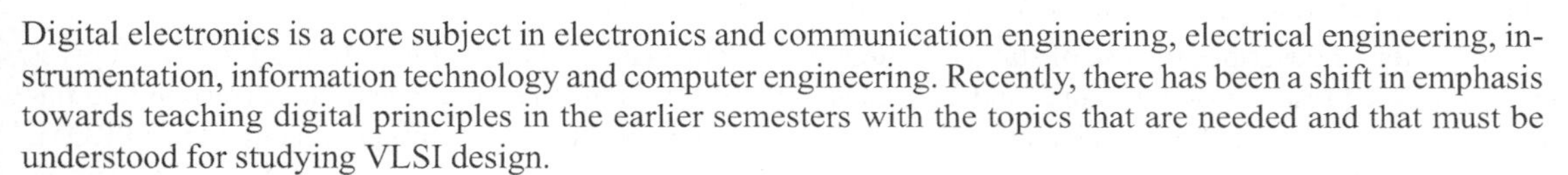

Digital electronics is a core subject in electronics and communication engineering, electrical engineering, instrumentation, information technology and computer engineering. Recently, there has been a shift in emphasis towards teaching digital principles in the earlier semesters with the topics that are needed and that must be understood for studying VLSI design.

An engineering student takes a course in digital principles at a very early stage in his or her academic life. In addition, he or she may not be from an English-medium school, and therefore, may face great difficulties in comprehending the subject and in perfecting their problem-solving skills. This present book is targeted at such students.

This book fulfils the requirements of students by detailing the fundamental building blocks of digital design with a large number of examples and prepares the students to study VLSI circuit design in their eighth or ninth semesters. Each chapter focuses on a single aspect and highlights points that the reader must remember. The solved examples and practice exercises provided in each chapter will enable a reader to master the principles of digital circuits and design easily.

The organization of the book is as follows:

Chapter 1 the concepts of logic 1 and 0 and the positive and negative logic assignments to the circuit conditions. It explains the standard ways by which the 1's and 0's describe the circuit conditions. It provides a comparison of digital versus analog circuits. It also offers many exemplary circuits. A reader will thus easily comprehend the representation of a circuit input or output condition in terms of a 1 or a 0.

Chapter 2 explains the decimal, octal, binary, hexadecimal, and BCD number systems. It shows how to convert a number from one system to another. It provides many examples and practice exercises that help the reader to comprehend the various number systems and the conversion of a number from one system to another.

Chapter 3 covers the core concepts of binary arithmetic and how to use the unsigned, signed, and two's complement numbers. It explains addition, subtraction, multiplication and division of binary numbers using numerical examples.

Chapter 4 explains Boolean expressions and laws, and how to use these to simplify a logic circuit. It explains the steps that are required for obtaining minterms or maxterms in a two or three or four variables' Boolean expression for obtaining the SOP and POS formats, respectively.

Chapter 5 has an innovative presentation of Karnaugh maps and their importance in simplifying a logic circuit for implementing a Boolean expression originally in an SOP or POS format. It describes how to find the prime implicants (minterms or maxterms). It includes advanced topics such as—multi-output minimization, Quine-McCluskey and hypercube-representation-based computer minimization techniques.

Chapter 6 describes the different types and families of logic gates—RTL, DTL, TTL, I2L, ECL, HTL, NMOS and CMOS and the circuits for the NOTs, NANDs or NORs. The provided examples clarify the parameters of these gates; for example, fan-outs, propagation delays, voltage levels and current levels.

Chapter 7 describes interfacing of logic circuits between same or different families and types. In-depth explanation of the interfacing concept is provided through examples.

Chapter 8 covers the three different types of the logic gates—open collector, open drain and tri-state.

Chapter 9 explains problem formulation for the combinational circuit of the gates and implementation of the design. It clarifies the meaning of a combination circuit in contrast to sequential circuits.

Chapter 10 explains the designs of half adder, full adder and subtractor. It then explains the important circuits—the decoder, encoder, multiplexer and demultiplexer, which are the building blocks of many combinational circuits. Examples and practice exercises help the students improve their problem solving skills in *combinational circuit design*.

Chapter 11 explains the design of code converters, comparators for finding equality, greater-than and less-than, parity generators and multi bit AND, OR, XOR, and NOT operation circuits.

Chapter 12 describes the use of standard integrated circuits (ICs) and use of ROMs and PROMs. It gives an innovative presentation by figures and examples of the PROMs. Each example and the corresponding figure shows which are the fused and intact OR links that implement a particular combination circuit in a PROM.

Chapter 13 describes the use of PAL, PROM and PLAs. An innovative presentation of the PAL, PROM and PLA is presented through figures and examples. The fused and intact AND or AND-OR links in a PAL or PLA are explained clearly to the reader.

Chapter 14 initiates the description of sequential circuits. It describes the SR, D, JK, and master slave sequential circuits through circuits, timing diagrams and state tables. It gives the characteristic equations for these. A reader will learn by examples how to draw the timing diagrams for the change of states in these flip-flops and latches.

Chapter 15 gives the advanced digital principles—Mealy machines and Moore machine sequential circuits, state minimization, drawing the excitation, transition and state tables and showing the state diagrams. Sequential circuits are used in many applications. Must-learn concepts are described in a student-friendly manner. The large number of tables and diagrams in the examples will enable the reader to hone his or her state minimization problem solving skill.

Chapter 16 describes the practical sequential-circuit building blocks—registers and counters. The common designs, registers and counters are explained.

Chapter 17 describes the RAM memory and the interfacing of several memory chips using the decoders. A systematic approach that is helpful when designing a decoding circuit is described. A reader will learn to design the decoding circuit with multiple types of memory chips.

Chapter 18 features an advanced topic—asynchronous sequential circuits and their analysis in fundamental mode. It explains the races and race free assignments. A reader will improve his or her asynchronous race free design skills by learning from the tables, diagrams and examples given in the chapter.

Chapter 19 covers the hazards present in the logic circuits—static, dynamic and essential. It explains how to identify these and then obtain a hazard-free design.

Chapter 20 describes ADC, DAC, analog switches, analog MUXs and deMUXs. These circuits enable interfacing the digital world with analog skills.

Chapter 21 gives an overview of the advanced programmable logic devices and the FPGAs that are programmed for specific applications using the computer.

Every effort has been made to provide precise information and correct solutions for the examples in the book. Despite this, errors may be still present. The author will be grateful to the readers for pointing these out to him.

The author will be grateful for any suggestions. These can be sent to professor@rajkamal.org. Students, please note that the PowerPoint slides and solutions to the exercises will be available on the Web soon. Queries are heartily welcome and can be sent through a query-sending link at the author's Web site http://www.rajkamal.org.

Raj Kamal

ACKNOWLEDGEMENTS

I am grateful to Dr M. S. Sodha, a renowned teacher, and Kalvivallal T. Kalasalingam, Chairman of Arulmigu Kalasalingam College of Engineering, Krishnankoil.

I am grateful to our Vice Chancellor Mr C. S. Chadha for his full support and cooperation. I am also thankful to my colleagues, particularly the young faculty of Computer and Electronics and the laboratory colleagues at the university.

I would like to acknowledge my gratitude to Dr Chelliah Thangaraj, Dr S. Radhakrishnan and Prof. G. Sudhakar of Arulmigu Kalasalingam College of Engineering, Krishnankoil.

I am especially thankful to Ms. S. Alagu and Ms. Suganthi Lakshmanan for thoroughly reading the manuscript and Mr. Annathurai for making the AutoCAD drawings, and Mr. S. Murugan and Mr. Manickandan for assistance.

Finally, I would like to thank my family members—Ms. Sushil Mittal, Shalin Mittal, Needhi Mittal, Dr Shilpi Kondaskar, Dr Atul Kondaskar and baby Arushi Kondaskar—for their full cooperation and for encouraging me to write this book.

CHAPTER 1

Basic Digital Concepts

OBJECTIVE

In this chapter, we shall learn the concept of a digital circuit and its logic states, '1' and '0'. We shall also understand the differences between analog and digital circuits.

BASIC CONCEPTS

1.1 CONCEPTS OF '1's, '0's

An electronic circuit in which a state switches (changes) between two distinct states when there is a change in the input states or conditions is called a digital circuit. A state of a circuit means either a distinct region of output or input voltages or currents or frequencies or phases or conditions of a circuit. One of the regions is represented by a logic 'true' or 1 or 'high' or 'yes' state and the other region is represented by a logic 'false' or 0 or 'low' or 'no' state.

1.1.1 Positive Logic

Let us assume, a logic state 1 is represented by a voltage, V between 5 V and 2.8 V and a state 0 by V between 0.8 V and 0 V. (This is the case of *positive logic*, when the lower voltage represents 'low' or 0 state and the higher voltage represents 'high' or 1 state).

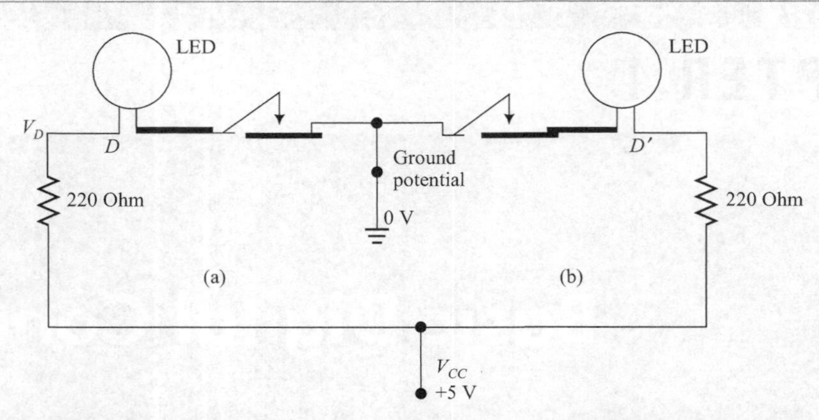

FIGURE 1.1 (a) A digital circuit consisting of a simple switch. (b) Another digital circuit consisting of a simple switch.

(A) Consider a simple circuit given in Figure 1.1(a). When the switch is pressed, the voltage at D is +5 V and when released, it is ~0 V. Logic state is 'true' or 1 when the switch is pressed and 'false' or 0 when released.

(B) Now consider the circuit given in Figure 1.1(b). When the switch is pressed, the voltage at D' is +0 V and when released, it is ~5 V. Logic state is 'true' or 1 when switch is released and 'false' or 0 when pressed.

1.1.2 Negative Logic

Now let us assume, a logic state 1 represented by a voltage, V between 0.8 V and 0 V and a state 0 by V between 5 V and 2.8 V. (This is called the case of '*negative logic*', when a lower voltage represents 'high' or 1 state and the higher voltage represents 'low' or 0 state.)

(A) Consider the circuit given in Figure 1.1(a). When the switch is pressed, the voltage at D is +5 V and when released, it is ~0 V. Logic state is 0 when the switch is pressed and when released 1.

(B) Now consider, the circuit given in Figure 1.1(b). When the switch is pressed, the voltage at D' is ~ 0 V and when released, it is ~5 V. Logic state is 0 when the switch is released and 1 when pressed.

1.1.3 Popular Representations of the Digital Circuits

Table 1.1 shows two states, 1 and 0 in a few popular digital circuits.

1.2 ANALOG VS. DIGITAL CIRCUITS

Table 1.2 gives the advantages and disadvantages of a digital circuit vis-a-vis an analog circuit.

TABLE 1.1 1 and 0 state in a few popular digital circuits

Name of the corresponding representative digital circuit	Electronic state when logic state is called 1 in a digital circuit	Electronic state when logic state is called 0 in a digital circuit
TTL*	High +ve voltage (2.8 V to 5 V) between output and ground and low current ~μA sourcing from a circuit due to a 'OFF' output-stage transistor in a non-conducting state.	Low ~0 voltage (0 V to 0.8 V) between output (or input) and ground and high current ~mA sinking into a circuit due to the 'ON' output-stage transistor in a conducting state.
High Speed CMOS	High +ve voltage (+5 V ± to 3.3 V) between output (or input) and circuit supply ground.	Low + ve voltage (+1.6 V to +0 V) between output (or input) and circuit supply ground.
RS232C	High –ve voltage (–3 V to –25 V) between output (or input) and ground and 'low current' through a circuit.	High +ve voltage (+3 V to +25 V) between output (or input) and ground and 'low current' through a circuit.
Modem	Higher ~1240 Hz frequency at the output or input circuit.	Lower ~1040 Hz frequency at the output or input circuit.
Teletype loop	Higher ~16 to 20 mA current from the output to an input circuit.	Lower ~0 to 4 mA current from the output to an input circuit.

*Worst TTL case is 2.4 V ± 0.4 V for logic 1 and 0.4 V ± 0.4 V for logic 0. Note: Different families of logic circuits have different noise immunity than 0.4 V. The level of voltage changes that will not affect the input or output logic state is defined as the noise immunity.

TABLE 1.2 Advantages and disadvantages of a digital circuit vis-a-vis an analog circuit

Digital circuit		Analog circuit	
Advantages	Disadvantages	Advantages	Disadvantages
1. More close to logic and number systems and a logic state has a definiteness and thus the circuit inputs and outputs have preciseness and reliability.	1. Physical values are reflected by a complex digital to analog conversion circuit.	1. More close to physical system and physical values.	1. Logical values are reflected by a complex analog to digital conversion circuit.
2. Capabilities of logical decisions and arithmetic, logic and Boolean operations.	2. The circuits have high complexities. To represent a big decimal number, a large number of components are needed.	2. Capabilities of representing directly a physical value. For example, the value of temperature, wind, speed, etc can be represented by a corresponding voltage or current or frequency or phase.	2. A circuit shows deviation with time due to circuit's temperature changes or physical condition changes.

Table 1.2 Contd.

3. Noise in voltage or current does not matter because an output or input is measured in terms of its logic state(s), not in terms of a value of the voltage or current or frequency or phase.	3. Slower speed due to greater number of components needed to represent a state. For example, a bipolar junction transistor as inverter will act faster than a circuit in a TTL inverter.	3. Noise in voltage or current does matter because an output or input is measured in terms of a value of voltage or current or frequency or phase.	3. Faster speed due to lesser number of components needed to represent a physical value.

■ EXAMPLES

Example 1.1

Using the circuit in Figure 1.1, fill the Table 1.3 given below: (Write logic state 1 when $V_D > 2.8$ V and less than 5 V and 0 when $V < 0.8$ V. Write * when indeterminate).

TABLE 1.3

V_D	Logic state at *D*	V_D	Logic state at *D*
5 V	1	1 V	*
4 V	1	0.75 V	0
3.5 V	1	0.5 V	0
2 V	*	~0 V	0

Example 1.2

Consider a circuit in Figure 1.2. The circuit consists of two switches, SL and SU. One is at the ground floor and the other at the first floor. When the lamp is ON let us assume logic state = True and when the lamp is OFF let us assume logic state = False. Find the logic state in different conditions of the switches and show these in Table 1.4.

TABLE 1.4

SL	SU	Lamp	Logic state of the lamp
ON	ON	ON	1
OFF	OFF	OFF	0
ON	OFF	OFF	0
OFF	ON	OFF	0

Example 1.3

Consider the circuit in Figure 1.3. It consists of diodes, fused diodes (unconnected) and the resistances in the digital circuit outputs. It shows four inputs, Y_0, Y_1, Y_2 and Y_3. What will the inputs at D_0, D_1 when the output $Y_0 = 1$? Let 1 mean any voltage output between 2.8 V and 5 V and 0 mean any voltage between 0 and 0.8 V. Let 1 mean any input voltage between 2.4 V and 5 V. (Note: Assume worst-case 1 to be 2.4 V $\pm$ 0.4 V and let worst-case 0 be

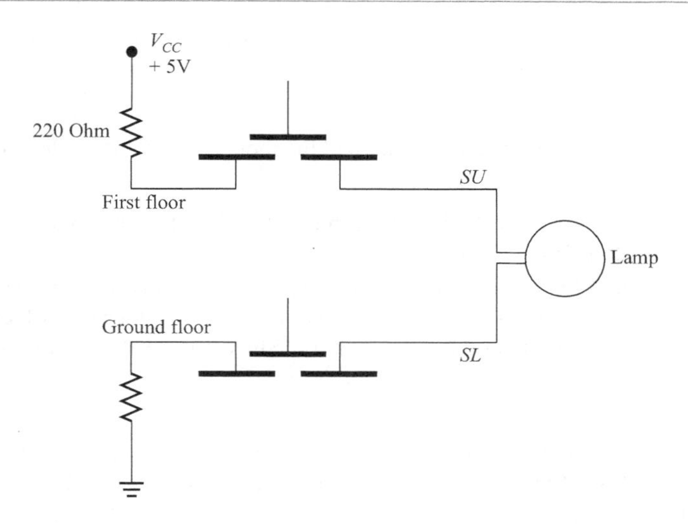

FIGURE 1.2 A circuit consisting of two switches *SL* and *SU* at the ground floor and first floor. When the lamp is on let us assume logic state = 1 and when the lamp is on let us assume logic state = 1.

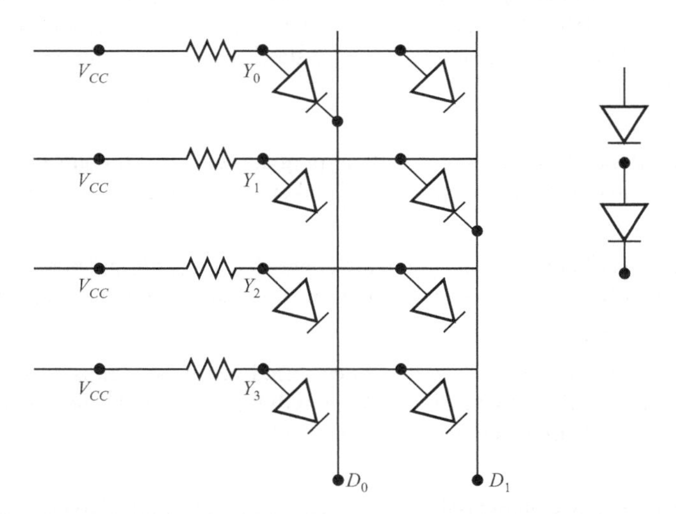

FIGURE 1.3 A digital circuit with four inputs, Y_0, Y_1, Y_2 and Y_3 and two outputs, *D*0 and *D*1. Here, 1 means output voltage between 2.8 V and 5 V and 0 mean any output voltage between 0 and 0.8 V and 1 also means any input voltage between 2.4 V and 5 V. Supply operational voltage = ~ 5.0 V.

0.4 V $\pm$ 0.4 V. Assume except one all other three inputs are unconnected. Supply operational voltage, V_{CC} = 5.0 $\pm$ 0.25 V = 5.0 $\pm$ 0.25 V). A dot represents a junction.

Solution

(i) When *Y*0 = 1, the diode connected to *D*0 will be forward biased above the diode threshold voltage. Assume threshold voltage = 0.35 V, then voltage at *D*0 will be just 0.35 V less than the voltage at *Y*0. Since *Y*0 = 1 (> 2.8 V) therefore input at *D*0 = 1 (> 2.4 V).

(ii) When $Y0 = 1$, because the diode is not connected to $D1$, the voltage at $D1$ will be ~0 V, therefore $D1 = 0$ (< 0.8 V).

Example 1.4

Consider the circuit in Figure 1.3. What will the inputs $D0$, $D1$ when the output $Y1 = 1$?

Solution

(i) When the $Y1 = 1$, the diode is not connected to $D0$, then the voltage at $D0$ will be ~0 V, therefore $D0 = 0$ (< 0.8 V).

(i) When the $Y1 = 1$, the diode connected to $D1$ will be forward biased above the diode threshold voltage. Assume threshold voltage = 0.35 V, then the voltage at $D1$ will be just 0.35 V less than the voltage at $Y1$. Since $Y1 = 1$ (> 2.8 V) therefore $D1 = 1$ (>2.8 V).

Example 1.5

Consider the circuit in Figure 1.3. What will the input $D0$ when the $Y1 = 1$ but the resistance $R0$ from $D0$ is fused (unconnected) or is of very high value (>> 1 MΩ)? (Note: Ω is a symbol for Ohm).

Solution

When the $R > 1$ MΩ the input, the $V_{CC} = 5$ V is not connected to $D0$, then the voltage at $D0$ and $Y1$ cannot be defined because there is no connection to the potential source through R. $D0$ is at the indeterminate (indefinable) logic state. A state can only be defined if R is not disconnected.

Example 1.6

Consider third the thick pin in an electrical plug at the mains. It connects to the electrical earth connection of the laboratory. A wire is taken from it. Can it be called at logic state 0?

Solution

No, because it is a hanging connection, not connected to the ground potential and logic circuit power supply. Existance of merely the 0 V, does not mean logic state = 0.

Example 1.7

The circuit in Figure 1.3 consists of diodes, fused diodes and resistances. It shows four inputs, $Y0$, $Y1$, $Y2$ and $Y3$ to the diodes. What will the states at $D0$ and $D1$ when the supply voltage to the logic circuit suddenly falls from 5 V to 2.5 V?

Solution

Between 2.8 V and 5 V output only, state is 1 as per the definition given for the digital circuit in Figure 1.3. Since supply voltage is now 2.5 V, the input is 1 from the worst-case consideration ($V > 2.4$ V). Circuit will function improperly as a digital circuit whenever the supply voltage to the logic circuit falls such that an input $Y0$ or $Y1$ or $Y2$ or $Y3$ is less than 2.4 V.

Example 1.8

Assume that the circuit operational voltage $V_{DD} = 5.0 \pm 0.25$ V. Let us assume that in a logic circuit, 1 means any voltage between $(2/3)V_{DD}$ and V_{DD}, and 0 mean any voltage between $(1/3)V_{DD}$ and V_{SS}. Assume that the circuit is same as in Figure 1.3. What will $D0$ and $D1$

when the output $Y0 = 1$ (>3.7 V)? Assume other outputs unconnected. (Assume voltages are with respect to V_{SS} and V_{SS} connects to supply ground potential).

Solution

(i) When $Y0 = 1$, the diode connected to $D0$ will be forward biased above the diode threshold voltage. Assume threshold voltage = 0.35 V, then the voltage at $D0$ will be just 0.35 V less than the voltage at $Y0$. Since $Y0 = 1$ (> 3.7 V) therefore the input $D0 = 1$ (> 3.3 V). Note: When the $Y0 = 1$ such that < 3.7 V but > 3.3 V, it will function improperly as a digital circuit because voltage drop across the diode will make the input $D0$ below 3.33 V and thus less than 2/3 V_{DD}.

(ii) When $Y0 = 1$, the diode is not connected to $D1$, then voltage at $D1$ will be ~0 V, therefore $D1 = 0$ (< 1.67 V).

Example 1.9

Let us assume that in a logic circuit, state 1 defines by a frequency 2250 Hz and 0 by frequency 2050 Hz. Assume that only $Y0$ is connected and $Y1$, $Y2$ and $Y3$ are not connected. Assume that the circuit is same as in Figure 1.3. What will the $D0$ and $D1$ when the $Y0 = 1$ (2250 Hz)?

Solution

(i) When $Y0 = 1$ (means R connected to a 2250 Hz 1 V source in place of V_{CC}), the diode connected to $D0$ will be forward biased when the input is above the diode threshold voltage. Frequency at $D0$ will therefore also be 2250 Hz. Since input $Y0 =$ 1 (2250 Hz) therefore output $D0 = 1$ (2250 Hz).

(ii) When $Y0 = 1$, the diode is not connected to $D1$, then voltage at $D1$ will be indeterminate as no signal of either 2250 Hz or 2050 Hz is appearing at $D1$.

The circuit will function improperly and is not a digital circuit as output $D1$ is indefinable. It just behaves as an analog circuit.

Example 1.10

Circuit in Figure 1.3 consists of diodes, fused diodes and resistances. It shows four inputs, $Y0$, $Y1$, $Y2$ and $Y3$. What will the $D0$, $D1$ when the $Y0 = 1$? Assume a digital circuit with a negative logic in which 1 means any voltage input between 0 and 0.4 V and 0 mean any voltage output between 2.8 V and 5 V. Assume other outputs unconnected. Supply operational voltage $V_{CC} = 5.0 \pm 0.25$.

Solution

(i) When $Y0 = 1$ (< 0.4 V), the diode connected to $D0$ will not be forward biased as the diode is at V below threshold voltage. Since $Y0 = 1$ (< 0.4 V) therefore input $D0 = 1$ (< 0.8 V).

(ii) When $Y0 = 1$(< 0.8 V), because the diode is not connected to $D1$, the voltage at $D1$ will be ~ 0 V, therefore $D1 = 1$ (< 0.8 V).

The logic states at $D0$ and $D1$ are (1, 1) in case of negative logic and (0, 1) in case of positive logic.

EXERCISES

1. Using the circuit in Figure 1.1, fill the Table 1.5 given below: [Assume negative logic circuit and write logic state 1 when $V < 0.8$ V and 0 when $V > 2.8$ V to supply voltage. Write * when indeterminate.]

TABLE 1.5

V_D	Logic state at D	V_D	Logic state at D
5 V		1 V	
4 V		0.75 V	
3.5 V		0.5 V	
2 V		~0 V	

2. Redraw the circuit in Figure 1.2 so that outputs are as per Table1.6 below:

TABLE 1.6

SL	SU	Lamp	Logic state of the Lamp
ON	ON	ON	1
OFF	OFF	OFF	0
ON	OFF	ON	1
OFF	ON	ON	1

3. Consider the circuit in Figure 1.3. What will the states at $D0$, $D1$ when the input $Y2 = 1$?
4. Using a pencil draw in a row horizontally the seven small-unfilled circles, like the one in character small *o*. Repeat this nine times vertically such that circles are in nine rows and seven columns. Now erase the circles in each row such that English character A becomes visible. Write the sequence of 0s and 1s in all the nine rows. Assume that a circle means 1 and no circle means 0. (Hint: The first row will have only one circle in the center filled for the character A. Hence the first row sequence answer will be 0001000).
5. Using the circuit in Figure 1.3, fill the Table 1.7 given below: (Use negative logic).

TABLE 1.7

Y	$D1$	$D0$
$Y0 = 1$		
$Y1 = 1$		
$Y2 = 1$		
$Y3 = 1$		

6. Let us assume that for the logic states in a circuit,
   (i) 1 defines by an output signal voltage 5 V and
   (ii) 0 defines by the output signal voltage 0.4 V.

   Assume output logic state changes every ms in the in sequence 1, 1, 0, 1, 1, 0, 0 and 0. Show the waveform within the period 0 to 8 ms.
7. Let us assume that in the logic signals from a circuit,
   (iii) 1 means the output signal is of voltage –12 V and
   (iv) 0 means the output signal is of voltage +12 V.

   Assume output logic changes every ms in the in sequence 1, 1, 0, 1, 1, 0, 0 and 0. Show the waveform within the period 0 to 8 ms.
8. Let us assume that in the logic signals from a modem,
   (v) 1 means the output signal is of frequency 2 kHz and
   (vi) 0 means the output signal is of frequency 1 kHz.

   Assume output signal is a sine wave. Show the waveform within the period 0 to 8 ms when digital bits are in sequence 1, 1, 0, 1, 1, 0, 0 and 0.
9. Let us assume that in the logic signals from a modem,
   (vii) 1 means phase of the output signal is +90° and
   (viii) 0 means phase of the output signal is –90°.

   Assume that the output signal is a sine wave of frequency = 1 kHz. When the logic state is 1 between 0 to 1 ms, 2 to 3 ms and 4 to 5 ms, and logic state is 0 between 1 to 2 ms, 3 to 4 ms and 5 to 6 ms, show the output waveform between 0 to 6 ms on a graph paper.
10. Let us assume that in the 1 kHz signals from a modem,
    (i) when the logic pair of bits = 0 and 1 means the phase of the output signal is 0°,
    (ii) when the logic pair of bits = 0 and 0 means the phase of the output signal is +90°,
    (iii) when the logic pair of bits = 1 and 0 means the phase of the output signal is +180°, and
    (iv) when the logic pair of bits = 1 and 1 means the phase of the output signal is –90°.

    Show the waveform within the period 0 to 4 ms when the digital bits are in the sequence 1, 1, 0, 1, 1, 0, 0 and 0.

QUESTIONS

1. Define a digital circuit.
2. Define analog circuit.
3. What does 1 means in a logic circuit of TTL family? (Hint: refer Table 1.1).
4. What does 0 mean in a logic circuit of TTL family? (Hint: refer Table 1.1).

5. What does 1 mean in a negative logic circuit operating at 5 V?
6. What does 1 mean in a RS232C logic circuit? (Hint: refer Table 1.1).
7. What does 1 mean in a logic circuit of a modem operating at 1040 Hz and 1240 Hz? (Hint: refer Table 1.1).
8. Draw a +5 V circuit in which when the switch is pressed, output is 1 and when release a the output is 0.
9. Draw a +5 V negative logic circuit in which when the switch is pressed, output is 1 and when release the output is 0.
10. What does 0 means in a logic circuit of a modem operating at 2050 Hz and 2250 Hz? (Hint: refer Example 1.9).

CHAPTER 2

Number Systems

OBJECTIVE

In this chapter, we shall learn the concept of various number systems: binary, octal, decimal, hexadecimal, and conversion from one to another.

BASIC CONCEPTS

2.1 DECIMAL NUMBER

A decimal number is a number based upon ten values: 0, 1, 2, 3, 4, 5, 6, 7, 8 and 9. Suppose there are twelve apples. How can we represent twelve? One is that take one for each apple and write 111111111111. Easier way is write 2 at the right most position and 1 at the left of 2 and assume that *weight (w) of the left side digit is always ten times* ($w_d = 10$) compared to the neighboring one on the right side. Now, the number 12 is twelve represented by only two digits, 12 one digit is at right position ($p = 0$) and other is at the left position ($p = 1$). We get a formula as under.

Value, $$12 = 1 \times w_d^1 + 2 \times w_d^0 \qquad ...(2.1)$$

Here $p = 0$ and 1 counting from right to left and basic weight $w_d = 10$. [A basic weight, w is also called *base*. It is the ratio of the weights of two successive places, (left side weight/ right side weight) in a pair of digits].

Now, suppose there are hundred apples. It again becomes difficult to represent by one 1 continuing hundred times from rightmost to leftmost place. However, if we use the assumption that the weight of the left side digit is always ten times compared to the one on the right side of it, which means we assume that base = 10, then 100 represents the hundred apples. We get the similar formula as above. Here $p = 2$, 1 and 0, at leftmost, middle and right most places, respectively.

Total number, $100 = 1 \times w_d^p + 0 \times w_d^p + 0 \times w_d^p$ ...(2.2)

where $w_d = 10$.

If you put by mistake an addition 0 on the right, the number of apples will become ten times more.

Total number, $1000 = 1 \times w_d^p + 0 \times w_d^p + 0 \times w_d^p + 0 \times w_d^p$ ...(2.3)

where a place $p = 3, 2, 1$ and 0 from leftmost to rightmost and $w_d = 10$.

2.2 BINARY NUMBER

A binary number is a number based upon only two numbers (0 and 1). [Base $w_d = 2$]. It is unlike decimal number based upon ten numbers, 0, 1, 2, 3, 4, 5, 6, 7, 8 and 9. Advantage of two-number system is its analogy with a digital circuit, which is based upon the two distinct states (1 or 0). **A two-number system is a powerful way of representation.** It will be clear from the following story:

A king had a smart minister. Once the king asked the minister that he wanted to gift him something and he may ask it. If I agree to your request than I will definitely give it to you. Minister politely said, Sir, it will be impossible for you to give the gift in spite of your easily agreeing for it, so please withdraw your gift offer. King said, I don't withdraw an offer once made.

Minister politely said, Sir, please take a chessboard with 64 squares on it. Just give one grain of rice for the first square and just double the number of grains for each succeeding square. King said no problem, he agreed to give the gift. Minister asked the king to think again. King said no need for re-thinking. King started ordering the grains. Rice in the whole kingdom were exhausted but still the higher number of squares could not be filled. Practically the minister asked the number of grains by the following formula:

Number of grains $= 1 \times w_0 + 1 \times w_a^p + 1 \times w_a^p + \cdots + 1 \times w_b^p + 1 \times w_b^p + 1 \times w_b^p$ ...(2.4)

where $p = 63, 62, 61, \ldots, 2, 1$ and 0 from left to right. There are total 64 values for p for 64 digits due to 64 squares on the chessboard. Also, base $w_b = 2$ as in binary system because minister asked only two times the grain at the next place of the square.

A Golden Rule to Remember — Sum of the Weights

A general formula to get a number, N is as under:

$$N = \text{sum of the } y_p.(\text{Base})^p$$
$$= y_{p\max-1} \cdot w_0^{p\,max-1} + y_{p\max-2} \cdot w_0^{p\,max-2}$$
$$+ y_{p\max-3} \cdot w_0^{p\,max-3} + \ldots + y_2 \cdot w_0^2 + y_1 \cdot w_0^1 + y_0 \cdot w_0^0 \quad \ldots(2.5)$$

where p = place of the digit = 0, 1, 2, ..., $(pmax - 3)$, $(pmax - 2)$ or $(pmax - 1)$, from right to left, and $p$max is maximum the number of places used in the representation and base (basic weight). w_0 = 10 for decimal numbers, 8 for octal numbers, 2 for binary numbers and 16 for hexadecimal numbers. The $y_0, y_1, y_2, ...$ are the digits. The w_0^p is weight at the place p. In case of decimal numbers, each digit = 0 or 1 or 3 or ... or $(w_0 - 1)$ at the right-most, left first from that, left second from that and so on till left $(p\text{max} - 1)^{th}$ position.

2.3 USE OF BINARY NUMBERS TO REPRESENT A NUMBER

A binary number is a number based upon two values, 0 and 1. A digital circuit is extremely suitable to deal with binary numbers. Just as a number has definiteness (preciseness), a state of a digital circuit has definiteness. As a digital circuit is based upon the two distinct states, 1 and 0, the binary numbers (called bits) are used for a number.

Let us assume there are 16 apples. In a binary system, 0 will represent first apple, the second by 1, the third by 10, the fourth by 11 and so on. The last apple will be represented by 1111. Table 2.1 shows the numbers in a binary system starting from 0, that represents first, second third, fourth, fourteenth, fifteenth and sixteenth apple starting from 0. Why does 1111 represents sixteenth apple in binary number system? [Binary means a number with base 2. Use formula, binary number value = $2^0 . 1 + 2^1 . 1 + 2^2 . 1 + 2^3 . 1$. This equals 15. If we start counting from 0, sixteenth apple is actually 15th. By convention, we may optionally write b as the subscript after a binary number as shown in last row of the table. This is to distinguish it from the decimal numbers.

TABLE 2.1

Decimal Value	Four bit binary representation	Eight bit binary representation	Sixteen bit binary representation
0	0000_b	$0000\ 0000_b$	$00000000\ 0000\ 0000_b$
1	0001_b	$0000\ 0001_b$	$00000000\ 0000\ 0001_b$
2	0010_b	$0000\ 0010_b$	$00000000\ 0000\ 0010_b$
3	0011_b	$0000\ 0011_b$	$00000000\ 0000\ 0011_b$
.	.	.	.
.	.	.	.
.	.	.	.
13	1101_b	$0000\ 1101_b$	$00000000\ 0000\ 1101_b$
14	1110_b	$0000\ 1110_b$	$00000000\ 0000\ 1110_b$
15	1111_b	$0000\ 1111_b$	$00000000\ 0000\ 1111_b$

Points to Remember

(1) A binary number is chosen as a four-bit number in case its decimal value lies between 0 and 15. [It means between 0 and $(2)^4 - 1$].

(2) A binary number is chosen as an eight-bit number in case its decimal value lies between 0 and 255. [It means between 0 and $(2)^8 - 1$].

(3) A binary number is chosen as a sixteen-bit number in case its decimal value lies between 0 and 65535. [It means between 0 and $(2)^{16} - 1$].

(4) A binary number is chosen as a thirty-two bit number in case its decimal value lies between 0 and (65536 × 65536 – 1). [It means 0 and $(2)^{32} - 1$].

(5) A binary number is chosen as a sixty-four bit number in case its decimal value lies between 0 and (65536 × 65536 × 65536 × 65536 – 1). [It means between 0 and $(2)^{32} - 1$].

(6) A general formula to get the total. N is again as under with only w_0 changed

$$N = y_{p\max-1} \cdot w_0^{p\max-1} + y_{p\max-2} \cdot w_0^{p\max-2} + y_{p\max-3} \cdot w_0^{p\max-3} + ... + y_2 \cdot w_0^2 + y_1 \cdot w_0 + y_0 \cdot w_0^0 \quad ...(2.6)$$

where $p = p\max - 1, p\max - 2, p\max - 3, ..., 2, 1$ and 0, and $p\max$ is maximum the number of places used in the representation and $w_0 = 2$. The y_0, y_1, y_2, ... are the digit = 0 or 1 at the right-most, left first from that, left second from that and so on.

2.4 USE OF OCTAL NUMBERS TO REPRESENT A NUMBER

An octal number is a number based upon eight values, 0, 1, 2, 3, 4, 5, 6, and 7.

Let us assume there are 16 bananas. In the octal system, 0 will represent first banana, second banana by 1, third by 2, fourth by 3 and so on. 17 will represent the last banana. Table 2.2 shows the numbers in a octal system starting from 0, that represents first, second third, fourth, fourteenth, fifteenth and sixteenth banana starting from 1. Why does 17_8 represent the sixteenth banana in octal number system? (Octal means a number with base 8. Use formula, octal number value = $8^0 . 7 + 8^1 . 1$. This equals 15. If we start counting from 0, sixteenth banana is actually the 15th). By convention we write 8 as the subscript after an octal number. This is to distinguish it from a decimal number.

TABLE 2.2

Decimal value	Two digit octal representation	Four digit octal representation	Eight digit octal representation
0	00_8	0000_8	$0000\ 0000_8$
1	01_8	0001_8	$0000\ 0001_8$
2	02_8	0002_8	$0000\ 0002_8$
3	03_8	0003_8	$0000\ 0003_8$
.	.	.	.
.	.	.	.
.	.	.	.
13	15_8	0015_8	$0000\ 0015_8$
14	16_8	0016_8	$0000\ 0016_8$
15	17_8	0017_8	$0000\ 0017_8$

Points to Remember

(1) An octal number is chosen as a four-digit number in case its decimal value lies between 0 and 4095_a. (0 and $(8)^4 - 1$).

(2) An octal number is chosen as a two-digit number in case its decimal value lies between 0 and 63. (0 and $(8)^2 - 1$).

(3) A general formula to get the total N is as under with only w_0 changed:

$$N = y_{p\max-1} x\, w_8^{p\max-1} + y_{p\max-2} x\, w_8^{p\max-2} + y_{p\max-3} x\, w_8^{p\max-3} + \cdots + y_2 x\, w_8^2 + y_1 x\, w_8^1 + y_0 x\, w_8^0 \quad \text{...(2.7)}$$

where a place $p = p\max - 1, p\max - 2, p\max - 3, ..., 2, 1$ and 0, and $p\max$ is maximum the number of places used in the representation and $w_8 = 8$. The y_0, y_1, y_2, ... are the digits with each digit is either 0 or 1 or 6 or maximum 7, at the right-most, left first from that, left second from that and so on.

2.5 USE OF DECIMAL NUMBERS TO REPRESENT A NUMBER

A decimal number is a number based upon ten values 0, 1, 2, 3, 4, 5, 6, 7, 8 and 9.

Let us assume there are 16 bananas. In decimal system, 0 will represent first banana, second banana by 1, third by 2, fourth by 3 and so on. 15 will represent last banana. Table 2.3 shows the numbers in decimal system starting from 0, that represents first, second third, fourth, fourteenth, fifteenth and sixteenth banana starting from 0. Why does the15 represent sixteenth banana in decimal number system? [Decimal means a number with base 10. Using the formula (4), decimal number value = $10^0.5 + 10^1.1$. This equals 15. If we start counting from 0, sixteenth banana is actually the 15th.] By convention, we may optionally write d as the subscript after a decimal number as shown in the table. This is to distinguish it from octal or hexadecimal numbers.

TABLE 2.3

Decimal value	Two digit decimal representation	Four digit decimal representation	Eight digit decimal representation
0	00_d	0000_d	$0000\ 0000_d$
1	01_d	0001_d	$0000\ 0001_d$
2	02_d	0002_d	$0000\ 0002_d$
3	03_d	0003_d	$0000\ 0003_d$
.	.	.	.
.	.	.	.
.	.	.	.
13	13_d	0013_d	$0000\ 0013_d$
14	14_d	0014_d	$0000\ 0014_d$
15	15_d	0015_d	$0000\ 0015_d$

2.6 USE OF HEXADECIMAL NUMBERS TO REPRESENT A NUMBER

A hexadecimal number is a number based upon sixteen values; 0, 1, 2, 3, 4, 5, 6, 7, 8, 9, *A*, *B*, *C*, *D*, *E* and *F*.

Let us assume there are 21 bananas. In the hexadecimal system, 0 will represent first banana, second banana by 1, ninth by 8, fourteenth by *D*, fifteenth by *E* and sixteenth one by *F*. Last one by 14H. Table 2.4 shows the numbers in the hexadecimal system starting from 0, that represents the first, second, tenth, eleventh, fourteenth and fifteenth to 21st banana starting from 1st. Why does 14H represents the last banana when counted from 0 in the hexadecimal number system? (Hexadecimal means a number with base 16. Using the formula, hexadecimal octal number value = $16^0.4 + 16^1.1 = 4 + 16$. This equals 20.). By convention, we write H after a hexadecimal number. This is to distinguish it from the decimal number.

TABLE 2.4

Decimal value	Two digit hexadecimal representation	Four digit hexadecimal representation	Eight digit hexadecimal representation
0	00H	0000H	0000 0000H
1	01H	0001H	0000 0001H
.			
.			
9	09H		
10	0AH	000AH	0000 000AH
.		.	.
.			.
13	0DH	000DH	0000 000DH
14	0EH	000EH	0000 000EH
15	0FH	000FH	0000 000FH
16	10H	0010H	0000 0010H
17	11H	0011H	0000 0011H
18	12H	0012H	0000 0012H
19	13H	0013H	0000 0013H
20	14H	0014H	0000 0014H

Note: (1) Some times dollar sign is used before a hexadecimal number. Then 0014H will be written as $0014. (2) Some times 0x (zero followed by small case x) is used before a hexadecimal number. Then 0014H or $0014 will be written as 0x0014. (3) Some times, H is written as small case h in the subscript.

Points to Remember

(1) A hexadecimal number is chosen as a four hex-digit number in case its decimal value lies between 0 and 65536_d. (0 and $(16)^4 - 1$).

(2) A hexadecimal number is chosen as a two hex-digit number in case its decimal value lies between 0 and 255_d [0 and $(16)^2 - 1$].

(3) A hexadecimal number is chosen as a eight hex-digit number in case its decimal value lies between 0 and $(65536 \times 65536 - 1)_d$ [0 and $(16)^8 - 1$].

(4) A general formula to get the total N is again same as under with only w_0 changed:

$$N = y_{p\max-1} \cdot w_b^{p\max-1} + y_{p\max-2} \cdot w_b^{p\max-2} + y_{p\max-3} \cdot w_b^{p\max-3}$$

$$+ \cdots + y_2 \,.\, w_b^2 + y_1 \,.\, w_b^1 + y_0 \,.\, w_b^0 \qquad \text{...(2.8)}$$

where a place $p = p\max - 1, p\max - 2, p\max - 3, ..., 2, 1$ and 0, and $p\max$ is maximum the number of places used in the representation and $w_b = 16$. The $y_0, y_1, y_2, ...$ are the digits with a digit = 0 or 1.. or E or F, at the right most, left first from that, left second from that and so on.

2.7 USE OF BCD NUMBER TO REPRESENT A DECIMAL NUMBER

BCD format is a representation in which each decimal digit replaces a four-bit binary number.

A decimal number cannot store in a computer except in binary format. Therefore, it can be stored in BCD format. 0_d is 0000_{bcd}. 1_d is 0001_{bcd}. 9_d is 1001_{bcd}. 38_d is 0011 1000_{bcd}.

Let us assume there are 16 bananas. In the BCD system, when 0000 will represent first banana, second banana by 0001, the third by 0010, fourth by 0011 and so on. 0001 0101 will represent the last banana. Table 2.5 shows the numbers in the BCD system starting from 0000, that represents the first, second, third, fourth, fourteenth, fifteenth and sixteenth banana starting from 0. Why does 0001 0101 represent the sixteenth banana in the decimal number system? (BCD means each decimal number is represented as a four-bit binary number (called nibble). By convention, we may optionally write *bcd* as the subscript after a BCD number as shown in last row of the table. This is to distinguish it from the octal or hexadecimal number.

TABLE 2.5

Decimal value	Two digit BCD representation	Four digit BCD representation
0_d	0000 0000_{bcd}	0000 0000 0000 0000_{bcd}
1_d	0000 0001_{bcd}	0000 0000 0000 0001_{bcd}
2_d	0000 0002_{bcd}	0000 0000 0000 0002_{bcd}
3_d	0000 0003_{bcd}	0000 0000 0000 0003_{bcd}
.	.	.
.	.	.
.	.	.
13_d	000 1 0003_{bcd}	0000 0000 0001 0003_{bcd}
14_d	000 1 0004_{bcd}	0000 0000 0001 0003_{bcd}
15_d	000 1 0005_{bcd}	0000 0000 0001 0005_{bcd}

Points to Remember in the Number Systems

1. Use of any digit other than 0 and 1 is illegal in the binary system.
2. Use of any digit other than 0, 1, 2, 3, 4, 5, 6, 7, 8 or 9 is illegal in the decimal system.
3. Use of any digit other than 0, 1, 2, 3, 4, 5, 6, or 7 is illegal in the octal system.
4. Use of any digit other than between 0 and 9 or any character other than *A*, *B*, *C*, *D*, *E* or *F* is illegal in the hexadecimal system.
5. Use of any nibble other than a nibble between 000 and 1001 is illegal in the BCD system. (A nibble is a sequence of 4-bits).

2.8 FRACTIONAL NUMBERS

Fractional numbers are frequently needed in arithmetic calculation. How do we represent 1.5 in binary or octal number systems? How do we represent 2.333 in binary or octal number systems? The dot is also known as a decimal point in the decimal number system.

Answer is we represent fraction numbers also by using a variation of the golden rule given in Section 2.2. Weights of the digits before the point (integer part) are taken in the same way as mentioned in equation (2.5) and weight of digits after dot (fraction part) are successively reduced by the same weight factor as earlier. The golden rule to remember is again the sum of the weights in case of fractional number also. *Weights increase by basic weight factor as we go left from the dot in the integer part and weights decrease by basic weight factor as we go right from the dot in the fractional part.* Equations (9) and (10) gives the new golden rule to be applied.

A Golden Rule to Remember — Sum of the Weights in Case of Fractional Numbers

A general formula to get the number, N is as under:

Integer Part = sum of the y_p.(Base)^p before the dot

$$= [y_{p\max-1} \cdot w_0^{p\max-1} + y_{p\max-2} \cdot w_0^{p\max-2} + y_{p\max-3} \cdot w_0^{p\max-3} + \dots + y^2 \cdot w_0^2 + y_1 \cdot w_0^1 + y_0 \cdot w_0^0] \quad \text{...(2.9)}$$

and

Fractional Part = sum of the y_q.(Base)^q after the dot

$$= (y_{-1} \cdot w_0^{-1} + y_{-2} \cdot w_0^{-2} + y_{-3} \cdot w_0^{-3} + \dots + y_{-q\max+2} \cdot w_0^{-q\max+2} + y_{-q\max+1} \cdot w_0^{-q\max+1} + y_{-q\max+1} \cdot w_0^{-q\max}) \quad \text{...(2.10)}$$

where p = place of the digit before the dot (integer part) = 0, 1, 2, ..., ($p max$ – 3), ($p max$ – 2) or ($p max$ – 1), from right to left, and $p max$ is maximum the number of places used in the representation, q = place of the digit after the dot (fractional part) = – 1, – 2, – 3, ..., (– $q max$ + 2), (– $q max$ + 1) or (– $q max$), from left to right, and $q max$ is maximum the number of places used in the representation after the dot and base (basic weight) w_0 = 10 for decimal number, 8 for octal number, 2 for binary number and 16 for hexadecimal numbers . The y_0, y_1, y_2, ... are the digits. In case of decimal numbers, each digit being 0 or 1 or 3 or ...... or (w_0 – 1) at the right-most (– $q$max)th position, left first from that, left second from that and so on till left most (p max – 1)th position.

Following Tables 2.6 and 2.7 helps in understanding the concept. For a binary fractional number, we find the values of y for w_0 = 2 and find whether y_{-1} = 1 or 0, y_{-2} = 1 or 0, y_{-3} = 1 or 0, y_{-4} = 1 or 0, y_{-5} = 1 or 0, y_{-6} = 1 or 0, ... so that the sum of the terms in right hand side of Equation (10) equals the fractional part.

TABLE 2.6

Fractional decimal	Formula for basic weight, w_0 = 2	Binary representation
0.5_d	.5 = 1/2 = (1×2^{-1}) So, y_{-1} = 1	0.10000000_b
0.25_d	.25 = 1/4 = (1×2^{-2}) So, y_{-2} = 1	0.01000000_b
0.125_d	.25 = 1/8 = (1×2^{-3}) So, y_{-3} = 1	0.00100000_b
0.0625_d	$.0625_d$ = 1/16 = (1×2^{-4}) So, y_{-4} = 1	0.00010000_b
0.03125_d	$.03125_d$ = 1/32 = (1×2^{-5}) So, y_{-5} = 1	0.00001000_b

Table 2.6 Contd.

0.015625_d	$.015625_d = 1/64 = (1 \times 2^{-6})$ So, $y_{-6} = 1$	0.00000100_b
0.0078125_d	$.0078125_d = 1/128 = (1 \times 2^{-7})$ So, $y_{-7} = 1$	0.00000010_b
0.00390625_d	$.001953125_d = 1/256 = (1 \times 2^{-8})$ So, $y_{-8} = 1$	0.00000001_b
0.001953125_d	$.001953125_d = 1/512 = (1 \times 2^{-9})$ So, $y_{-9} = 1$	0.000000001000_b
0.0009765625_d	$.0009765625_d = 1/1024 = (1 \times 2^{-10})$ So, $y_{-10} = 1$	0.000000000100_b

TABLE 2.7

Fractional decimal	Formula for basic weight, $w_0 = 8$	Octal representation
0.125_d	$.125 = 1/8 = (1 \times 8^{-1})$ So, $y_{-1} = 1$	0.1000_{octal}
0.250_d	$.25 = 2 \times .125 = (2 \times 8^{-1})$ So, $y_{-1} = 2$	0.2000_{octal}
0.5_d	$.5 = 4 \times 1/8 = (4 \times 8^{-1})$ So, $y_{-1} = 4$	0.4000_{octal}
0.625_d	$.625 = 5 \times 1/8 = (5 \times 8^{-1})$ So, $y_{-1} = 5$	0.5000_{octal}
0.875_d	$.875_d = 7 \times 1/8 = (7 \times 8^{-1})$ So, $y_{-1} = 7$	0.7000_{octal}
0.015625_d	$.015625_d = 1/64 = (1 \times 8^{-2})$ So, $y_{-1} = 0$ and $y_{-2} = 1$	0.0100_{octal}
0.001953125_d	$.001953125_d = 1/512 = (1 \times 8^{-3})$ So, $y_{-1} = 0$, $y_{-2} = 0$ and $y_{-3} = 1$	0.0010_{octal}
0.00390625_d	$.00390625_d = 2 \times 1/512 = (2 \times 8^{-3})$ So, $y_{-1} = 0$, $y_{-2} = 0$ and $y_{-3} = 2$	0.0020_{octal}

For an octal fractional number, we find the values of y for $w_0 = 8$ and find whether $y_{-1} =$ 1 or 0 or ... or 7, $y_{-2} = 0$ or 1 or ... 7, $y_{-3} = 0$ or 1 ... or 7, $y_{-4} = 0$ or 1 ... or 7 so that the sum of the terms in right hand side of Equation (10) equals the fractional part.

Examples 2.15 and 2.16 will explain the use of the Tables 2.6 and 2.7 respectively.

EXAMPLES

Example 2.1

Conversion from binary number to decimal number:

Here we use the formula value = sum of the $(\text{base})^{\text{digit place}}$ multiplied by a digit in each term. Also take rightmost digit place = 0. Refer equation (2.6). Sum is performed from rightmost to leftmost digit place.

Suppose base of a number is 2 and binary number is 0110 1001. Leftmost place is 7th in case of 8-bit binary number and 15th in case of 16 bit binary number.

Solution

At 0th digit place, digit = 1.

At 1st digit place, digit = 0.

At 2nd digit place, digit = 0.

At 3rd digit place, digit = 1.

At 4th digit place, digit = 0.

At 5th digit place, digit = 1.

At 6th digit place, digit = 1.

At 7th digit place, digit = 0.

Therefore, value (in decimal) =

$(2)^0.1$ +

$(2)^1.0$ +

$(2)^2.0$ +

$(2)^3.1$ +

$(2)^4.0$ +

$(2)^5.1$+

$(2)^6.1$ +

$(2)^7.0$ =

$(64 + 32 + 8 + 1) = 105_d$

Corresponding decimal value is 105.

Example 2.2

Conversion from decimal number to binary number:

Let the required binary number be of 8 bits. It means it will have rightmost 0th digit-place to left-most 7th place.

Let the decimal number be 137_d. We start dividing 137_d first by 2^7, then by 2^6, then by 2^5, then by 2^4, then by 2^3, then 2^2, then by 2^1 and finally by 2^0 (It means by 128, 64, 32, 16, 8, 4, 2 and 1, successively). The quotients provide the binary number's 7th digit-place, 6th digit-place, ... and rightmost 0th place, respectively.

Solution

At 7th digit place, digit = $137/2^7 = 137/128 = 1$ (remainder = 9)

At 6th digit place, digit = $9/2^6 = 9/64 = 0$ (remainder = 9)

At 5th digit place, digit = $9/2^5 = 9/32 = 0$ (remainder = 9)

At 4th digit place, digit = $9/2^4 = 9/16 = 0$ (remainder = 9)

At 3rd digit place, digit = $9/2^3 = 9/8 = 1$ (remainder = 1)

At 2nd digit place, digit = $1/2^2 = 1/4 = 0$ (remainder =1).

At 1st digit place, digit = $1/2^1 = 1/2 = 0$ (remainder = 1).

At 0th digit place, digit = $1/2^0 = 1/1 = 1$ (remainder = 0).

Therefore, corresponding to decimal value 137_d binary number = 1000 1001.

Example 2.3

Conversion from decimal number to octal number:

Let the required octal number be of 4 digits. It means it will have rightmost 0th digit-place to left-most 3rd place.

Let the decimal number be 649_d. We divide 649_d first by 8^3, then by 8^2, then by 8^1, and finally by 8^0. (It means by 512, 64, 8 and 1, successively). The quotients provide the octal number's 3rd digit place, 2nd digit place,.... and rightmost 0th place, respectively.

Solution

At 3rd digit place, octal-digit = $649/8^3 = 649/512 = 1$ (remainder = 137)

At 2nd digit place, octal-digit = $137/8^2 = 137/64 = 2$ (remainder = 9).

At 1st digit place, octal-digit = $9/8^1 = 9/8 = 1$ (remainder = 1).

At 0th digit place, octal-digit = $1/8^0 = 1/1 = 1$ (remainder = 0).

Therefore, corresponding to decimal value 649_d the octal number = 1211_8.

Example 2.4

Conversion from binary number to octal number:

Let the required octal number be of 4 digits. It means it will have rightmost 0th digit-place to left-most 3rd place.

Let the binary number be 1100 1001. It is easier if we first convert it to decimal. The decimal number is 201_d using the formula in equation (6), $= 1 \times 2^7 + 1 \times 2^6 + 0 \times 2^5 + 0 \times 2^4 + 1 \times 2^3 + 0 \times 2^2 + 0 \times 2^1 + 1 \times 2^0$.

We therefore divide 201_d first by 8^3, then by 8^2, then by 8^1, and finally by 8^0. (It means by 512, 64, 8 and 1, successively). The quotients provide the octal number's 3rd digit place, 2nd digit place,.... and rightmost 0th place, respectively.

Solution

At 3rd digit place, octal-digit = $201/8^3 = 201/512 = 0$ (remainder = 201)

At 2nd digit place, octal-digit = $201/8^2 = 201/64 = 3$ (remainder = 9).

At 1st digit place, octal-digit = $9/8^1 = 9/8 = 1$ (remainder = 1).

At 0th digit place, octal-digit = $1/8^0 = 1/1 = 1$ (remainder = 0).

Therefore, corresponding to binary value 1100 1001, decimal value is 201_d and the octal number = 0311_8.

Example 2.5

Conversion from binary number to hexadecimal number:

Let the required hexadecimal number be of 4 digits. It means it will have rightmost 0th digit place to left-most 3rd place.

Let the binary number be 0100 0010 1101 0011. Each nibble (the sequence of 4 bits) can be directly converted to hex-digits as follows.

Solution

At 3rd digit place, hex-digit = $0100_b = 4_d = 4_h$

At 2nd digit place, hex digit = $0010_b = 2_d = 2_h$

At 1st digit place, octal-digit = $1101_b = 13_d = D_h$

At 0th digit place, octal-digit = $0011_b = 3_d = 3_h$

Therefore, corresponding to 0100 0010 1101 0011_b the hexadecimal number = $42D3_h$.

Example 2.6

Conversion from hexadecimal number to binary number:

Let the required hexadecimal number be of 4 digits. Each hex-digit has a nibble (set of four bits). The binary number will be from rightmost 0th digit-place to left-most 15th place.

Let the hexadecimal number be 0x5A9F (using prefix 0x convention in place of *h* subscript as post fix). Each hex-digit can be directly converted to a nibble.

Solution

At 15th down to 12th bit place, bits are 0101 as hex-digit = 5

At 11th down to 8th bit place, bits are 1010 as hex-digit = *A*

At 7th down to 4th bit place, bits are 1001 as hex-digit = 9

At 3rd down to 0th bit place, bits are 1111 as hex-digit = *F*

Therefore, corresponding to 0x5A9F hexadecimal number the binary number *i* = 0101 1010 1001 1111.

Example 2.7

Conversion from hexadecimal number to decimal number:

We use the formula in equation (2.8), which finds sum of the terms $(16)^{\text{digit place}}$ multiplied by a hex-digit. Take rightmost digit place = 0. Sum is performed from rightmost to leftmost digit place.

Suppose base of a number is 16 and hexadecimal number is of four hex digits, 6A0C H. Leftmost place is 3rd in case of 4-digit hex decimal number.

Solution

At 0th digit place, hex-digit = C = 12_d.

At 1st digit place, hex-digit = 0.

At 2nd digit place, hex-digit = *A* = 10_d.

At 3rd digit place, hex-digit = 6_d.

Therefore, value (in decimal) =

$(16)^0.12$ +

$(16)^1.0$ +

$(16)^2.10$ +

$(16)^3.6$

$= (1.\ 12 + 16.\ 0 + 256.\ 10 + 4096.\ 6) = (12 + 2560 + 24576) = 27148_d$.

The hex digits, 6A0C H equals 27148_d.

Example 2.8

Conversion from octal number to decimal number:

We use the formula in equation (2.7) value, which finds sum of $(8)^{\text{digit place}}$ multiplied by an octal digit in each term. Take rightmost digit-place = 0. Sum is performed from rightmost to leftmost digit place.

Suppose base of a number is 8 and octal number is of four octal digits, 6715_8. Leftmost place is 3rd in case of 4 digit octal number.

Solution

At 0th digit place, octal-digit = 5_8.

At 1st digit place, octal-digit = 1_8.

At 2nd digit place, octal -digit = 7_8.

At 3rd digit place, octal-digit = 6_8.

Therefore, value (in decimal) =

$(8)^0.5$ +

$(8)^1.1$ +

$(8)^2.7$ +

$(8)^3.6$

= (1. 5 + 8. 1 + 64. 7 + 512. 6) = (5 + 8 + 448 + 3072) = 3533_d. The octal number, 6715_8 equals 3533_d.

Example 2.9

Conversion from octal number to binary number.

We use two steps. First the formula in equation (2.7) for decimal value, which finds sum of $(8)^{\text{digit place}}$ multiplied by the octal-digit. Take rightmost digit place = 0. Sum is performed from rightmost to leftmost digit place. This gives the decimal number. The decimal number is then converted to the binary number.

Suppose base of a number is 8 and octal number is of four octal digits, 6715_8. We get the decimal number = 3533 from the above example. Now, it is converted to binary as follows.

Solution

At 15th digit place, digit = $3533/2^{15}$ = 3533/32768 = 0 (remainder = 3533)

At 14th digit place, digit = $3533/2^{14}$ = 3533/16384 = 0 (remainder = 3533)

At 13th digit place, digit =$3533/2^{13}$ = 3533/8192 = 0 (remainder = 3533)

At 12th digit place, digit = $3533/2^{12}$ = 3533/4096 = 0 (remainder = 3533)

At 11th digit place, digit = $3533/2^{11}$ = 3533/2048 = 1 (remainder = 1485)

At 10th digit place, digit = $1485/2^{10}$ = 1485/1024 = 1 (remainder = 461).

At 9th digit place, digit = $461/2^9$ = 461/512 = 0 (remainder = 461).

At 8th digit place, digit = $461/2^8$ = 461/256 = 1 (remainder = 205).

At 7th digit place, digit = $205/2^7$ = 205/128 = 1 (remainder = 77)

At 6th digit place, digit = $77/2^6$ = 77/64 = 1 (remainder = 13)

At 5th digit place, digit = $13/2^5$ = 13/32 = 0 (remainder = 13)

At 4th digit place, digit = $13/2^4$ = 13/16 = 0 (remainder = 13)

At 3rd digit place, digit = $13/2^3$ = 13/8 = 1 (remainder = 5)

At 2nd digit place, digit = $5/2^2$ = 5/4 = 1 (remainder = 1).

At 1st digit place, digit = $1/2^1$ = 1/2 = 0 (remainder = 1).

At 0th digit place, digit = $1/2^0$ = 1/1 = 1 (remainder = 0).

Therefore, corresponding to octal value 6715_8 = decimal value 3533_d, the binary number = 0000 1101 1100 1101.

Example 2.10

Conversion from decimal number to hexadecimal:

We use two steps. First convert the decimal number to binary number of 4 or 8 or 16 bits, depending up on whether the decimal number is less or equal to 15 ($= 2^4-1$), 255 ($= 2^8 -1$) or 65535 ($= 2^{16}-1$). Conversion is done as in Example 2.2. Now the binary number can be converted to hexadecimal easily by using each nibble for converting to a hex digit, as shown in Example 2.5.

Let decimal number be 56_d. This is more than 15 and less than 255. Hence 8-bit binary number or two hex-digit number can be used. Each four bit of a binary number correspond to a hex-digit.

Solution

Step 1. Convert 56_d to binary as follows:

At 7th digit place, digit = $56/2^7$ = 56/128 = 0 (remainder = 56)

At 6th digit place, digit = $56/2^6$ = 56/64 = 0 (remainder = 56)

At 5th digit place, digit = $56/2^5$ = 56/32 =1 (remainder = 24)

At 4th digit place, digit = $24/2^4$ = 24/16 = 1 (remainder = 8)

At 3rd digit place, digit = $8/2^3$ = 8/8 = 1 (remainder = 0)

At 2nd digit place, digit = $0/2^2$ = 0/4 = 0 (remainder = 0).

At 1st digit place, digit = $0/2^1$ = 0/2 = 0 (remainder = 0).

At 0th digit place, digit = $0/2^0$ = 0/1 = 0 (remainder = 0).

Binary number is 00111000_b. Each nibble (a set of 4 bits) can be directly converted to hex-digits as follows.

At 1st digit place, hex digit = $0011_b = 3_d = 3_h$

At 0th digit place, hex digit = $1000_b = 8_d = 8_h$

Therefore, corresponding to the decimal value 56_d binary number = 00111000_b, and the hexadecimal number = 38_h.

Example 2.11

Conversion from octal number to hexadecimal:

We use two steps. First convert the octal number to a decimal number, as in the Example 2.8. Now a decimal number can be converted to a hexadecimal as shown in Example 2.10.

Let octal number be 0356_8.

Solution

Step 1. Convert 0356_8 to a decimal as follows:

At 0th digit place, octal-digit = $6_h = 6_d$.

At 1st digit place, octal-digit = $5_h = 5_d$.

At 2nd digit place, octal-digit = $3_h = 3_d$.

At 3rd digit place, octal-digit = $0_h = 0_d$.

Therefore, value (in decimal) =

$(8)^0.6 +$

$(8)^1.5 +$

$(8)^2.3 +$

$(8)^3.0+$

$(1.\ 6 + 8.\ 5 + 64.\ 3 + 512\ .\ 0) = (6 + 40 + 192) = 238_d$.

Step 2. Convert 238_d to hexadecimal as follows:

At 7th digit place, digit = $238/2^7 = 238/128 = 1$ (remainder = 110)

At 6th digit place, digit = $110/2^6 = 110/64 = 1$ (remainder = 46)

At 5th digit place, digit = $46/2^5 = 46/32 = 1$ (remainder = 14)

At 4th digit place, digit = $14/2^4 = 14/16 = 0$ (remainder = 14)

At 3rd digit place, digit = $14/2^3 = 14/8 = 1$ (remainder = 6)

At 2nd digit place, digit = $6/2^2 = 6/4 = 1$ (remainder = 2).

At 1st digit place, digit = $2/2^1 = 2/2 = 1$ (remainder = 0).

At 0th digit place, digit = $0/2^0 = 0/1 = 0$ (remainder = 0).

Binary number is $1110\ 1110_b$. Each nibble (a set of 4 bits) can be directly converted to hex-digits as follows.

At 1st digit place, hex-digit = $1110_b = 14_d = E_h$

At 0th digit place, hex digit = $1110_b = 14_d = E_h$

Therefore, corresponding to the decimal value 238_d binary number = $1110\ 1110_{bb}$, and the hexadecimal number = EE_h.

Example 2.12

Conversion from hexadecimal number to octal number:

We use two steps. First convert the hexadecimal number to a decimal number, as in the Example 2.7. Now a decimal number can be converted to an octal as shown in Example 2.3.

Let the hexadecimal number be 56_h.

Solution

At 0th digit place, hex-digit = $6_h = 6_d$.

At 1st digit place, hex-digit = $5_h = 5_d$.

Therefore, number (in decimal) =

$(16)^0.6 +$

$(16)^1.5 +$

$(1.\ 6 + 16.\ 5) = 86_d$.

The hexadecimal number, 56_h equals 86_d. Now let us convert it to octal number as follows:

We start dividing 86_d first by 8^3, then by 8^2, then by 8^1, and finally by 8^0. (It means by 512, 64, 8 and 1, successively). The quotients provide the octal digit at 3rd digit place, 2nd digit place, ... and rightmost 0th place, respectively.

At 3rd digit place, octal-digit = $86/8^3 = 86/512 = 0$ (remainder = 86)

At 2nd digit place, octal-digit = $86/8^2 = 86/64 = 1$ (remainder = 22).

At 1st digit place, octal-digit = $22/8^1 = 22/8 = 2$ (remainder = 6).

At 0th digit place, octal-digit = $6/8^0 = 6/1 = 6$ (remainder = 0).

Therefore, corresponding to the decimal value 86_d the octal number = 0126_8.

Example 2.13

Conversion from a decimal number to an eight digit BCD number:

Let the decimal number be 123789. We use the following steps. First convert each decimal digit from right to left by the equivalent nibble (four-bit number). The more BCD digits, place the needed nibbles 0000 on the left. (Left side zeros does not effect a number, only the addition of right side zeros effects a number).

Solution

Step 1. Convert into BCD

At 5th digit place, digit = 1_d hence answer is 0001_b.

At 4th digit place, digit = 2_d hence answer is 0010_b.

At 3rd digit place, digit = 3_d hence answer is 0011_b.

At 2nd digit place, digit = 7_d hence answer is 0111_b.

At 1st digit place, digit = 8_d hence answer is 1000_b.

At 0th digit place, digit = 9_d hence answer is 1001_b.

BCD number with six digits is $0001\ 0010\ 0011\ 0111\ 1000\ 1001_b$.

Step 2. Convert into eight-digit BCD by putting additional two nibbles on the left to the six digit BCD.

Eight digit BCD number for 123789_d is $0000\ 0000\ 0001\ 0010\ 0011\ 0111\ 1000\ 1001_{bcd}$.

Example 2.14

Finding illegal representation in the followings:

$120A_d$ $1010\ 011_{bcd}$ 0208_{octal} $1010\ 2011_b$ $GC0A_h$

Solution

(1) $120A_d$ is illegal as A is not the digit permitted in the decimal system.

(2) $1010\ 011_{bcd}$ is illegal as a four-bit binary number can be used, not a three bit binary.

(3) 0208_{octal} is illegal as 8 is not the digit permitted in the octal system, only up to 7 permitted.

(4) $1010\ 2011_b$ is illegal as 2 is not the digit permitted in the binary system.

(5) $GC0A_h$ is illegal as G is not the digit permitted in the hex-decimal system.

Example 2.15 Finding the binary representation of the following fractional numbers:

0.5_d 1.5_d 2.3333_d 3.875_d 13.125_d 14.6666_d 1.58_d

Solution

We apply the integer and fractional parts formulae given in Equations (9) and (10) and find the values of y at various places before the dot and after the dot.

Let the decimal number fractional part equals $.333_d$. We start dividing $.333_d$ first by 2^{-1}, then by 2^{-2}, then by 2^{-3}, then by 2^{-4}, then by 2^{-5}, then 2^{-6}, then by 2^{-6} and so on successively till the sum equals the part or till the desired precision is obtained. (It means divide by .5, .25, .125, .0625, .03125, .015625, …and so on successively). The quotients provide the binary bits at -1st digit place, -2nd digit-place, ... and rightmost $-q\max^{th}$ place, respectively.

At $q = -1^{st}$ digit place, digit $= .333/2^{-1} = .333/0.5 = 0$ (remainder = .333)

At $q = -2^{nd}$ digit place, digit $= .333/2^{-2} = .333/0.25 = 1$ (remainder = .0833)

At $q = -3^{rd}$ digit place, digit $= .0833/2^{-3} = .0833/0.125 = 0$ (remainder = .0833)

At $q = -4^{th}$ digit place, digit $= .0833/2^{-4} = .0833/0.0625 = 1$ (remainder = .0208)

At $q = -5^{th}$ digit place, digit $= .0208/2^{-5} = .0208/0.03125 = 0$ (remainder = .0208)

At $q = -6^{th}$ digit place, digit $= .0208/2^{-6} = .0208/.015625 = 1$ (remainder = .005175)

At $q = -7^{th}$ digit place, digit $= .005175/2^{-7} = .005175/.0078125 = 0$ (remainder = .005175)

At $q = -8^{th}$ digit place, digit $= .005175/2^{-8} = .005175/.00390625 = 1$ (remainder = .00126875)

At $q = -9^{th}$ digit place, digit $= .00126875/2^{-9} = .00126875/.001953125 = 0$ (remainder = .00126875)

At $q = -10^{th}$ digit place, digit $= .00126875/2^{-10} = .00126875/0.0009765625 = 1$ (remainder = .0002921875).

Sum of the 10 terms is 0.3330078125. Remainder .0002921875 is insignificant so further operations for $q = -11$, $q = -12$ and lower stopped here.

Table 2.8 gives the final results for all the desired decimal fractional numbers: 0.5, 1.5, 2.333, 3.875, 13.125, 14.666 and 1.58.

TABLE 2.8

Decimal fractional number	Integer and fractional parts formulae for basic weight, $w_0 = 2$	Binary representation
0.5_d	$0. = [0 \times 2^0]$ and $.5 = [1 \times 2^{-1}]$	00.100000_b
1.5_d	$1 = [1 \times 2^0]$ and $.5 = [1 \times 2^{-1}]$	0001.100000_b
2.3333_d	$2 = 1 \times 2^1 + 0 \times 2^0$ and $.333 = 0 \times 2^{-1} + 1 \times 2^{-2} + 0 \times 2^{-3} + 1 \times 2^{-4} + 0 \times 2^{-5} + 1 \times 2^{-6} + 0 \times 2^{-7} + 1 \times 2^{-8} + 0 \times 2^{-9} + 1 \times 2^{-10}$	$10.\ 0101010101_b$

Table 2.8 Contd.

3.875_d	$3 = 1 \times 2^1 + 1 \times 2^0$ and $.875 = 1 \times 2^{-1} + 1 \times 2^{-2} + 1 \times 2^{-3}$	11.1110_b
13.125_d	$13 = 1 \times 2^3 + 1 \times 2^2 + 0 \times 2^1 + 1 \times 2^0$ and $.125 = 0 \times 2^{-1} + 0 \times 2^{-2} + 1 \times 2^{-3}$	1101.0010_b
14.6666_d	$14 = 1 \times 2^3 + 1 \times 2^2 + 1 \times 2^1 + 0 \times 2^0$ and $.666 = 1 \times 2^{-1} + 0 \times 2^{-2} + 1 \times 2^{-3} + 0 \times 2^{-4} + 1 \times 2^{-5} + 0 \times 2^{-6} + 1 \times 2^{-7} + 0 \times 2^{-8} + 1 \times 2^{-9} + 0 \times 2^{-10}$	1110.1010101010_b
1.58_d	$1 = 1 \times 2^0$ and $.58 = 1 \times 2^{-1} + 0 \times 2^{-2} + 0 \times 2^{-3} + 1 \times 2^{-4} + 0 \times 2^{-5} + 1 \times 2^{-6} + 0 \times 2^{-7} + 0 \times 2^{-8} + 0 \times 2^{-9} + 1 \times 2^{-10} + 1 \times 2^{-11} + 1 \times 2^{-12}$	01.100101000111_b

Example 2.16

Finding octal representation of the following fractional numbers:

0.5_d 1.5_d 2.333_d 3.875_d 13.125_d 14.666_d 1.58_d

Solution

We apply the integer part and fractional part formulae in Equations (9) and (10) and find the values of y at various places before the dot and after the dot.

Let the decimal number fractional part equals $.333_d$. We start dividing $.333_d$ first by 8^{-1}, then by 8^{-2}, then by 8^{-3}, then by 8^{-4} and so on successively till the sum of the terms in Equation (10) equals the part or till the desired precision in the sum is obtained (It means by .125, .015625, .001953125, .000244140625 and so on successively). The quotients provide the octal-digits at -1st digit place, -2nd digit place and so on up to $-q\max^{th}$ place.

At $q = -1^{st}$ digit place, digit $= .3333/8^{-1} = .333/.125 = 2$ (remainder $= .0833$)

At $q = -2^{nd}$ digit place, digit $= .0833/8^{-2} = .083/.015625 = 5$ (remainder $= .005175$)

At $q = -3^{rd}$ digit place, digit $= .005175/8^{-3} = .005175/0.001953125 = 2$ (remainder $=$ .00126875)

At $q = -4^{th}$ digit place, digit $= .00126875/8^{-4} = .00126875/0.000244140625 = 5$ (remainder $= .000048053125$). Since remainder is negligible compared to .333, we stop the further operations for higher precision at this point.

Table 2.9 gives the final results for all the desired decimal fractional numbers: 0.5, 1.5, 2.333, 3.875, 13.125, 14.666 and 1.58.

TABLE 2.9

Decimal fractional number	Integer and fractional parts formulae for basic weight, $w_0 = 8$	Octal representation
0.5_d	$0 = 0 \times 8^0$ and $.5 = 4 \times 8^{-1}$	0.4000_{octal}
1.5_d	$1 = 1 \times 8^0$ and $.5 = 4 \times 8^{-1}$	1.4000_{octal}
2.333_d	$2 = 2 \times 8^0$ and $.333 = 2 \times 8^{-1} + 5 \times 8^{-2} + 2 \times 8^{-3} + 5 \times 8^{-4}$	2.2525_{octal}
3.875_d	$3 = 3 \times 8^0$ and $.875 = 7 \times 8^{-1}$	3.7000_{octal}
13.125_d	$13 = 1 \times 8^1 + 5 \times 8^0$ and $.125 = 1 \times 8^{-1}$	15.1000_{octal}
14.6666_d	$14 = 1 \times 8^1 + 6 \times 8^0$ and $.6666 = 5 \times 8^{-1} + 2 \times 8^{-2} + 5 \times 8^{-3} + 2 \times 8^{-4}$	16.5252_{octal}
1.58_d	$1 = 1 \times 8^0$ and $.58 = 4 \times 8^{-1} + 5 \times 8^{-2} + 0 \times 8^{-3} + 7 \times 8^{-4}$	1.4507_{octal}

EXERCISES

1. In a 32-bit computer, what are the maximum and minimum possible binary numbers? Convert these into maximum and minimum possible positive decimal numbers.
2. Let us represent addresses in binary number system as follows, 0000000000000000, 0000000000000001, 0000000000000010, and so on, up to certain maximum as per following table. Calculate the number of addresses present in decimal. (Fill column 3 of the Table 2.10).

TABLE 2.10

Starting address in binary representation	Maximum address in binary Representation	Number of addresses available (Answer in decimal system)
0000 0000 0000 0000	0000 0000 0000 1111	
0000 0000 0000 0000	0000 0000 0111 1111	
0000 1111 0000 0000	0000 1111 1111 1111	
1111 0000 0000 0000	1111 1111 1111 1111	

3. Let us represent addresses in the hex-number system as followed: 0000000000000000, 0000000000000001, 0000000000000010, and so on, up to certain maximum as per following table. Calculate the number of addresses present in decimal. (Fill column 3 of the Table 2.11.)

TABLE 2.11

Starting address in hexadecimal representation	Maximum address in hexadecimal representation	Number of addresses available (Answer in decimal system)
0000H	1111H	
A000H	AFFFH	
0000 FFFFH	0001 FFFFH	
0000 FFFDH	0008 FFFFH	

4. Convert the octal numbers into binary, decimal, BCD and hexadecimal numbers in the Table 2.12.

TABLE 2.12

Number in octal	Decimal	16 bit binary	4 digit hexadecimal	4 digit BCD
3600_{octal}				
1200_{octal}				
0200_{octal}				
0777_{octal}				

5. Convert the decimal numbers into BCD, binary, octal and hexadecimal numbers in the Table 2.13.

TABLE 2.13

Number in decimal	4 digit BCD	8 digit octal	16 bit Binary	4 digit Hexadecimal
3600_d				
1200_d				
0200_d				
0777_{dl}				

6. Convert the BCD into the decimal, BCD, binary, octal and hexadecimal numbers in the Table 2.14.

TABLE 2.14

Number in BCD	4 digit decimal	8 digit octal	16 bit Binary	4 digit hexadecimal
1000011101000010_{bcd}				
1001011001100001_{bcd}				
0001011001101001_{bcd}				
0001001101010100_{bcd}				

7. Derive complete solutions by finding the quotients and remainders for 0.5, 1.5, 2.333, 3.875, 13.125, 14.666 and 1.58 as done for 0.333 conversion to binary in Example 2.15 and check that you obtain the results same as in Table 2.8.
8. Derive complete solutions by finding the quotients and remainders for 0.5, 1.5, 2.333, 3.875, 13.125, 14.666 and 1.58 as done for 0.333 conversion to octal in Example 2.16 and check that you obtain the results same as in Table 2.9.
9. Using Equations (9) and (10), convert the fractional numbers in binary and octal to the decimals. (Table 2.15)

TABLE 2.15

Fractional number in binary	Decimal number	Fractional number in octal	Decimal number
$0.1100\ 1100_b$		7.7_{octal}	
1001.011001100001_b		4.5670_{octal}	
0011.011001101001_b		15.1234_{octal}	
1111.001101010100_b		0.00067_{octal}	

QUESTIONS

1. Suppose you have a cheque for Rs. 10000/-. What is the number system, you are using? What is the base you are using? What are the weights of the digits 1, 0, 0, 0, 0, and 0?

2. Suppose you place an additional 0 after 10000 at the cheque. What are the weights of 1, 0, 0, 0, 0, 0 and 0 now?
3. Your salary is going to be 100000 per month. (We wish it to be so!). How will your salary store at the 64-bit computer in the binary format.
4. Find and remove illegally represented values from Table 2.16.

TABLE 2.16

Number in decimal	4 digit BCD	8 digit octal	16 bit binary	4 digit hexadecimal
3600_d	1100 0011	0600_{octal}	1100 0311	$B6AG_h$
$120A_d$	1010 0011	$120A_{octal}$	1010 0011	$1C0A_h$
0200_d	0200 0000	0200_{octal}	0100 0000	0000_h
0777_{dl}	111 0000	0778_{octal}	111 0001	0777_{hl}

5. Answer how did you find the illegal values in Table 2.16.
6. Why is 0.5252_{octal} twice of 0.2525_{octal} when 0.5050_{dl} is twice of 2525_d? (Hint: The remainder is 2 and carry is 1 when 5 multiplies by 2 in multiplication of the octal numbers and the remainder is 0 and carry is 1 when 5 multiplies by 2 in multiplication of the decimal numbers).

CHAPTER 3

Binary Arithmetic and Two's Complement Arithmetic

OBJECTIVE

In Chapter 2, we learnt binary numbers. We used unsigned (positive) numbers only. In this chapter, the concept of binary arithmetic numbers (positive and negative) and their operations will be discussed. We shall also learn the concept of two's complement numbers and their operations.

A 32-bit computer system, like a Pentium, can easily perform arithmetic operations on two binary numbers as big as four bytes (32 bits) [1 byte = 8 bits]. Nowadays, computers can simultaneously deal with an 8-byte word (a 64-bit number).

3.1 BINARY ADDITION ARITHMETIC

Let us assume that two arithmetic number A and B each of four bits, which are to be added. Let us assume bits in A are A_3, A_2, A_1 and A_0 and bits in B are B_3, B_2, B_1 and B_0. Let Cy means any previous operation carry and Cy' means carry for future use. Table 3.1 shows how the bits add and equation numbers that we will be using later.

Let A_3, A_2, A_1 and A_0 are 1, 1, 0 and 1, respectively. B_3, B_2, B_1 and B_0 are 0, 1, 1 and 1, respectively. Use of Table 3.1 for the binary additions can be seen as under:

1 1 0 1 (= Decimal 13)

Add 0 1 1 1 (= Decimal 7).

TABLE 3.1

Input bits *A*, *B* and Cy of previous operation Carry			Output bits Sum, *S* and Cy' carry for future operation		Equation number
A	B	Cy	*S*	Cy'	
0	0	0	0	0	...(3.1)
0	0	1	1	0	...(3.2)
0	1	0	1	0	...(3.3)
0	1	1	0	1	...(3.4)
1	0	0	1	0	...(3.5)
1	0	1	0	1	...(3.6)
1	1	0	0	1	...(3.7)
1	1	1	1	1	...(3.8)

Let us start from the right most place, $p = 0$.

Step 1: Right most place, $p = 0$ operation 1 + 1 + carry 0 = Sum 0 and carry Cy' = 1 Equation (3.7)

Step 2: Place $p = 1$ left to right most operation 0 + 1 + carry 1 = Sum 0 and Cy' = 1 Equation (3.4)

Step 3: Place $p = 2$, left to the place at step 2 operation 1 + 1 + carry 1 = Sum 1 and Cy' = 1 Equation (3.8)

Step 4: Place $p = 3$ left to place at step 3 operation 1 + 0 + carry 1 = Sum 0 and Cy' = 1 Equation (3.6)

Answer Sum = 0 1 0 0 [= Decimal 4] and carry =1 to p = 4. [Remember carry is fifth place from the right most place.]

Above answer is correct because the Cy' =1 at $p = 4$ has weight = 16 and 16 + 4 = 20 decimal. This is what was the answer expected.

In brief, the above operation can be written as follows:

Carry 1 1 1 1

1 1 0 1 [= Decimal 13.]

Add 0 1 1 1 [= Decimal 7.]

Sum = 0 1 0 0 and Carry at 5th place $p = 4$ is equal to 1. [= Decimal 20]

Points to Remember

- Sum of two bits 0 and 0 is 0 if the previous carry added is 0.
- Sum of two bits 0 and 0 is 1 if the previous carry added is 1.
- Sum of two bits 1 and 0 is 1 if the previous carry added is 0.
- Sum of two bits 1 and 0 is 0 if the previous carry added is 1.
- Sum of two bits 1 and 1 is 0 if the previous carry added is 0.
- Sum of two bits 1 and 1 is 1 if the previous carry added is 1.

3.2 ARITHMETIC NUMBER REPRESENTATIONS

Arithmetic binary numbers are of three types.

3.2.1 Unsigned Number

An unsigned number is very frequently used in arithmetic. For example, in defining number of apples, bananas, currency notes, people etc.

Points to Remember

Maximum four bit unsigned number is 15 [1111 binary = 15 decimal ($2^4 - 1$)].

Maximum 16 bit unsigned number is 65535 decimal as 1111 1111 1111 1111 binary = decimal ($2^{16} - 1$).

Maximum 32 bit unsigned number is 1111 1111 1111 1111 1111 1111 1111 1111 binary = 4294967695 decimal ($2^{32} - 1$).

3.2.2 Signed Magnitude Number

A signed magnitude number means that it can be positive and negative depending on its maximum significant bit (msb) = 0 or 1, respectively. *In a signed magnitude number, maximum significant bit (on left most side) represents sign*. Such a representation has been found impractical to implement arithmetic subtraction operations by the logic gates easily. It is therefore used very rarely for subtractions in a computer.

Points to Remember

A 4 bit signed magnitude number if 0111 will represent +7, as msb is 0 and if 1111 will represent –7, as msb is 1.

An 8 bit signed number if 0111 1111 will represent +127 as msb is 0 and if 1111 1111 will represent –127 as msb is 1.

3.2.3 Two's Complement

A signed number representation given in (ii) above has been found less practical to implement subtraction by the logic gates. To represent a negative number, we simply take an unsigned number of 8 bits or 16 bits or 32 bits and find two's complement number of 8 bits, 16 bits or 32 bits, respectively. A two's complement signed number representation given here has been found very easy to implement by the logic gates and lets us perform the arithmetic operations of subtraction easily.

To find two's complement, we first complement all the bits (that is find one's complement) and then perform an increment by 1 (Complementing a bit or one's complement means convert all 1s as 0s and all 0s as 1s).

Let us consider, for example, a four-bit number 0101 (+5 decimal). Now, if we complement all four bits of it, we get 1010. [A complement means a NOT logical operation individually on each bit.] Let us now increment by 1, which means add 0001. We get the final result, two's complement = 1011.

In brief, the above operation can be written as follows:

	0 1 0 1 [= decimal 5].
One's Complement	1 0 1 0
Add	0 0 0 1
Sum	1 0 1 1
Therefore,	
Two's Complement =	1 0 1 1 [= decimal –5].

Now 1011 can be said to represent – 5 (negative decimal five) as it's two's complement twice gives back the original number 0101 (+5) (if we say 'find two's complement of a number', it means multiply that number by –1 or negate that number. If we say find two's complement and then again find two's complement, it means a two's complement is found twice).

Two's complement of a 16-bit number is also found in same way. Number 0100 0000 0000 1000 will have two's complement = (1011 1111 1111 0111 + 0000 0000 0000 0001) = (1011 1111 1111 1000). Now let us again find the two's complement of 1011 1111 1111 1000 = (0100 0000 0000 0111+ 0000 0000 0000 0001) = (0100 0000 0000 1000).

In brief, the above operation can be written as follows:

	0100 0000 0000 1000 (= decimal 16392).
One's Complement	1011 1111 1111 0111
Add	0000 0000 0000 0001
Two's Complement	1011 1111 1111 1000 (= decimal – 16392).

We recover the same number back after finding two's complement two times as if when we negate and then negate again, we recover same original number back. This proves that a two's complement of a number is an arithmetical like negative of that original number i.e. it is from an arithmetic multiplication by –1.

It is because of the above property, that two's complement is an excellent way to represent (and process by digital circuits) negative values of a number. Two's complement number is easy to implement by the gates, as the complementing and incrementing circuits are very practical to make.

Four bit binary number represents one of the 16 distinct values, 0 to 15_d (Table 2.1). Hence four bits binary arithmetic number can also represent 16 values between –8 to +7 (eight negative and eight positive).

Eight bit binary numbers represent only 256 distinct values, 0 to 255_d. Hence eight bit binary arithmetic number can also represent 256 distinct values within –128 to 127.

An n-bit binary arithmetic number can be within -2^{n-1} and $(2^{n-1} - 1)$. Therefore, a 16-bit binary arithmetic number is within -2^{15} and $(2^{15} - 1)$. Also, a 32-bit binary arithmetic number is within -2^{31} and $(2^{31} - 1)$.

An important fact is therefore as under: the highest positive valid arithmetic number is when at msb there is 0 and all remaining bits are 1s. The highest positive 8 bits number is 0111 1111. The highest positive 16 bits number is 0111 11111111 1111.

The lowest negative valid arithmetic number is when at msb there is 1 and all bits are 0s. A lowest negative 8 bits number is 1000 0000. A lowest negative 16 bits number is 1000 0000 0000 0000.

An 8-bit arithmetic number is within 1000 0000 *to* 0111 1111.

A 16-bit arithmetic number is within 1000 0000 0000 0000 *to* 0111 1111 1111 1111.

Table 3.2 gives examples of two's complement numbers.

TABLE 3.2

Decimal value	Four bit binary two's complement representation	Eight bit binary two's complement representation	Sixteen bit binary two's complement representation
+1	0001_b	$0000\ 0001_b$	$00000000\ 0000\ 0001_b$
−1	1111_b	$1111\ 1111_b$	$11111111\ 1111\ 1111_b$
−2	1110_b	$1111\ 1110_b$	$11111111\ 1111\ 1110_b$
−3	1101_b	$1111\ 1101_b$	$11111111\ 1111\ 1101_b$
.	.	.	.
+8	Not Feasible*	$0000\ 1000_b$	$00000000\ 0000\ 1000_b$
−8	1000_b	$1111\ 1000_b$	$11111111\ 1111\ 1000_b$
−13	Not Feasible*	$1111\ 0011_b$	$11111111\ 1111\ 0011_b$
−14	Not Feasible*	$1111\ 0010_b$	$11111111\ 1111\ 0010_b$
.	.	.	.
+128	Not Feasible*	Not Feasible	$00000000\ 10000\ 0000_b$
−128	Not Feasible*	$1000\ 0000_b$	$11111111\ 1000\ 0000_b$
.	.		
−213	Not Feasible*	Not Feasible$	$11111111\ 0010\ 1011_b$
−214	Not Feasible*	Not Feasible$	$11111111\ 0010\ 1010_b$

*Four bits two's complement is defined down to −8 decimal, not −9, −10, ...

$*Eight bits two's complement is defined down to −128 decimal, not −129, −130, ...

Points to Remember

1. For representing the negative numbers and for arithmetic operations in a computer, use of two's complement number is a way.
2. A 4 bit signed two's complement number 0111 will represent +7 and then 1001 will be the two's complement and will represent −7.
3. An 8 bit signed two's complement number 0111 1111 will represent +127 and then 1000 0001 will be the two's complement and will represent −127.
4. Two's complement of a two's complement will return the original number because negation twice always returns the same original number. $x = [-(-x)]$
5. Addition of a number with its two's complement number will return all 0s because addition of x and $-x$ always gives 0. [0111 + 1001 = 0000.]
6. An n-bit binary arithmetic number lies within -2^{n-1} and $(2^{n-1} - 1)$ only.

7. The highest positive arithmetic number is when at msb there is 0 and all remaining bits are 1s.
8. The lowest negative arithmetic number is when at msb there is 1 and all bits are 0s.
9. An arithmetic numbers is the lowest number when at msb there is 1 and all bits are 0s and the highest number when at msb there is 0, and all remaining bits are 1s.

3.2.4 Sign Extended Two's Complement Number for the Arithmetic Operations on Bigger Numbers

Recall column 3 at rows 1 to 4 in Table 3.2. We shall note that if a four bit two's complement is found and then if an 8 bit complement is found, we simply extend the maximum significant bit (msb) at bit place $p = 3$ by copying it at p = 4, $p = 5$, $p = 6$ and $p = 7$.

Recall column 4 at rows 1 to 4 in Table 3.2. We shall also note that if an eight bits two's complement is found and then if a 16 bit complement is found, we simply extend the maximum significant bit (msb) at bit place $p = 7$ at column 3 by copying it at $p = 8$ up to $p = 15$.

Let two's complement of an 8 bit number is found as follows. Number 0000 1000 will have two's complement = (1111 0111 + 0000 0001) i.e. 1111 1000. Now if we have to find 16-bit two's complement number from an 8-bit number, we simple extend the sign by first finding the maximum significant bit (msb) and then extend it towards left successively to get the 16 bit number. Resulting number is 11111111 1111 1000. Also consider the following example:

Example

The 16 bit two's complement sign extension 0000 1000 of will be 0000 0000 0000 1000 because msb = 0. The 16 bit two's complement sign extended 1111 1000 is 1111 1111 1111 1000 because msb at b15 (15th place bit) = 1.

Remember the Following Arithmetic Operation

We simply extend the sign of maximum significant bit (msb) up to the new msb when converting an 8 bit arithmetic number to 16-bit or a 16 bit number to a 32-bit arithmetic number.

3.3 BINARY SUBTRACTION

Suppose we have to find $A - B$. We first represent A as a binary arithmetic number. Then find the two's complement of B. Add A with B's two's complement and we will get $A - B$. Table 3.3 shows this.

Columns 1 and 2 give the decimal values of A and B. Our goal is to find $A - B$. It is done in four steps. We first convert these into binary arithmetic numbers in steps 1 and 2. In step 3, we find negative of B by finding two's complement. Step 4: is binary addition of the 16-bit arithmetic numbers.

1. Step 1: Column 3 gives the binary representation of A.
2. Step 2: Column 4 gives the binary representation of B.

3. Step 3: Column 5 is obtained by two's complement of column 4 by first finding one's complement and then adding 1. It is thus binary representation of $-B$.
4. Step 4: Column 6 is obtained by binary addition of column 3 and 5. It is thus addition of A and $-B$.

TABLE 3.3

Decimal value A	Decimal value B	(Arithmetic binary) number A	(Arithmetic binary) number B	Two's complement (arithmetic binary) number $-B$	After addition of columns 3 and 5 $A-B$	Answer decimal
+128	+128	0000 0000 1000 0000	0000 0000 1000 0000	1111 1111 1000 0000	0000 0000 0000 0000	0
+128	–128	0000 0000 1000 0000	1111 1111 1000 0000	0000 0000 1000 0000	0000 0001 0000 0000	256
–128	+128	1111 1111 1000 0000	0000 0000 1000 0000	1111 1111 1000 0000	1111 1111 0000 0000	–256
+1020	+1017	0000 0011 1111 1100	0000 0011 1111 1001	1111 1100 0000 0111	0000 0000 0000 0011	3
+1020	–1017	0000 0011 1111 1100	1111 1100 0000 0111	0000 0011 1111 1001	0000 0111 1111 0101	2037
–1017	–1020	1111 1100 0000 0111	1111 1100 0000 0100	0000 0011 1111 1100	0000 0000 0000 0011	3
–1017	+1020	1111 1100 0000 0111	0000 0011 1111 1100	1111 1100 0000 0100	1111 1000 0000 1011	–2037

Decimal equivalent of column 6 is the answer, which is given in column 7. We find that provided we represent positive and negative number as the arithmetic numbers in two's complement representation, when we first find the two's complement of the value to be subtracted and then add it we get the answer, which is correct in each row.

Let us understand the steps in last row of the table, which is to find –1017 – 1020 (Subtract +1020 from –1017).

1. Step 1: 1111 1100 0000 0111 in last row column 3 is –1017.
2. Step 2: 0000 0011 1111 1100 in last row column 4 is +1020.
3. Step 3: 1111 1100 0000 0100 in last row column 5 is –1020.
4. Step 4: 1111 1000 0000 1011 in last row column 6 represents the answer –2037.

It can be seen as follows: If we take one's complement, we get 0000 0111 1111 0100. Now add 1 to it. We get 0000 0111 1111 0101. It is (1024 + 512 + 256 + 128 + 64 + 32 + 16 + 5) = 2037. Hence, 1111 1000 0000 1011 is –2037.

```
Carry   1111 1000 0000 100
        \\\\ \\\\ \\\\ \\\
        1111 1100 0000 0111    (= Decimal –1017).
        1111 1100 0000 0100    (= Decimal –1020).
        ===================
Add     1111 1000 0000 1011    (= Decimal –2037).
        ===================
```

3.4 BINARY MULTIPLICATION OF UNSIGNED NUMBERS

Let us consider the multiplication of two unsigned numbers, each of 4 bits.

Table 3.4 shows various steps to be sequentially executed before we obtain by an exercise the result of a multiplication of the 4 bits by 4 bits; X_3, X_2, X_1 and X_0 (for **X**) by Y_3, Y_2, Y_1 and Y_0 (for **Y**). The result is M_7, M_6, M_5, M_4, M_3, M_2, M_1 and M_0, an 8 bit number **M**. Table columns 3 to 10 correspond to places p of the A (multiplicand), B (multiplier) and M (result) unsigned numbers.

1. Shifting left means shifting the resulting bit to next higher significant place p.
2. Conventionally, the dot operator represents an AND operation. AND operations—

$$1.\,1 = 1 \quad ...(3.8)$$

$$1.\,0 = 0 \quad ...(3.9)$$

$$0.\,1 = 0 \quad ...(3.10)$$

$$0.\,0 = 0 \quad ...(3.11)$$

In the equations are used as follows: In Table 3.4, after the AND operation, $X_0.Y_0$ the resulting bit will be 1 if both bits X_0 and Y_0 are 1s else the resulting bit will be 0. In Table 3.4, after the AND operation on $X_3\,X_2\,X_1\,X_0.Y_0$ the resulting product P will be $X_3\,X_2\,X_1\,X_0$ if $Y_0 =$ 1. After the AND operation on $X_3\,X_2\,X_1\,X_0.Y_0$ the resulting product P will be 0000 if $Y_0 = 0$.

TABLE 3.4 Multiplication method for 4 bits by 4 bits of unsigned numbers x and y

Step No	Instruction steps	Resulting binary number	4 BITS → multiply by 4 bits →				X_3 Y_3	X_2 Y_2	X_1 Y_1	X_0 Y_0
			p = 7	p = 6	p = 5	p = 4	p = 3	p = 2	p = 1	p = 0
1.	AND WITH Y_0	$P0$					$X_3.Y_0$	$X_2.Y_0$	$X_1.Y_0$	$X_0.Y_0$
2.	AND WITH Y_1 and shift left	$P1$				$X_3.Y_1$	$X_2.Y_1$	$X_1.Y_1$	$X_0.Y_1$	
3.	Find Partial Sum, $P0 + P1$	$S0$			s5	s4	s3	s2	s1	s0
4.	AND WITH Y_2 and shift left	$P3$			$X_3.Y_2$	$X_2.Y_2$	$X_1.Y_2$	$X_0.Y_2$		
5.	Find Partial Sum, $S0 + P3$	$S1$		s'6	s'5	s'4	s'3	s'2	s'1	s'0
6.	AND WITH Y_3 and shift left	$P4$		$X_3.Y_3$	$X_2.Y_3$	$X_1.Y_3$	$X_0.Y_3$			
7.	Find Partial Sum, $S1 + P4$	$S2$	s''7	s''6	s''5	s''4	s''3	s''2	s''1	s''0
8.	RESULT from $S2$	**M = S2**	M_7	M_6	M_5	M_4	M_3	M_2	M_1	M_0

Now let us see the process of multiplication by exemplary multiplication (Table 3.5).

Let multiplicand be 10_d (1010_b)

Let multiplier be 13_d (1101_b).

It means $X_3 = 1, X_2 = 0, X_1 = 1$ and $X_0 = 0$ and means $Y_3 = 1$, $Y_2 = 1$, $Y_1 = 0$ and $Y_0 = 1$.

Table 3.5 shows the steps in multiplication.

Answer in last row is 1000 0010_b (= decimal 130) as expected.

In brief we can write above steps as under:

Step 1: $P0$	= x x x x 1 0 1 0
Step 2: $P1$	= x x x 0 0 0 0 _
Step 3: $S0$	= x x 0 0 1 0 1 0
Step 4: $P2$	= x x 1 0 1 0 _ _
Step 5: $S1$	= x 0 1 1 0 0 1 0
Step 6: $P3$	= x 1 0 1 0 _ _ _
Step 7: $S2$	= 1 0 0 0 0 0 1 0
M	= 1 0 0 0 0 0 1 0
	= Decimal 130.

TABLE 3.5 Steps in Multiplying 1010 with 1101

1.	AND WITH $Y_0 = 1$	$P0 = 1010_b$					$X_3.Y_0$	$X_2.Y_0$	$X_1.Y_0$	$X_0.Y_0$
							1	0	1	0
2.	AND WITH $Y_1 = 0$ and shift left	$P1$				$X_3.Y_1$	$X_2.Y_1$	$X_1.Y_1$	$X_0.Y_1$	
						0	0	0	0	
3.	Find Partial Sum, $P0 + P1$	$S0$			s5	s4	s3	s2	s1	s0
					0	0	1	0	1	0
4.	AND WITH $Y_2 = 1$ and shift left	$P3$			$X_3.Y_2$	$X_2.Y_2$	$X_1.Y_2$	$X_0.Y_2$		
					1	0	1	0		
5.	Find Partial Sum, $S0 + P3$	$S1$		s'6	s'5	s'4	s'3	s'2	s'1	s'0
				0	1	1	0	0	1	0
6.	AND WITH $Y_3 = 1$ and shift left	$P4$		$X_3.Y_3$	$X_2.Y_3$	$X_1.Y_3$	$X_0.Y_3$			
				1	0	1	0			
7.	Find Partial Sum, $S1 + P4$	$S2$	s'7	s'6	s"5	s"4	s"3	s"2	s"1	s"0
			1	0	0	0	0	0	1	0
8.	RESULT from $S2$	**M = S2**	M_7	M_6	M_5	M_4	M_3	M_2	M_1	M_0
			1	0	0	0	0	0	1	0

3.5 MULTIPLICATION METHOD WHEN MULTIPLIER IS 2^n WHERE n IS AN INTEGER

When multiplier is 2, we simply shift left once and we get the answer. Let the multiplicand be 1101 (decimal 13). If multiplier is 2, on shifting left once the result will be 11010 (decimal 26) as expected.

When multiplier is 4, we simply shift left twice as $4 = 2^2$, and we get the answer. Let multiplicand be 1101 (decimal 13). If multiplier is 4, on shifting left twice the result will be 110100 (decimal 52) as expected.

When multiplier is 8 we simply shift left thrice as $8 = 2^3$, and we get the answer. Let multiplicand be 1101 (decimal 13). If multiplier is 8, on shifting left thrice the result will be 1101000 (decimal 104) as expected.

> An unsigned number is shifted left n times for multiplication by 2^n to get the result.

3.6 MULTIPLICATION METHOD WHEN MULTIPLIER IS A SMALL NUMBER

When multiplier is a small number, n it may more efficient to add the multiplicand with itself that $n - 1$ times.

Let multiplicand be 1101 (decimal 13). Let multiplier be 3. Add multiplicand twice by itself. 1101 + 1101 + 1101 = 1000 1001 (decimal 39).

Let multiplicand be 1101 (decimal 13). Let multiplier is 5. Add multiplicand four times by itself. 1101 + 1101 + 1101 + 1101 + 1101 = 0100 0001 (decimal 65).

> When multiplier is a small number, perform binary addition of multiplicand by itself (multiplier – 1) times.

3.7 MULTIPLICATION METHOD FOR *n*-BIT BY *N*-BIT SIGNED NUMBERS, *X* AND *Y*

When we multiply the binary two's complement format arithmetic number, then process is as under:

1. Check if multiplicand X has msb = 0 if yes, take $X = X'$ as such. If no then let a = msb of $X = 1$.
2. If $a = 1$, find two's complement of X and make that equal to X'.
3. Check if multiplier Y has msb = 0 if yes, take $Y = Y'$ as such. If no then let b = msb of $Y = 1$.
4. If $b = 1$, find two's complement of Y and make that equal to Y'.
5. Multiply X' with the multiplier Y' as steps described in Table 3.4 and find M.
6. If $a = b = 1$, then result is $M' = M$.
7. If $a = b = 0$, then also result $M' = M$.
8. If either $a = 0$ and b = 1 or $a = 1$ and $b = 0$, then find two's complement of M and make that as result equal to M'.

Note that for an 8 bit signed arithmetic multiplicand or multiplier, the multiplication can only be if each is between –128 and + 127. Result will be two's complement format 16-bit arithmetic-number. Note that for 16-bit signed arithmetic multiplicand or multiplier, the multiplication can only be between –32768 and + 32767. Result will be two's complement format 32-bit arithmetic-number.

3.8 BINARY ARITHMETIC DIVISION BY SUCCESSIVE SUBTRACTION METHOD

Let dividend be X and divisor be Y. When we divide the unsigned format number (integers non fractional numbers), then process to get the resulting quotient Q and remainder R is as under:

> 1. Set the initial quotient = 0000.
> 2. Check if $X < Y$, if yes, then Q is unchanged and $R = X$. Stop the process.

3. If $X > Y$, increment the quotient (new Q in first cycle is = 0001, second cycle it will be 0010).
4. Find $X - Y$ using two's complement arithmetic and get the R.
5. Set $X = R$.
6. Repeat steps 2 to 5 till $X < Y$.
7. Now Q is the result for the quotient and R = finally left X is the final remainder.

Let X = 0111 and Y = 0011. Here X > Y.

Step 1: Q = 0000

Step 2: $X > Y$ and so go to next step and when X becomes $\leq Y$, then stop.

Step 3: Q = 0001

Step 4: Find $X - Y$ = 0111 – 0011 = 0111 + 1101 = 0100. R = 0100.

Step 5: $X = R$ = 0100.

Step 6: Repeat steps 2 to 5. We get Q = 0001 + 1 = 0010 and R = 0001.

Answer is Q = 0010 (decimal 2) and R = 0001 (decimal 1) as expected from division of 7 by 3.

EXAMPLES

Example 3.1

Binary addition of two 8-bit numbers 1010 1010 and 1100 1100.

Solution

	1 0 1 0 1 0 1 0 (= decimal 172)
Add	1 1 0 0 1 1 0 0 (= decimal 204)

Let us start from the right most place, $p = 0$.

Step 1: Right most place, $p = 0$ operation 0 + 0 + carry 0 = Sum 0 and Cy' = 0

Step 2: Place $p = 1$ operation 1 + 0 + carry 0 = Sum 1 and Cy' = 0

Step 3: Place $p = 2$ operation 0 + 1 + carry 0 = Sum 1 and Cy' = 0

Step 4: Place $p = 3$ operation 1 + 1 + carry 0 = Sum 0 and Cy' = 1

Step 5: Place, $p = 4$, the operation is 0 + 0 + carry 1 = Sum 1 and Cy' = 0

Step 6: Place $p = 5$ operation is 1 + 0 + carry 0 = Sum 1 and Cy' = 0

Step 7: Place $p = 6$ operation is 0 + 1 + carry 0 = Sum 1 and Cy' = 0

Step 8: Place $p = 7$ operation is 1 + 1 + carry 0 = Sum 0 and Cy' = 1

Answer Sum = 0111 0110 (= decimal 118) and carry =1 to $p = 8$. (remember it is ninth place with significance = 256).

Expected answer was 374 decimal, which meant carry 1 with significance of 256 and 118 in case of 8-bit addition.

Example 3.2 The numbers given in Table 3.6 are the valid unsigned numbers.

Solution

TABLE 3.6

16 bit Number	8 bit Number
0000 0000 0000 0000	1111 0000
1111 0000 0000 0111	1111 0000
0000 1111 0000 1111	1111 0000
1111 1111 11111 111	0000 1111

Example 3.3 The numbers given in Table 3.7 columns 1 and 4 are the valid signed magnitude numbers and Table 3.7 columns 2 and 3 gives the sign and decimal values, respectively. Table 3.7 columns 5 and 6 also gives the sign and decimal values, respectively.

Solution

We take the msb (maximum significant bit). If it is 1 the sign is assigned negative. If it is 0 the sign is assigned positive. Then the decimal value is found from the remaining bits other than the msb bit. Therefore, in column 4 of row 3, the 1111 1111, sign is –ve because msb = 1 and the magnitude of the remaining 7-bits is 111 1111 = 127. Also, in column 4 of row 2, the 0111 1111, sign is +ve because msb = 0 and the magnitude of the remaining 7-bits is 111 1111 = 127.

TABLE 3.7

16 bit number	Sign	Decimal value	8 bit number	Sign	Decimal value
0000 0000 0000 1111	+	15	0000 1100	+	12
1111 0000 0000 0000	–	28672	0111 1111	+	127
0000 1111 0000 0001	+	3841	1111 1111	–	127
1111 1111 1111 1111	–	32767	1111 0000	–	112

Example 3.4 Find the two's complement of positive numbers given in Table 3.8 column 1.

Solution

1. Step 1: We take the binary number in column 2.
2. Step 2: We then find one's complement by changing all 1s by 0s and all 0s by 1s. One's complement is now is as per column 3.
3. Step 3: Now add 0000 0001 to get two's complement in column 4.
4. Step 4: Now we verify by adding column 2 with column 4 and check whether result of binary addition is all bits = 0s or not.

TABLE 3.8

Decimal value	Step 1: Find 8 bit number	Step 2: Find one's complement	Step 3: Add 0000 0001 give two's complement number	Decimal value	Step 4: Verify by binary addition of columns 2 and 4
14	0000 1110	1111 0001	1111 0010	–14	00000000
65	0100 0001	1011 1110	1011 1111	–65	00000000
112	0111 0000	1000 1111	1001 0000	–112	00000000
113	0111 0001	1000 1110	1000 1111	–113	00000000

Example 3.5

Find whether a given number in columns 1 and 3 of Table 3.9 is a valid positive number in two's complement format.

Solution

We look at the msb, it should be 0 for a valid positive number in two's complement arithmetic representation.

TABLE 3.9

8 bit Number	Is it a valid +ve number?	16 bit number	Is it a valid +ve number?
0000 1110	Yes	1111 0010 1101 0010	No
0100 0001	Yes	0011 1111 0011 1111	Yes
1111 0000	No	0001 0000 1111 0000	Yes
1111 1001	No	1000 1111 0001 0000	No

Example 3.6

Find whether the given numbers in columns 1 and 3 of Table 3.10 are the valid negative numbers in two's complement format.

Solution

We look at the msb, it should be 1 for a valid negative number in two's complement arithmetic representation.

TABLE 3.10

8 bit number	Is it valid –ve number?	16 bit number	Is it valid –ve number?
0000 1110	No	1111 0010 1101 0010	Yes
0100 0001	No	0011 1111 0011 1111	No
1111 0000	Yes	0001 0000 1111 0000	No
1111 1001	Yes	1000 1111 0001 0000	Yes

Example 3.7 Find sign extended 16-bit two's complement format arithmetic number for the 8-bit numbers in Table 3.11.

Solution

We look at the msb, if it is 0, we extend 0s from place $p = 8$ to $p = 15$. If it is 1, we extend 1s from place $p = 8$ to $p = 15$.

TABLE 3.11

8 bit Signed number	Is it valid +ve or –ve number?	16 bit sign extended number	Is it valid +ve or –ve sign extended number?
0000 1110	+ve	0000 0000 0000 1110	+ve
0100 0001	+ve	0000 0000 0100 0001	+ve
1111 0000	–ve	1111 1111 1111 0000	–ve
1111 1001	–ve	1111 1111 1111 1001	–ve

Example 3.8 Subtract a negative number 1111 1111 1111 0000 (decimal – 16) in two's complement format by another negative number, 1111 0111 1111 1001 (decimal – 2055).

Solution

The steps to find 1111 1111 1111 0000 (decimal = –16) – 1111 0111 1111 1001 (decimal = – 2055) are as follows:

1. Step 1: Take the number as such 1111 1111 1111 0000.
2. Step 2: Find two's complement of 1111 0111 1111 1001 by first finding one's complement and the incrementing it by 1. One's complement is 0000 1000 0000 0110. Adding 1 gives 0000 1000 0000 0111.
3. Step 3: Binary addition of the followings:

```
Carry     1111 00 0000 0000 0
          \\\\ \\ \\\\ \\\\ \\

A =        1111 1111 1111 0000 (decimal = –16).
–B =       0000 1000 0000 0111 (decimal = +2055).
          ====================
           0000 0111 1111 0111
```

Answer is 0000 0111 1111 0111 (decimal = 2039). Leave the last carry.

4. Step 4: Answer is 0000 0111 1111 0111 (decimal = 2039).

Example 3.9 Ideally a 16 bit number when multiplied by a 16 bit number, the result can be up to 32 bits. Test whether answer after multiplication will be a 16 bit number or higher when we multiply a negative number 1111 1111 1111 0000 (decimal –16) in two's complement format by another negative number, 1111 0111 1111 1001 (decimal –2055).

Solution

$$A = 1111\ 1111\ 1111\ 0000 = -16_d = -2^4.$$

$$B = 1111\ 0111\ 1111\ 1001 = -2055_d = -(2^{11} + 2^2 + 2^1 + 2^0)$$

Magnitude of number $A = 2^4$ and magnitude of number $B > 2^{11}$. Answer after multiplication will be more than 2^{15}. Multiplication of two negative numbers is a positive number. Maximum positive number is $+ (2^{15} - 1)$. Hence multiplication is not feasible without sign extension of A and B to 20 bits or higher.

Example 3.10

Multiply a negative number 1111 1111 1110 0011 (decimal – 29] in two's complement format by another negative number, 1111 1011 1111 1001 (decimal – 1031).

Solution

The steps in multiplications are as under:

Step A: X = 1111 1111 1110 0011. It is a negative number as msb of $X = 1$.

Take two's complement

X' = 0000 0000 0001 1101 (decimal 29).

Step B: Y = 1111 1011 1111 1001. It is negative number as msb of $Y = 1$

Take two's complement

Y' = 0000 0100 0000 0111 (decimal 1031).

Step C: Instead of multiplying X' and Y', let us multiply Y' and X' because X' has smaller number of significant bits (= 5 only) (this is an intelligent move to reduce the number of steps in multiplication).

Multiplicand = 0000 0100 0000 0111

Multiplier = 0000 0000 0001 1101

= xxxx xxxx xxx1 1101 (considering up to significant bits only).

Step 1: $P0$ = xxxx xxxx xxxx xxxx 0000 0100 0000 0111

Step 2: $P1$ = xxxx xxxx xxxx xxx0 0000 0000 0000 000_

Step 3: $S0$ = xxxx xxxx xxxx xxx0 0000 0100 0000 0111

Step 4: $P2$ = xxxx xxxx xxxx xx00 0001 0000 0001 11__

Step 5: $S1$ = xxxx xxxx xxxx xx00 0001 0100 0010 0011

Step 6: $P3$ = xxxx xxxx xxxx x000 0010 0000 0011 1___

Step 7: $S2$ = xxxx xxxx xxxx x000 0011 0100 0101 1011

Step 8: $P4$ = xxxx xxxx xxxx 0000 0100 0000 0111 ____

Step 9: $S3$ = xxxx xxxx xxxx 0000 0111 0100 1100 1011

Since msb of both X and Y were 1, therefore the answer is positive. Therefore, further negation is not to be done here. The answer is M = 0000 0000 0000 0000 0111 0100 1100 1011 = decimal 29899 as expected from 1031×29.

Example 3.11

Multiply a negative number 1111 1111 1111 0000 (decimal –16) in two's complement format by another negative number, 1111 1011 1111 1001 (decimal –1031).

Solution

The steps in multiplications are as under:

Step A: X = 1111 1111 1111 0000. It is a negative number as msb of $X = 1$

Take two's complement

$X' = 0000\ 0000\ 0001\ 0000$

Step B: $Y = 1111\ 1011\ 1111\ 1001$. It is a negative number as msb of $Y = 1$

Take two's complement

$Y' = 0000\ 0100\ 000\ 0111$

Step C: Instead of multiplying X' and Y', let us multiply Y' and X' because X' has a smaller number of significant bits (= 5 only),

Multiplicand = 0000 0100 0000 0111

Multiplier = 0000 0000 0001 0000

= xxxx xxxx xxx1 0000 (considering up to significant bits only).

Step 1: $P0$ = x x x x xxxx xxxx xxxx 0000 0100 0000 0111

Shift left four times (as multiplier is 2^4), an intelligent move to reduce the number of steps in multiplication.

Step 2: $P1$ = x x x x xxxx xxxx xxx0 0000 1000 0000 1110

Step 3: $P2$ = x x x x xxxx xxxx xx00 0001 0000 0001 1100

Step 4: $P3$ = x x x x xxxx xxxx x000 0010 0000 0011 1000

Step 5: $P4$ = x x x x xxxx xxxx 0000 0100 0000 0111 0000

Since msb of both X and Y were 1, therefore the answer is positive. Therefore, further negation is not to be done.

Answer = 0000 0000 0000 0000 0100 0000 0111 0000 = decimal 16496.

Example 3.12 Divide 0000 0100 0000 (decimal + 64) by 0000 0010 0001 (decimal 33).

Solution

Let $X = 0000\ 0100\ 0000$ and $Y = 0000\ 0010\ 0001$. Here $X > Y$.

Step 1: $Q = 0000$

Step 2: $X > \text{Y}$ and so go to next step.

Step 3: $Q = 0001$

Step 4: Find $X - Y = 0000\ 0100\ 0000 - 0000\ 0010\ 0001 = 0000\ 1000\ 0000 + 1111\ 1101\ 1111 = R = 0000\ 0001\ 1111$. (Two's complement of 0000 0010 0001 is 1111 1101 1110 + 0000 0000 0001 = 1111 1101 1111).

Step 5: $X = R = 0001\ 1111$.

Step 6: Since $X < Y$ now, we get $Q = 0001$ and $R = 0001\ 1111$.

Answer is $Q = 0001$ (decimal 1) and $R = 11111$ (decimal 31) as expected from division of 64 by 33.

EXERCISES

1. Do binary addition of the rows and columns of arithmetic numbers in the Table 3.12.

TABLE 3.12

16 bit Number	8 bit Number	Result of binary addition of the row	
0000 0000	0000 1111	Cy = 0	0000 1111
1111 0000	1000 1111	Cy = 1	0111 1111
0000 1111	1111 0000	Cy = 0	1111 1111
1111 1111	1111 1111	Cy = 1	1111 1110
Column Total = ______	Column Total = ______	Column Total = ______	

2. The numbers given in Table 3.13 are the valid unsigned numbers. Find their decimal values.

TABLE 3.13

16 bit Number	Decimal value	8 bit Number	Decimal value
0000 0000 0000 0000		1111 0000	
1111 0000 0000 0111		1111 0000	
0000 1111 0000 1111		1111 0000	
1111 1111 1111 1111		0000 1111	

3. The numbers given in Table 3.14 are the valid signed magnitude numbers. Find their decimal values.

TABLE 3.14

16 bit Signed magnitude number	Decimal value	8 bit Signed magnitude number	Decimal value
0000 0000 0000 0000		1111 0000	
1111 0000 0000 0111		1111 0000	
0000 1111 0000 1111		1111 0000	
1111 1111 1111 1111		0000 1111	

4. The numbers given in Table 3.15 are the valid arithmetic two's complement numbers. Find their decimal values.

TABLE 3.15

16 bit arithmetic two's complement number	Decimal value	8 bits arithmetic two's complement Number	Decimal value
0000 0000 0000 0000		1111 0000	
1111 0000 0000 0111		1111 0000	
0000 1111 0000 1111		1111 0000	
1111 1111 1111 1111		0000 1111	

5. The numbers in columns 1 and 4 given in Table 3.16 are the decimal numbers. Find in case feasible a corresponding binary number. Enter * in case it is not possible to find valid number with given length of the bits.

TABLE 3.16

Decimal value	8 bits arithmetic two's complement number	8 bits Signed magnitude number	Decimal value	16 bits arithmetic two's complement number	16 bits unsigned number
120			192		
–120			3072		
127			44000		
–75			29899		
33			+32678		
–33			–32678		
–128			–40720		
192			–24000		

* Means not possible to find a valid number with given length of the bits.

6. Find sign extended two's complement format arithmetic-number from the given decimal numbers in Table 3.17.

TABLE 3.17

Decimal	8 bits arithmetic number	16 bits sign extended number	Decimal	16 bits arithmetic number	32 bits sign extended number
–112			11200		
79			–65		
–65			–32678		
36			+32767		

7. Subtract a negative number in column 1 or 3 by another number in column 2 or 4 of Table 3.18. (Fill all columns of the rows.)

TABLE 3.18

Decimal value *A*	Decimal value *B*	(Arithmetic binary) number *A*	(Arithmetic binary) number *B*	Two's complement (Arithmetic binary) number –*B*	After addition of columns 3 and 5 *A* – *B*	Answer (Decimal)
+128	+128					
+128	–128					
		1110 1111	0000 0000			
		1000 0001	1000 0000			
		0000 0011	0000 0011			
		1111 1100	1111 1001			

8. Multiply the numbers A and B in the Table 3.19.

TABLE 3.19

Multiplicand decimal value *A*	Multiplier decimal value *B*	(Arithmetic binary) number *A*	(Arithmetic binary) number *B*	Result *M* in binary arithmetic format	Result *M* in decimal format	Is binary multiplication result verified
+128	+128					
+128	–128					
		1110 1111	0000 0000			
		1000 0001	1000 0000			
		0100 0011	0010 0011			
		1111 1100	1111 1001			

9. Multiply the numbers *A* with *B* = 2, 16, 64, 32768 in the Table 3.20. Verify that the results in all rows at the last columns are same as expected.

TABLE 3.20

Multiplicand decimal value *A*	Multiplier decimal value *B*	(Arithmetic binary) number *A*	(Arithmetic binary) positive numbers *A*' and *B*' multiply	Result M' = *A*' × *B*' in binary arithmetic format by shift left method	From *M*' result *M* = *A* × *B* in binary arithmetic format	*M* in decimal
+128	2					
+128	16					
–128	64					
32568	32768					
65767	–2					
32767	–16					
32	–64					
256	–32768					

10. Divide a number 0111 1011 1111 1001 by divisor 0111 1111 1110 0011.
11. Fill the following Table 3.21.

TABLE 3.21

Number *A*	Number *B*	Binary addition of *A* and *B*	Subtraction of *A* with *B*	Multiplication of *A* and *B*	Division of *A* and *B* *Q* and *R*
1100 0011	1100 0111				
1010 0011	1010 0011				
0100 0000	0100 0000				
111 0000	111 0001				

Q = Quotient and *R* = Remainder

QUESTIONS

1. What is the range of 16 bit unsigned numbers?
2. What is the range of 16 bit signed magnitude numbers? Prove that 0 decimal has two representations?
3. What is the range of 16 bit signed two's complement numbers? Prove that there is only one representation for 0 decimal.
4. Prove by two examples that two's complement of a number taken twice returns the original number.
5. Prove by two examples that binary addition of a number and its two's complement always gives all 0s, provided the numbers are valid within the given length of the bits.
6. Show by an example that we can subtract both positive and negative numbers by two's complement arithmetic.
7. Your scholarship is 10000 per month. Find annual scholarship by binary multiplication.
8. Your scholarship is 10000 per month. Find annual scholarship by binary multiplication using left shift method (Hint: Find 8 and 4 months amount by left shifts thrice and twice. Add these two results, you get annual salary).
9. Show that for a positive number, just right shift by one place gives the division by 2.
10. A lady fashion designer attends one customer every 50 minutes. Find time taken by 10 customers using standard partial product and sums method.
11. A lady fashion designer attends one customer every 50 minutes. Find time taken by 10 customers using left shift based multiplication method (Hint: Find time spent for 2 customers by left shifting once. Find time spent on 8 customers by left shifting the result twice. Add these two results, you get the 10 customers time).
12. Find sum of a series $1 + 2 + 3 + 4 + \cdots + 99 + 100$ using binary multiplication approach (Hint: Use the standard arithmetic series sum formulas: Sum $= n(n+1)/2$).

13. Your four months total salary is 40000. Find monthly salary by binary division using right shift method.
14. Your study hours per month are 400. Find number of minutes you study per day in a 30-day month using binary arithmetic division.

CHAPTER 4

Boolean Algebra and Theorems, Minterms and Maxterms

OBJECTIVE

In Chapter 3, we discussed binary signed, unsigned and signed magnitude numbers and binary arithmetic operations. In this chapter, we shall learn logic operations and the concept of truth table. We shall learn Boolean algebraic laws and theorems. Topics such as sum of products (SOPs) and product of sums (POSs) and minterms and maxterms will also be studied.

Six digital-electronics logic-gates are used as basic element in a simple as well as complex circuit. These are NOT, NAND, AND, OR, NOR and XOR. An interesting combination is NOT-XOR.

4.1 THE NOT, AND, OR LOGIC OPERATIONS

Figure 4.1 gives the logic symbols as well as the truth tables of NOT, NAND, AND, OR gates.

4.1.1 The NOT Logic Operation

NOT operation [Figure 4.2(a)] means that the output is the complement of the input. If input is logic 1, an output F is logic 0 and if input A is logic 0, the output is logic 1. In other words,

$$F = \overline{A} \qquad ...(4.1)$$

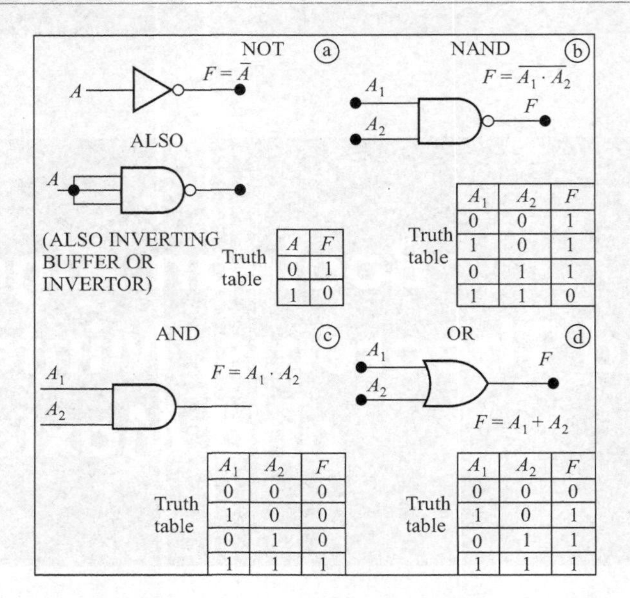

FIGURE 4.1 The logic symbols in a circuit and the truth tables of NOT, NAND, AND, OR operations.

'A' is a Boolean variable, that represents an input. It can have two values, 1 or 0. 'F' is also a Boolean variable, that represents an output. It have two states, 1 or 0. Bar over A denotes a NOT logic operation on A.

NOT in a circuit is generally represented by a triangle followed by a bubble or a bubble followed by a triangle (refer Figure 4.1(a)).

Point to Remember

NOT gate unique property is that output is 1 if input is at 0 logic state and output is 0 if input is at 1 logic state.

4.1.2 The AND Logic Operation

The unique property of AND is that its output is 0 unless all the inputs to it are at the logic 1s. It is represented by the symbol in Figure 4.1(c). Two-input (A_1 and A_2) AND-gate has the following way of writing its operations for an output F.

$$F = A_1 \cdot A_2 \qquad ...(4.2)$$

F, A_1 and A_2 are the Boolean variable representations and can either take the state = 1 or 0. Dot between two states indicates AND logic operation using these. Operation is such that when both inputs are 1s, the output is 1 else it is 0.

Three-input (A_1, A_2 and A_3) AND gate has the following way of writing its operations for the output F.

$$F = A_1 \cdot A_2 \cdot A_3 \qquad ...(4.3)$$

Operation is again such that when all three inputs are 1s, the output is 1 else it is 0.

Point to Remember

AND gate unique property is that output is 1 only if all of the inputs are at 1.

We will note later that AND symbol differs from NAND only by the omission of the bubble (circle) in that. We can note the differences in symbols and truth tables of NOT and AND in Figures 4.1(a) and (c).

Point to Remember

A truth table has 2^n rows for 2^n combination of n inputs. It gives in each of its row m outputs for a given combination. It gives the logic output(s) after the logical operations under different possible conditions of the input(s).

4.1.3 The OR Logic Operation

An OR operation means that the output is 0 only if all the inputs are 0. It is represented by the symbol in Figure 4.1(d). If any of the inputs is 1, the output F is 1. A two input OR gate has the following way of showing the operation.

$$F = A_1 + A_2 \quad ...(4.4)$$

Three inputs OR gate has the following way of showing the operation.

$$F = A_1 + A_2 + A_3 \quad ...(4.5)$$

Sign of + between the two logic states indicates an OR logic operation.

4.2 THE NAND AND NOR LOGIC OPERATIONS

4.2.1 NAND Gate

NAND gate property is that output is 1 if any of the input is at 0 logic state. Let us consider two inputs with the states A_1 and A_2 at a NAND gate.

$$\text{(Output) } F = \overline{A_1 \cdot A_2} \quad ...(4.6)$$

Bar denotes a NOT logic operation after the dot operation, $A_1.A_2$. The meaning of dot in $A_1.A_2$ is AND operation, which is explained in section 4.1.2.

Let us consider three inputs with the states A_1, A_2 and A_3 at the NAND gate.

$$\text{(Output) } F = \overline{A_1 \cdot A_2 \cdot A_3} \quad ...(4.7)$$

1. NAND gate with all inputs made common gives us the NOT operation. [Figure 4.1(a)]
2. NAND gate preceded by a NOT gate gives us AND gate.

Points to Remember

NAND gate unique property is that output is 1 if any of the input is at 0 logic state. NAND operation is also a NOT operation after an AND operation.

Figure 4.2 gives the logic symbols as well as truth tables of NOR, XOR NOT-XOR and NOT-NOT (Buffer) gates.

4.2.2 NOR Gate

An OR circuit followed a NOT circuit gives a NOR gate and is shown in Figure 4.2(a). Its unique property is that its output is 0 if any of the inputs is 1. A two input NOR has

$$F = A_1 + A_2 \qquad ...(4.8)$$

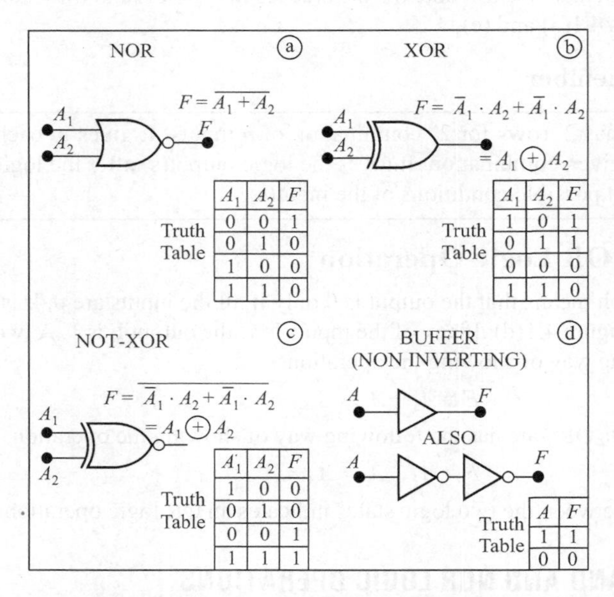

FIGURE 4.2 The logic symbols as well as truth tables of NOR, XOR NOT-XOR and Buffer gates.

4.3 THE XOR, NOT-XOR, NOT-NOT LOGIC OPERATIONS

4.3.1 XOR Logic Operation

An XOR gate (Figure 4.2(b)) is called Exclusive OR gate. Its unique property is that the output is 1 only if odd numbers of the inputs at it are 1s. The output—

$$F = A_1 \cdot \overline{A}_2 + \overline{A}_1 \cdot A_2 \text{ or } A_1 \oplus A_2 \qquad ...(4.9)$$

Exclusive OR operation symbol is ⊕ (Equation (4.9)) XOR symbol in a logic circuit is as shown in Figure 4.2(b). XOR is important in the circuits for addition of two binary numbers. (Section 3.1)

4.3.2 NOT-XOR (XNOR) Logic Operation

A NOT-XOR gate (Figure 4.2 (c)) has a unique property that the output is 0 only if odd numbers of the input are 1s. It is like a NOR gate but differs in the output when even number of its inputs are 1s.

4.3.3 NOT-NOT Logic Operation

A NOT-NOT (also called a BUFFER gate) (Figure 4.2(d)) simply gives same state at output as that at the input.

Its output $F = \overline{\overline{A}}$ instead of $F = \overline{A}$ in the NOT logic operation.

Its symbol in a circuit is shown in Figure 4.2(d) along with its truth table.

4.4 BOOLEAN ALGEBRAIC RULES (FOR OUTPUTS FROM THE INPUTS)

George Boole in 1854 developed mathematics now referred as Boolean algebra. It differs from usual mathematics. It is based upon the logic. For example, $A + A = A$ because true or true remains (remember that a + stands for OR logic). Two truths cannot negate to false. In Boolean algebra, there are only two numbers 1 and 0 corresponding to true and false or high and low or OFF and ON. Inversion (complementation) of true is false, and vice versa. In digital electronic logic circuits design, OR the algebraic rules are highly useful. The following laws (rules) can be said to be associated with the Boolean algebra.

4.4.1 OR Rules

The OR laws are described by following equations:

$$A + 1 = 1 \quad \text{...(4.10a)}$$

$$A + 0 = A \quad \text{...(4.10b)}$$

$$A + A = A \quad \text{...(4.10c)}$$

$$A + \overline{A} = 1 \quad \text{...(4.10d)}$$

An OR operation is denoted by a plus sign. OR laws means (i) any number (0 or 1) is a first input to an OR gate and another number at the second input is 1 then answer is 1, (ii) if another is 0 then answer is same as first input and (iii) If two inputs to an OR gate complements then output is '1'. These can be observed in an inset of the truth table in Figure 4.1(d).

4.4.2 AND Rules

Let us recall an AND operation (Equations (3.8) to (3.11)). It is denoted by the dot sign. True and true make true. True and false make false. False and false also remain false.

$$A \cdot 1 = A \quad \text{...(4.11a)}$$

$$A \cdot 0 = 0 \quad \text{...(4.11b)}$$

$$A \cdot A = A \quad \text{...(4.11c)}$$

$$A \cdot \overline{A} = 0 \quad \text{...(4.11d)}$$

These rules can be observed in an inset of the truth table in Figure 4.1(c).

A useful Boolean rule, which derives from equations (4.10a), (4.10b) and (4.11a), is as follows:

$$X + X \cdot Y = X \quad \text{...(4.11d)}$$

4.4.3 NOT Rules (Rules of Complementation)

A NOT operation is denoted by putting a bar over a number. A NOT true means false. A NOT false means true.

$$\overline{0} = 1 \quad \text{...(4.12a)}$$

$$\overline{1} = 0 \quad \text{...(4.12b)}$$

$$\overline{\overline{A}} = A \quad \text{...(4.12c)}$$

Equation (4.12c) means that if A is inverted (complemented) and then again inverted, we get the original number (Figure 4.2(d)). Let us refer to an inset of the truth table in Figure 4.1(a) for equations (4.12a and b).

4.5 BOOLEAN ALGEBRAIC LAWS

4.5.1 Commutative Laws

These laws mean that order of a logical operation is immaterial.

$$A_1 + A_2 = A_2 + A_1 \quad \text{...(4.13a)}$$

$$A_1 \cdot A_2 = A_2 \cdot A_1 \quad \text{...(4.13b)}$$

4.5.2 Associative Laws

These Laws allow a grouping of the Boolean variables.

$$X + (Y + Z) = (X + Y) + Z \quad \text{...(4.14a)}$$

$$X \cdot (Y \cdot Z) = (X \cdot Y) \cdot Z \quad \text{...(4.14b)}$$

4.5.3 Distributive Laws

These laws simplify the problems in the logic designs.

$$X \cdot (Y + Z) = (X \cdot Y) + (X \cdot Z) \quad \text{...(4.15a)}$$

$$X + (Y \cdot Z) = (X + Y) \cdot (X + Z) = (X + Y) \cdot (X + Y + Z) \quad \text{...(4.15b)}$$

$$X + (\overline{X} \cdot Y) = X + Y \quad \text{...(4.15c)}$$

The last two equations are typical to Boolean algebra, and are not followed in the usual algebra.

4.6 DEMORGAN THEOREMS

First theorem shows an equivalence of a NOR gate with an AND gate having bubbled inputs (Figure 4.3(a)), and is given by the equation:

$$\overline{A_1 + A_2} = \overline{A_1} \cdot \overline{A_2} \quad \text{...(4.16a)}$$

Point to Remember

> First DeMorgan theorem states, complement of two or more variables and then AND operation on these is equivalent to NOR operation on these variables. (NOR means complement of two or more variables OR).

Second theorem shows an equivalence of a NAND gate with an OR having bubbled inputs as shown in Figure 4.3(b) and is given by the equation:

$$\overline{A_1 \cdot A_2} = \overline{A_1} + \overline{A_2} \quad \text{...(4.16b)}$$

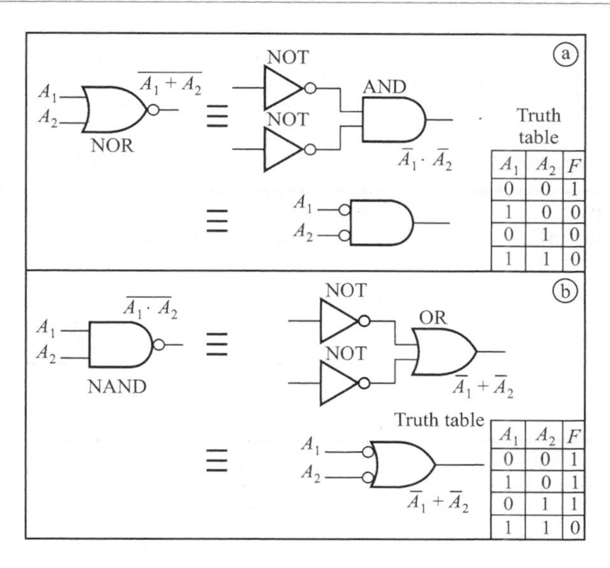

FIGURE 4.3 (a) DeMorgan first theorem showing an equivalence of NOR gate (b) DeMorgan second theorem showing an equivalence of NAND gate.

Point to Remember

> Second DeMorgan theorem states that complement of two or more variables and then OR operation on these is equivalent to a NAND operation on these variables (NAND means complement of two or more variables AND).

In fact both the equations (4.16a and 4.16b) also hold for the cases of the multiple (more than two) inputs.

$$\overline{A_1 + A_2 + A_3 + \ldots} = \overline{A_1} \cdot \overline{A_2} \cdot \overline{A_3} \ldots \qquad \text{...(4.17a)}$$

$$\overline{A_1 . A_2 . A_3 \ldots} = \overline{A_1} + \overline{A_2} + \overline{A_3} \ldots \qquad \text{...(4.17b)}$$

These theorems find wide use in the digital logic circuit's design. A circuit is implementable by one single basic logic gate used as a basic building gate. (A house is constructed easily using the identical bricks). Similar is the criteria for choosing only the NANDs or the NORs as basic building units in the digital ICs.

Simple Formulae to Remember for the Applications

Tables 4.1 and 4.2 give methods using three steps for introducing a complement over whole term and for removal of complete bar over the whole term, respectively. We note the following from DeMorgan theorems in Equations (4.16) and (4.17), if we use three steps:

1. complement each term on right hand side, then
2. convert the dot (AND) operation to + (OR) operation or + (OR) operation to dot (AND) operation, dot sign to + sign and vice vera and then
3. complement the remaining expression on the whole, we get the left hand side terms.

> We can remove bar over the entire expression, by complementing each variable under it and then changing sign if + to a dot and if dot then to a +.

TABLE 4.1 Three step method for introducing complement over whole term in a Boolean expression

Exemplary term	Step 1: complementation	Step 2: Sign change	Step 3: Complementation over whole term
$\overline{C} + D + \overline{E}$	$C + \overline{D} + E$	$C \cdot \overline{D} \cdot E$	$\overline{C \cdot \overline{D} \cdot E}$
$(A + B) + \overline{C}$	$\overline{(A + B)} + C$	$\overline{(A + B)} \cdot C$	$\overline{\overline{(A + B)} \cdot C}$

TABLE 4.2 Removal of complete bar over the whole term in a Boolean expression

Exemplary term	Step 1: Complementation	Step 2: Sign Sign change	Step 3: Remove
$\overline{\overline{C} \cdot D \cdot E}$	$\overline{C \cdot \overline{D} \cdot \overline{E}}$	$\overline{C + \overline{D} + \overline{E}}$	$C + \overline{D} + \overline{E}$
$\overline{\overline{A} + C + D}$	$\overline{A + \overline{C} + \overline{D}}$	$\overline{A \cdot \overline{C} \cdot \overline{D}}$	$A \cdot \overline{C} \cdot \overline{D}$

4.7 THE SUM OF THE PRODUCTS (SOPs) AS PER BOOLEAN EXPRESSION AND MINTERMS

An output from the logic gates can be represented as the sum of the Minterms. Advantage of using the SOP form is that the functions of any combinatin of logic gates can be represented by the AND gates at the inputs and an OR gate at the outputs.

4.7.1 SOPs for Two Variables (Two Inputs) Case

Consider a two variable case, *A* and *B* in Table 4.3. Advantage of using the two variable SOP form is that functions of any two input logic gate functions or a truth table be represented by four ANDs at the inputs and an OR at an output.

For two input case, an output *S* for an XOR gate can be written by *S*0 as follows: (Table 4.3 shows how to select the minterms for the XOR, AND, OR and NAND gate outputs.)

$S0 = \overline{A} \cdot B + A \cdot \overline{B}$; ($S0 = 1$ when $A = 0$ and $B = 1$, and $S0 = 1$ when $A = 1$ and $B = 0$). In the table, for an AND the ($S1 = 1$ when $A = 1$ and $B = 1$). In the table, for an OR and ($S2 = 1$ when $A = 0$ and $B = 1$ or $A = 0$ and $B = 1$ $A = 1$ and $B = 1$).

From fourth column in Table 4.3 for an XOR we find that only second and third minterm are contributing in giving the output $S0 = 1$. There is simple way of writing the above SOP terms as follows:

$$S0 = mn1 + mn2 = \Sigma m\ (1, 2)$$

Therefore, the SOP for two inputs AND is $S1 = A \cdot B$. From fifth column in Table 4.3, we find that only fourth minterm is contributing in giving the output = 1

$$S1 = (m(3))$$

The SOP for two inputs OR is $S2 = \overline{A} \cdot B + A \cdot \overline{B} + A \cdot B$. From sixth column in Table 4.3, we find that second, third and fourth minterm are contributing in giving the output = 1. Therefore, the S output is as follows:

$$S2 = \Sigma m(1, 2, 3)$$

The SOP for two inputs NAND is $S3 = \overline{A} \cdot \overline{B} + \overline{A} \cdot B + A \cdot \overline{B}$. From seventh column in Table 4.3, we find that first, second and third minterm are contributing in giving the output = 1. Therefore, the S output is as follows:

$$S1 = \Sigma m(0, 1, 2).$$

TABLE 4.3

A Input	B Input	Minterm and Boolean function	Minterm selected when output required XOR *S0*	AND *S1*	OR *S2*	NAND *S3*
0	0	$mn0 = \overline{A} \cdot \overline{B}$	0	0	0	1
0	1	$mn1 = \overline{A} \cdot B$	1	0	1	1
1	0	$mn2 = A \cdot \overline{B}$	1	0	1	1
1	1	$mn3 = A \cdot B$	0	1	1	0

Note: Columns 1, 2, 4 are as per truth table for XOR in Figure 4.2(b). Columns 5, 6 and 7 are as per truth table of AND, OR and NAND.

4.7.2 SOPs for Three Variables (Three Inputs) Case

Consider a three-variables case, A, B and C (Table 4.4). Advantage of using SOP an form is that functions of any three input logic gate functions can be represented by maximum eight ANDs at the inputs and eight input OR at an outputs. Using column 6, the SOP for a logic circuit defined by $S1$ minterms can be written as follows:

$$S1' = \overline{A} \cdot B \cdot \overline{C} + A \cdot \overline{B} \cdot \overline{C} + A \cdot B \cdot C = \Sigma m(2, 4, 7) \quad ...(4.18)$$

The SOP output using column 7 for the $S2$ minterms can be written as follows:

$$S2' = \overline{A} \cdot B \cdot C + A \cdot \overline{B} \cdot \overline{C} + A \cdot \overline{B} \cdot C + AB\overline{C} = \Sigma m(3, 4, 5, 6) \quad ...(4.19)$$

The SOP using column 8 for the $S3'$ minterms can be written as follows:

$$S3' = \overline{A} \cdot \overline{B} \cdot C + \overline{A} \cdot B \cdot \overline{C} + \overline{A} \cdot B \cdot C + ABC = \Sigma m\ (1, 2, 3, 7) \quad ...(4.20)$$

The SOP using column 9 for the $S4'$ minterms can be written as follows:

$$S4' = A \cdot B \cdot C = \Sigma m\ (7) \quad ...(4.21)$$

TABLE 4.4

A	*B*	*C*	Minterm	Boolean function	Minterm selected when output required *S1'* (Eq. 4.18)	*S2'* (Eq. 4.19)	*S3'* (Eq. 4.20)	*S4'* (Eq. 4.21)
0	0	0	*mn0*	$\overline{A} \cdot \overline{B} \cdot \overline{C}$	0	0	0	0
0	0	1	*mn1*	$\overline{A} \cdot \overline{B} \cdot C$	0	0	1	0

Table 4.4 Contd.

0	1	0	*mn*2	$\overline{A}\cdot B\cdot\overline{C}$	1	0	1	0
0	1	1	*mn*3	$\overline{A}\cdot B\cdot C$	0	1	1	0
1	0	0	*mn*4	$A\cdot\overline{B}\cdot\overline{C}$	1	1	0	0
1	0	1	*mn*5	$A\cdot\overline{B}\cdot C$	0	1	0	0
1	1	0	*mn*6	$A\cdot B\cdot\overline{C}$	0	1	0	0
1	1	1	*mn*7	$A\cdot B\cdot C$	1	0	1	1

Note: Columns 1, 2, 3 and 6 define the truth table for *S*1′.

4.7.3 SOPs for Four Variables (Four Inputs) Case

Consider a four -variables case, *A*, *B*, *C* and *D* in Table 4.5. Advantage of using SOP form 4-inputs in logic circuit design is that functions of any four input logic gate functions can be represented by maximum sixteen ANDs at the inputs and sixteen input OR at the outputs. Using column 7, the SOP for the S5 minterms can be written as some of minterms mn2, mn10 and mn15 as follows:

$$S5 = \overline{A}\cdot\overline{B}\cdot C\cdot\overline{D} + A\cdot\overline{B}\cdot C\cdot\overline{D} + A\cdot B\cdot C\cdot D = \Sigma m\,(2, 10, 15) \qquad ...(4.22)$$

Above equation shows that the logic circuit to implement truth table in Table 4.5 can be drawn as shown in Figure 4.4 using 3 four-input NANDs and one three-input OR gate.

TABLE 4.5

A	*B*	*C*	*D*	Minterm	Boolean function	Minterm selected when output
						S5 Eq. 4.22
0	*0*	0	0	*mn*0	$\overline{A}\cdot\overline{B}\cdot\overline{C}\cdot\overline{D}$	0
0	*0*	0	1	*mn*1	$\overline{A}\cdot\overline{B}\cdot C\cdot\overline{D}$	0
0	*0*	1	0	*mn*2	$\overline{A}\cdot\overline{B}\cdot C\cdot\overline{D}$	1
0	*0*	1	1	*mn*3	$\overline{A}\cdot\overline{B}\cdot C\cdot D$	0
0	*1*	0	0	*mn*4	$\overline{A}\cdot B\cdot\overline{C}\cdot\overline{D}$	0
0	*1*	0	1	*mn*5	$\overline{A}\cdot B\cdot\overline{C}\cdot D$	0
0	*1*	1	0	*mn*5	$\overline{A}\cdot B\cdot C\cdot\overline{D}$	0
0	*1*	1	1	*mn*7	$\overline{A}\cdot B\cdot C\cdot D$	0
1	*0*	0	0	*mn*8	$A\cdot\overline{B}\cdot\overline{C}\cdot\overline{D}$	0
1	*0*	0	1	*mn*9	$A\cdot B\cdot\overline{C}\cdot\overline{D}$	0
1	*0*	1	0	*mn*10	$A\cdot\overline{B}\cdot C\cdot\overline{D}$	1
1	*0*	1	1	*mn*11	$A\cdot\overline{B}\cdot C\cdot D$	0
1	*1*	0	0	*mn*12	$A\cdot B\cdot\overline{C}\cdot\overline{D}$	0
1	*1*	0	1	*mn*13	$A\cdot B\cdot\overline{C}\cdot D$	0
1	*1*	1	0	*mn*14	$A\cdot B\cdot C\cdot\overline{D}$	0
1	*1*	1	1	*mn*15	$A\cdot B\cdot C\cdot D$	1

Note: Columns 1, 2, 3, 4, 8 define the truth table for outputs as per Boolean expression in Equation (4.22).

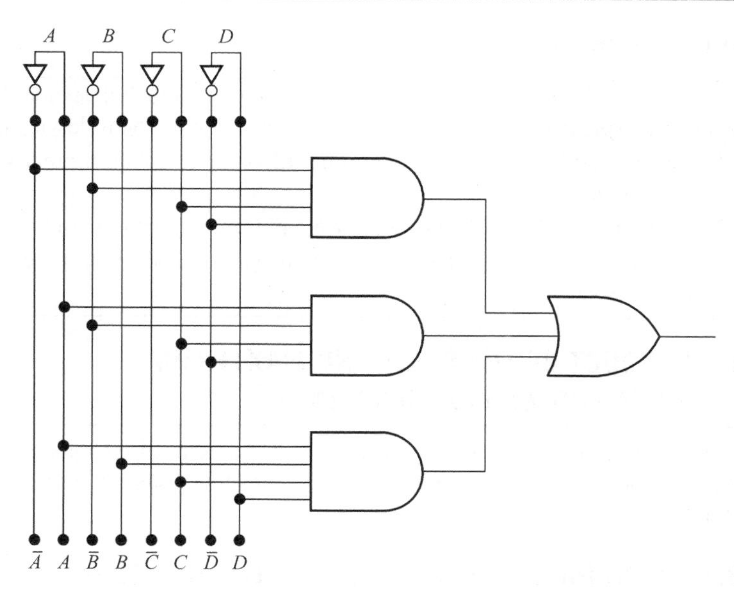

FIGURE 4.4 A logic circuit to implement Boolean expression (Eq. (4.22), truth table in Table 4.5) using the three four-input NANDs and one three-input OR gate.

Points to Remember

A simple way for laying the complement sign over a variable in m^{th} minterm of an n-variable SOP is as follows.

1. Expand m in the n-variable binary form.
2. At a place where the binary number is 0, put the complement sign and else no sign.

Let minterm is 5th and $S = \Sigma m\ (5)$ and SOP is in four variable form.

1. Expand m: 5 as 0101.
2. Lay the complement signs at places wherever it is 0 in m: $\overline{A} \cdot B \cdot \overline{C} \cdot D$.

Let minterm is 7th and $S = \Sigma m\ (7)$ and SOP is in four variable form.

1. Expand m: 1 as 0111.
2. Put complement signs at places wherever it is 0: $\overline{A} \cdot B \cdot C \cdot D$.

4.7.4 Conversion of a Boolean Expression or Truth Table Outputs into the Standard SOP Format

Suppose in a four variable SOP, there is a term with two variables only, SOP = $C \cdot D$. We perform AND operation with $(\overline{A} + A) \cdot (\overline{B} + B)$. (Using OR rule in equation (4.10d) that OR with a complement of the same variable or term with itself is always 1).

$$\text{SOP} = C \cdot D = (\overline{A} + A) \cdot (\overline{B} + B) \cdot C \cdot D = (\overline{A} + A)(\overline{B} \cdot C \cdot D + B \cdot C \cdot D)$$
$$= \overline{A} \cdot \overline{B} \cdot C \cdot D + A \cdot \overline{B} \cdot C \cdot D + A \cdot \overline{B} \cdot C \cdot D + A \cdot B \cdot C \cdot D$$
$$= \Sigma m(3, 7, 11, 15)$$

Points to Remember

When finding the minterms and converting to an n-variable SOP standard format, *in any term of the expression if a variable is not present*, then do the followings:

1. Perform additional AND operation with a term containing that variable and its complement.
2. Continue ANDing till all n variable are present in each term of the n-variable SOP.
3. Repeat the process for each term that has a missing variable in the Boolean expression.

4.8 PRODUCT OF THE SUMS AND MAXTERMS FOR A BOOLEAN EXPRESSION

An output from the logic gates can be represented as sum of the maxterms. Advantage of using POS form is that functions of any logic gates functions can be represented by OR at the inputs and an AND gate at the outputs.

4.8.1 POS for Two Variables (Two Inputs) Case

Consider a two variable case—A and B in Table 4.6. Advantage of using z-input variables POS form is that functions of any two inputs logic gates functions can be represented by four ORs at the inputs and an AND at the output.

Table 4.6 shows how to select the maxterms or XOR, AND, OR and NAND gates outputs.

TABLE 4.6

A	B	Maxterm and Boolean function	Maxterm selected when output required XOR $\overline{P0}$	AND $\overline{P1}$	OR $\overline{P2}$	NAND $\overline{P3}$
0	0	$Mx0 = A + B$	0	0	0	1
0	1	$Mx1 = A + \overline{B}$	1	0	1	1
1	0	$Mx2 = \overline{A} + B$	1	0	1	1
1	1	$Mx3 = \overline{A} + \overline{B}$	0	1	1	0

For two input case, an output P for an XOR gate can be written as follows: $\overline{P_0} = (A + B) \cdot (\overline{A} + \overline{B})$. The $\overline{P_0} = 0$ when $A = 0$ and $B = 0$ and when $A = 1$ and $B = 1$. From fourth column in Table 4.6, we find that only first and fourth maxterms are contributing in giving the output $\overline{P} = 0$. Also in maxterm notations, it can be written as follows:

$$\overline{P_0} = \Pi Mx\ (0, 3)$$

The POS for two-input AND is

$$\overline{P_1} = \Pi(A + B). (A + \overline{B}) \cdot (\overline{A} + B)$$

From fifth column in Table 4.6, we find that first, second and third maxterms are contributing in giving the output = 0

$$\overline{P_1} = \Pi Mx\,(0, 1, 2)$$

The POS for two inputs OR is

$\overline{P_2} = (A + B)$. From sixth column in Table 4.6, we find that only first maxterm is contributing in giving the output = 0.

$$\overline{P_2} = \Pi\, Mx\,(0)$$

The POS for two inputs NAND is

$\overline{P_3} = (\overline{A} + \overline{B})$. From seventh column in Table 4.6, we find that only fourth maxterm is contributing in giving the output = 0.

$$\overline{P_3} = \Pi\, Mx\,(3)$$

4.8.2 POS for Three Variables (Three Inputs) Case

Consider a three-variables case, A, B and C in Table 4.7. Advantage of using 3 input variables POS form is that functions of any three inputs logic gates functions can be represented by maximum eight ORs at the inputs and the AND at an output. Using column 6, the POS for the $P1'$ maxterms can be written as follows:

$$\overline{P1'} = (A + \overline{B} + C) \cdot (\overline{A} + B + C) \cdot (\overline{A} + \overline{B} + \overline{C}) = \Pi M(2, 4, 7) \quad ...(4.23)$$

The POS using column 7 for the $\overline{P2'}$ maxterms can be written as follows:

$$\overline{P2'} = (A + \overline{B} + \overline{C}) \cdot (\overline{A} + B + C) \cdot (\overline{A} + B + \overline{C}) \cdot (\overline{A} + \overline{B} + C) = \Pi M(3, 4, 5, 6) \quad ...(4.24)$$

The POS using column 8 for the $\overline{P3'}$ maxterms can be written as follows:

$$\overline{P3'} = (A + B + \overline{C}) \cdot (A + \overline{B} + C) \cdot (A + \overline{B} + \overline{C}) \cdot (\overline{A} + \overline{B} + \overline{C}) = \Pi M(1, 2, 3, 7) \quad ...(4.25)$$

The POS using column 9 for the $\overline{P4'}$ maxterm can be written as follows:

$$\overline{P4'} = (\overline{A} + \overline{B} + \overline{C}) = M(7) \quad ...(4.26)$$

TABLE 4.7

A	B	C	Maxterm		Maxterm selected when output required			
					$\overline{P1'}$ Eq. 4.23	$\overline{P2'}$ Eq. 4.24	$\overline{P3'}$ Eq. 4.25	$\overline{P4'}$ Eq. 4.26
0	0	0	Mx0	$A + B + C$	1	1	1	1
0	0	1	Mx1	$A + B + \overline{C}$	1	1	0	1
0	1	0	Mx2	$A + \overline{B} + C$	0	1	0	1
0	1	1	Mx3	$A + \overline{B} + \overline{C}$	1	0	0	1
1	0	0	Mx4	$\overline{A} + B + C$	0	0	1	1
1	0	1	Mx5	$\overline{A} + B + \overline{C}$	1	0	1	1
1	1	0	Mx6	$\overline{A} + \overline{B} + C$	1	0	1	1
1	1	1	Mx7	$\overline{A} + \overline{B} + \overline{C}$	0	1	0	0

4.8.3 POS for Four Variables (Four Inputs) Case

Consider a four-variables case, A, B, C and D in Table 4.8. Advantage of using 4-input variables POS form is that functions of any four inputs logic gates functions can be represented by maximum sixteen ORs at the inputs and the AND at an output. Using column 7, the POS for the $\overline{P5}$ maxterms can be written as follows:

$$\overline{P5} = (A + B + \overline{C} + D) \cdot (\overline{A} + B + \overline{C} + D) \cdot (\overline{A} + \overline{B} + \overline{C} + \overline{D}) = \Pi M(2, 10, 15) \quad ...(4.27)$$

TABLE 4.8

A	B	C	D	Maxterm	Boolean Function	Maxterm selected when output required
						$\overline{P5}$ (Eq. 4.27)
0	0	0	0	Mx0	$A + B + C + D$	1
0	0	0	1	Mx1	$A + B + C + \overline{D}$	1
0	0	1	0	Mx2	$A + B + \overline{C} + D$	0
0	0	1	1	Mx3	$A + B + \overline{C} + \overline{D}$	1
0	1	0	0	Mx4	$A + \overline{B} + C + D$	1
0	1	0	1	Mx5	$A + \overline{B} + C + \overline{D}$	1
0	1	1	0	Mx6	$A + \overline{B} + \overline{C} + D$	1
0	1	1	1	Mx7	$A + \overline{B} + \overline{C} + \overline{D}$	1
1	0	0	0	Mx8	$\overline{A} + B + C + D$	1
1	0	0	1	Mx9	$\overline{A} + B + C + \overline{D}$	1
1	0	1	0	Mx10	$\overline{A} + B + \overline{C} + D$	0
1	0	1	1	Mx11	$\overline{A} + B + \overline{C} + \overline{D}$	1
1	1	0	0	Mx12	$\overline{A} + \overline{B} + C + D$	1
1	1	0	1	Mx13	$\overline{A} + \overline{B} + C + \overline{D}$	1
1	1	1	0	Mx14	$\overline{A} + \overline{B} + \overline{C} + D$	1
1	1	1	1	Mx15	$\overline{A} + \overline{B} + \overline{C} + \overline{D}$	0

Points to Remember

A simple way for laying the complement sign over a variable in an Mth maxterm of an n-variable POS is to as follows.

1. Expand M in n-variable binary form.
2. At a ever place where the binary number is 1 put the complement sign else no sign such.

Let maxterm is 11th and $\overline{P} = \Sigma M(11)$ and POS is in four variable form.

1. Expand M 11: 1011.
2. Lay complement signs at places where ever it is 1 in M: POS = $\overline{A} + B + \overline{C} + \overline{D}$

Let maxterm is 6th and $\overline{P} = \Sigma M(6)$ and POS is in four variable form.

1. Expand M 6: 0110.
2. Lay the complement signs at places where ever it is 1: $A + \overline{B} + \overline{C} + D$.

4.8.4 Conversion of a Boolean Expression into Standard POS Format

When finding the maxterms and converting to an n-variable POS standard format, *in any maxterm of the expression if a variable is not present*, then do the followings:

1. Perform additional OR operation in the maxterm with a term containing that variable ANDed with complement of that and get two POS maxterms using a distributive law $X + (Y \cdot Z) = (X + Y) \cdot (X + Z)$ with Y and Z as variable and its complement, respectively.
2. Continue ORing till all n variable are present in each term of the n-variable POS.
3. Repeat the process for each term that has a missing variable in the Boolean expression.

The above steps are clarified below:

Assume that a standard POS in four-variable format is to be found for the following:

$$\overline{P} = (A + C + D) \cdot (A + \overline{B} + C + D) \text{ (Missing } B \text{ in first maxterm).}$$

First maxterm = $(A + \overline{B} \cdot B + C + D)$. Use ANDing rule that AND with the complement is 0, and ORing with 0 has no effect on an expression. Refer equations (4.11d) and (4.10b).

First term = $(A + \overline{B} + C + D)(A + B + C + D)$ (Using distributive law $X + (Y \cdot Z) = (X + Y) \cdot (X + Z)$ in equation (4.15b) by taking $X = A + C + D$, $Y = \overline{B}$ and $Z = B$).

Two terms in the last expression are in standard POS form of the first maxterm with variable B not present. Therefore,

$$\overline{P} = M(u) \cdot M(0) \cdot M(u)$$
$$= \Pi M(0, 4) \text{ using Equation (4.11c)}$$

■ EXAMPLES

Example 4.1

Use DeMorgan theorem to simplify $F = \overline{A + B} + \overline{C \cdot \overline{D} \cdot E}$. (Note: This is an example to remove the bar over the individual terms).

Solution

Use a *remove whole bar* formula for both terms as follows:

Step 1: Complement each variable and change sign: R.H.S. First term = $\overline{A} \cdot \overline{B}$ Second term = $\overline{C} + D + \overline{E}$.

R.H.S = $\overline{A} \cdot \overline{B} + \overline{C} + D + \overline{E}$

Example 4.2

Use DeMorgan theorem to simplify $F = \overline{A}\cdot\overline{B} + \overline{C}\cdot\overline{D}$. (Note: This is an example to place the bar over the terms).

Solution

Use a three step formula for both terms as follows:

Steps 1, 2 and 3: Complement, change sign and put bar over whole:

$$\overline{A+B} + \overline{C+D}$$

For further simplification, use the three step formula once again as follows:

Steps 1, 2 and 3: $\overline{(A+B)\cdot(C+D)}$

Example 4.3

Prove that $F = \overline{A}.B + A.\overline{B}$ is exclusive OR operation and it equals

$$= \overline{\overline{\overline{(A\cdot B)}\cdot A}\cdot\overline{\overline{(A\cdot B)}\cdot B}}$$

Solution

There is a dot (AND) operation between $\overline{A}$ and B. It means that A has to be 0 when $B = 1$ to get logic state 1 from the first term. There is a dot (AND) operation between A and B. It means that B has to be 0 when $A = 1$ to get logic-state 1 from the second term. XOR gate gives answer 1 only when two variable A and B are distinct from each other. Hence, it is proved that the $\overline{A}\cdot B + A\cdot\overline{B}$ is an XOR operation.

Now we prove the second part.

Use a *remove whole bar* formula for topmost bar as follows:

Step A: Complement each variable or term and convert sign:

$$(\overline{A\cdot B}).A + (\overline{A\cdot B})\cdot B$$

Use a remove whole bar formula for the bar as follows:

Step B: Complement each variable or term and convert sign:

$$(\overline{A} + \overline{B})\cdot A + (\overline{A} + \overline{B})\cdot B$$

Use distributive law [Eq. (4.15a)] to separately AND the A in first term and B in the second term as follows:

$$\overline{A}\cdot A + \overline{B}\cdot A + \overline{A}\cdot B + \overline{B}\cdot B$$

Use ANDing rule described by Equation (4.11d) that an AND between a variable and its complement is always 0. Hence, first and fourth terms are 0s. The ORs with 0 has no effect [Eq. (4.10b)]. Therefore, first and last terms can be removed

We get the second and third terms as follows:

$$A\cdot\overline{B} + \overline{A}\cdot B.$$

Note: Commutative law [Eq. (4.13b)] gives $\overline{B}\cdot A = A\cdot\overline{B}$.

Hence the L.H.S of expression for F is same as R.H.S. for F after simplification.

Example 4.4

Prove that for constructing XOR from NANDs we need four NAND gates.

Solution

We first prove the expression in Example 4.3. We have four bars over the ANDs. [$(\overline{A\cdot B})$ is counted once only.] Therefore, we need four NANDs for constructing an XOR gate.

Example 4.5

Prove $X+(\overline{X}\cdot Y)=X+Y$ a distributive law in Equation (4.15c).

Solution

Let $F=\overline{\overline{X}\cdot(\overline{\overline{X}\cdot Y})}$ (Using a three step formula for DeMorgan theorem application).

Solve for $\overline{F}$ first as follows:

$\overline{F}=\overline{X}\cdot(X+\overline{Y})$ (Using the remove whole bar formula on the second term within the bracket).

$=\overline{X}\cdot X+\overline{X}\cdot\overline{Y}$ (Use distributive law in Equation (4.15a)).

$=\overline{X}\cdot\overline{Y}$ (Use ANDing rule described by Equation (4.11d) that AND between a variable and its complement is always 0 and OR rule that OR of A with 0 gives A (Equation 4.10b)).

$\overline{F}=\overline{X+Y}$ (Using remove whole bar formula on the second term within the bracket).

$F=X+Y$ (Taking complement of L.H.S. and R.H.S. both)

Example 4.6

Prove $X+(Y\cdot Z)=(X+Y)\cdot(X+Z)=(X+Y)\cdot(X+Y+Z)$ a distributive law mentioned in Equation (4.15b). Use DeMorgan theorems.

Solution

It means

R.H.S $=\overline{(\overline{X+Y})+(\overline{X+Z})}$ (Using a three step formula for DeMorgan theorem application).

$=\overline{(\overline{X+Y})+\overline{X}\cdot\overline{Z}}$ (Using remove whole bar formula on the second term within the bracket).

$=\overline{(\overline{X}\cdot\overline{Y})+(\overline{X}\cdot\overline{Z})}$ (Using remove whole bar formula on the first term).

$=\overline{(\overline{Y}+\overline{Z})\cdot\overline{X}}$ (Using the distributive law in Equation (4.15a)). ...(4.28a)

L.H.S $=X+Y\cdot Z$

$=X+\overline{\overline{Z}+\overline{Y}}$ (Using deMorgan Theorem)

$=\overline{\overline{X}\cdot(\overline{Z}+\overline{Y})}$ (Using deMorgan Theorem) ...(4.28b)

L.H.S. = R.H.S. because Equation (4.28a) equals Equation (4.28b)

We proved that $X+(Y\cdot Z)=(X+Y)\cdot(X+Z)$.

It also means it is proved $(X+Y)+(Y\cdot Z)=(X+Y+Y)\cdot(X+Y+Z)=(X+Y)\cdot(X+Y+Z)$ (Replace X by $X+Y$ on both sides. Because if a law is true for X then it should also be true for $X+Y$).

Example 4.7

Prove $X+(Y\cdot Z)=(X+Y)\cdot(X+Z)$ a distributive law [Equation (4.15b)] by using AND and OR Boolean rules and Boolean laws.

Solution

Alternative solution of the problem in Example 4.6 is as follows:

$$(X+Y)\cdot(X+Z)$$

$= X \cdot X + X \cdot Z + Y \cdot X + Y \cdot Z$ (Use distributive law in Equation 4.15a).
$= X + X \cdot Z + Y \cdot X + Y \cdot Z$ (Using ANDing rule Equation (4.11c)).
$= X(1 + Z) + Y \cdot X + Y \cdot Z$ (Using the distributive law and taking $X = X.1$ due to the rule that AND with 1 returns the same term. [Equation (4.11a)]
$= X + Y \cdot X + Y \cdot Z$ (Using OR rule in equation (4.10a). OR with 1 give 1).
$= X(1 + Y) + Y \cdot Z$ (Using as above the equation (4.11a)).
$= X + Y \cdot Z$ (Using as above the equation (4.10a)).

Example 4.8 Simplify $A \cdot C + A \cdot (C + B) + C \cdot (C + B)$ using Boolean rules. and draw the simplest possible logic circuit.

Solution

Expression $= A \cdot C + A \cdot (C + B) + C \cdot (C + B)$
$= A \cdot C + A \cdot C + A \cdot B + C \cdot C + C \cdot B$ (Using distributive law in equation 4.15a))
$= A \cdot C + A \cdot B + C \cdot C + C \cdot B$ (Using OR rule that OR with itself gives the same term; equation (4.10c). So $A.C + A.C = A.C$)
$= A \cdot C + A \cdot B + C + C \cdot B$. (Using rule that AND operation by itself gives the same term. Equation (4.11c))
$= A \cdot C + A \cdot B + (1 + B) \cdot C$ (Using distributive law and taking $C = C \cdot 1$ due to the rule that AND with 1 returns the same term. Equation (4.11a))
$= A \cdot C + A \cdot B + C$ (Using OR rule in equation (4.10a). OR with 1 give 1)
$= A \cdot B + C \cdot (A + 1)$. (Using distributive law between first and third term)
$= AB + C$ (Using OR rule in Equation (4.10a). OR with 1 give 1)

Figure 4.5 shows the simplest logic circuit for $A \cdot C + A \cdot (C + B) + C(C + B)$.

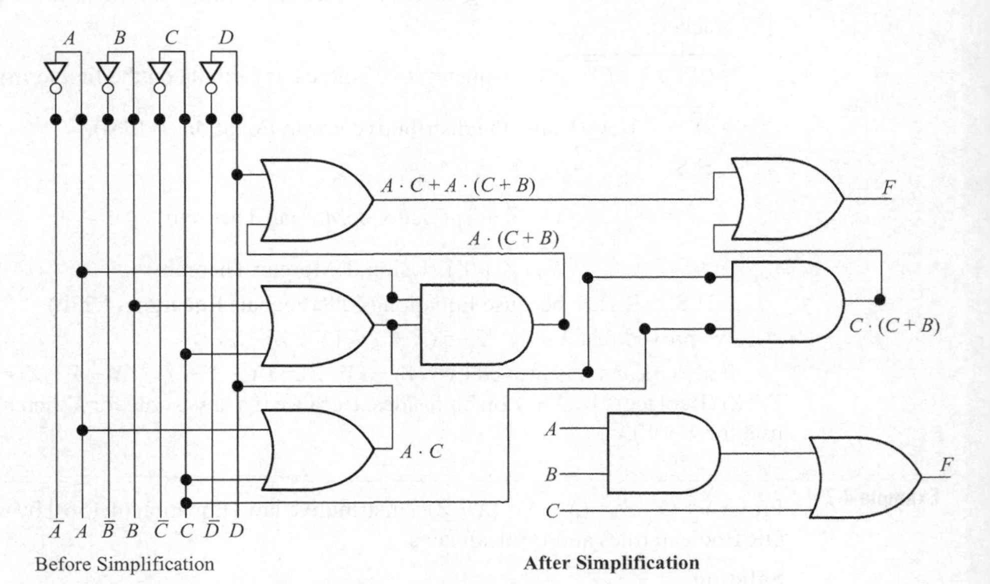

FIGURE 4.5 $A \cdot C + A \cdot (C + B) + C(C + B)$ after its simplification as $A.B + C$.

Example 4.9 Find minterms and give SOP form of a combinational logic circuit that has the truth table as in Table 4.9.

TABLE 4.9

Term number	*A*	*B*	*C*	*D*	Output *S*1
0	0	0	0	0	1
1	0	0	0	1	0
2	0	0	1	0	0
3	0	0	1	1	0
4	0	1	0	0	1
5	0	1	0	1	0
6	0	1	1	0	0
7	0	1	1	1	0
8	1	0	0	0	1
9	1	0	0	1	0
10	1	0	1	0	0
11	1	0	1	1	0
12	1	1	0	0	0
13	1	1	0	1	0
14	1	1	1	0	0
15	1	1	1	1	1

Solution

We note that output is 1 in line number 0, 4, 8 and 15. There are four minterms present

Therefore $\text{SOP} = \Sigma m\,(0, 4, 8, 15)$

$$= \overline{A} \cdot \overline{B} \cdot \overline{C} \cdot \overline{D} + \overline{A} \cdot B \cdot \overline{C} \cdot \overline{D} + A \cdot \overline{B} \cdot \overline{C} \cdot \overline{D} + A \cdot B \cdot C \cdot D$$

We use a simple way for laying the complement signs at the 0th, 4th, 8th and 15th terms. In an mth minterm in an n-variable SOP it is as follows.

1. Expand $m = 0, 4, 8$ and 15 in n variable binary form. 0000, 0100, 1000, 1111.
2. Where ever the binary number is 0 put the complement sign over the variable else leave as such.

Example 4.10 Find maxterms and give POS form of a combinational logic circuit that has the truth table as in Table 4.10.

TABLE 4.10

Term number	*A*	*B*	*C*	*D*	Output *P*1
0	0	0	0	0	0
1	0	0	0	1	1
2	0	0	1	0	1
3	0	0	1	1	1
4	0	1	0	0	0
5	0	1	0	1	1

Table 4.10 Contd.

6	*0*	*1*	1	0	1
7	*0*	*1*	1	1	1
8	*1*	*0*	0	0	0
9	*1*	*0*	0	1	1
10	*1*	*0*	1	0	1
11	*1*	*0*	1	1	1
12	*1*	*1*	0	0	1
13	*1*	*1*	0	1	1
14	*1*	*1*	1	0	1
15	*1*	*1*	1	1	1

Solution

We note that output is 0 in line number 0, 4, and 8. There are thus three maxterms present

Therefore $\text{POS} = \Sigma M\,(0, 4, 8)$

$$= (A + B + C + D) \cdot (A + \overline{B} + C + D) \cdot (\overline{A} + B + C + D)$$

We use a simple way for putting complement signs at the 0th, 4th, and 8th. In an Mth maxterm in a four variable POS is to as follows.

1. Expand $M = 0$, 4 and 8 in four variable binary form 0000, 0100, 1000
2. When binary number is 1 lay the complement sign over the variable else leave as such.

Example 4.11

Convert $A \cdot B \cdot C + A \cdot D$ expression into standard SOP format.

Solution

Since there are four variables *A*, *B*, *C* and *D* in the expression, we have to obtain four variable SOP for $ABC + AD$

First term $= A \cdot B \cdot C \cdot (\overline{D} + D)$ Using rule that ORing a variable with its complement is always 1. [Equation (4.10d)]

$$= A \cdot B \cdot C \cdot \overline{D} + A \cdot B \cdot C \cdot D$$

Second Term $= A \cdot D$

$$= A \cdot (B + \overline{B}) \cdot D$$

First for missing variable *B*, we use a rule that ORing a variable with its complement is always 1. [Equation (4.10d)]

$$= A \cdot \overline{B} \cdot D + A \cdot B \cdot D$$

Now for missing variable *C* in both terms, we use a rule that ORing a variable with its complement is always 1. [Equation (4.10d)]

$$= A \cdot \overline{B} \cdot (\overline{C} + C) \cdot D + A \cdot B \cdot (\overline{C} + C) \cdot D$$

$$= A \cdot \overline{B} \cdot \overline{C} \cdot D + A \cdot \overline{B} \cdot C \cdot D + A \cdot B \cdot C \cdot D + A \cdot B \cdot \overline{C} \cdot D.$$

$A \cdot B \cdot C + A \cdot D =$ First term SOP standard form + Second term SOP standard form

$$= A \cdot B \cdot C \cdot \overline{D} + A \cdot B \cdot C \cdot D + A \cdot \overline{B} \cdot \overline{C} \cdot D + A \cdot \overline{B} \cdot C \cdot D + A \cdot B \cdot C \cdot D + A \cdot B \cdot \overline{C} \cdot D.$$

$$= m(14) + m(15) + m(9) + m(11) + m(15) + m(13)$$

$$= \Sigma m(9, 11, 13, 14, 15)$$

Example 4.12 Convert $(A + B + C) \cdot (A + D)$ expression into standard POS format.

Solution

Since there are four-variables A, B, C and D in the expression, we have to obtain four variable POS.

$$(A + B + C) \cdot (A + D)$$

First term $= (A + B + C + \overline{D} \cdot D)$

Using rule that ANDing a variable with its complement is always 0; [Equation (4.11d)] Further ORing with 0 has no effect.

$= (A + B + C + \overline{D}) \cdot (A + B + C + D)$ [Using distributive law in equation (4.15b) that $X + Y \cdot Z = (X + Y) \cdot (X + Z)$.]

Second Term $= A \cdot D$

$= (A + B \cdot \overline{B} + D)$

Using rule that ANDing variable B with its complement is always 0; equation (4.11d). Further ORing with 0 has no effect.

$= (A + \overline{B} + D) \cdot (A + B + D)$ [Using distributive law in equation (4.15b) that $X + Y \cdot Z = (X + Y)(X + Z)$.]

$= (A + \overline{B} + \overline{C} \cdot C + D) \cdot (A + B + \overline{C} \cdot C + D)$

Using rule that ANDing variable C with its complement is always 0; [Equation (4.11d).] Further ORing with 0 has no effect.

$= (A + \overline{B} + \overline{C} + D) \cdot (A + \overline{B} + C + D) \cdot (A + B + \overline{C} + D) \cdot (A + B + C + D)$

[Using distributive law in equation (4.15b) that $X + Y \cdot Z = (X + Y)(X + Z)$.]

$(A + B + C).(A + D) =$ (First term POS standard form).(Second term POS standard form)

$= (A + B + C + \overline{D}) \cdot (A + B + C + D) \cdot (A + \overline{B} + \overline{C} + D) \cdot (A + \overline{B} + C + D) \cdot (A + B + \overline{C} + D) \cdot (A + B + C + D).$

$= (A + \overline{B} + \overline{C} + D) \cdot (A + \overline{B} + C + D) \cdot (A + B + \overline{C} + D) \cdot (A + B + C + D) \cdot (A + B + C + \overline{D})$

[Simplifying using rule that ANDing with itself returns the same variable or term; equation (4.11c).]

$= \Pi M(0, 1, 2, 4, 6)$ using Table 4.8.

EXERCISES

1. Use DeMorgan theorem to simplify $F = \overline{A + A \cdot B} + \overline{C \cdot D \cdot E}$.
   (Hint: Use a remove whole bar formula for both terms).
2. Use DeMorgan theorem to simplify $F = \overline{A} \cdot \overline{B} \cdot \overline{C} + \overline{D}$.
   (Hint: Use three step formula for both terms).

3. Prove that $\overline{F} = \overline{A} \cdot B + A \cdot \overline{B} + A \cdot B$ is NAND operation using DeMorgan theorems.
4. Construct XOR from NORs
5. Prove that $(A + B) \cdot (C + D) = \overline{A} \cdot \overline{B} + \overline{(C + D)}$.
6. Solve $\overline{A} \cdot \overline{B} + \overline{C} \cdot \overline{D} + E$ applying DeMorgan theorems to each term.
7. Prove $A + (B \cdot C \cdot D) = (A + B) \cdot (A + C)\,(A + D)$. Use DeMorgan theorems.
8. Prove $A + (B \cdot C \cdot D) = (A + B) \cdot (A + C)\,(A + D)$ by using AND and OR Boolean rules and laws.
9. Simplify $A \cdot B \cdot C + A(C + B) + C(C + B)$ using Boolean rules and draw the simplest possible logic circuit.
10. Find minterms and give SOP forms for $S1$, $S2$ and $S3$ of three combinational logic circuits that has the truth table outputs as in Table 4.11.

TABLE 4.11

Term number	A	B	C	D	Required output		
					S1	S2	S3
0	*0*	*0*	0	0	0	1	0
1	*0*	*0*	0	1	1	0	0
2	*0*	*0*	1	0	0	0	0
3	*0*	*0*	1	1	0	0	1
4	*0*	*1*	0	0	0	1	0
5	*0*	*1*	0	1	1	0	0
6	*0*	*1*	1	0	0	0	0
7	*0*	*1*	1	1	0	0	1
8	*1*	*0*	0	0	0	0	0
9	*1*	*0*	0	1	1	0	0
10	*1*	*0*	1	0	0	0	0
11	*1*	*0*	1	1	0	0	1
12	*1*	*1*	0	0	0	1	0
13	*1*	*1*	0	1	1	0	0
14	*1*	*1*	1	0	0	0	0
15	*1*	*1*	1	1	0	0	1

11. Find maxterms and give three POS form $P1$, $P2$ and $P3$ of three combinational logic circuits that has the truth table as in Table 4.12.

TABLE 4.12

Term number	A	B	C	D	Required output		
					$\overline{P1}$	$\overline{P2}$	$\overline{P3}$
0	*0*	*0*	0	0	1	1	0
1	*0*	*0*	0	1	1	0	1
2	*0*	*0*	1	0	0	1	1
3	*0*	*0*	1	1	1	1	1

Table 4.12 Contd.

4	*0*	*1*	0	0	1	1	1
5	*0*	*1*	0	1	1	1	0
6	*0*	*1*	1	0	1	0	1
7	*0*	*1*	1	1	0	1	1
8	*1*	*0*	0	0	1	1	1
9	*1*	*0*	0	1	1	1	1
10	*1*	*0*	1	0	1	1	0
11	*1*	*0*	1	1	1	0	1
12	*1*	*1*	0	0	0	1	1
13	*1*	*1*	0	1	1	1	0
14	*1*	*1*	1	0	1	0	1
15	*1*	*1*	1	1	0	1	1

12. Convert SOP = AB expression into four variable standard SOP format.
13. Convert $(\overline{A} + \overline{B} + C)$. expression into standard four variables POS format four variables.
14. Convert SOP = $\overline{AB} + \overline{BC} + \overline{AC}$ expression into three variable standard SOP format after simplifying using DeMorgan Theorem.
15. Convert POS $(A + B + \overline{C \cdot D})$ expression into standard four variables POS format four variables.
16. Construct SOP expression and POS expression for a four input NAND.

QUESTIONS

1. Give the OR and NOT rules in Boolean algebra?
2. Write the AND rules in Boolean algebra?
3. Explain associative, commutative and distributive laws with two examples each for their uses.
4. What is the advantage of simplifying using Boolean rules and laws?
5. Why do we use NANDs in fabricating a complex TTL based logic circuits?
6. Why do we use NORs in fabricating a complex CMOS based logic circuits?
7. How do we make a maximum four input circuit using ANDs and ORs using SOPs?
8. How do we make a maximum four input circuit using ORs and ANDs using POs?
9. How do we convert POS form to SOP form? Explain by an exemplary four-variable case.
10. How do we convert SOP form to POS form? Explain by an exemplary four-variable case.

CHAPTER 5

Karnaugh Map and Minimization Procedures

OBJECTIVE

In Chapter 4, we studied Boolean algebraic rules, laws and theorems and their uses in simplifying a logic circuit. We also discussed the concept of SOPs and POSs to write Boolean expression in terms of a standard format for implementation by AND gates and OR gates. In this chapter, we shall learn how to develop a Karnaugh map, tabulate the minterms, and minimize a circuit using the map. Computer-aided minimization procedure will also be studied.

5.1 THE THREE-VARIABLE KARNAUGH MAP AND TABLES

5.1.1 Karnaugh Map from the Truth Table

Recall the truth tables at the insets in Figures 4.1(a) to (d) and 4.2(a) to (d) and in Tables 4.3 to 4.12. We note that truth table has all possible inputs at the columns on the left side and the required or observed output(s) at the right hand side.

Table 5.1 shows an unfilled Karnaugh map. A three variable Karnaugh map is a two-dimensional map built from a truth table with cells arranged as per Table 5.1. Since number of rows in a three variable (three inputs) truth table are 8, the map has 8 cells; two cells horizontal and four cells vertical. It can also be vice versa.

Points to Remember

> Row 1st has cells for AB as 00. It corresponds $\overline{A}.\overline{B}$.
> Row 2nd has cells for AB as 01. It corresponds $\overline{A}.B$
> Row 3rd has cells for AB as 11 (Complement of row 1st). It corresponds $A.B$.
> Row 4th has cells for AB as 10. (Complement of row 2nd). It corresponds $A.\overline{B}$.
> Column 1st has cells for $C = 0$.
> Column 2nd has cells for $C = 1$.

TABLE 5.1 Three variable Karnaugh Map

AB \ C		$\overline{C}$ 0	C 1
$\overline{A}\,\overline{B}$	00		
$\overline{A}B$	01		
AB	11		
$A\overline{B}$	10		

Let in a three variable truth table, when $A = 0$, $B = 1$, $C = 0$, the output = 1. We place 1 in the 2nd row and column 1st. This is because 2nd row is for AB as 01 and column 1st is for $C = 0$.

When $A = 1$, $B = 1$, $C = 1$, we place 1 in 3rd row column 2nd. This is because 3rd row is for AB 11 and column 2nd is for $C = 1$.

Table 5.2 is a filled map with the above states of A, B and C at the truth table.

TABLE 5.2 Three variable Karnaugh Map for $S = 1$ an output only when $A = 0$, $B = 1$, $C = 0$, and $A = 1$, $B = 1$, $C = 1$ at a truth table of a given logic circuit

AB \ C		$\overline{C}$ $\overline{0}$	C 1
$\overline{A}\,\overline{B}$	00		
$\overline{A}B$	01	1	
AB	11		1
$\overline{A}\,\overline{B}$	10		

Remember the Following Operations

> 1. When output is 1 for a given combination of A, B and C, we place 1 at the corresponding cell.
> 2. Complete the step 1 for all the rows of truth table with output states = 1.

5.1.2 Karnaugh Map from the Minterms in a SOP

Recall Sections 4.7.1 to 4.7.3. Recall the Tables 4.3 to 4.5. A truth table for the output from a logic circuit can also be expressed as SOP minterms. Therefore, Karnaugh map can also be made from the minterms corresponding to an output.

Table 5.1 showed an unfilled Karnaugh map. A three variable Karnaugh map can be a two-dimensional map built from the minterms for a truth table. The cells are arranged as per Table 5.3. It shows the minterm numbers corresponding to the cells. Since maximum number of minterms in a three variables (three inputs) SOP can be 8, the map has 8 cells; two cells horizontal and four cells vertical (it can also be vice versa).

Points to Remember

Row 1st has cells for AB as 00. It corresponds $\overline{A}.\overline{B}$ part containing the minterms $m(0)$ and $m(1)$.

Row 2nd has cells for AB as 01. It corresponds $A.B$ containing minterms $m(2)$ and $m(3)$.

Row 3rd has cells for AB as 11 (complement of row 1st). It corresponds $A.B$ containing minterms $m(6)$ and $m(7)$.

Row 4th has cells for AB as 10 (complement of row 2nd). It corresponds $A.B$ containing minterms $m(4)$ and $m(5)$.

Column 1st has cells for $C = 0$.

Column 2nd has cells for $C = 1$.

TABLE 5.3 Three variable Karnaugh Map

AB \ C		$\overline{C}$ 0	C 1
$\overline{A}\,\overline{B}$	00	$m(0)$	$m(1)$
$\overline{A}B$	01	$m(2)$	$m(3)$
AB	11	$m(6)$	$m(7)$
$A\overline{B}$	10	$m(4)$	$m(5)$

Let in a three variable minterm, $\overline{A}BC$ is present and $A.\overline{B}.\overline{C}$ is present in a particular three input logic circuit output under different input states. It means $S = \Sigma m$ (3, 4). It means terms $m(3)$ and $m(4)$ are present. Therefore we place 1 in 2nd row column 2nd and place 1 in row 4th column 1.

Table 5.4 is a filled map with the above SOP output and above set of minterms.

TABLE 5.4 Three variable Karnaugh Map for $S = 1$ an output when SOP = Σm (3, 4) = $\overline{A}.B.C + A.\overline{B}.\overline{C}$

AB \ C		$\overline{C}$ 0	C 1
$\overline{A}\,\overline{B}$	00		
$\overline{A}B$	01		1
AB	11		
$A\overline{B}$	10	1	

Remember the Following Operation

1. When minterm is present in a SOP expression, we place 1 at the corresponding cell for the term using Table 5.3.
2. Complete the step 1 for all the rows for all minterms present in SOP of a Boolean expression for the logic circuit.

5.1.3 Karnaugh Map from the Maxterms IN A POS

Recall Sections 4.7.4 to 4.7.6. Recall the Tables 4.6 to 4.8. A truth table for the output from a logic circuit can also be expressed as POS maxterms. Therefore, Karnaugh map can also be made from the maxterms corresponding to an output.

Table 5.5 shows an unfilled Karnaugh map. A three variable Karnaugh map can be a two-dimensional map built from the maxterms for a truth table. The cells are arranged as per Table 5.5. It shows the maxterm numbers corresponding to the cells. Since maximum number of minterms in a three variables (three inputs) POS can be 8, the map has 8 cells; two cells horizontal and four cells vertical (It can also be vice versa).

Points to Remember

Row 1st has cells for AB as 00. It corresponds $(A + B)$ part containing the maxterm.

Row 2nd has cells for AB as 01. It corresponds $(A + \overline{B})$ containing minterm.

Row 3rd has cells for AB as 11. (complement of row 1st). It corresponds $(\overline{A} + \overline{B})$ containing minterm.

Row 4th has cells for AB as 10. (complement of row 2nd). It corresponds $(\overline{A} + B)$ containing minterm.

Column 1st has cells for $C = 0$.

Column 2nd has cells for $C = 1$.

TABLE 5.5 Three variable Karnaugh Map

AB \ C		C 0	$\overline{C}$ 1
$A + B$	00	$M(0)$	$M(1)$
$A + \overline{B}$	01	$M(2)$	$M(3)$
$\overline{A} + \overline{B}$	11	$M(6)$	$M(7)$
$\overline{A} + B$	10	$M(4)$	$M(5)$

Note: For map of maxterms, we just replace the dot (AND) sign operation with + (OR) sign and replace each variable A, B and C by the respective complements.

Let in a three variable map the maxterm, $(A + \overline{B} + \overline{C})$ is present and the $(\overline{A} + B + \overline{C})$ is present in a particular three input logic circuit output under different input states. It means $\overline{P} = \Pi\, M(3, 5)$. [$M(3)$ and $M(5)$ are present.] Therefore, we place 0 in 2nd row column 2nd and place 0 in row 4th column 2.

Table 5.6 is a filled map with the above POS output and above set of Maxterms in the cells.

TABLE 5.6 Three variable Karnaugh Map for $\overline{P}$ = an output when POS = $\Pi M(3, 5) = (A + \overline{B} + \overline{C}) . (\overline{A} + B + \overline{C})$

C / AB		C 0	$\overline{C}$ 1
$A + B$	00		
$A + \overline{B}$	01		0
$\overline{A} + \overline{B}$	11		
$\overline{A} + B$	10		0

Note: For map of Maxterms, we just fill 0s at the map using cells for maxterms defined in Table 5.3.

Remember the Following Operations

1. When maxterm is present in a POS expression, we place 0 at the corresponding cell for the term.
2. Complete the step 1 for all the rows for the maxterms of a Boolean expression or logic circuit corresponding to that.
3. For map of the POS maxterms from minterms K-map, we just consider a dot (AND) sign operation replaced with the + (OR) sign and each variable A, B and C just replaced by their respective complements. Also instead of placing 1s, we place 0s in POS Karnaugh maps (Compare maps Tables 5.5 with 5.3 and expands in Tables 5.6 and 5.4.).

5.2 FOUR VARIABLE KARNAUGH MAP AND TABLES

5.2.1 Karnaugh Map from the Truth Table

From Sections 5.1.1 to 5.1.3, we have learnt that for a three variable (input) case, a Karnaugh map is another way of representation of the truth table and the circuit logic outputs under various logic states of the inputs. Now consider the ***four variables*** (inputs) case.

Table 5.7 shows an unfilled four-variable Karnaugh map. A four variable Karnaugh map is a two-dimensional map built from a truth table with cells arranged as per Table 5.7. Since number of rows in a four variables (four inputs) truth table are 16, the map has 16 cells; four cells horizontal and four cells vertical.

Points to Remember

Row 1st has cells for AB as 00. It corresponds to $\overline{A}.\overline{B}$.

Row 2nd has cells for AB as 01. It corresponds to $\overline{A}.B$.

Row 3rd has cells for AB as 11. (Complement of row 1st). It corresponds to $A.B$.

Row 4th has cells for AB as 10. (Complement of row 2nd). It corresponds to $A.\overline{B}$.

Column 1st has cells for CD as 00. It corresponds to $\overline{C}.\overline{D}$.

Column 2nd has cells for CD as 01. It corresponds to $\overline{C}.D$.

Column 3rd has cells for CD as 11. (Complement of column 1st). It corresponds to $C.D$.
Column 4th has cells for CD as 10. (Complement of column 2nd). It corresponds to $C.\overline{D}$.

TABLE 5.7 Four variable Karnaugh Map

AB \ CD		$\overline{C}\,\overline{D}$ 00	$\overline{C}D$ 01	CD 11	$C\overline{D}$ 10
$\overline{A}\,\overline{B}$	00				
$\overline{A}B$	01				
AB	11				
$A\overline{B}$	10				

Let in a four variable truth table there are four rows in which output = 1,

1. When $A = 0, B = 0, C = 0, D = 1$, the output =1. We place 1 in the 1st row and column 2nd. This is because 1st row is for AB as 00 and column 2nd is for $CD = 01$.
2. When $A = 0, B = 0, C = 1, D = 1$, the output = 1. We place 1 in the 1st row and column 3rd. This is because 1st row is for AB as 00 and column 3rd is for $CD = 11$.
3. When $A = 0, B = 1, C = 0, D = 1$, the output = 1. We place 1 in the 2nd row and column 2nd. This is because 2nd row is for AB as 01 and column 2nd is for $CD = 01$.
4. When $A = 0, B = 1, C = 1, D = 1$, the output = 1. We place 1 in the 2nd row and column 3rd. This is because 2nd row is for AB as 01 and column 3rd is for $CD = 11$.

Table 5.8 is a filled map with the above states at a truth table.

TABLE 5.8 Four variable Karnaugh Map for output $S = A$ logic circuit in which the output = 1 when A, B, C, and D = 0001, 0011, 0101 and 0111, respectively

AB \ CD		$\overline{C}\,\overline{D}$ 00	$\overline{C}D$ 01	CD 11	$C\overline{D}$ 10
$\overline{A}\,\overline{B}$	00		1	1	
$\overline{A}B$	01		1	1	
AB	11				
$A\overline{B}$	10				

Points to Remember

1. When output is 1 for a given combination of A, B, C and D, we place 1 at the corresponding cell.
2. Complete the step 1 for all the rows of truth table with outputs = 1.

5.2.2 Karnaugh Map from the Minterms in an SOP

Four variables (input) Karnaugh map can also be made from the minterms corresponding to an output expression.

Table 5.9 shows an unfilled Karnaugh map. A four variable Karnaugh map can be a two-dimensional map built from the minterms for a truth table. The cells are arranged as per

Table 5.9. It shows the minterm numbers corresponding to the cells. Since maximum number of minterms in a four variables (four inputs) SOP can be 16, the map has 16 cells, four cells horizontal and four cells vertical.

Remember

Row 1st has cells for *AB* as 00. It corresponds $\overline{A}.\overline{B}$ part containing the minterm $m(0)$ to $m(2)$.

Row 2nd has cells for *AB* as 01. It corresponds $\overline{A}.B$ containing minterms $m(4)$ to $m(7)$.

Row 3rd has has cells for *AB* as 11 (Complement of row 1st). It corresponds $A.B$ containing minterms $m(12)$ to $m(15)$.

Row 4th has cells for *AB* as 10. (Complement of row 2nd). It corresponds $A.\overline{B}$ containing minterms $m(8)$ to $m(11)$.

Column 1st has cells for *CD* as 00. It corresponds $C.D$ part containing the minterms.

Column 2nd has cells for *CD* as 01. It corresponds to $C.D$ containing minterms.

Column 3rd has cells for *CD* as 11. (Complement of column 1st). It corresponds to $C.D$ containing minterms.

Column 4th cells for *CD* as 10. (Complement of column 2nd). It corresponds to $C.\overline{D}$ containing minterms.

TABLE 5.9 Four variables Karnaugh Map with minterms of the SOP Expression

AB \ CD		$\overline{C}\,\overline{D}$ 00	$\overline{C}D$ 01	CD 11	$C\overline{D}$ 10
$\overline{A}\,\overline{B}$	00	$m(0)$	$m(1)$	$m(3)$	$m(2)$
$\overline{A}B$	01	$m(4)$	$m(5)$	$m(7)$	$m(6)$
AB	11	$m(12)$	$m(13)$	$m(15)$	$m(14)$
$A\overline{B}$	10	$m(8)$	$m(9)$	$m(11)$	$m(10)$

Let in a four variable the minterms—$\overline{A}.\overline{B}.C.\overline{D}$ is present, $\overline{A}.\overline{B}.C.D$ is present, $\overline{A}.B.C.D$ is present, $\overline{A}.B.C.\overline{D}$ is present, $A.B.C.\overline{D}$ is present and $A.B.C.D$ is present in a particular four inputs logic circuit output under different input states. It means $S = \Sigma m$ (2, 3, 6, 7, 14, 15). It means 6 terms 2, 3, 6, 7, 14 and 15 are present. There we place 1s as shown in six cells of Table 5.10.

Table 5.10 is a filled map with the above SOP output and above set of six minterms.

TABLE 5.10 Four variable Karnaugh Map for S = 1–an output when SOP = Σm (2, 3, 6, 7, 14,15)

AB \ CD		$\overline{C}\,\overline{D}$ 00	$\overline{C}D$ 01	CD 11	$C\overline{D}$ 10
$\overline{A}\,\overline{B}$	00			1	1
$\overline{A}B$	01			1	1
AB	11			1	1
$A\overline{B}$	10				

Remember the Following Operations

1. When minterm is present in an SOP expression, we place 1 at the corresponding cell for the term.
2. Complete the step 1 for all the rows for the minterms of a Boolean expression or logic circuit corresponding to that.

5.2.3 Karnaugh Map from the Maxterms in a POS

Recall Sections 4.7.4 to 4.7.6. Recall the Tables 4.6 to 4.8. A four-input truth table for the output from a logic circuit can also be expressed as maximum 16 POS maxterms. Therefore, Karnaugh map can also be made from the maxterms corresponding to a four input logic circuit.

Table 5.7 showed an unfilled Karnaugh map. A four variable Karnaugh map can be a two-dimensional map built from the 16 maxterms for a four input truth table. The cells are arranged as per Table 5.11. It shows the maxterm numbers corresponding to the cells. Since maximum number of minterms in a four variables (four inputs) POS can be 16, the map has 16 cells; four cells horizontal and four cells vertical.

TABLE 5.11 Four variables Karnaugh Map for maximum 16 Maxterms in a POS expression

A + B \ C + D		$C + D$ 00	$C + \overline{D}$ 01	$\overline{C} + \overline{D}$ 11	$\overline{C} + D$ 10
$A + B$	00	$M(0)$	$M(1)$	$M(3)$	$M(2)$
$A + \overline{B}$	01	$M(4)$	$M(5)$	$M(7)$	$M(6)$
$\overline{A} + \overline{B}$	11	$M(12)$	$M(13)$	$M(15)$	$M(14)$
$\overline{A} + B$	10	$M(8)$	$M(9)$	$M(11)$	$M(10)$

Note: For map of maxterms, we just replace in Table 5.9 dot (AND) sign operation with + (OR) sign and replace each variable *A*, *B*, *C* and *D* by their respective complements.

Points to Remember

Row 1st has cells for *AB* as 00. It corresponds $A + B$ part containing the maxterms $M(0)$ to $M(3)$.

Row 2nd has cells for *AB* as 01. It corresponds $A + \overline{B}$ containing maxterm $M(4)$ to $M(7)$.

Row 3rd has cells for *AB* as 11. (Complement of row 1st). It corresponds $\overline{A} + \overline{B}$ containing maxterms $M(12)$ to $M(15)$.

Row 4th has cells for *AB* as 10. (Complement of row 2nd). It corresponds $A + B$ containing maxterms $M(8)$ to $M(11)$.

Column 1st has cells for *CD* as 00. It corresponds $C + D$ part containing maxterms.

Column 2nd has cells for *CD* as 01. It corresponds $C + \overline{D}$ containing maxterms.

Column 3rd has cells for *CD* as 11. (Complement of column 1st). It corresponds $\overline{C} + \overline{D}$ containing maxterms.

Column 4th has cells for *CD* as 10. (Complement of column 2nd). It corresponds $\overline{C} + D$ containing maxterms.

Let in a four variable Maxterms present are 0, 1, 2, 3 and 10 and 11—$(A+B+C+D)$ is present, $(A+B+C+\overline{D})$ is present, $(A+B+\overline{C}+\overline{D})$ is present, $(A+B+\overline{C}+D)$ is present, $(\overline{A}+B+\overline{C}+D)$ is present and $(\overline{A}+B+\overline{C}+\overline{D})$ is present in a particular four input logic circuit output under different input states. It means $\overline{P} = \Pi M\,(0, 1, 2, 3, 10, 11)$. We place 0s as per Table 5.12.

Table 5.12 is a filled map with the above POS output and above set of six maxterms.

TABLE 5.12 Four variable Karnaugh Map for $\overline{P} = 0$ an output when POS = ΠM (0, 1, 2, 3, 10, 11)

A + B \ C + D		$C + D$ 00	$C + \overline{D}$ 01	$\overline{C} + \overline{D}$ 11	$\overline{C} + D$ 10
$A + B$	00	0	0	0	0
$A + \overline{B}$	01				
$\overline{A} + \overline{B}$	11				
$\overline{A} + B$	10			0	0

Note: For map of maxterms, we just fill 0s at the map using Table 5.11.

Remember the Following Operations

1. When maxterm is present in a POS expression, we place 0 at the corresponding cell for the term.
2. Complete the step 1 for all the rows for the maxterms of a Boolean expression 0 or logic circuit corresponding to that.
3. From a map for SOP, for a map of the POS maxterms, we just consider a dot (AND) sign operation replaced with the + (OR) sign and each variable A, B, C and D just replaced by their respective complements. Also instead of placing 1s, we place 0s in POS Karnaugh maps.

5.3 FIVE AND SIX VARIABLE KARNAUGH MAPS AND TABLES

When the number of variables are five, we can make two maps, each with four variables, B, C, D and E with fifth variable $A = 0$ and $A = 1$, respectively. One map is considered as a upper layer map and other is considered as a lower layer map, when performing the minimization from the cell adjacencies. Table 5.13 shows a five variable Karnaugh map set.

Table 5.14 shows how to construct a six variable map. A six variable 64 cells Karnaugh map can be made a pair of two five variable map one upper for $A = 0$ and other lower for $A = 1$ each with 32 cells. It can also be made from a quad of four 4-variable maps; left upper for $A = 0$, $B = 0$, right upper for $A = 0$, $B = 1$, left lower for $A = 1$, $B = 1$ and right lower for $A = 1$ and $B = 0$, respectively. We can place one layer over another of four variable map in three dimension in the order, left upper, right upper, left lower and right lower, respectively. Total maximum minterms can be 64 for a six variable map.

TABLE 5.13 Five variables Karanaugh Map set with 32 minterms of the SOP expression and upper layer for $A = 0$ on the left side and lower layer for $A = 1$ on the right side

← $A = 0$ →

DE		$\overline{D}\,\overline{E}$	$\overline{D}E$	DE	$D\overline{E}$
BC		00	01	11	10
BC	00	$m(0)$	$m(1)$	$m(3)$	$m(2)$
BC	01	$m(4)$	$m(5)$	$m(7)$	$m(6)$
BC	11	$m(12)$	$m(13)$	$m(15)$	$m(14)$
BC	10	$m(8)$	$m(9)$	$m(11)$	$m(10)$

← $A = 1$ →

DE		$\overline{D}\,\overline{E}$	$\overline{D}E$	DE	$D\overline{E}$
BC		00	01	11	10
BC	00	$m(16)$	$m(17)$	$m(19)$	$m(18)$
BC	01	$m(20)$	$m(21)$	$m(23)$	$m(22)$
BC	11	$m(28)$	$m(29)$	$m(31)$	$m(30)$
BC	10	$m(24)$	$m(25)$	$m(27)$	$m(26)$

TABLE 5.14 Six-variable Karnaugh Map set with 64 minterms of the SOP expression and upper layer for $A = 0$ on the upper side and for $A = 1$ on the lower side

← $A = 0, B = 0$ →

EF		$\overline{E}\,\overline{F}$	$\overline{E}F$	EF	$E\overline{F}$
CD		00	01	11	10
$\overline{C}\,\overline{D}$	00	$m(0)$	$m(1)$	$m(3)$	$m(2)$
$\overline{C}D$	01	$m(4)$	$m(5)$	$m(7)$	$m(6)$
CD	11	$m(12)$	$m(13)$	$m(15)$	$m(14)$
$C\overline{D}$	10	$m(8)$	$m(9)$	$m(11)$	$m(10)$

← $A = 0, B = 1$ →

EF		$\overline{E}\,\overline{F}$	$\overline{E}F$	EF	$E\overline{F}$
CD		00	01	11	10
$\overline{C}\,\overline{D}$	00	$m(16)$	$m(17)$	$m(19)$	$m(18)$
$\overline{C}D$	01	$m(20)$	$m(21)$	$m(23)$	$m(22)$
CD	11	$m(28)$	$m(29)$	$m(31)$	$m(30)$
$C\overline{D}$	10	$m(24)$	$m(25)$	$m(27)$	$m(26)$

← $A = 1, B = 1$ →

EF		$\overline{E}\,\overline{F}$	$\overline{E}F$	EF	$E\overline{F}$
CD		00	01	11	10
$\overline{C}\,\overline{D}$	00	$m(48)$	$m(49)$	$m(51)$	$m(50)$
$\overline{C}D$	01	$m(52)$	$m(53)$	$m(55)$	$m(54)$
CD	11	$m(60)$	$m(61)$	$m(63)$	$m(62)$
$C\overline{D}$	10	$m(56)$	$m(57)$	$m(59)$	$m(58)$

← $A = 1, B = 0$ →

EF		$\overline{E}\,\overline{F}$	$\overline{E}F$	EF	$E\overline{F}$
CD		00	01	11	10
$\overline{C}\,\overline{D}$	00	$m(32)$	$m(33)$	$m(35)$	$m(34)$
$\overline{C}D$	01	$m(36)$	$m(37)$	$m(39)$	$m(38)$
CD	11	$m(44)$	$m(45)$	$m(47)$	$m(46)$
$C\overline{D}$	10	$m(40)$	$m(41)$	$m(43)$	$m(42)$

5.4 AN IMPORTANT FEATURE IN THE DESIGN OF A KARNAUGH MAP

5.4.1 Only Single Variable Changes into Its Complement in a Pair of Adjacent Cells

We note an important observation in the Karnaugh map designs (Tables 5.1 to 5.13). In a cell adjacent to any cell, whether on left or right or on up or down (not diagonal) has only a single-variable changes into it's complement.

In three-variable map, first row first column cell only $\overline{C}$ changes to C on moving to second column. In fact, each row; $\overline{C}$ is changing C in the right side adjacent cell (Tables 5.1 to 5.4).

In cells of row 1 and row 2, each column, $\overline{B}$ is changing to B at the down side adjacent cell. From row 2 to 3, $\overline{A}$ is changing to A at the down side cell. From 3 to 4, B is changing to $\overline{B}$ at the down side cell.

In four variable map also, there is only variable change in adjacent cell to any cell, whether on left or right or on up or down (not diagonal). (Tables 5.7 to 5.12). A single-variable change occurs into its complement.

Table 5.15 left side upper portion shows the adjacent pairs ($\bar A B\bar C D$ and $\bar A BCD$) and ($\bar A BC\bar D$ and $\bar A\bar B C\bar D$).

> Therefore, we can form in a specific case a pair of adjacent cells having 1s at left or right or at up or down (not diagonal) and adjacency condition is that only one variable is distinct by its complement.

TABLE 5.15 Adjacent cell pairs, adjacent four cells, adjacent eight cells in four different four variable Karnaugh Maps with 16 minterms each in the SOP expression

	CD	$\bar C\bar D$	$\bar C D$	CD	$C\bar D$		CD	$\bar C\bar D$	$\bar C D$	CD	$C\bar D$
AB		00	01	11	10	AB		00	01	11	10
$\bar A\bar B$	00				1	AB	00	1	1		1
$\bar A B$	01		1	1	1	AB	01	1	1		1
AB	11					AB	11				1
$A\bar B$	10	1				AB	10	1	1	1	1

	CD	$\bar C\bar D$	$\bar C D$	CD	$C\bar D$		CD	$\bar C\bar D$	$\bar C D$	CD	$C\bar D$
AB		00	01	11	10	$\bar A\bar B$		00	01	11	10
$\bar A\bar B$	00		1	1		$\bar A B$	00	1	1	1	1
$\bar A B$	01	1	1	1	1	AB	01		1		
AB	11	1	1	1	1	AB	11	1	1		1
$A\bar B$	10		1	1		$A\bar B$	10		1		1

Note: $\bar A\bar B\bar C\bar D$ cell is also taken as adjacents to $AB\bar C\bar D$ cell, $\bar A\bar B\bar C\bar D$ and $\overline{ABCD}$ is a pair cell as there is only one variable complementation on going to adjacent cell.

5.4.2 Only Two Variables Change into Their Complements in Adjacent Cells in a Square or Column of Four Cells

In four adjacent cells, whether horizontally or vertically or in a square with four cells, two variables change into their complement.

In three-variable map (Table 5.3), in a set (1st row 1st column, 1st row 2nd column, 2nd row 1st column, 2nd row 2nd column) two variables $\bar C$ and $\bar B$ change to C and B in the square with four cells and $\bar A$ is same for all the four cells. In four cells of first column, two variables, A and B change into their complements and $\bar C$ remains same. In second column C remains same.

In four variable map also, there are only two variable changes in a square or column change within adjacent cell to any cell, whether on left or right or on up or down (not diagonal) and other two remain identical (Table 5.5 to 5.8). The first column cell and last column cell in a row are considered adjacent cells.

Similary, first row cell and last row cell in a column are also the adjacent cells.

Therefore, we can form in a specific case either a square or column of four adjacent cells having 1s in all and only two variables are distinct by their complements and remaining ones remain unchanged. Adjacency condition is that only two variable changes occur within the adjacent cells.

Table 5.15 right side upper portion shows examples of the square and column of the four adjacent cells.

5.4.3 Three Variables change into Their Complements in Adjacent Cells in a Box of Eight Adjacent Cells

In eight adjacent cells in a box, only three variables have their complements also and remaining variables (present in four variable or higher variable maps) remain common to all the cells (Table 5.1 to 5.8).

Therefore, we can form in a specific case a 8-cell box of adjacent cells having 1s in all and only three variables are distinct by their complements and remaining one(s) remains unchanged.

Table 5.15 left side lower shows examples of the box of adjacent eight cells and right lower shows the adjacent 4 cells and 2 cells.

5.4.4 First and Last Columns for First and Last Rows and Purpose of Deciding Adjacency in a Karnaugh Map

We note an important observation in the Karnaugh map design. (Tables 5.1 to 5.13).

1. Two cells, one each at the upper most row and the lower most row can also be considered as adjacent if we wrap the map in a horizontal axis cylindrical form and there is only a single-variable that changes into the complement when we consider two cells of the same column in the upper most and lower most rows.
2. Two cells, one each at the left most column and the right most column can also be considered as adjacent if we wrap the map in a vertical axis cylindrical form and there is only a single-variable that changes into it complement when we consider two cells of the same row in the left most and rightmost columns.
3. Four cells distributed at the upper rows and lower rows can also be considered as adjacent if we wrap the map in a horizontal axis cylindrical form and there are only two variable(s) that change into it the complements in these cells. Other variable(s) remain common in these cells.
4. Four cells distributed at the left-most column and the right-most column can also be considered as adjacent if we wrap the map in vertical axis cylindrical form and there

are only two variables that change into their complements in these cells. Other variable(s) remain common in these cells.

5. Eight cells distributed in the upper rows and lower rows can also be considered as adjacent if we wrap the map in a horizontal axis cylindrical form and there are three variables that change into the complements when we consider four cells at same column(s) in the upper most and lower most rows.
6. Eight cells distributed in the left most column and right most column can also be considered as adjacent if we wrap the map in a vertical axis cylindrical form and there are only three variables that change into the complements when we consider four cells at the same row(s) in left and right columns.

Table 5.16 shows wrapping adjacencies between the two cells, four cells and eight cells at four Karnaugh maps.

TABLE 5.16 Wrapping adjacencies of three cell pairs (left upper), two four cell quads (right upper), two eight cell octets (left lower) and two wrapping adjacencies (right-lower) at the four different four variable Karnaugh Maps with 16 minterms each for the SOP expressions

	CD	$\overline{C}\,\overline{D}$	$\overline{C}D$	CD	$C\overline{D}$
AB		00	01	11	10
$\overline{A}B$	00				1
$A\overline{B}$	01	1			1
$\overline{A}B$	11				
$A\overline{B}$	10	1			1

	CD	$\overline{C}\,\overline{D}$	$\overline{C}D$	CD	$C\overline{D}$
CD		00	01	11	10
AB	00	1	1		1
$\overline{A}B$	01	1			1
$\overline{A}B$	11				
AB	10	1	1	1	

	CD	$\overline{C}\,\overline{D}$	$\overline{C}D$	CD	$C\overline{D}$
AB		00	01	11	10
$\overline{A}\,\overline{B}$	00	1	1	1	1
$\overline{A}B$	01	1			1
AB	11	1			1
$A\overline{B}$	10	1	1	1	1

	CD	$\overline{C}\,\overline{D}$	$\overline{C}D$	CD	$C\overline{D}$
AB		00	01	11	10
$\overline{A}\,\overline{B}$	00	1	1	1	
$\overline{A}B$	01				
AB	11				
$A\overline{B}$	10	1	1	1	

5.4.5 Use of Don't Care (or Unspecified) Input Conditions for Purpose of Deciding Adjacencies in a Karnaugh Map

Don't care condition means that a combination of input states do occur and whether the outputs for those states taken as 1s or 0s, it does not matter. We can thus place 1s at the corresponding minterm places in SOP form of the Karnaugh map. The 1s are placed only at the places where it leads to make or improve adjacencies. We can also place 0s at the corresponding maxterm places in POS form of the Karnaugh map. The 0s are paced only at the places where it leads to make or improve adjacencies. Better way is that we place an '*x*' for the don't care input states. Use it at places where it leads to simplification and leave it where it does not lead to simplification. The improved adjacencies lead to simplification of the circuit design.

We can take x for determining the adjacent pairs or quads or octets. Table 5.17 shows the use of don't care in selecting conditions the adjacencies. Assume that $\overline{A}.B.\overline{C}.\overline{D}$ and $\overline{A}.B.C.D$ and $A.\overline{B}.\overline{C}.D$ are don't care input states [($A = 0$, $B = 1$, $C = 0$ and $D = 0$), $A = 0$, B, C and $D = 1$), and ($A = 1$, $B = 0$, $C = 0$ and $D = 1$)].

We can use the two don't care conditions to make two octets and leave and ignore third one x at 2nd row and third column place in the map.

TABLE 5.17 Use of Don't care input combinations for determining the adjacencies

AB \ CD		$\overline{C}\,\overline{D}$ 00	$\overline{C}D$ 01	CD 11	$C\overline{D}$ 10
$\overline{A}\,\overline{B}$	00	1	1	1	1
$\overline{A}B$	01	x		x	1
AB	11	1			1
$A\overline{B}$	10	1	x	1	1

Now one octect is by columns 1 and 4 adjacencies and other octect is by rows 1 and 4 adjacencies.

5.5 SIMPLIFICATION OF LOGIC CIRCUIT REALIZATION BY MINIMIZATION USING ADJACENCIES

5.5.1 Minimization of a Karnaugh Map Using Pairs of Adjacent Cells

In Sections 5.4.1 and 5.4.4, we learnt that pair of cells has one of the variable as the complements of each other. Hence only common variable in the pair of terms needs to be retained and that variable removed after simplification of two terms as one. This follows from the application of the Boolean OR and AND rules; $X + X = 1$ and $X.1 = X$.

A pair simplifies by reducing two minterms as one in an SOP. The procedure for removing one variable and simplifying two terms in SOP into one can be understood as follows by an exemplary Karnaugh map in Table 5.18.

Table 5.18 shows the adjacent pairs of cells.

There are five places where there are 1s. Map corresponds to five minterms (Table 5.3) and following SOP expression.

$$S = \overline{A}.\overline{B}.\overline{C}.\overline{D} + A.\overline{B}.\overline{C}.\overline{D} + \overline{A}.B.C.\overline{D} + A.B.C.\overline{D} + A.\overline{B}.C.\overline{D} \qquad ...(5.1)$$

First and second terms are for the adjacent cells (assumed wrapped map) and have 'A' variable as complements. Hence, A is removed from first two terms and $\overline{B}.\overline{C}.\overline{D}$ is left.

TABLE 5.18 A Karnaugh Map with two pairs of adjacent cells

AB \ CD		$\overline{C}\,\overline{D}$ 00	$\overline{C}\,D$ 01	CD 11	$C\overline{D}$ 10
$\overline{A}\,\overline{B}$	00	1			
$\overline{A}\,B$	01				1
AB	11				1
$A\overline{B}$	10	1			1

Third and fourth terms are for the adjacent cells at the middle of the last column. These have A variable as complements. Hence, A is removed from these two terms and $B.C.\overline{D}$ is left. Fifth term has a wrapping adjacency pair with $A\,\overline{B}\,\overline{C}\,\overline{D}$ pair. Therefore, it can't be removed but can be simplified as $A.\overline{B}.\overline{D}$. (a 3-variable term).

Simplified Boolean expression becomes as follows.

$$S = \overline{B}.\overline{C}.\overline{D} + B.C.\overline{D} + A.B.\overline{D}$$

5.5.2 Minimization of a Karnaugh Map Using Quads of Four Adjacent Cells

In Sections 5.4.2 and 5.4.4, we learnt that quad of four adjacent cells has two of the variables as the complements of each other. Hence only common variables in four terms needs to be retained and those two variables are removed for the simplification of logic circuit. Thus three minterms can be reduced from the four and only one is then needed for designing a circuit. It follows from Boolean OR and AND rules on both of them; $\overline{X} + X = 1$, $\overline{Y} + Y = 1$, $X.1 = X$ and $Y.1 = Y$.

Table 5.19 shows the adjacent quads of four adjacent cells.

This procedure can be understood as follows by an exemplary Karnaugh map in Table 5.19. There are nine places where there are 1s. Map corresponds to nine minterms and following SOP expression.

$$S = \overline{A}.\overline{B}.\overline{C}.D + \overline{A}.B.\overline{C}.\overline{D} + \overline{A}.B.\overline{C}.D + \overline{A}.B.C.\overline{D} + A.B.\overline{C}.\overline{D} + A.B.\overline{C}.D + A.B.C.\overline{D} + A.\overline{B}.\overline{C}.D + A.\overline{B}.C.\overline{D} \quad ...(5.2)$$

TABLE 5.19 A Karnaugh Map with two quads of four adjacent cells

AB \ CD		$\overline{C}\,\overline{D}$ 00	$\overline{C}\,D$ 01	CD 11	$C\overline{D}$ 10
$\overline{A}\,\overline{B}$	00		1		
$\overline{A}\,B$	01	1	1		1
AB	11	1	1		1
$A\overline{B}$	10		1		1

Four terms (1st, 3rd, 6th and 8th in the SOP Equation) for second column adjacent cells have A and B variables as well as their complements. Hence, A and B are removed from first four terms and $\overline{C}.D$ is left. Four terms are also for the four adjacent cells (after wrapping the map into cylinder with a vertical axis). These have A and C variables as well as their complements. Hence, A and C are removed from these four terms and $B.\overline{D}$ is left. Ninth term has a pair with the seventh and B can be removed.

Simplified Boolean expression becomes as follows.

$$S = \overline{C}.D + B.\overline{D} + A.\overline{B}.C.\overline{D} = \overline{C}.D + B.\overline{D} + A.C.\overline{D} \qquad ...(5.3)$$

5.5.3 Minimization of a Karnaugh Map Using Octet of Eight Adjacent Cells

In Section 5.4.3, we learnt that octet of eight adjacent cells has three variable as well as their complements present. Hence only common variable(s) in eight terms needs to be retained and those three variables are removed for the simplification of logic circuit. It follows from Boolean OR and AND rules on both of them; $X + X = 1$, $Y + Y = 1$, $Z + Z = 1$, $X.1 = X$, $Y.1 = Y$ and $Z.1 = Z$. The number of minterms reduces from 8 to 1 only. This procedure can be understood as follows by an exemplary Karnaugh map in Table 5.20.

Table 5.20 shows an octet of adjacent eight cells.

There are nine places where there are 1s. Map corresponds to nine minterms and following SOP expression.

$$S = \overline{A}.\overline{B}.\overline{C}.\overline{D} + \overline{A}.\overline{B}.C.D + \overline{A}.\overline{B}.C.\overline{D} + \overline{A}.B.\overline{C}.\overline{D}$$
$$+ \overline{A}.B.C.\overline{D} + A.B.\overline{C}.\overline{D} + A.B.C.\overline{D} + A.\overline{B}.\overline{C}.\overline{D} + A.\overline{B}.C.\overline{D} \qquad ...(5.4)$$

TABLE 5.20 A Karnaugh Map with an octet of eight adjacent cells

AB \ CD		$\overline{C}\,\overline{D}$ 00	$\overline{C}D$ 01	CD 11	$C\overline{D}$ 10
$\overline{A}\,\overline{B}$	00	1		1	1
$\overline{A}B$	01	1			1
AB	11	1			1
$A\overline{B}$	10	1			1

Eight terms (all except 2nd Equation (5.4)) for first and fourth column adjacent cells (from wrapping adjacencies) have A, B and C variables as well as their complements. Hence, A, B and C are removed from the eight terms and only $\overline{D}$ is left. Second term has a pair with third. Therefore, it can be simplified. Answer after simplification is as follows:

$$S = \overline{D} + \overline{A}.\overline{B}.C.D = \overline{D} + \overline{A}.\overline{B}.C \qquad ...(5.5)$$

5.5.4 Minimization of a Karnaugh Map Using Offset Adjacencies and Diagonal Adjacencies

Consider the following map in Table 5.21.

TABLE 5.21 A Karnaugh map with an offset adjacency and a diagonal adjacency

AB \ CD		$\overline{C}\,\overline{D}$ 00	$\overline{C}\,D$ 01	CD 11	$C\overline{D}$ 10
$\overline{A}\,\overline{B}$	00	1		1	
$\overline{A}B$	01	1		1	
AB	11		1		
$A\overline{B}$	10	1			

We note that rows 3 and 4 have a diagonal adjacency. $\overline{C}$ and A are common variables between and two minterms sum will have the term $B.\overline{D}$ and $\overline{B}.D$ after Boolean simplification. The operation equals an XOR operation. Hence, simplification gives $(A.\overline{C}).(B.\text{XOR}.D)$.

We also note that columns 1 and 3 have an offset adjacency between the four cells. $\overline{A}$ variable is common between these. B variable has its complement B and is removable. So answer is $\overline{A}.(\overline{C}.\overline{D} + C.D) = \overline{A}(C.\text{XNOR}.D)$. [The XNOR logic gate and truth table was given in Figure 4.2(c).]

Simplified expression from the map is as follows:

$$S = (A.\overline{C})\,(B.\text{XOR}.D) + \overline{A}.(C.\text{XNOR}.D). \qquad ...(5.6)$$

When XOR and XNOR form of simplifications are needed then the diagonal and offset adjacencies, respectively, are used.

5.5.5 Minimization by Finding Prime Implicants

The definitions and technique for minimizing a Boolean SOP expression or truth table (also called finding **prime implicants**) are as follows—

1. A variable in complemented as well as in un-complemented format is used in a Boolean expression. It can also be called a **literal**.
2. A product term (minterm) present in a function (Boolean expression) can also be called **implicant** and its function = 1. An implicant is implemented by the AND gates at the inputs (first level).
3. A **cover** means a set of all implicants that contains all the implicants (minterms) whose function values are 1s and that are needed to complete the map.
4. A function is implemented by ANDs in all the *implicants* at the *cover* and using the OR gates at the second level.

5. An AND (first level)-OR (second level) circuit can be converted to an NAND — NAND circuit at both first and second levels. (Refer use of DeMorgan Theorem).
6. A **prime implicant** means an implicant, which can't be further simplified into another implicant with less number of literals in it. It can't be ORed with another implicant to get less number literals (variables).
7. *Prime implicants* give a simpler minimized circuit than the circuits using *implicants*.

1. Convert a Boolean expression into standard SOP format or add more truth table rows till table has 2^n rows for n-inputs (variables to be used for making the map) with output marked as x. (x means don't care) for added rows.
2. Use x as 1 when forming pairs, quads and octets. Remove x from the other remaining cells at the map.
3. If $n = 3$, construct a three variable Karnaugh map and put the 1s at the places corresponding to the minterms present (implicants) or truth table rows corresponding to the output = 1.
4. If $n = 4$ or 5 or 6, construct a 4 or 5 or 6 variable Karnaugh map and put the 1s at the places corresponding to the minterms present (implicants) or truth table rows corresponding to the output = 1.
5. Find the adjacency cell octets of 1s in the map. Consider wrapping adjacencies also. It will simplify the eight terms for an octet into one term with three variables removed.
6. Find the adjacency cell quads of 1s in the map. Consider wrapping adjacencies also. It will simplify the four terms for a quad into one term with two variables removed.
7. Find the adjacency of cell pairs of 1s in the map. Consider wrapping adjacencies also. It will simplify the two terms for a pair into one term with two variables removed.
8. Find the diagonal adjacency(s) also in case XOR gate(s) are also to be used.
9. Find the offset adjacency(s) also in case XNOR gate(s) are also to be used.

Steps for finding the prime implicants, called minimization technique (minimizing a Boolean POS expression or truth table) using POS based is as follows Karnaugh Map:

1. Convert a Boolean expression into Standard POS format or add more rows in the truth table rows in incompletely specified till it has 2^n rows for n-inputs (variables to be used for making the map) with output marked as x. (x means don't care).
2. A sum term (maxterm) present in a function (Boolean expression) gives the output = 0 by its function = 0. A maxterm present in the map is implemented by the OR gates at the inputs (first level).
3. A cover means a set of all the maxterms whose function values are 0s and that are needed to complete the map.
4. A function is implemented by ORs in all the maxterms present and using the AND gates at the second level.
5. An OR (first level)-AND (second level) circuit can be converted to an NOR– NOR circuit at both first and second levels (Refer use of DeMorgan Theorem).

6. If $n = 3$, construct a three variable Karnaugh map and put the 0s at the places corresponding to the maxterms present or find those truth table rows that correspond to the output = 0.
7. If $n = 4$ or 5 or 6, construct a 4 or 5 or 6 variable Karnaugh map and put the 1s at the places corresponding to the maxterms present or truth table rows corresponding to the output = 0.
8. Use x as 0 when forming pairs, quads and octets. Remove x from the other remaining cells at the map.
9. Find the adjacency cell octets of 0s in the map. Consider wrapping adjacencies also. It will simplify the eight terms for an octet into one term with three variables removed.
10. Find the adjacency cell quads of 1s in the map. Consider wrapping adjacencies also. It will simplify the four terms for a quad into one term with two variables removed.
11. Find the adjacency cell pairs of 0s in the map. Consider wrapping adjacencies also. It will simplify the two terms for a pair into one term with 2 variables removed.

5.6 DRAWING OF LOGIC CIRCUIT USING AND-OR GATES, OR-AND GATES, NAND'S ONLY, NOR'S ONLY

Karnaugh map in SOP function corresponds to 1. We can use the minterm for each cell by for making a circuit from AND gates. Later on all the outputs of AND gates are ORed. For example, we can use Table 5.19 Karnaugh map for the minterms and we get an AND – OR circuit (at first and second level, respectively). This is demonstrated in Figure 5.1.

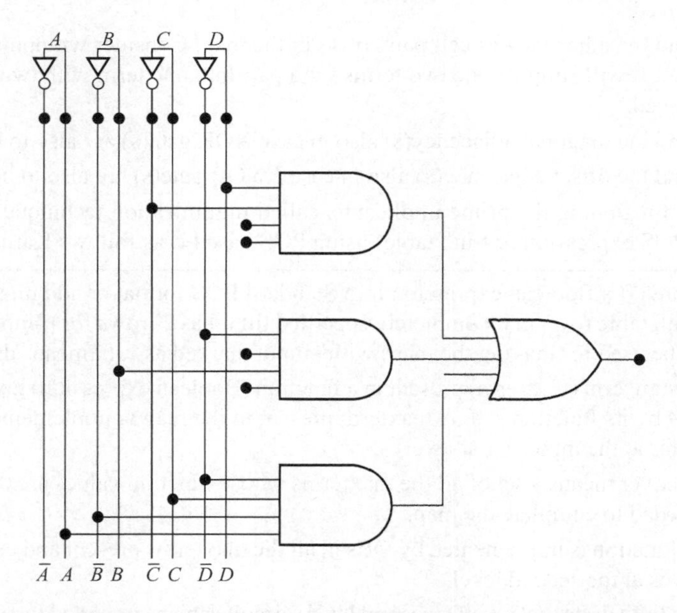

FIGURE 5.1 An AND-OR circuit Representation of Karnaugh map in Table 5.19 after simplification.

Karnaugh map in POS function corresponds to 0. We can use the maxterm for each cell by for making a circuit from OR gates. Later on all the outputs of OR gates are ANDed. For example, we can use Table 5.12 Karnaugh map maxterms and we get an OR–AND circuit (at first and second level, respectively). This is demonstrated in Figure 5.2.

Using DeMorgan theorem, an AND-OR circuit can be converted to NANDs only circuit. This is demonstrated in Figure 5.3. Using DeMorgan theorem, an OR-NAND circuit can be converted to NORs only circuit. This is demonstrated in Figure 5.4.

5.7 REPRESENTATIONS OF A FUNCTION (COVER) FOR A COMPUTER-AIDED MINIMIZATION FOR SIMPLIFYING THE LOGIC CIRCUITS

When simplifying and finding the prime implicants manually and up to fewer variables (six), the Karnaugh maps are suitable. For complex circuits, the computer-aided minimization is used.

5.7.1 Representation in Cube Format for Computer-aided Minimization

A cube has three axes, X-axis, Y-axis and Z-axis. Figure 5.5 shows a cube. It has 8 vertices. The origin vertex coordinates are 000, to represent $\overline{A}.\overline{B}.\overline{C}$. The other 3 vertices along the axes has coordinates are 100, 010, 001 (corresponding to $A.\overline{B}.\overline{C}$, $\overline{A}.B.\overline{C}$ and $\overline{A}.\overline{B}.C$). In three X-Y, Y-Z and Z-X planes, the other vertices are 110, 011, 101 (corresponding to $A.B.\overline{C}$, $\overline{A}.B.C$ and $A.\overline{B}.C$). One vertex along the cue diagonal is 111 (corresponding to $A.B.C$). Thus there are eight coordinates, each corresponding to a cube vertex and each corresponding to a minterm (an implicant) in SOP.

Each vertex is marked if the function value = 1 of the corresponding implicant. Two marked vertices are joined together if these are along an axis. Joined marked vertices are shown by dark or different coloured axis in the cube. Figure 5.5 shows the cubical representation for a cover with the implicants given by following Equation (5.7). It is for SOP function, S corresponding to Karnaugh map in Table 5.22.

$$S = \overline{A}.B.C + A.\overline{B}.\overline{C} + A.\overline{B}.C = \Sigma m\,(4, 3, 5)$$
$$= (011,\ 100,\ 101) \qquad \text{...(5.7)}$$

Now, pair 100, 101 can be written as $10x$.

The coordinate of mid-point of a dark Z-axis starting from 100 is (1, 0, 0.5). Let us write it as $10x$. Similarly we can write and find other dark axes mid points. For example, X-axis mid point is $x00$, when it starts from 000.

If XY plane passing through 000 has all its line as dark, its mid point can be written as $xx0$. If XY plane passing through 001 has all its line as dark, its mid point can be written as $xx1$. Similarly we can write mid point of all planes having the dark axes at the cube.

Mid point of the cube can be written as (x, x, x) when all the eight vertices are marked and all the cube axes are dark.

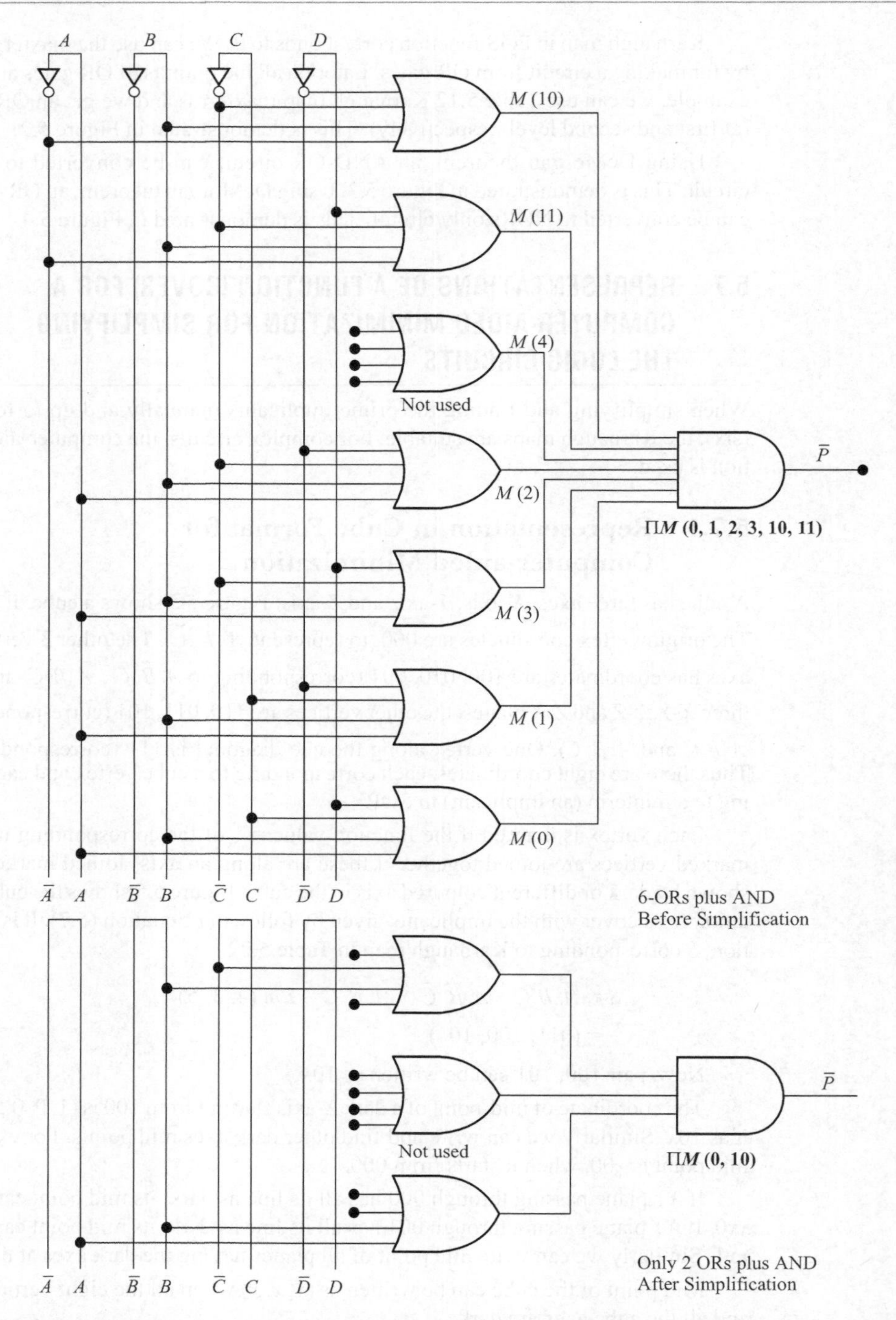

FIGURE 5.2 Using two quads, taking (2, 3, 10, 11) wrapping adjacencies also in account, the OR-AND circuit a representation of Karnaugh map of Table 5.12 before and after simplification.

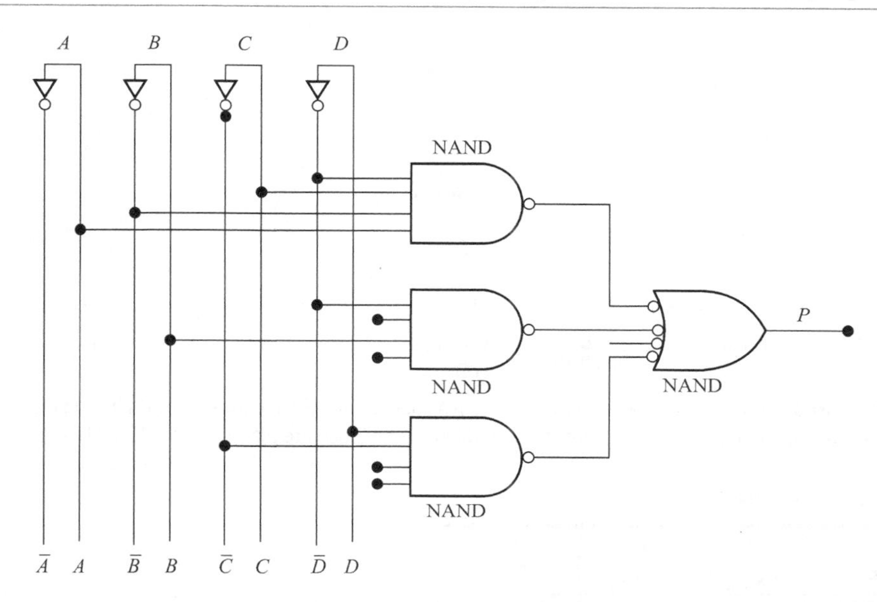

FIGURE 5.3 Representation of AND-OR circuit in Figure 5.1 into NANDs only circuit. [OR gate with all bubbled input is equivalent to NAND (deMorgan Theorem)].

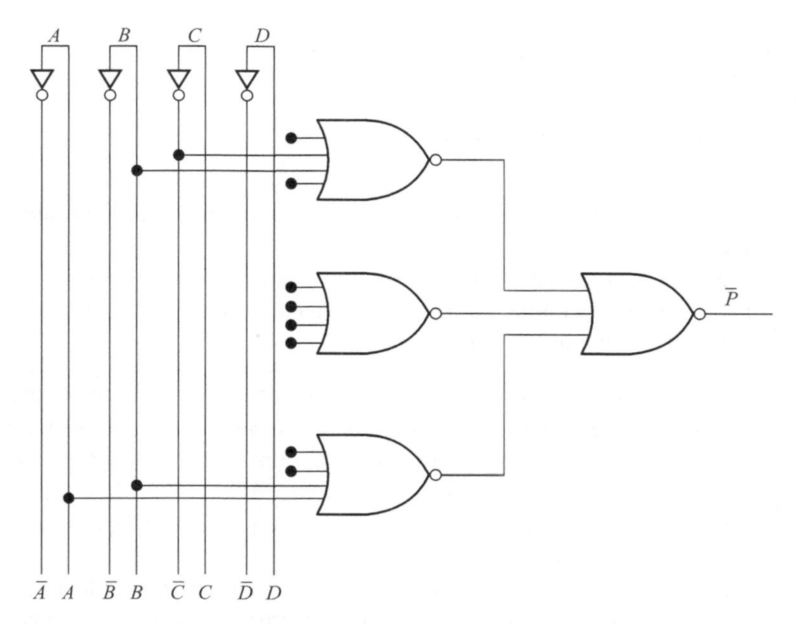

FIGURE 5.4 Representation of an OR-AND circuit in Figure 5.2 by NORs only.

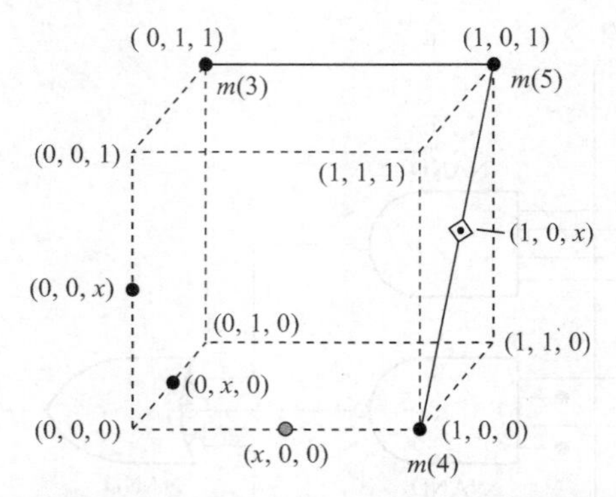

FIGURE 5.5 Representation of implicants (minterms) in a cube format. Each vertex, which is having the function value = 1 of the implicant, is marked. Marked two vertices are joined together if these are along an axis. (Dark lines in the figure.)

TABLE 5.22 Three variable Karnaugh Map

AB \ C		$\overline{C}$ 0	C 1
$\overline{A}\,\overline{B}$	00		
$\overline{A}B$	01		1
AB	11		
$A\overline{B}$	10	1	1

The equation (5.7) is rewritten as

$$S = \{10x, 011\} \quad ...(5.8)$$

(When the pair of vertices 100, 101 are written by its axis mid-point as $10x$ in Figure 5.5)

We can see below that finding prime implicants becomes very easy using the representation as a cube if we follow following steps and these steps can also be programmed and minimization is possible by that program.

1. Write the implicants of a function, S is terms of the coordinates of the vertices.
2. Find a pair in which X coordinates only change and are 0 and 1. Reduce it to one term with coordinate $= x$. Do the same with Y and Z. For example, convert the pair (1, 0, 1) and (1, 1, 1) into $(1, x, 1)$.
3. Continue till all pairs are taken care (It is equivalent to replacing a pair of vertices by an axis mid-point).
4. Again pair the newly found coordinates with one of its coordinate marked as x also. For example, pair $(0, x, 0)$ and $(0, x, 1)$ into $(0, x, x)$ (It is equivalent to replacing a pair of axes mid-points by a plane mid-point).
5. Again pair the newly found coordinates with two of its coordinates marked as x also. For example, pair $(x, x, 0)$ and $(x, x, 1)$ into (x, x, x). (It is equivalent to replacing a pair of planes mid-points at a the cube mid point).

Assume that a circuit is that its Boolean expression in SOP format has minterms 4, 5, 6 and 7. Therefore, S can be expressed in a cube coordinates' format as follows:

$$S = \{(111, 110, 101, 100) \quad ...(5.9)$$

The steps which are implemented by a computer-based minimization, are as follows:

$$S = (1, 1, 1), (1, 1, 0), (1, 0, 1), (1, 0, 0)$$

$$= ((1, 1, x), (1, 0, x)\}$$

(Pairing 1st and 2nd and 3rd and 4th set of coordinates. It is equivalent to replacing a pair of vertex coordinates by an axial mid-point).

$$= (1, x, x) \quad ...(5.10)$$

(Pairing 1st and 2nd.) It is equivalent to replacing a pair of axial mid-points by a plane mid-point.

The minimized function is now

$$S = A \quad ...(5.11)$$

The A is prime implicant of the four implicants of Equation (5.9). The logic circuit is just output equal to input A. $S = \overline{A}$ for minterms $\Sigma m(0, 1, 2, 3)$ case of implicants, then simplified minimized circuit is just a not operation on A.

5.7.2 Representation in Four-Dimensional Hypercube Formats for a Computer-aided Minimization

A four dimensional hyper cube has four axes, $A1$-axis, $A2$-axis, $A3$ and $A4$-axis. It will have 16 vertices. The origin vertex coordinates are 0000. The other coordinates will be 0001, 0010, 0011, 0100, 0101, 0110, 0111, 1000, 1001, 1010, 1011, 1100, 1101, 1110 and 0111.

The coordinate of mid-point on a darkened axis from 1000 to 1001 is (1, 0, 0, 0.5) (An axis is darkened if there are vertices falling on it). Let us write it as $100x$. Similarly, we can write, we can find other dark axes mid-points. For example, $A1$-axis mid-point is $x000$, when it starts from 0000 to 1000.

If $A1$-$A2$ plane passing through 0000 has all its line as dark, its mid point can be written as $xx00$. If $A1$-$A2$ plane passing through 0001 has all its line as dark, its mid point can be written as $xx01$. We can similarly write mid-point of all planes with the dark axes in a four dimensional cube.

Mid-point of the four dimensional cube can be written as (x, x, x, x) when all the 16 vertices are marked and all the axes are dark.

Consider the following four literal (variable) SOP equation:

$$S = \{\overline{A}.\overline{B}.\overline{C}.\overline{D} + \overline{A}.\overline{B}.C.D + \overline{A}.B.\overline{C}.\overline{D} + \overline{A}.B.C.D + A.\overline{B}.\overline{C}.\overline{D}$$
$$+ A.B.\overline{C}.D\} = \Sigma m\ (0, 3, 4, 7, 8, 13) \quad ...(5.12)$$

Its Karnaugh map representation is same as given above in Table 5.21. This equation gives the implicants in four dimensional cube format as follows.

$$S = \{0000, 0011, 0100, 0111, 1000, 1101\} \quad ...(5.13)$$

We can see below that finding prime implicants becomes very easy using the representation as a four dimensional cube if we follow following steps and these steps can also be programmed and minimization is possible by that program.

1. Write the implicants of a function, S is terms of the coordinates of the four dimensional cube vertices.
2. Find a pair in which one coordinate only change from 0 to 1. Reduce the pair to one term with coordinate = x. For example, convert the pair (1, 0, 0, 1) and (1, 1, 0, 1) into (1, x, 0, 1).
3. Continue till all pairs are taken care. It is equivalent to replacing a pair of vertex coordinates by an axis mid-point.
4. Again pair the newly found coordinates with one of its coordinate being x. For example, pair (0, x, 0, 0) and (0, x, 1, 0) into (0, x, x, 0). It is equivalent to replacing a pair of axes mid-points by a plane mid-point.
5. Again pair the newly found coordinates with two of its coordinates being x. For example, pair (x, x, 0, 0) and (x, x, 1, 0) into (x, x, x, 0). It is equivalent to replacing a pair of plane mid-points by an internal cube mid-point to a four dimensional cube.
6. Again pair the newly found coordinates with three of its coordinates being x. For example, pair (x, x, x, 0) and (x, x, x, 1) into (x, x, x, x). It is equivalent to replacing a pair of axes mid-points by the four dimensional cube mid-point.

Exemplary steps in computer based minimization (finding prime implicants) of Equation (5.13) based on four-dimensional cube are as follows

$$S = \{(0, 0, 0, 0), (0, 0, 1, 1), (0, 1, 0, 0), (0, 1, 1, 1), (1, 0, 0, 0), (1, 1, 0, 1)\}$$
$$= \{(0, x, 0, 0), (0, x, 1, 1), (1, 0, 0, 0), (1, 1, 0, 1)\} \quad ...(5.14)$$

(Pairing 1st and 3rd and 2nd and 4th set of coordinates).

The minimized function has four prime implicants:

$$\overline{A}.\overline{C}.\overline{D} + \overline{A}.C.D + A.\overline{B}.\overline{C}.\overline{D} + A.B.\overline{C}.D \quad ...(5.15)$$

The AND-OR logic circuit is now minimum when using the Equation (5.15).

5.7.3 Representation in Hypercube (Multi-dimensional cube) Formats for Computer-aided Minimization

A hypercube (multidimensional dimensional cube) has n axes, A_1 axis, A_2 axis, A_3, A_4 axis, ... up to A_n axis and is used for n literals (variables). It will have 2^n vertices that corresponds to 2^n minterms. The coordinates are easily written using 2^n sets of n-bit binary numbers between 0 and $2^n - 1$.

Let U and V are two cubes (or hypercubes) with coordinates u_i and v_i, respectively.

We can see below that finding prime implicants becomes very easy using the representation as a hypercube also if we follow following steps and these steps can also be programmed and minimization is possible by that program.

1. Write the implicants of a function, S is terms of the coordinates of the n dimensional hyper cube vertices.
2. Find a pair in which one of the coordinates change and that are 0 and 1. Reduce pair to one term with that coordinate = x. For example, convert the pair (1, ..., 0, 0, 1) and (1, ..., 1, 0, 1) into (1, ..., x, 0, 1).
3. Continue till all pairs are taken care. It is equivalent to replacing a pair of vertices coordinates by the axis mid-point.

4. Again pair the newly found coordinates with one of its coordinate as x also. For example, (0, ... x, 0, 0) and (0, ... x, 1, 0) into (0, ... x, x, 0). It is equivalent to replacing a pair of axes mid-points by a plane mid-point.
5. Again pair the newly found coordinates with two of its coordinates as x also. For example, pair (x, x, ..., 0, 0) and (x, x, ..., 1, 0) into (x, x, ... x, 0). It is equivalent to replacing a pair of axes mid-points by an internal cube mid-point to a four dimensional cube when $n = 4$.
6. Again pair the newly found coordinates with three of its coordinates as x also. For example, pair (x, x, 1, ... x) and (x, x, 0, ..., x) into (x, x, x, ..., x). It is equivalent to replacing a pair of axes mid-points by the hypercube mid-point.
7. Continue searching and forming more and more pairs till we reach mid-point of ($n - 1$) dimension cube.

We notice that computer based minimization approach is analogous, whether we use three cube (three variable) or four-cube or hypercube format. We write a cover (an SOP function) as the function of a set of coordinates on a hypercube before running the program for the steps given above.

5.8 MULTI-OUTPUT SIMPLIFICATION

Karnaugh map simplifies and gives reduced number of minterms or maxterms (product terms or sum terms) for an output of a combinational circuit. Often there are multiple outputs. For example, a binary to—segment decoder or 4-bit gray code converter. Let $Y0$, $Y1$, ..., $Yn - 1$ are the outputs for a set of input variable $X0$, $X1$, ... $Xm - 1$. A truth table of these will have m columns for the inputs and n columns for the outputs. Number of rows will be equal to number of possible input combinations 2^m.

One method is to write Karnaugh map for the each and fabricate the separate circuit for each output. This does not minimize the cost as there are several terms that may be common and can be implemented by a common AND-OR sub-array. Other methods are as follows:

Method 1: Finding Common Set of Quads or Diads

Consider the Figures 5.6 (a) and (b). It shows implementation using the AND-ORs circuits for the two Karnaugh maps for two outputs $Y0$ and $Y1$ in Tables 5.23 and 5.24, respectively. The tables show a dashed square envelope for common quad of the adjacent cells in the two maps. The Boolean expressions after Karnaugh minimization are $Y_0 = \overline{A}.D$ and $Y_1 = \overline{A}.D + B.\overline{C}.\overline{D}$.

TABLE 5.23 Karnaugh Map for $Y_0 = \overline{A}.D$

$\overline{A}\,\overline{B}$ \ $\overline{C}\,\overline{D}$		$\overline{C}\,\overline{D}$ 00	$\overline{C}D$ 01	CD 11	$C\overline{D}$ 10
$\overline{A}\,\overline{B}$	00		1	1	
$\overline{A}B$	01		1	1	
AB	11				
$A\overline{B}$	10				

TABLE 5.24 Karnaugh Map for $Y_1 = \overline{A}.D + B.C.D$

AB \ $\overline{C}\,\overline{D}$		$\overline{C}\,\overline{D}$ 00	$\overline{C}D$ 01	CD 11	$C\overline{D}$ 10
$\overline{A}\,\overline{B}$	00		1	1	
$\overline{A}B$	01	1	1	1	
AB	11	1			
$A\overline{B}$	10				

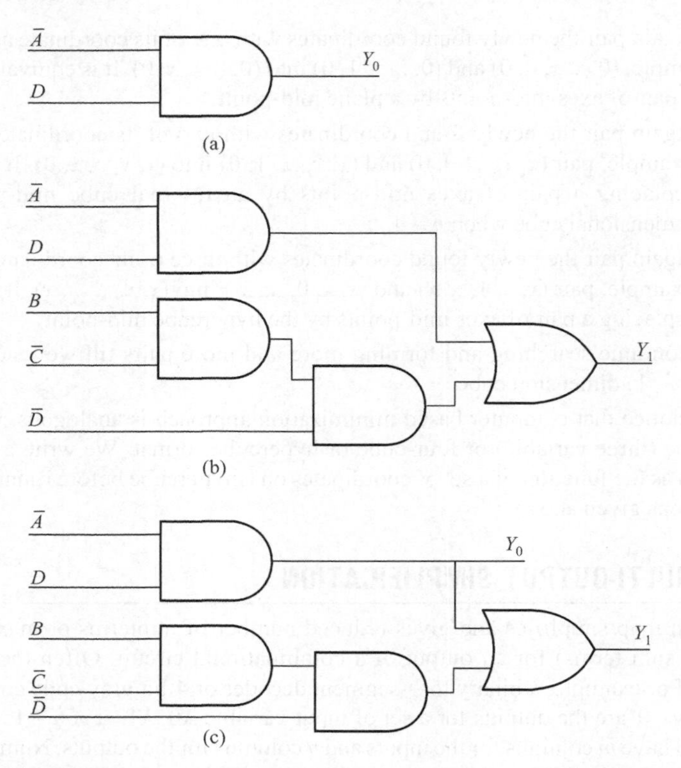

FIGURE 5.6 (a) AND-OR Array for Y_0 (b) AND-OR Array for Y_1 (c) Multi-output implementation using AND-OR sub-arrays.

Since $\overline{A}.D$ quad (pair of four adjacent cells) is common in both, the simplified low cost circuit is as per Figure 5.6 (c). Number of gates reduces by two in the common implementation.

Method 2: Finding a Common Group of Terms

Consider the Figures 5.7 (a) to (e). It shows implementation using the AND-ORs circuits for the two Karnaugh maps for two outputs Y_0 and Y_1 in Tables 5.25 and 5.26, respectively. The tables show *a* # sign marked term in each. These terms form a group with a common term in the two maps. The use of this common term in one group gives an implementation, which minimizes with lesser number of gates than when implementing by common diads or quads.

The Boolean expressions after Karnaugh minimization are $Y_0 = \overline{A}.\overline{B}.C + \overline{A}.C.D$ and $Y_1 = \overline{A}.B.C + A.C.\overline{D}$. $Y_0 = \Sigma m\ (2, 3, 7)$ and $Y_1 = \Sigma m\ (2, 6, 7)$ [from Table 5.9].

Since no diad (pair of two adjacent cells) is common in expression for *Y*0 and *Y*1 in both, total four ANDs of three inputs (Figures 5.7(a) and (b)) and six two-input NANDs will be needed in Figures 5.7(c) and (d). A simplified low cost circuit is as per Figure 5.7(e). Number of gates reduces by one in the common implementation. We need five two-input

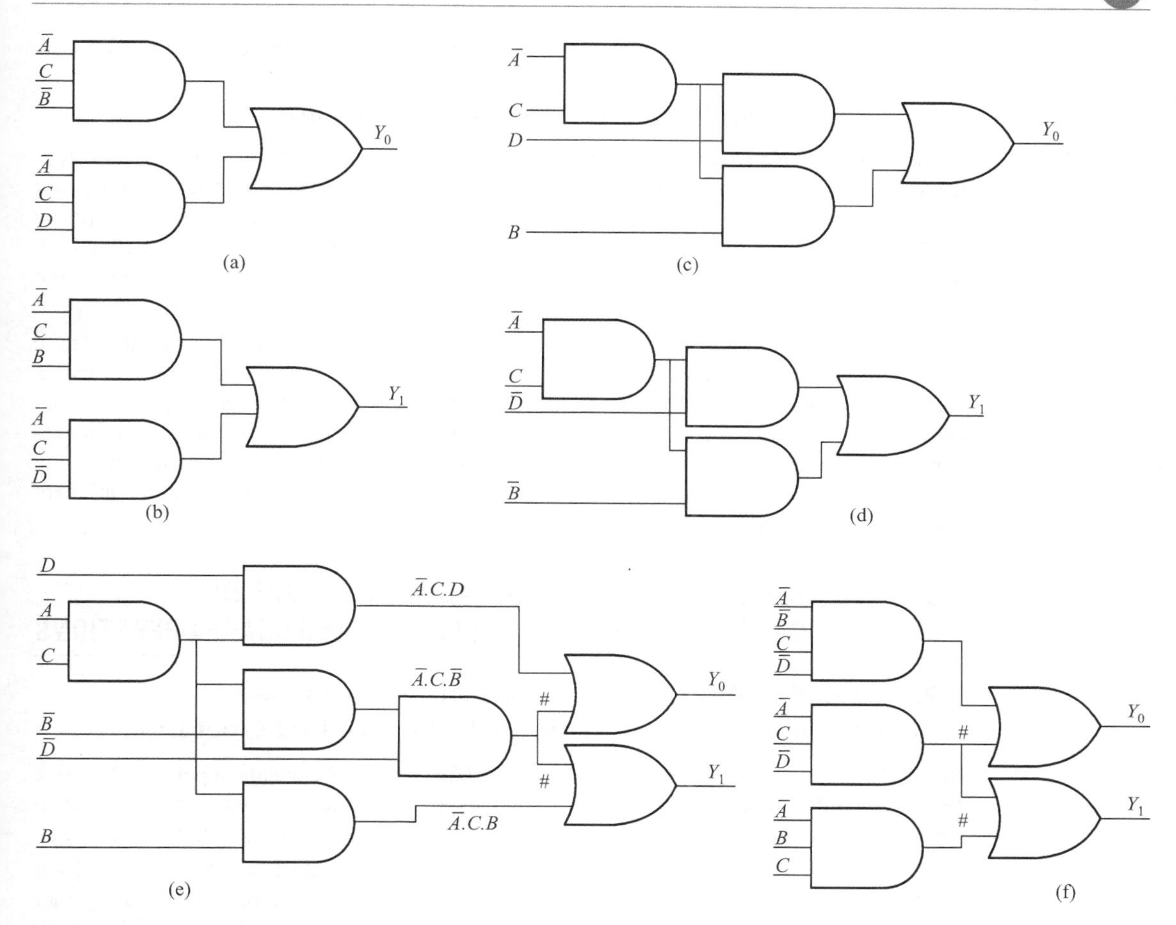

FIGURE 5.7 (a) and (b) 3-input AND-OR Array for Y_0, Y_1 (c) and (d) Two input AND-OR Array for Y_0, Y_1 (c) and (d) Multi-output implementation for Y_1 (e) A joint implement (f) simplification of AND-OR arrange in joint implementation.

TABLE 5.25 Karnaugh Map for $Y_0 = \bar{A}.\bar{B}.C + \bar{A}.C.D$

$\bar{C}\bar{D}$ / $\bar{A}\bar{B}$		$\bar{C}\bar{D}$ 00	$\bar{C}D$ 01	CD 11	$C\bar{D}$ 10
$\bar{A}\bar{B}$	00			1	1#
$\bar{A}B$	01			1	
AB	11				
$A\bar{B}$	10				

TABLE 5.26 Karnaugh Map for $Y_1 = \bar{A}.B.C + \bar{A}.C.\bar{D}$

$\bar{C}\bar{D}$ / AB		$\bar{C}\bar{D}$ 00	$\bar{C}D$ 01	CD 11	$C\bar{D}$ 10
$\bar{A}\bar{B}$	00				1#
$\bar{A}B$	01			1	1
AB	11				
$A\bar{B}$	10				

ANDs or alternatively one four input and two three input gates (Figure 5.7(f)).

5.8.1 Prime Implicants for Multi-Outputs Case

Prime implicant of a Boolean function is a product term for the function, which has no other term with lesser literals to represent the function. A prime implicants for multi-outputs case is a prime implicant of either one of the individual function or of their products. A prime implicant for multi-outputs Y_0 and Y_1 case is a prime implicant of either Y_0 or Y_1 or $Y_0.Y_1$. A prime implicants for multi-outputs Y_0, Y_1 and Y_2 case is a prime implicant of any of the seven functions either (a) Y_0 or Y_1 or Y_2 or (b) $Y_0.Y_1$ or $Y_1.Y_2$ or $Y_2.Y_0$ or (c) $Y_0.Y_1.Y_2$.

Cost minimization has two approaches either by reducing the number of gates or the number of inputs to the gates. Cost minimization is for a sum of the set of prime-implicants such that all the prime implicants of either that function or its product are present. That means that the cost minimization for Y_i is either done using the prime-implicants expression for Y_i or for a product function of Y_i. Reason for this is when considering Y_i the other function values do not matter. They correspond to don't care condition and don't care condition can be taken as 1 in the product.

5.9 TWO OUTPUTS SIMPLIFICATION—COMPUTER-BASED PRIME IMPLICANTS USING STAR PRODUCT AND SHARP OPERATIONS

5.9.1 Combination of Two Cubes Differing in One Variable into One Cube — A Star Product Operation

Recall the pairing operation described in sections 5.7.1 to 5.7.3 Consider a pair in which one coordinate only change and are 0 and 1 in two terms. We reduce it to one term with coordinate $= x$. For example, we convert the pair (1, ..., 0, 0, 1) and (1, ..., 1, 0, 1) into (1, ..., x, 0, 1).

Since conversion to a single x basically means a removal of one of the variable or reducing the dimension of the cube. Therefore, two cubes can also be combined together into one cube. Let us assume that there are two m-dimensional cubes, U and V with m-variables each. Let the variables are $u_1, u_2, ..., u_m$ on first hyper cube and $v_1, v_2, ..., v_m$ on second hyper cube. A variable, u or v is either 1 or x or 0. x means a don't care condition.

Now star-product operation $U * V$ is as follows:

For each pair of u_i and v_i,

(i) $u_i.v_i = 0$ if both are 0.

(ii) $u_i.v_i = 0$ if $u = x$ and $v = 0$.

(iii) $u_i.v_i = 0$ if $u = 0$ and $v = x$.

(iv) $u_i.v_i = 1$ if both are 1.

(v) $u_i.v_i = 1$ if $u = x$ and $v = 1$.

(vi) $u_i.v_i = 1$ if $u = 1$ and $v = x$.

(vii) $u_i.v_i = \varepsilon$ if $u = 1$ and $v = 0$.

(viii) $u_i.v_i = \varepsilon$ if $u = 0$ and $v = 1$.

(ix) $u_i.v_i = x$ if $u = x$ and $v = x$.

The $i = 1, 2, ..., m$. Star product operation means a cube W with coordinates W_i. W is obtained by a star-product operation $U * V$ using expressions (i) to (ix) for each value of i.

(1) $W = \varepsilon$, if $u_i.v_i = \varepsilon$ for greater than one value of i between 1 and m.

(2) $W_i = u_i.v_i$ if $u_i.v_i = 0$ or 1 or x, where W_i is i^{th} coordinate of W.

(3) $w_i = x_i$ if $u_i.v_i = \varepsilon$.

Points to Remember

1. If $W = \varepsilon$, it means that U and V can't be combined into W.
2. Two cubes can be combined into one cube if they differ in one (not none or more) of the variable i between $i = 1$ to m.
3. W is discarded for minimization if they differ but the W is included in both U and V. For example, $U_i = \{x11)$ and $V_i = \{1, 1, x\}$. Now $W_i = \{111\}$ from (v) and (v_i) above, which means W_i is included in U_i and V_i both.
4. W can be included for minimization if they differ but the W is not included in either U or V but not in both. For example, $U_i = \{x, 0,1\}$ and $V_i = \{1, x, 0\}$. Now $W_i = \{1, 0, x\}$, which means a W is included in V but not included in u.

5.9.2 Finding Essential Prime Implicants Using Two Cubes — A Sharp Operation

Recall section 5.4.5. Refer inclusion of 1 at a cell in the map from a don't care condition(s) (incomplete specification case) to complete a Karnaugh map quad or pair or octet. Consider a pair of vertex on a hypercube in which one coordinate is x and other is either 0 or 1 in two vertices. We consider it as one vertex with coordinate = x replaced by either by 1 or 0 as follows. For example, convert the pair (1, ..., x, 0, 1) and (1, ..., 1, 0, 1) into (1, ..., 0, 0, 1) or convert the pair (1, ..., x, 0, 1) and (1, ..., 0, 0, 1) into (1, ..., 1, 0, 1).

Since conversion to a single x basically means a removal of one of the variable or reducing the dimension of the cube.

Let us assume that there are two m-dimensional cubes, U and V with m- variables. Let the variables are $u_1, u_2, ..., u_m$ on first cube and $v_1, v_2, ..., v_m$ on second cube. A variable, u or v is either 1 or x or 0.

Now sharp operation $U \# V$ means the following:

For each pair of u_i and v_i,

(i) $u_i \# v_i = \varsigma$ if both are 0.

(ii) $u_i \# v_i = 1$ if $u = x$ and $v = 0$.

(iii) $u_i \# v_i = \varsigma$ if $u = 0$ and $v = x$.

(iv) $u_i \# v_i = \varsigma$ if both are 1.

(v) $u_i \# v_i = 0$ if $u = x$ and $v = 1$.

(vi) $u_i \# v_i = \varsigma$ if $u = 1$ and $v = x$.

(vii) $u_i \# v_i = \varepsilon$ if $u = 1$ and $v = 0$

(viii) $u_i \# v_i = \varepsilon$ if $u = 0$ and $v = 1$.

(ix) $u_i \# v_i = \varsigma$ if $u = x$ and $v = x$.

The i = 1, 2, ..., m. Sharp operation means a cube W obtained by sharp operation $U \# V$ defined as follows:

(1) $W = U$, if $u_i \# v_i = \varepsilon$ for some value of i between 1 and m (For some i, the variable is complementary in U and V); and defined by (vii) and (viii) relationships.

(2) $W = \varepsilon$ if $u_i \# v = \varsigma$ for all values of i.

(3) $W = \underset{\cdot}{U}(u_1, u_2, ..., u_i, ..., u_m)$ for cases in which $u = x$ and v_i = either 0 or 1 ($\underset{\cdot}{U}$ is a special sign for the union. Union is for all value of i where these conditions for u and v exist).

Note the important points as following:

1. $W = \underset{\cdot}{U}$, it means that U and V differs at least in one variable and that variable(s) is complement of each other in U and V.
2. $W = \varepsilon$ means that cube U fully covers V. For all i, both u and v either 0s, or 0 and x, or x and 1, or 1 and x, or x and x.
3. $W = \underset{\cdot}{U}$ is a new union where that part, which was not covered in $\underset{\cdot}{U}$, is included now. When $u_i = x$ and is 0, and $v_i = 1$, the u did not include v_i. For a value of i, when $u_i = x$ and is 1, and $v_i = 0$, the u_i did not include v_i. It is now included in the union.
4. By using the sharp operation, all essential prime implicants can be grouped. This will result in minimum possible cover (implementation by AND-OR circuits.)

5.9.3 Computer-Based Minimization Method to find Minimum Required Cover (SOP function implicants)

Following are the steps for a computer-based minimization to get the minimum required cover (SOP function essential prime implicants).

1. Step 1: Specify the coordinated of marked vertices, which correspond to the given set of SOP minterms (implements). Let us call it as SET A.
2. Step 2: Specify the coordinated of marked vertices, which correspond to the given set of SOP minterms (implements) after including the don't care conditions) (unspecified table rows or unspecified input conditions). Let us call it as SET A'.
3. Step 3: Find prime implicants using the Star-product operation in the SET A' implicants.
4. Step 4: Find union after the Sharp operation to find essential prime implicants and thus the minimum cover for A'.
5. Step 5: Write the least cost (minimized) SOP expression.
6. Step 6: Make the AND-OR logic circuit for the terms.

5.10 COMPUTER-BASED MINIMIZATION—QUINE-McCLUSKEY METHOD

It has been observed in Sections 5.1 to 5.6 that a Karnaugh map can be used to get the minimization by forming diads, quads and octets that remove the one, two and three, variables respectively. The basis for the removal is the Boolean OR operation $X.\overline{Y} + X.Y = X$, where Y is a single variable and X is a single variable or a multivariable product term. Therefore, two product terms $\overline{A}.C.D.\overline{B} + \overline{A}.C.D.B$ will reduce to $\overline{A}.C.D$ because they differ in one variable value B.

$X = A.C.D$ and $Y = B$ in above example). In Karnaugh map two adjacent cell pairs differ in one variable value. That is why a variable reduces on pairing adjacent or wrapping-adjacency showing 1s in the SOP terms of Karnaugh map.

There are two problems in Karnaugh map approach. First is how to work with the map in case of increase in the number of variables visualization becomes more and more cumbersome. Second difficulty is a adoptability to a computer based minimization. To take care of these two problems, one procedure of using cube and hyper cube vertices coordinates for the minterms was described in Section 5.9. A popular procedure is the procedure known as Quine-McCluskey method. It also provides the answer to above two problems.

5.10.1 Quine-McCluskey Method of finding Prime Implicants

Step 1: Write the expression for a Boolean function Y in terms of the product term. $Y = \Sigma m(1, 3, 5, 6, 7, 10, 14, 15) = \overline{A}.\overline{B}.\overline{C}.D + \overline{A}.\overline{B}.C.D + \overline{A}.B.\overline{C}.D + \overline{A}.B.C.\overline{D} + \overline{A}.B.C.D + A.\overline{B}.C.\overline{D} + A.B.C.\overline{D} + A.B.C.D$ is an exemplary expression undertaken here for explaining the steps. Let an index represent the number of 1s in a SOP term for Y. List all the terms with index = 0 in one list, index = 1 in next list, index = 2 in next list, index = 4 in the next list and so on. For the four variable case, the maximum value of index = 4 and the indices are 0, 1, 2, 3 and 4. Separate each list by line at the end of the list. Columns 1 to 4 of Table 5.27 shows the five lists for the example under description.

There are four list-end lines in the four cells to show five separate lists. There is no term in the list with index 0. (Note: Logic of this step can be understood as follows. Basically a list having additional 1 compared to previous list is the one having the potential to have the pairing terms within the lists for the terms which can be paired to satisfy the expression $X.\overline{Y} + X.Y = X$).

Step 2a: Start from index 0 list and find a pair of minterm in the index 1 list in cycle 1. Continue to find pair upto lists 0 and 5 pairing. Index 0 list is empty. Therefore, no pairs. Switch to step 2b with no action.

Step 2b: Start cycl3 2 from index 1 and find a pair of minterm in the index 1 list and next list with index = 2. (The pair is the one, satisfies that $X.\overline{Y} + X.Y = X$). Indices 1 and 2 lists are not empty. Therefore, the action is to be for finding the pairs after comparisons. In column-5 two lists' comparison is done (between pair of minterms 1 and 3, 1 and 5, 1 and 6, and 1 and 10). Pairs 1 and 3 and 1 and 5 differ in one variable each for third and second places (taking from left), respectively. (Recall Table 5.9 Karnaugh map also, the minterms $m(1)$ and $m(3)$, and $m(1)$ and $m(5)$ are the adjacent cell pairs.) Write up a list of SOP pairs in next column (column 6). Refer entries 00–1 and 0–01. (0001 + 0011) of column 3 gives 00–1 and (0001 + 0101) gives 0–01 in the second column. The dash sign means removed variable after using the OR relation. Successful pairs are check marked. Put check marks in column 2 sign at 1, 3 and 5 in Table 5.27. Draw a line below the list 1 in columns 5 and 8.

Step 2c: Start from index 2 and find a pair of minterms in the index 2 list and next list with index = 3. Indices 2 and 3 lists are not empty. Therefore, the action is to be for finding the pairs after comparisons. In list 2 column-5 the two lists' comparison is given after OR operations between pair of minterms (3, 7), (3, 14), (5, 7), (5, 14), (6, 7), (6, 14), (10, 7) and (10, 14). Pairs 7a(3, 7), 7b(5, 7), 7c(6, 7), 14a(6, 14), 14b(10, 14) are satisfying $X.Y + X.Y =$

TABLE 5.27 Finding the prime implicants using the Quine-McCluskey method

List	Minterms	Binary form and Index		Comparison cycle 2		New index and list no.		Comparison cycle 3		Comparison cycle 4
1	~	~	0	3^a	|00–1	1	1	7*d*	0––1	Not required
	–	~	0	5^a	0–01					
2	1‘	0001	1	7^a	0–11			15^c	–11–	
3	3%	0011	2	7^a	01–1					
	5%	0101	2	7^c	011–	2	2			
	6%	0110	2	14^a	–110					
	10%	1010	2	14^b	1–10					
4	7^	0111	3	15^a	–111					
	14^	1110	3	15^b	111–	3	2			
5	15\	1111	4							

(a)~ means index 0 correspond to no minterm. (b) ‘ means index = 1 because only at one place there is 1. (c) % means index = 2 because only at two places there are 1s. (d) ^ means index = 3 because only at three places there are 1s. (e) \ means index = 4 because only at four places there are 1s. (f) | Dash sign here means variable *C* removed from the minterm on simplification between *m*(1) and *m*(3). The term now left is $\overline{A}\,\overline{B}\,D$, represented by 00–1. (g) 3^a and 5^a means 3 and 1 and 5 and 1 paired in step 2b. Cycle 1 does not exists as no pair of list 0 with others.

X expression. Recall Table 5.9 Karnaugh map also, the minterms m(6) and (14), and (10) and (14) are the adjacent cell pairs, and so on. We can verify adjacency from the Table 5.9 for all these five pairs at the two lists. Put a check mark on 3, 5, 6, 10, 7 and 14 and carry remaining result of sum of product terms in the pairs to next column (column 6) second list. Successful pairs are check marked. Refer entries 0–11 and 01–1. (0011 + 0111) gives 0–11 and (0101 + 0111) gives 01–1. The dash sign means removed variable after using the OR relation described above. Draw a line below new list 2 at the last cell in column 5 to 8.

Step 2d: Start from index 3 and find a pair of minterms in the index 3 list and next list with index = 4. Indices 3 and 4 lists are not empty. Therefore, the action is to be for finding the pairs after comparisons. In column-5 two lists' comparison is given between pair of minterms (7, 15) and (14, 15). Both pairs are satisfying $X.\overline{Y} + X.Y = X$ expression. Recall Table 5.9 Karnaugh map also, the minterms *m*(14) and (15), and (7) and (15) are the adjacent cell pairs. Write the result of sum of product terms in the pairs in next column (column 6). In the present step, the column 6 list 3 entries are –111 and 111–. (0111 + 1111) gives –111 and (1110 + 1111) gives 111- (15a and 15b, respectively). The dash sign means a removed variable after using the OR relation described above. Draw a line below new list at the last cells columns 5 to 8.

Carry Step further for the five variable cases, Step 2f for six variable case and so on. Since the example is for four variables case, in step 2 steps 2a, 2b, and 2c suffice. Move to next Step 3. Let us number the three lists (column 8). Marking '*a*' or '*b*' or '*c*' signifies that these are the terms of the lists at the second cycle of the actions at the Steps 2a to 2d. A number specified on the left of the dashed minterms sorted out in the Step 2s corresponds to the minterm number presuming – sign as 1. For example, 00-1 is marked as 3a. If next time at cycle 3 also if 00-1 occurs then it is marked as 3b. Labeling of minterms is done here for better understanding of each of the steps described here.

Operations in steps 2a, 2b, 2c, ... are called cycle 2 operations.

General Instructions for Step 3 Cycle 3

1. Compare all pairs between (i) first and second lists of column 6, (i) second and third and so on. In present example, the three lists are present at column 6. Therefore, only Steps 3a and 3b are required. Compare the pairs in column 6 but in between two adjacent lists only.
2. Only those pairs in which the dash(es) are occurring at the same places needs to be considered. For example, pair 3^a and 7^a need not to be considered as dashes are at the third and second places from most most bit places, in 00–1 and 0–01, respectively.

Step 3a: Start from index 1 as index 0 term is not present in column 6. Find a pair of minterms in the corresponding list and next list with index = 2. (Index is now counted after ignoring the dash(es). Pairs should also have the dashes at the same place. Indices 1 and 2 lists in the column are not empty. Therefore, the action is to be for finding the pairs after comparisons. In column-9 two lists' comparison is given between the pair of minterms (3a, 7b) and (5a, 7a). Pair (3a, 7b) only is satisfying $X.\overline{Y} + X.Y = X$ expression. Put the check marks on 3a, 5a, 7b and 7a terms. Carry remaining result of sum of product terms in the pairs to next cycle first list. We find that both pairs give same term 0 – – 1. Refer third cycle entry 0 – – 1. Entries 0 – – 1 is because (00–1 + 01–1) gives 0– –1 and (0 – 01 + 0 – 11) gives 0 – – 1. Two dash signs means two removed variables after using the OR relation described above in cycles 2 and 3. Draw a line below new list at the last cell in columns 9 and 10.

Step 3b: Start from index 2. Find a pair of minterms in the new index 2 list and next list with index = 3 (index is now counted after ignoring the dash(es). The pair is the one, which satisfies the $X.\overline{Y} + X.Y = X$. Pairs should also have the dashes at the same place. Indices 2 and 3 lists 2 and 3 are not empty. Therefore, the action is to be for finding the pairs after the comparisons. In column-9 two lists' comparison is done between pair of minterms (7c, 15b) and (14a, 15a), which only are satisfying $X.\overline{Y} + X.Y = X$ expression. Put the check marks on 7c, 15b, 14a and 15a terms. Carry remaining result of sum of product terms in the pairs to next cycle list. We find that both pairs give the same term –11–. Refer third cycle entry –11–. Entries –11– is because (011– + 111–) gives –11– and (–110 + –111)) gives – 11–. Two dash signs means two removed variables after using the OR relation described above, operated in cycle 1 and cycle 2. Draw a line below new list at the last cell in columns 9 and 10.

Out of nineth entries in column 6, the eigth are marked (paired in cycle 3 for generating column 10). Continuation of the further cycle 4 steps are not required as both the lists in column 10 have one entry each and the pair between cannot be formed because the dashes are at the different places (0 – – 1, – 11 –).

Collect all the unmarked entries from (unpaired entry) columns 3, 6 and 10. Column 6 has left one uncheck marked entry $m(10)$[1–10] and column 10 has two entries $m(1)$ and $m(6)$ [– 11 – and 0 – – 1] unpaired. An unchecked entry corresponds to a prime implicant.

Three prime implicants are therefore as follows:

$$Y = \Sigma m\ (1, 3, 5, 6, 7, 10, 14, 15) = \Sigma m(1, 6, 10)\ \overline{A}.\overline{C}.D + \overline{A}.D + B.C$$

The above process looks tedious, but easily implements using a computer program. For multiple variables, this will be the best option.

5.10.2 Finding Minimal Sum from the Prime Implicants for an Output

Exemplary prime implicants table was shown in Table 5.28. We assume don't care variable as 1. Minimal sum is sum of prime implicants, which are absolutely necessary, else the Boolean function itself modifies.

First check whether each row has at least one term present and check that whether removal of a row still leaves label y in at least one column. If removal of a row is possible then remove provided a row still leaves label y in at least one column (for example, assume there is y at $m7$ also. Then middle row can be removed because $m6 + m7 = m6$ and it removes the redundant term).

We therefore get on summing all the prime implicants at the table the minimal sum expression. In the following expression, there is no redundancys. Hence, the answer is follows:

$$Y = A.C.\overline{D} + A.\overline{D} + BC$$

It is also called *irredundant disjoint* expression. None of the row of Table 5.26 can be removed; otherwise the function will be modified.

TABLE 5.28 Prime implicants table from the Quine-McCluskey method for finding redundancy and then obtaining minimal sum expression (irredundant disjoint expression)

	m0	m1	m2-m5	m6	m7	m8	m9	m10	m14	m15
BC				y						
$\overline{A}D$		y								
$AC\overline{D}$								y		

y shows the prime implicants (1–10, 0– – 1, –11–) in the Table 5.27 example.

5.10.3 Finding Minimal Sum for the Multi-Output Case using Quine-McCluskey Method

Quine-McCluskey method can be applied to multiple outputs case as follows. We have to find three sets of prime implicant table for two- outputs case, for Y_0, Y_1 and Y_0Y_1.

Let $Y_0, Y_1, ..., Y_{n-1}$ are the outputs for a set of input variable $X_0, X_1, ... X_{m-1}$. A truth table of these will have m columns for the inputs and n columns for the outputs. Number of rows will be number of possible input combinations 2^m. Now suppose $n = 2$ and $m = 3$. Therefore, the Boolean functions are Y_0 and Y_1, which depends on X_0, X_1 and X_2 only.

Let us consider a *labeled product* term. Let us understand a labeled product term by an example. Suppose in the truth table, for $X_0X_1X_2 = 001$, the $Y_0 = 1$ and $Y_1 = 0$, then labeled product term is $001Y_0-$. It means that Y_0 term exists for 001 input and Y_1 does not exist for 001 input. For $X_0X_1X_2 = 000$, if $Y_0 = 1$ and $Y_1 = 1$, then labeled product term is $000Y_0Y_1$. For $X_0X_1X_2 = 010$, if $Y_0 = 0$ and $Y_1 = 1$, then labeled product term is $001–Y_1$.

Now find the prime implicants using Quine-McCluskey method for the five-variable case in place of four input variables A, B, C and D case considered in section 5.10.1 (Note: a general case will be $(m + n)$ variable case).

After finding prime implicants, make three prime implicants table in place of one table. One table is for the variable $X_0X_1X_2Y_0$-, other for $X_0X_1X_2$-Y_1 and another for $X_0X_1X_2Y_0Y_1$. Prime implicants in third table will correspond to the common prime implicants so that the terms can be used in Y_0 and Y_1 both. Prime implicants in first table will correspond to the terms that can be used in Y_0. Prime implicants in second table will correspond to the Y_1.

EXAMPLES

Example 5.1 From the given three inputs, A, B, and C truth table (Table 5.29), construct a Karnaugh map and then construct the SOP functions based on map for output S. Simplify the result and find a Boolean expression.

TABLE 5.29 Truth table

A	B	C	Output S
0	0	0	1
0	0	1	0
0	1	0	1
0	1	1	0
1	0	0	1
1	0	1	0
1	1	0	1
1	1	1	0

Solution

The 0th, 2nd, 4th and 6th row of truth table have the output = 1. We put the 1s at the 000, 010, 100 and 110 cells in the Karnaugh map (Table 5.1). (Four minterms (implicants) are the ones for which $S = 1$). We can use Table 5.3 to get the SOP—

$$S = \Sigma m\,(0, 2, 4, 6)$$

TABLE 5.30 Map from truth table and corresponding minterms

AB \ C		$\overline{C}$ 0	C 1
$\overline{A}.\overline{B}$	00	1	
$\overline{A}.B$	01	1	
$A.B$	11	1	
$A.\overline{B}$	10	1	

AB \ C	0	
00	$m(0)$	
01	$m(2)$	
11	$m(6)$	
10	$m(4)$	

Table 5.30 shows the map.

There is a quad containing four adjacent columns in first column. Hence two variables can be removed out of three. Only $\overline{C}$ is left and

$$S = \overline{C}.$$

Example 5.2 From the given four inputs, A, B, C and D truth table (Table 5.31) construct a Karnaugh map, construct the SOP functions based on the Karnaugh map for an SOP output S. There is no specifications given for $A = 1$, $B = 0$, $C = 1$, and $D = 1$.

TABLE 5.31 Truth table

A	B	C	D	S
0	0	0	0	0
0	0	0	1	0
0	0	1	0	0
0	0	1	1	1
0	1	0	0	0
0	1	0	1	1
0	1	1	0	0
0	1	1	1	1
1	0	0	0	0
1	0	0	1	1
1	0	1	0	0
1	1	0	0	0
1	1	0	1	1
1	1	1	0	0
1	1	1	1	0

Solution

If we count from 0th row, the 3rd, 5th, 7th, 9th and 13th row of truth table have the output = 1. Now use the map in Table 5.7 as template to fill 1s from the truth table. We put the 1s at the 0011, 0101, 0111, 1001 and 1101 cells in the Karnaugh map (Table 5.32). Eleventh row is not present and we assume it as don't care condition cell. We put the x in the 1011 cell. Five minterms (implicants) are the ones for which $S = 1$. Using Table 5.9, we find 1s at $m3$, $m5$, $m7$, $m9$, $m13$.

$$S = \Sigma m\,(3, 5, 7, 9, 13)]$$

TABLE 5.32 Four variable Karnaugh Map and adjacencies

AB \ CD		$\overline{C}\,\overline{D}$ 00	$\overline{C}D$ 01	CD 11	$C\overline{D}$ 10
$\overline{A}\,\overline{B}$	00			1	
$\overline{A}B$	01		1	1	
AB	11		1		
$A\overline{B}$	10		1	x	

Wrapping Adjacency

We have two pairs of the adjacent cells ($m5$, $m7$) and ($m9$, $m13$). One additional-pair exits assuming x (don't care) as 1 at last row. This additional-pair exits on considering a wrapping adjacency. We can remove one variable each. Hence answer is as follows:

$$S = A.\overline{C}.D + \overline{A}.B.D + \overline{B}.C.D$$

Example 5.3

From the given Karnaugh map in Table 5.33, find the minterms present at the SOP.

Solution

TABLE 5.33 Three variable Karnaugh Map and minterms

AB \ C	$\overline{C}$ 0	C 1
00		1
01		1
11	1	
10		

AB \ C	0	1
00	–	$m(1)$
01	–	$m(3)$
11	$m(6)$	–
10	–	–

There are 1s at the cells, 001, 011, and 110. Hence 1st, 3rd and 6th minterm is present (using Table 5.3). Hence $m(1)$, $m(3)$ and $m(6)$ are present in the SOP—

$$S = \Sigma m(1, 3, 6)$$

Example 5.4

From the given maxterms in Table 5.34, find the Karnaugh map. Conditions for $A = 1$ are not specified and are to be taken as don't care. Find the POS expression also.

TABLE 5.34 Maxterms and POS output function

A	B	C	D	Maxterm	POS Output function
0	0	0	0	$Mx0$ $A + B + C + D$	1
0	0	0	1	$Mx1$ $A + D + C + \overline{D}$	1
0	0	1	0	$Mx2$ $A + B + \overline{C} + D$	0
0	0	1	1	$Mx3$ $A + B + \overline{C} + \overline{D}$	1
0	1	0	0	$Mx4$ $A + \overline{B} + C + D$	0
0	1	0	1	$Mx5$ $A + \overline{B} + C + \overline{D}$	1
0	1	1	0	$Mx5$ $A + \overline{B} + \overline{C} + D$	0
0	1	1	1	$Mx7$ $A + \overline{B} + \overline{C} + \overline{D}$	1

Solution

There are 0s at $M(2)$, $M(4)$ and $M(6)$. Hence, we put the 0s in the cells at 0010, 0100, 0110 as alone in filling Table 5.12 (Section 5.2.3). We also put x for the cell positions with $\overline{A} = 1$. Therefore, the map is as per Table 5.35. We have

$\overline{P} = \Pi M(2, 4, 6)$ (from the Table 5.11) if we do not consider don't care conditions. Else $\overline{P} = \Pi M(2, 4, 6, 8, 9, 10, 11, 12, 13, 14, 15)$ by assuming 0s at the places of x.

TABLE 5.35 Four variables Karnaugh Map for maximum 16 Maxterms in a POS expression–$(A + B + \overline{C} + D).(A + \overline{B} + \overline{C} + D).(A + \overline{B} + C + D)$ when not considering don't care terms

$A + B$ \ $C + D$		$C + D$ 00	$C + \overline{D}$ 01	$\overline{C} + \overline{D}$ 11	$\overline{C} + D$ 10
$A + B$	00				0
$A + \overline{B}$	01	0			0
$\overline{A} + \overline{B}$	11	x	x	x	x
$\overline{A} + B$	10	x	x	x	x

Note: $\overline{P} = \overline{A}.(A + \overline{C} + D). (A + \overline{B} + D))$ on placing 0s for x.

Example 5.5 Show the octets, quads and pair of adjacent cells in Table 5.36 below and hence simplify the given implicants (minterms) for function $S = 1$ and obtain prime implicants.

TABLE 5.36 Use of don't care input combinations for determining the adjacencies

AB \ CD		00	01	11	10
–	00	1	1	1	1
–	01	x		x	1
–	11	1			1
–	10	1	x	1	1

AB \ CD		$C + D$ 00	$C + \overline{D}$ 01	$\overline{C} + D$ 11	$\overline{C} + D$ 10
–	00	1	1	1	1
–	01	x		x	1
–	11	1			1
–	10	1	x	1	1

Solution

The map shows two octets consisting of (1st and 4th rows) and (1st and 4th columns) considering wrapping adjacencies and taking don't care conditions as 1. The map also shows a quad in first and in second rows.

Hence $S = B + \overline{C} \cdot \overline{A}$.

But the second term $\overline{A} \cdot \overline{C}$ is not essential, because x can as well be taken as 0. Therefore, the essential (prime) implicant is only one term in S, and

$S = B$ is the least cost AND-OR circuit.

One AND has all inputs = B.

Example 5.6

Find Karnaugh map of $X + \overline{Y}.Z$ in POS standard format (POS cover) and verify the answer by minimizing the map.

Solution

Recall Boolean rule; $X + \overline{Y}.Z = (X + \overline{Y}).(X + Z)$ (Equation 4.15b). We have to first convert it into standard POS format.

This can be written as $\overline{P} = (X + \overline{Y} + Z.\overline{Z}).(X + Y.\overline{Y} + Z)$. [Using ANDing rule that AND with the complement is 0 and ORing rule that OR with 0 has no effect in a Boolean operation.

$$\overline{P} = (X + \overline{Y} + \overline{Z}).(X + \overline{Y} + Z).(X + Y + Z).(X + \overline{Y} + Z)$$

(After expanding by using the rule $X + Y.Z = (X + Y).(X + Z)$ for both the ANDing terms). Using ANDing rule that $X.X = X$ for the second and fourth term in P:

$$= (X + \overline{Y} + \overline{Z}).(X + \overline{Y} + Z).(X + Y + Z)$$

The POS form therefore ΠM (0, 3, 2) (First term in $M(3)$, second term is $M(2)$ and third is $M(0)$. We have to put 0s in cells 000, 011 and 010 for forming the POS Karnaugh map).

Now recall Table 5.5 and put 0s at appropriate Maxterm positions and we get map as shown in Table 5.37.

TABLE 5.37 Three variable Karnaugh Map

XY \ Z		Z 0	$\overline{Z}$ 1
$X + Y$	00	0	
$X + \overline{Y}$	01	0	0

Table 5.37 is the desired map. We can verify that the map is correct as follows. There are two adjacent cell pairs. From pair in first column, Y is removable. So $(X + Z)$, which is common is left. From pair in second row, Z is removable. So $(X + \overline{Y})$, which is common is left. Using distribution rule $X + Y.Z = (X + Y).(X + Z)$, we get the simplest circuit $X + \overline{Y}.Z$. This is from where we started.

Example 5.7

Give Karnaugh map of $X + Y.Z$ in a SOP cover (standard format) and verify the answer by minimizing the map.

Solution

Recall ORing rule; $Y + \overline{Y} = 1$. (Equation 4.10d) We have the expression $X + Y.Z$. We first convert it into standard SOP format using this rule. Therefore, SOP S is as follows:

$$S = X.(\overline{Y} + Y).(\overline{Z} + Z) + (X + \overline{X}).Y.Z$$
$$= X.\overline{Y}.\overline{Z} + X.Y.\overline{Z} + X.\overline{Y}.Z + X.Y.Z + \overline{X}.Y.Z + X.Y.Z$$
$$= X.\overline{Y}.\overline{Z} + X.Y.\overline{Z} + X.\overline{Y}.Z + \overline{X}.Y.Z + X.Y.Z$$

(Remove fourth term $X.Y.Z$ occurring twice using rule $X + X = X$).

$= \Sigma m(3, 4, 5, 6, 7)$. (Above expression first term is $m(4)$, second is $m(6)$, third $m(5)$, fourth $m(3)$ and fifth $m(7)$).

We have to put 1s in cells 011, 100, 101, 110 and 111 for forming the SOP Karnaugh map.

Now recall Table 5.5 and put 1s at appropriate minterm positions (cell positions 011, 100, 101, 110 and 111) and get map as shown in Table 5.36.

TABLE 5.38 Three variable Karnaugh Map

XY \ Z	$\overline{Z}$ 0	Z 1
00		
01		1
11	1	1
10	1	1

(Diad: cells 01-1 and 11-1; Quad: rows 11 and 10)

Table 5.38 is desired the Map. We can verify that map is correct as follows. There are two adjacencies; one quad of 4 cells and one pair of cells. From quad in last two rows, Y and Z are removable. So X, which is common is left. From pair in 2nd column, X is removable. So $(Y.Z)$, which is common is left. We get simplest circuit as $X + Y.Z$. This is from where we started.

Example 5.8

From the given Karnaugh map in Table 5.39, find the simplified expression in terms of XOR and XNOR. logic circuits.

TABLE 5.39 Diagonal and offset adjacencies at Karnaugh Map

	AB \ CD	$\overline{C}\,\overline{D}$ 00	$\overline{C}D$ 01	CD 11	$C\overline{D}$ 10
–	00			1	1
XOR	01	1			
–	11		1	1	1
–	10	1			

(XOR: diagonal/offset adjacencies of 1s in cells 0100, 1101, 1000; XNOR: 1s in columns 11 and 10 of rows 00 and 11)

Solution

We find that between first row last two columns and third row last two columns, there are 4-cell with offset of one row. Therefore, we can write the simplified term as XNOR gates. (A XNOR B) $C.\overline{D}$ + ($\overline{A}$.XNOR.CD). This is equal to (A XNOR B) C. (From ORing rule $D + \overline{D} = 1$).

We also find that between 2nd and 3rd rows first two columns, there is a 2-cell pair having diagonal adjacent. $B.\overline{C}$ is common. We can write remaining simplified term as XOR gates, (A XOR D) $B.\overline{C}$.

We also find that between 3rd and 4th rows first two columns, there is another 2-cell with diagonal adjacent. $A.\overline{C}$ is common. Therefore, we can write the simplified term as XOR gates. (B XOR D) $A.\overline{C}$).

Therefore, S = (A XNOR B).C + (A XOR D) $B.\overline{C}$ + (B XOR D) $A.\overline{C}$.

Example 5.9

Represent $S = \overline{A}.\overline{B}.C. + \overline{A}.B.C. + A.\overline{B}.C$ in a cube form.

Solution

To write in terms of minterms, we put A, B and C as 1. From $\overline{1}\ \overline{1}\ 1, \overline{1}11$ and $1\ \overline{1}\ 1$ in the expression and evaluate the binary value in decimal. We find $S = \Sigma m\,(1, 3, 5)$ and the cube coordinates are (001, 011, 101) because $\overline{1} = 0$.

Answer is a cube with vertices at the coordinates (0, 0, 1), (0, 1, 1) and (1, 0, 1) marked. Also we show an axis joining (0, 0, 1), (0, 1, 1) as dark with mid point labeled as (0 x 1). This is because only Y middle coordinate is differing.

Also show another axis joining (0, 0, 1), (1, 0, 1) as dark with mid point labeled as (x, 0, 1).

Now two axis with mid points are (0 x 1) and (x, 0, 1). This will give simplified logic circuit as $(\overline{A} + \overline{B}).C$ from [$\overline{A}\,C$ is term from (0 x 1) and $\overline{B}\,C$ from (x 0 1)].

Example 5.10

Verify the simplification of $S = \overline{A}.\overline{B}.C + \overline{A}.B.C + A.\overline{B}.C$ as $(\overline{A} + \overline{B}).C$ from cube form in Example 5.9 by using Karnaugh map approach.

Solution

To write in terms of minterms, we put A, B and C as 1. From 111, 111, 111 and 111 in the expression and evaluate the binary value in decimal. $S = \Sigma m(1, 3, 5)$. Karnaugh map cells, which have 1s, are 001, 011 and 010. We get the map as per Table 5.40 using the Table 5.3.

TABLE 5.40

AB \ C	$\overline{C}$ 0	C 1
00		1
01		1
11		
10		1

Adjacency

Wrapping Adjacency

We have two pair of adjacent cells (one is with wrapping adjacency). From each pair we remove one variable and take the common two variables each.

We get $S = \overline{A}.C + \overline{B}.C = (\overline{A} + \overline{B}).C$

We get the same answer as by the cube form representation in Example 5.9.

EXERCISES

1. From the given three inputs, *A*, *B*, and *C* truth table in Table 5.41, construct the SOP functions based Karnaugh map for an output *S*.

TABLE 5.41 Three variable truth table

A	B	C	Output S
0	0	0	x
0	0	1	1
0	1	0	0
0	1	1	1
1	0	0	0
1	0	1	1
1	1	0	1
1	1	1	0

2. From the given four inputs, *A*, *B*, *C* and *D* truth table in Table 5.42, construct the SOP functions based Karnaugh map for output *S* (i) for SOP (ii) POS standard forms.

TABLE 5.42 Four variable truth table

A	B	C	D	S
0	0	0	0	1
0	0	0	1	0
0	0	1	0	
0	0	1	1	1
0	1	0	0	
0	1	0	1	1
0	1	1	0	1
0	1	1	1	1
1	0	0	0	1
1	0	0	1	
1	0	1	0	
1	0	1	1	x
1	1	0	0	
1	1	0	1	
1	1	1	0	
1	1	1	1	x

3. From the given Karnaugh map in Table 5.43, find the minterms present. Also minimize by two adjacent pairs and another adjacent pair oafter wrapping adjacency.

TABLE 5.43 Three variable Karnaugh Map

C / AB	$\overline{C}$ 0	C 1
00		1
01	1	
11		1
10	1	1

4. From the given Boolean expression $A.B + \overline{A}.C + B + A.D + \overline{A}.\overline{E} + D$, find the Karnaugh maps after converting the expression in both SOP and POS standard forms.
5. From the given Boolean expression $A.B.D + A.C.D + C.\overline{D} + A.D + D$ find the POS and SOP Karnaugh maps.
6. From the given Boolean expression $A.B.C + A.B.C + B.C$ find the Karnaugh maps after converting the expression in both SOP and POS standard forms.
7. Consider the following Table 5.44 for conversion of a decimal number to Gray code (a code in which next number have only one bit place changing). Assuming each code bit as a $S1$, $S2$, $S3$ and $S4$, draw four three-variable Karnaugh maps and simplify as much as possible (find prime implicants).

TABLE 5.44

Decimal value	4-bit Binary representation	Gray code representation
0	0000_2	0 0 0 0
1	0001_2	0 0 0 1
2	0010_2	0 0 1 1
3	0011_2	0 0 1 0
4	0100_2	0 1 1 0
5	0101_2	0 1 1 1
6	0110_2	0 1 0 1
7	0111_2	0 1 0 0
8	1000_2	1 1 0 0
9	1001_2	1 1 0 1
10	1010_2	1 1 1 1
11	1011_2	1 1 1 0
12	1100_2	1 0 1 0
13	1101_2	1 0 1 1
14	1110_2	1 0 0 1
15	1111_2	1 0 0 0

8. Consider the following Table 5.45 for conversion of a decimal number to excess-3 code (a code in which next number in binary form is three more than the binary value for the decimal value). Assuming each code bit as a $S1$, $S2$, $S3$ and $S4$, draw four three-variable Karnaugh maps and simplify as much as possible (find prime implicants).

TABLE 5.45

Decimal value	4-bit Binary representation	Excess-3 code representation
0	0000_2	0 0 1 1
1	0001_2	0 1 0 0
2	0010_2	0 1 0 1
3	0011_2	0 1 1 0
4	0100_2	0 1 1 1
5	0101_2	1 0 0 0
6	0110_2	1 0 0 1
7	0111_2	1 0 1 0
8	1000_2	1 0 1 1
9	1001_2	1 1 0 0

9. From the given minterms $S = \Sigma m(1, 7, 8, 9, 10)$ find the Karnaugh map-for four and five variable forms.
10. From the given maxterms $\overline{P} = \Pi M(1, 2, 7, 9)$ find the Karnaugh map for four variable forms.
11. From the given minterms $S = \Sigma m(1, 6, 11, 18)$ find the Karnaugh map for five variables forms.
12. From the given maxterms $\overline{P} = \Sigma M(2, 7, 9, 19, 42, 47)$ find the Karnaugh map six variable forms.
13. From the given maxterms $P = \Pi M(2, 7, 11, 9)$ find the Karnaugh map and draw the only NORs based circuit.
14. From the given minterms $S = \Sigma m(2, 7, 11, 9)$ find the Karnaugh map and draw the only NANDs based circuit.
15. Draw a Karnaugh map of four variable showing only wrapping adjacencies and having an octet, a quad and a pair of adjacent cells.
16. Simplify after first converting (A XOR B) XOR C to standard SOP form and then using Karnaugh map make a circuit with the NANDs only.
17. Simplify after first converting $(A.\overline{B}.C) + (\overline{A}.C.D)$ to standard POS form and then using Karnaugh map to make a circuit with the NORs only.
18. Show the octets, quad and the pairs of adjacent cells in Karnaugh map in Table 5.46 and hence simplify the given implicants (minterms) for function $S = 1$.

TABLE 5.46 Four variable Karnaugh Map

CD / AB	— 00	— 01	— 11	— 10
00	1	1	1	1
01				
11	1			1
10	1	x	1	1

19. Give Karnaugh map of $A + B.C.D$ in POS standard format (POS cover) and verify the answer by minimizing the map.
20. Give Karnaugh map of $A + B.C.D$ in SOP standard format (SOP cover) and verify the answer by minimizing the map.
21. From the given Karnaugh map in Table 5.47, find the simplified expression in terms of XOR and XNOR logic circuits.

TABLE 5.47 A Karnaugh Map, which includes wrapping and diagonal adjacencies

AB \ CD		CD 00	CD 01	CD 11	CD 10
$\overline{A}\,\overline{B}$	00		1		
$\overline{A}B$	01	1		1	1
AB	11				
$A\overline{B}$	10	1		1	1

22. Show simplification of $S = \overline{A}.B.C.\overline{D} + A.B.C.\overline{D} + \overline{A}.\overline{B}.C$ from Karnaugh map and by a 4-dimensional cube representation approach and draw logic circuit using AND-OR gates at first and second levels.
23. Draw $S = A.B.\overline{C}.\overline{D} + A.B.C.D.E$ Karnaugh map and hypercube representation approach and draw logic circuit using AND-OR gates at first and second levels.
24. Simplification of $S = A.B.\overline{C}.D + A.B.C.D + A.B.\overline{C}$ from Karnaugh map and by cube representation approach and draw logic circuit using NANDs.
25. Simplification of $S = \overline{A}.B.C.\overline{D} + \overline{A}.B.C.\overline{D} + A.B.C$ from Karnaugh map and by cube representation approach and draw logic circuit using NORs.
26. Using Karnaugh maps for multiple- outputs given in Tables 5.48 and 5.49 draw the AND-OR arrays minimize and find the minimal expressions for multiple outputs Y_0 and Y_1.

TABLE 5.48 Karnaugh Map for Y_0

$\overline{A}\,\overline{B}$ \ $\overline{C}\,\overline{D}$		$\overline{C}\,\overline{D}$ 00	$\overline{C}D$ 01	CD 11	$C\overline{D}$ 10
$\overline{A}\,\overline{B}$	00		1		
$\overline{A}B$	01		1		
AB	11	1			
$A\overline{B}$	10	1			

TABLE 5.49 Karnaugh Map for Y_1

AB \ $\overline{C}\,\overline{D}$		$\overline{C}\,\overline{D}$ 00	$\overline{C}D$ 01	CD 11	$C\overline{D}$ 10
$\overline{A}\,\overline{B}$	00	1	1	1	
$\overline{A}B$	01	1	1	1	
AB	11	1			
$A\overline{B}$	10				

27. Using Karnaugh maps for multiple-outputs given in Tables 5.50 and 5.51 draw the AND-OR arrays minimize and find the minimal expressions for multiple outputs Y_0 and Y_1.

TABLE 5.50 Karnaugh Map for Y_0

$\bar{C}\bar{D}$ / $\bar{A}B$		$\bar{C}\bar{D}$ 00	$\bar{C}D$ 01	CD 11	$C\bar{D}$ 10
$\bar{A}\bar{B}$	00	1			
$\bar{A}B$	01	1		1	1
AB	11			1	
$A\bar{B}$	10				

TABLE 5.51 Karnaugh Map for Y_1

$\bar{C}\bar{D}$ / AB		$\bar{C}\bar{D}$ 00	$\bar{C}D$ 01	CD 11	$C\bar{D}$ 10
$\bar{A}\bar{B}$	00	1			
$\bar{A}B$	01	1			1
AB	11			1	1

28. Minimize using $Y = \Sigma m(2, 3, 6, 7, 10, 14, 15)$ using Quine-McCluskey method.
29. Minimize and find minimal expression for $Y = \Sigma m(2, 4, 8, 9, 10, 11)$ using Quine-McCluskey method.
30. Using minimize and find the minimal expressions for multiple outputs $Y_0 = \Sigma m(2, 3, 6, 7, 10, 14, 15)$ and $Y_1 = \Sigma m(2, 4, 6, 7, 10, 11, 14)$ using Quine-McCluskey method.
31. From a prime implicants in Table 5.52, use the Quine-McCluskey method for finding redundancy and then obtains minimal sum expression (irredundant disjoint expression).

TABLE 5.52

	$m0$	$m1$	$m2$-$m5$	$m6$	$m7$	$m8$	$m9$	$m10$	$m13$	$m15$
BC				Y						
$\bar{A}D$		Y								
$AC\bar{D}$								Y		
$\bar{A}BCD$					Y					
$AB\bar{C}D$									Y	

QUESTIONS

1. How will you construct a three-variable Karnaugh map from a given three inputs A, B, and C truth table?
2. How will you construct a four-variable Karnaugh map from a given three inputs A, B, and C truth table?
3. Explain with two examples each, method of construction a Karnaugh map for output S (i) for SOP (ii) POS standard forms.
4. How will you find the minterms present from a Karnaugh map?
5. How will you draw logic circuit from the Karnaugh maps after converting the expression in (i) SOP and (ii) POS standard forms?
6. How will you find the wrapping adjacencies between first and last rows of a Karnaugh map?

7. How will you find the wrapping adjacencies between first and last columns of a Karnaugh map?
8. Show use of Karnaugh map for a Gray code converter.
9. How do you find the octets, quads and pairs in a Karnaugh map?
10. How do you minimize a Karnaugh map from the adjacencies?
11. How will you make a logic circuit after minimizing a SOP based Karnaugh map?
12. How will you make a logic circuit after minimizing a POS based Karnaugh map?
13. How do you from a Karnaugh map draw the only NORs based circuit?
14. How do you from a Karnaugh map draw the only NANDs based circuit?
15. How do you from a Karnaugh map draw the XNORs and XORs based circuits?
16. How do find the hypercube vertex coordinates from the five variable minterms?
17. Explain the star-product operation on the terms of a cube when minimizing using a computer program.
18. List the steps for a computer based minimizing method.
19. Describe Quine-McCluskey method to find prime implicants in five variable case.
20. Describe Quine-McCluskey method to find prime implicants in four input variables for the three multiple output expressions.

CHAPTER 6

Logic Gates

OBJECTIVE

In this chapter, we shall study RTL, DTL, TTL, ECL, ICL, HTL, NMOS and CMOS logic gates; the difference between the various families of gates and their speeds; propagation delay; operating frequency; power dissipated per gate; supply-voltage levels; and operational-voltage levels that define logic states 1 and 0.

In Chapter 4, we learnt six digital electronics logic operations used as basic operations in the complex circuits. These are NOT, NAND, AND, OR, NOR and XOR. We also learnt that an interesting combination is NOT-XOR (recall the Figures 4.1 and 4.2). There are the different types of circuits of the logic gates for implementing these operations; RTL, DTL, TTL, ECL, ICL, HTL, NMOS and CMOS. How are these circuits made? We shall learn this in this chapter. How does the different family of gates differ? These questions are answered in this chapter.

Let us first revise and learn the following concepts to have ease in learning the topics of this chapter.

6.1 REVISION OF THE IMPORTANT GATES

(A) NOT

NOT gate has the unique property that its output is 1 if the input is at 0 logic level and the output is 0 if input is at 1 logic level.

NOT logic is used when an output is desired to be the complement of the input. If all inputs of NAND gates are joined then that will also act as a NOT gate. NOT gate is also called inverting logic circuit. It is also called a complementing circuit.

Two DeMorgan theorems help a digital circuit designer to implement all the other logic gates with the help of either NOR gates only or NAND gates only. A NOT gate is implementable by a NAND or a NOR by joining all of their inputs.

(B) AND

A NAND gate followed by a NOT gate gives us an AND gate. Its symbol differs from NAND only by the omission of a bubble (circle). Its unique property is that its output is 0 unless all the inputs to it are at the logic 1. Two inputs AND gate has the following way of writing its operations.

$$F = A_1.A_2 \qquad ...(6.1)$$

Dot between two states indicate AND logic operation using these.

Remember

AND gate unique property is that output is 1 only if all of the inputs are at 1 logic level.

(C) OR

A NOR gate followed by a NOT gate (made by the common input NOR) gives us OR gate. Its symbol differs from the NOR only by the omission of a bubble (circle).

OR gate unique property is that output is '1' if any of the inputs are at '1' logic level.

(D) NOR

Its unique property is that its output is 0 if any of its inputs is 1. A NOR gate is a basic building block for other types of the logic gates other than TTLs (Section 6.4.3). In the TTL circuits, a NOR is fabricated in an IC by the several NANDs.

(E) XOR

An XOR gate (is called 'Exclusive OR' gate). Its unique property is that the output is 1 only if odd numbers of the inputs at it are 1s.

(F) NOT-XOR

A NOT-XOR gate has a unique property that the output is 0 only if odd numbers of the inputs are 1. It is like a NOR gate but differs in the output when even number of its inputs are 1.

(G) BUFFER (Non Inverting)

A BUFFER gate simply gives same state at output as that at the input. It can also be made by connecting two inverting circuits (NOTs) in series. It gives one advantage that we can connect to an output a greater number of other gates logic inputs with a non-inverting buffer in between than without the BUFFER, of course, at the expense of an additional propagation delay time which depends upon speed of the BUFFER logic gate internal circuit.

6.2 DIODE CIRCUIT

A *p-n* junction diode, shown by a triangle with a line for the *p* and *n* end, respectively, is used in the logic circuits as follows. (refer Figure 6.1 (a)). It is used in making logic circuits due to the following factors.

1. When a silicon p-end connects through a resistance ~2200 Ohm (~ 2.2 kΩ) to the supply of 5 V it will flow (5 V – 0.7 V)/2.2 kΩ] = ~ 2 mA current and its *p*-end will be at ~0.7 V (called threshold voltage) with respect to *n*-end connected to the supply ground. This *p*-end at low voltage now represents the logic state 0. (refer Figure 6.1 (b)).
2. When the *p*-end connects through a resistance ~2200 Ohm (~ 2.2 kΩ) to the supply of 5 V but the and *n*-end also connects to a high voltage ~ 2.4 V to 5 V, the current through the diode and resistance circuit will be negligibly small and the diode voltage is below the threshold. The *p*-end will also be at high voltage with respect to the supply ground. The both p-end and n-end now represents the logic state 1. (refer Figure 6.1 (c)).

Figures 6.1(d) and (e) shows how the two diodes can be used to create two-input AND and OR gate, respectively, and can also be used to give functionally the AND and OR operations within another logic circuit.

6.3 BIPOLAR JUNCTION TRANSISTORS AND MOSFETS

6.3.1 *N-P-N* Transistor Common Emitter Circuit

An *n-p-n* junction transistor, also called *n-p-n* bipolar junction transistor (*n-p-n* BJT) can be used in logic circuits as follows: Figure 6.2 (a) shows an *n-p-n* BJT symbol and its common emitter circuit. It also shows the D.C. state currents I_C, I_B, I_E through (i) V_{CC} supply-collector resistance, (ii) base-emitter and (iii) emitter, and V_{EE}^- supply ground, respectively. It also shows the D.C. state voltages V_R, V_{BC}, V_{BE} across (i) V_{CC}^+ supply and collector resistance, (ii) collector and base and (iii) base and emitter and V_{EE}^- supply ground, respectively.

When collector-end connects through a resistance between 0.6 kΩ to ~ 4 kΩ) to the supply of 5 V it will flow the current as per the base current I_B at that instant and the logic state (output) at the collector shall also depend on that. This circuit in its normal mode of operation is called inverting current amplifier and a small change in base current causes a large change in collector current. It is a called current switching transistor circuit when it is operated in one of the two—*cut-off* mode or *saturation* mode. In logic circuits, its use as a current switching circuit is of interest. The cut-off and saturation modes are switched by the base current changes such that the logic outputs are 1 and 0, when the logic inputs are 0 and 1, respectively.

1. **Cut-Off Mode Operation:** Base emitter *p-n* junction is reverse biased ($0 < V_{BE} < V_{BE\,(on)}$) and there is negligible current I_B flowing across it. The base collector-*p* junction is also reversed biased ($0 < V_{BC}$) and there is negligible current I_C through it. Voltage drop through *R* is negligible due to this transistor and collector is thus at high voltage of supply voltage ~ 5 V. The logic output represents 1 till other circuits or logic gates connected to through R_B decrease this voltage at the collector below a limit. Note: V_{BE} is base-emitter bias and $V_{BE\,(on)}$ is the bias needed to make base-

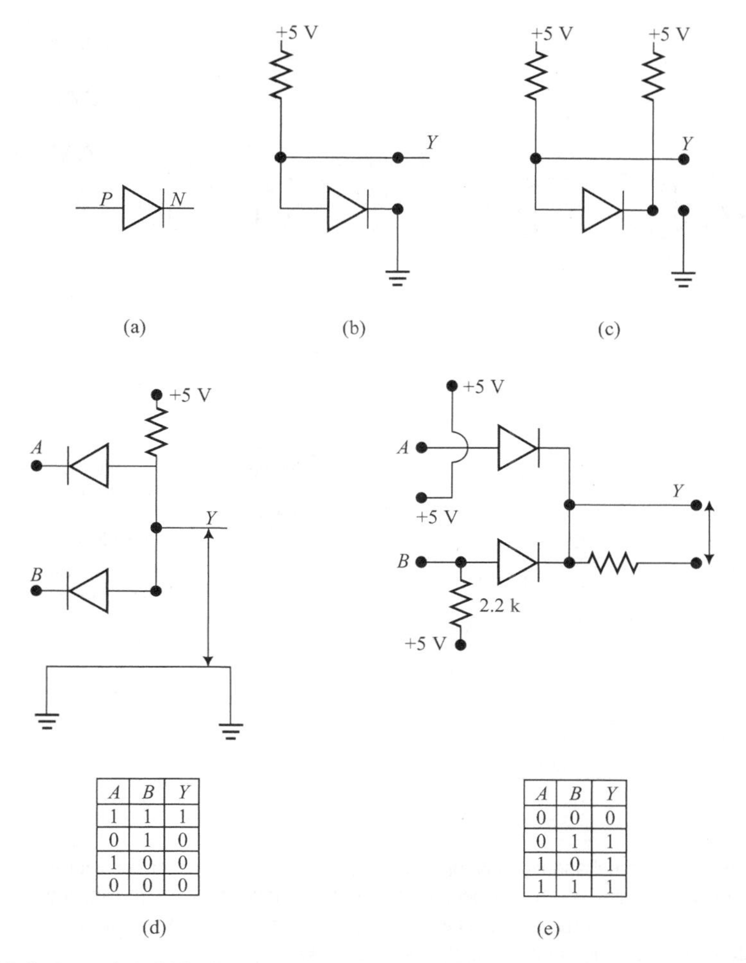

A	B	Y
1	1	1
0	1	0
1	0	0
0	0	0

A	B	Y
0	0	0
0	1	1
1	0	1
1	1	1

FIGURE 6.1 (a) A diode symbol. (b) A silicon diode at logic 0 (c) The diode at logic 1 (d) Two diodes creating a two-input AND (Truth table in inset). (e) Two diodes creating a two-input OR (Truth table in inset).

emitter junction forward biased. We must remember $V_{CE\,(sat)} = \sim 0.5$ V, $V_{BE\,(cutoff)} <$ 0.7 V in cut-off mode of operation. (refer Figure 6.2 (b)).

2. **Saturation Transistor Circuit Operation:** Base emitter *p-n* junction is forward biased ($V_{BE\,(sat)} = \sim 0.7$ V) and there is significant current I_B flowing across it. The base collector *p-n* junction is also forward biased ($V_{CE\,(sat)} = \sim 0.2$ V, base current large enough to give $I_C < \beta I_B$, where β is the forward current gain). There is significant current I_C (~ mA) through it. Voltage drop through R is large due to I_C this transistor and collector is thus at very low voltage ($V_{CE\,(sat)}$ ~0.2 V). The logic output represents 0 till other circuits or logic gates connected to through this transistor may also sink the current through it. We must remember, that when $I_C \approx \beta I_B$, the $V_{CE\,(sat)} =$ ~ 0.2 V, $V_{BE\,(ON)} = V_{BE\,(sat)} = \sim 0.7$ V and $V_{BC\,(sat)} = V_{BC\,(ON)} = \sim 0.5$ V and there is saturation mode of operation (refer Figure 6.1 (c)).

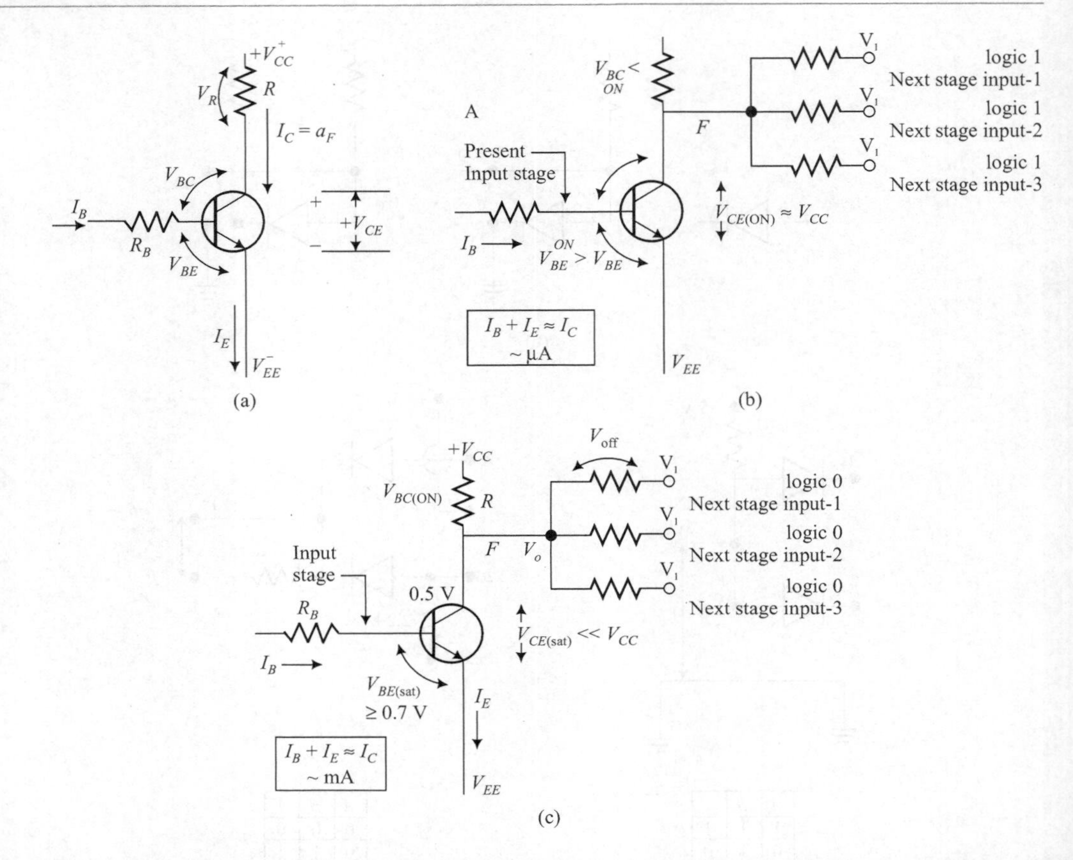

FIGURE 6.2 (a) The *n-p-n* BJT symbol and its common emitter circuit (b) The transistor collector negligible current corresponding to the OFF state of the transistor, which give functionally the 1 logic output state = 1 from the transistor collector (c) The transistor collector high current (~ mA) corresponding to the saturation state of the transistor, which give functionally the 0 logic output states, respectively, for giving the input(s) 0 to another logic circuit(s).

6.3.1.1 Important Equations for the Circuit

Important equations are $I_B = (V_i - V_{BE\,(ON)})/R_B$ ...(6.2)

where V_i is input voltage applied to R_B. This is because potential difference R_B equals the term in the bracket. In both saturation and normal modes, equation (6.2) holds. Also,

$$I_C = \beta I_B = \beta(V_i - V_{BE(ON)})/R_B \quad ...(6.3)$$

The output at F is given by

$$V_o = V_{CC} - I_C.R = V_{CC} - R.\beta(V_i - V_{BE\,(ON)})/R_B \quad ...(6.4)$$

When the V_i increases the V_o decreases till $V_o = V_{CE\,(sat)}$.

6.3.1.2 Important Feature of the *n-p-n* Common Emitter Circuit in Application in the Logic Circuits, and the Operational Voltage Levels that Define Logic States 1 and 0

1. V_i for the first stage can vary within a range provided the transistor remains in cut-off mode and the output is constant (~ V_{CC}), which decreases with the next stage loading circuit. We can assign a range for V_O (between the V_{OH} maximum = (~ $V^+{}_{CC}$) and V_{OH} minimum) and logic is said to be 1. The next stage(s) V_i can be between V_{OH} and (V_{OH}) minimum permitted drop in— between F and next stage(s) (Right side circuit of Figure 6.2 (c)).
2. V_I can vary within a range provided the transistor remains in saturation mode and the output is constant (~0.2V), which changes due to the currents from the next stage loading circuit(s). We can assign a range for V_O (V_{OL} = V_{OL} maximum to supply ground, V^-_{EE}) and logic is said to be 0. The next stage(s) V_I can be between supply ground V^-_{EE} and (V_{OL} maximum + permitted drop in between *F* and next stage(s).

6.3.2 MOSFET Circuits

MOSFET (Metal-Oxide Semiconductor Field Effect Transistor) logic circuits are most used circuits today. These are small and simple to fabricate on VLSI (very large scale integrated) chips. All memories, microprocessors, and computers therefore use these.

6.3.2.1 MOSFET Circuit Operation

A MOSFET is a three-layered structure; a metal (gate), a very thin oxide layer and a semiconductor (channel), in four types: *n*-channel, enhancement mode *n*-channel depletion mode, *p*-channel and enhancement mode *p*-channel depletion mode. (Figures 6.3(a) to (d) gives the symbol and gate-characteristics of four types of MOSFETs).

Enhancement mode *n*-channel MOSFET shows the characteristics as follows. When voltage between gate and source, V_{GS} = 0, the current I_D between the drain and source is 0 and is independent of voltage between drain and source, V_{DS}.

Small V_{DS} Region: MOSFET is OFF and is in cutoff state below a positive threshold Voltage $V^T{}_{GS}$. After a threshold, when $V_{GS} > V^T{}_{GS}$, a channel is formed (established for I_D flow) between drain and source and the I_D increases and linearly varies with V_{DS}. MOSFET is said to be in ON state. Further, when the V_{GS} increases, the slope of change in I_D is steeper and steeper with respect to V_{DS}. Channel resistance decreases steeply with ($V_{GS} - V^T{}_{GS}$). MOSFET is said to be in ON state. Channel width (along with electrons flow) is constant is function of ($V_{GS} - V^T{}_{GS}$) (Figure 6.3(a)) left side logic 0 region characteristic).

Constant I_D with V_{DS} Region: Now, $V_{DS} = V_{GS} + V_{GD}$. Increase in V_{DS} and therefore V_{GD} causes the channel narrowing, decrease in channel resistance. At a voltage called pinch-off voltage, V^T_{GS}, the channel resistance near drain is 0. At a given $V_{GS} = V_{GSS}$, the I_D does not change significantly with V_{DS} in ON state when $V_{DS} > V^T_{GS}$. The variation is no longer linear but is non-linear with V_{DS} and becomes nearly constant. Current I_{DSS} continues to flow in this region. (Figure 6.3(a) right side logic 1 region characteristic).

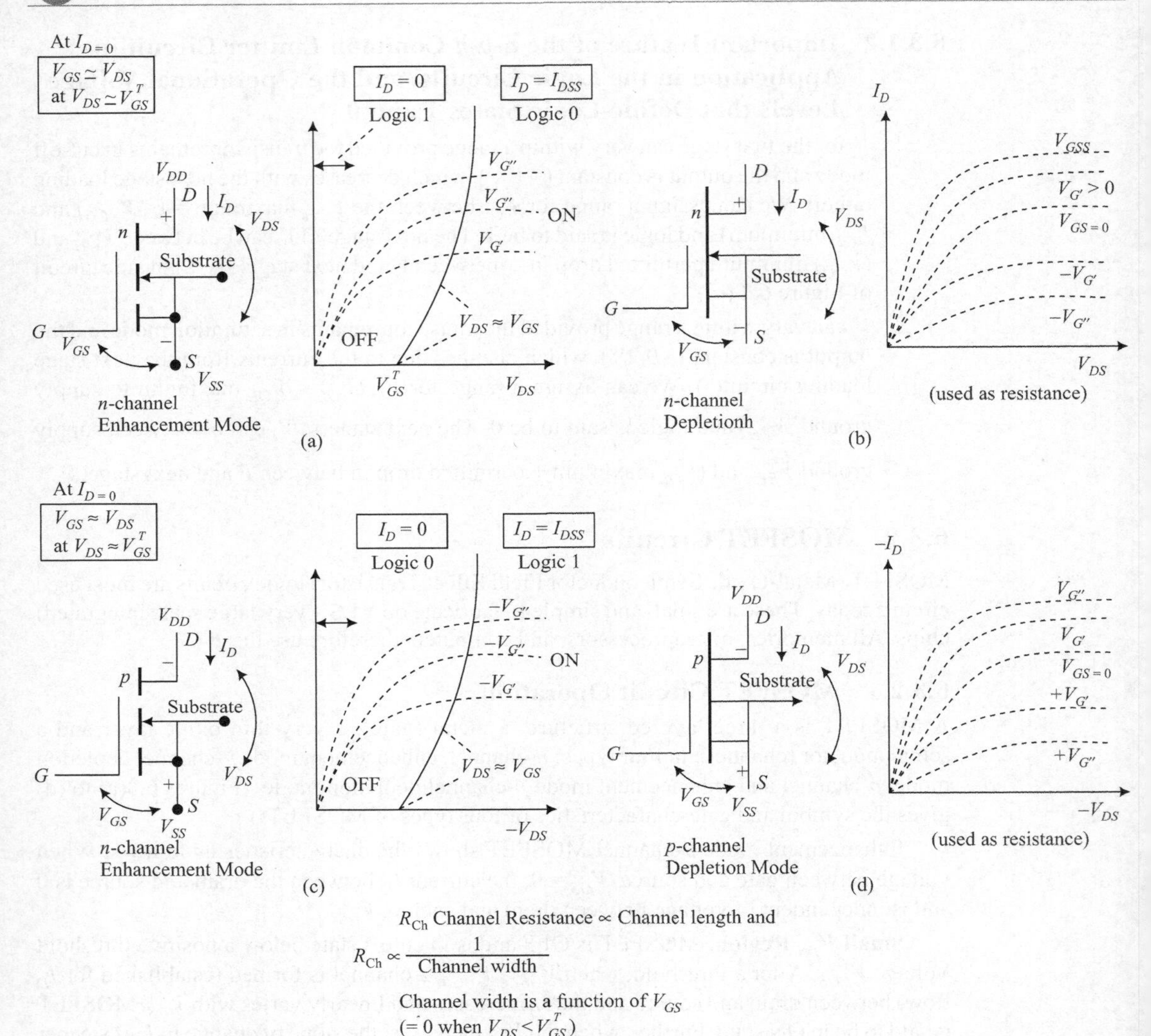

FIGURE 6.3 (a) to (d) Symbols of four types of MOSFETs. Also shown are the linear region characteristics for *n*-channel enhancement, *n*-channel depletion, *p*-channel depletion, and *p*-channel enhancement respectively. (e) Change in R_{Ch}

Enhancement mode needs $V_{GS} > V^T{}_{GS}$. A *depletion mode n-channel MOSFET* conducts at $V_{GS} = 0$ also and conduction starts from a negative threshold gate-source voltage, $V_{GS} > -V^T_{GS}$. MOSFET is OFF and is in cutoff state below this voltage. When voltage between gate and source, $V_{GS} = 0$, the current I_D between the drain and source is 0 and is independent of voltage between drain and source, V_{DS} (Figure 6.3(b)).

Enhancement mode *p*-channel MOSFET and Depletion mode *p*-channel MOSFET works similar to *n*-channel MOSFETs in these modes. The current is carried by holes; therefore the direction of current and voltages is reverse in *p*-channel with respect to *n*-channel.

For turning the MOSFET in ON state, an enhancement mode p-channel needs $V_{GS} < -V_{GS}^{T}$. Depletion mode p-channel needs $V_{GS} < +V_{GS}^{T}$. (Figure 6.3(c) and (d))

Enhancement mode MOSFET as a Resistance: If we join the drain and gate, $V_{GS} = V_{DS}$. When $I_D = 0$, $V_{DS} = V^{T}{}_{GS}$. The resistance is non-linear. Very high at $V_{DS} = V^{T}{}_{GS}$ and drops sharply with V_{DS} change. (Figure 6.3(e))

Depletion mode MOSFET as a Resistance: If we join the drain and gate, $V_{GS} = V_{DS}$. When $I_D = 0$, $V_{DS} = -V^{T}{}_{GS}$. The resistance is non-linear. Very high at $V_{DS} = -V^{T}{}_{GS}$ and drops sharply with V_{DS} change. At $V_{GS} = 0$ also it is not very high.

6.3.2.2 MOSFET Circuit Features

Static region resistance between input at gate and the channel is very high and the MOSFETs when at cut-off show very high input D.C. impedance. (There is only capacitive impedance between gate and source.) When at saturation, MOSFET (ON region) resistance between the drain and source is low (when drain source voltage is below the pinch off). The saturated region is above a pinch-off voltage, (the resistance between drain and source is more than between a BJT's collector and emitter).

6.4 RTL, DTL, TTL LOGIC GATES

Section 6.3 explained how a BJT transistor or a MOSFET can be used in two regions cut-off region and saturation region. This fact can be used to design the logic gates. Followings are the logic gates built on this concept.

6.4.1 Resistor–Transistor Logic (RTL)

Both input and output stage circuits in a NOR logic gate is shown in the Figure 6.4.

1. **Input Stage:** Two inputs to two n-p-n transistors through two 450 Ohm resistances.
2. **Output Stage for the next Input Stage:** The common output F connects the collector (s) of the transistors with the input stages given to other resistances at the next stage RTL gates.

Logic Operation

When input voltage at logic input A is low, the transistor T_A does not conduct (in cutoff stage). Similarly, when input B is low, the transistor T_B does not conduct. Output of these transistors is from their collectors, which are joined together and given supply voltage of 3.6 V through a 640 Ohm resistance. When both T_A and T_B do not conduct, the output at F is high voltage, close to the supply voltage of 3.6 V ($V_o = V_{CC} - I_C.R$).

When input voltage at logic input A is high, the transistor T_A conducts. Similarly, when input B is high, the transistor T_B conducts. Output of these transistors is from their collectors, which are joined together and given supply voltage of 3.6 V through a 640 ohm resistance. Therefore, when either of T_A and T_B conducts, the output at F is low voltage, closer to the supply ground voltage ($V_{CE\,(sat)} = \sim 0.2$ V). F connects to M number R-T stage (resistance transistor stage) transistors, $T_1, T_2 \ldots T_M$. ($M = 4$ in the figure)

NOR gate gives output = 1 when its all inputs are equal to 0. It gives output = 0, when any of the input = 1. Therefore, the circuit of Figure 6.4 works as two-input NOR gate.

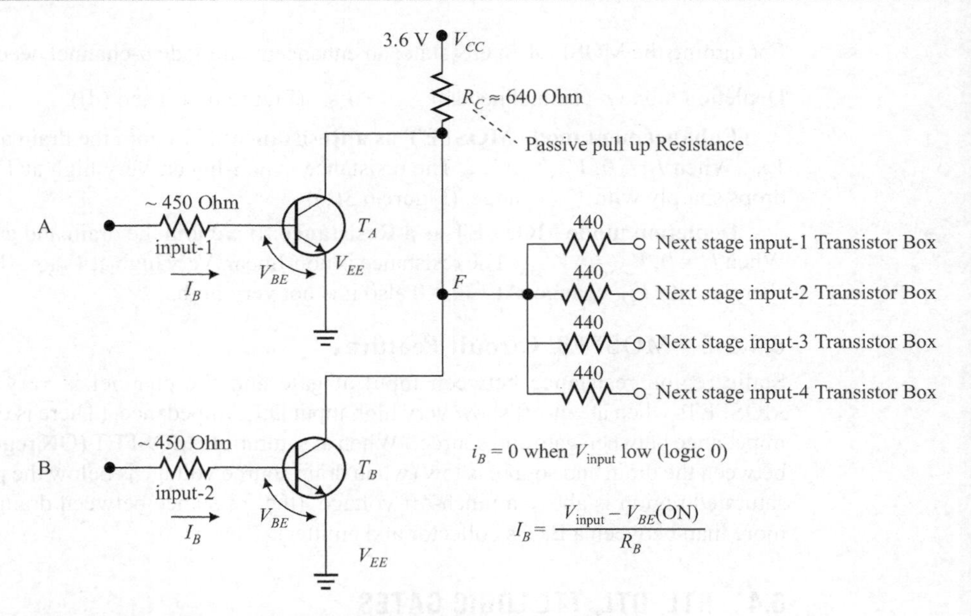

FIGURE 6.4 An RTL circuit for a NOR gate the output of which given to four gates.

Case of Input Stage Output Connected to m Output Stage Transistors as Load

When the output from common collector junction is to *m* number *R-T* stage (resistance transistor stage) transistors, each drives through 450 Ohm resistance as the load, the current from input stage 640 ohm resistance will be divided into *m* parts.

Base-Emitter voltage, when a transistor is ON (conducting in saturation region), is $V_{BE\ (sat)} \approx 0.8$V. Base current for each output stage transistor T_j(j = 1, will be $I_B = I_B = (V'_i - V_{BE\ (ON)})/R_B = (1/m)\ (3.6\ \text{V} - 0.8\text{V})\ /\ (640\ \text{Ohm} + (450/m)\ \text{Ohm}) = V'_i$ is the input to T_j. If $m = 4$, I_B will be 0.93 mA.

(***Point, you must note and practice:*** When solving a numerical problem, at each stage don't forget to put the units at numerator and denominator both).

The collector current when this transistor is conducting (is in saturation) will be $I_C = (3.6\text{V} - 0.2\ \text{V})/640\ \text{ohm} = 5.3$ mA. A T_j transistor gain h_{fe} must be equal or greater than I_{C0} (saturation stage current)/I_B. For $m = 4$, $h_{fe} >= (5.3\ \text{mA}/0.93\ \text{mA})$, $h_{fe} >= 5.8$. [h_{fe} = small signal current gain for the common emitter configuration, and is assumed here close to the D.C. current gain β in equations (6.2) to (6.4)].

We can have more output stage transistors (j can be higher) if h_{fe} is higher. More output stage transistors means more output stages can be driven from the output *F*.

For logic output of 1 at F, the output voltage depends on *m and m maximum permissible number depends on the* h_{fe} value of T_j transistor. Higher h_{fe} permits higher *m*.

For logic output 0 at *F*, the output voltage does not depends on *m*. This is because of the transistor T_A or T_B conducting and in saturation. The output at *F* is ~ 0.2 V when input at *A* or *B* >= 0.5 V (cut-off voltage). Cut-off voltage is the voltage, below which the transistor cut-off the current and above which the transistor is in saturation and is conducting.

Definition of Fan Out

> Number of logic gates at the next stage(s) that can be loaded to a given logic gate output so that voltages for each of the possible logic state remain within the defined limits (refer for example, for 1 between V_{OH} minimum and V_{CC}^{+} and for 0 between V_{OL} maximum and V_{EE}^{-}).

Note: V_{OH} is the voltage for next stage when logic = 1, V_{OL} when = 0, V_{OH} minimum and V_{OL} maximum are the permissible limiting values as per logic definition.

Propagation Delay

Definition of Propagation Delay

> Propagation delay for a logic output from a logic gate means the time interval between change in a defined reference point input voltage and reflection of its effect at the output. It can also be defined as the time interval between changes in a defined logic level input and reflection of its effect at the output logic level.

(Note: The slightly different values of the delays are obtained when the input change from 0 to 1 and change from 1 to 0. We can take average propagation delay. Also a propagation delay is also subject to variations in power supply and temperature. We can then define a statistical deviation and an average).

Calculation of Propagation Delay: Let base-emitter capacitance = C nF (nF means nanoFarad). If $m = 4$, the total capacitance all T_j in parallel being = 4.C. Resistance = 640 ohm + (450/m) Ohm = 752.5 Ohm. Propagation delay = (640.m + 450) C ns. (nF × Ohm = ns).

Definition of Noise Margin

> Noise margins for the logic outputs 1 and 0 means the permitted worst case voltage levels variations of the logic output 1 and 0, respectively, when an output from a stage is the input at the next stage(s). The margins are permitted due to expected internal temperature variations and power supply variations. Permitted noise margins reflects the digital circuit immunity and worst case performance consistency in the presence of an induced noise.

Calculation of the Noise Margins: For logic state 0, the output at F can be ~0.2 V and maximum 0.5 V, else the T_j will start conducting and go in saturation. Hence noise margin of logic state 0 in RTL based logic circuit is 0.3 V.

We have seen that the output voltage depends on m. For logic state 1, the output at F can be calculated as follows. Collector current at T_j = 5.3 mA for each transistor T_j base current is 5.3 mA/h_{fe} = 0.265 mA assuming h_{fe} = 20.

For $m = 4$, total base current needed from $F = m$. 0.265 mA = 1.06 mA. Voltage at F = voltage drop between T collector and emitter + total base current multiplied by total base resistance (450/m) Ohm. Thus voltage at $F \geq 0.8$ V + 1.06 mA. (450/m) Ohm ≥ 0.92 V.

Available output voltage at F can be calculated as follows:

V_O = 3.6 V – voltage drop at (450/m) ohm collector resistance + voltage drop at 0.8 V base-emitter. = 3.6 V – (640 ohm/(640 ohm + 450/m ohm). (0.8) V = 1.2 V for $m = 4$.

Hence the logic 1 at F can be between the 1.2 V and 0.92 V for $m = 4$ and $h_{fe} = 20$. For logic 1, noise margin will be 0.28 V when $m = 4$ and $h_{fe} = 20$.

Wired Logic (Wired OR or Implied OR) Connection

If outputs F and, F' at two RTL gates are made common (connected), the output can be considered as AND operation between the logic outputs. Because when both the outputs correspond to cutoff stages of the transistors, the output will remain unaffected and will be 1. When any of the outputs correspond to saturation condition ~0.2 V, the output from common point will become 0.2 V. If A, B are the inputs at one RTL NOR gate and C, D are inputs at another, NOR the output will be as follows:

$$F = \overline{(A + B)}.\overline{(C + D)} = \overline{(A + B + C + D)} \text{ (Using DeMorgan theorem).}$$

6.4.2 Diode–Transistor Logic (DTL)

A DTL circuit in Figure 6.5 works as three input NAND gate.

1. **Input Stage:** It is based on the applying of two or more inputs to two or more n-ends of the p-n junction diodes in place of passive ~450 Ohm resistance in an RTL circuit (Figure 6.4). The p-ends are common and connect to a 5000 Ohm resistance R_D, which connects to supply voltage of 5 V. The common point P also connects to a transistor-base through two forward biased diodes. The transistor (T) base also connects to the emitter through 5000 Ohm R_B connect. The emitter of T connects to supply ground. The collector of T connects to supply through a resistance R_C of 2200 Ohm.
2. **Output Stage for the next Input Stage:** A common output from the transistor T at F is given to other diodes at the other next stage(s) logic gates (DTL gates), which will get the input from the transistor, T.

Both input stage diodes and transistor form a NAND gate and next input stage circuit(s). The DTL based NAND and its connection to next stages is shown in the Figure 6.5.

6.4.2.1 Logic Operation for the Output at F

If input A or B or C is low ~0.2 V (near supply ground), the diode in the path of A or B or C, respectively will start conducting. Current through the diode will be approximately equal to (5 V – 0.7 V – 0.2 V)/(5000) Ohm = ~0.8 mA. (Voltage drop across a conducting p-n junction diode is threshold voltage $V_D^T \approx$ ~0.7 V). Voltage drop of 0.9 V (V_{in} + 0.2) connects to the base of T through 2-diode pair, each needing threshold voltage of 0.7 V and total 1.4 V to turn ON and conduct which is much more than 0.9 V available. Hence the transistor base is at ~0 V, (below cutoff voltage). When T is not conducting the output $F = 1$, high (~5 V). T does not conduct when any (or both of the inputs) is low because voltage at the base drops below the cut-off voltage needed at the base.

If all inputs A, B and C are high (> 0.7 V), the voltage at common p-ends will start exceeding 1.4 V and the 2-diode pair to the base will start conducting. When each input A, B and C exceeds 1.4 V and the voltage at the common p-ends exceeds (1.4 V + $V_{BE\,(\text{ON})}$) = 2.1 V, the base-emitter junction starts conducting. When each input exceeds 1.4 V, the diode at the input stops conduction and when exceeds 2.1 V, becomes reverse biased. $V_{BE\,(\text{ON})}$ remains at 0.7 V. If transistor T base-emitter current exceeds a limit, the T goes in saturation

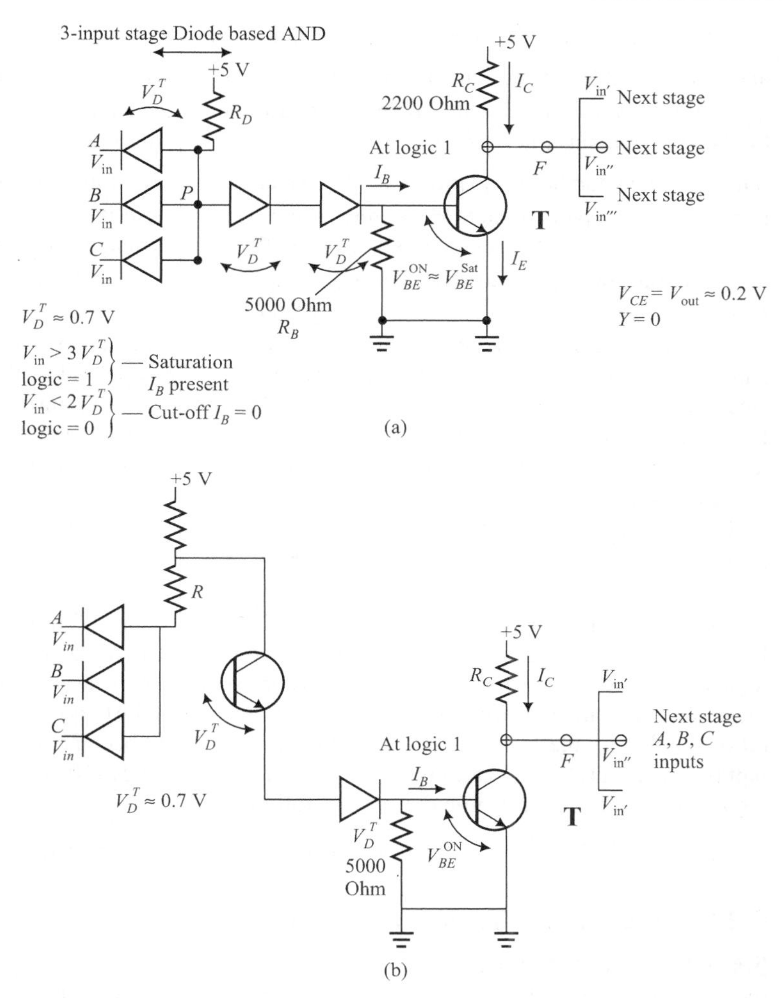

FIGURES 6.5 (a) The DTL based NAND and its connection to next stages (b) Modified DTL circuit by replacing 2-diode combination by one diode and one transistor.

mode and it will start conducting current I_C through R and $V_{CE} \approx 0.2$ V. Therefore, $F = 0$ when A and B both 1.

Property of a NAND is that its output is 0 when both the inputs are 1. Therefore, the circuit in Figure 6.5 works as a three-input NAND gate.

6.4.2.2 Case of Input Stage Output *F* Connected to *m* Output Stage Transistors as Load

When the output from common collector junction connects m number D-T stage (diode-transistor logic stage) transistors, each drives through one input diode and 2-series diodes (similar to first stage T), the current from input stage is 0 from the F when all the next stage inputs are high.

When a transistor T is conducting (logic 0 at F), the V_{CE} = ~0.2 V (in saturation stage) and V_{BE}^{sat} = ~0.8 V. The currents from each output stage diode with the transistors T_j(j = 1, ..., j) will be $m.I_C$. The T will remain in saturation until the condition, $mI_C < \beta I_B$, (Refer Section 6.3) remains true. I_B = (2.1V)/5000 Ohm = 0.4 mA (the condition is also called *current-sink logic* condition). [3 V_{DT}^{T} = 2.1 V] We can have more input D-T stages (j can be higher) if h_{fe} (≈ β) is higher. More input stages means more input stages can be driven through the output F. For logic output of 1 at F, the output voltage does not depend on m.

Calculation of Fan out: Number of logic gates at the next stage(s) that can be loaded to a given logic gate output is the fan-out. Fan-out $m = \beta I_B/I_C$.

Calculation of Propagation Delay: Let base-emitter capacitance = C nF (nF means nanoFarad). If m = 4, the total capacitance being all T_j in parallel = 4 C. Resistance is very small between base and emitter in logic 1 state. Therefore, transistor turn-on delay is small. Resistance in logic 0 state is 5000 ohm, therefore turn-off propagation delay = (5000) mC ns. (nF × Ohm = ns). Typically, the turn-on delay is 30 ns and turn-off delay is 80 ns.

Increasing Fan-out by a Modified DTL circuit: We have seen that I_B = 0.4 mA in 2-diode base-input based DTL. Figure 6.2(b) shows a modified DTL circuit by replacing 2-diode combination by one diode and one transistor. This improves the fan-out due to increase in I_B.

Wired Logic (Wired AND or Implied AND) Connection: If outputs F and F' at two DTL NAND gates connected, the output can be considered as AND operation between the logic outputs. Because when both the outputs correspond to cutoff stages of the transistors, the output will remain unaffected and will be 1. When any of the outputs correspond to saturation condition ~0.2 V, the output from common point will become 0.2 V. If A, B are the inputs at one DTL NAND gate and C, D are inputs at another, NAND the output Y on joining F and F' at common terminal will be as follows:

$$Y = \overline{(A.B)}.\overline{(C.D)} = \overline{(A.B + C.D)} \text{ (Using DeMorgan theorem).}$$

6.4.3 Transistor–Transistor Logic (TTL)

6.4.3.1 TTL NAND Gate

Since 1964, the digital electronic circuits based on the TTL have been introduced. TTL means that the circuit is based-upon transistor-transistor logic (TTL). The transistors in it are bipolar junction transistors (BJTs). In the TTL digital circuit, a voltage level V_{CC}^{+} of about +5 V with respect to the ground potential (GND) defines 'high' *i.e.* 1, and the level below about 0.65 V with respect to the V_{EE}^{-} GND defines 'low' *i.e.* 0 (Table 1.1). A typical TTL digital circuit is shown in Figure 6.6(a). Figure 6.6(b) shows passive pull up for using the circuit for the wired OR logic TTL output Figure 6.6(c) shows an open collector gate output. Figure 6.6(d) shows the symbol of four inputs NAND gate.

Figure 6.6(a) shows the TTL NAND gate electronic circuit using two transistors T and T'. Figure 6.6(b) gives the symbol of 4-input NAND gate shown in figure 6.6(a). [NAND has state 1 most of the times as its output F is 0 only when with all the inputs (for example, $A1$ and $A2$) are 1.] A NAND gate can be said to be basic building block of the all-digital TTL logic gates and other digital circuits.

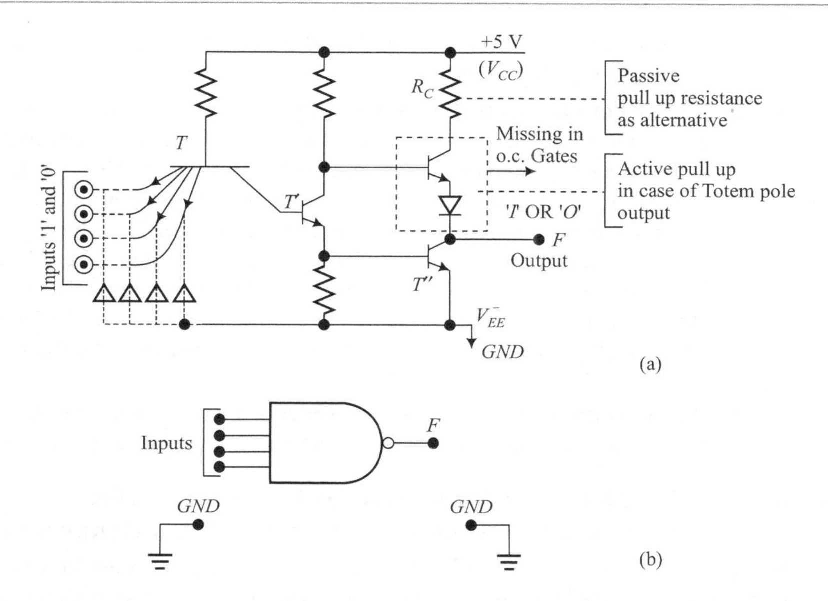

FIGURE 6.6 (a) TTL NAND Circuit with Active pull up (Totem Pole Output) (b) The symbols of four inputs NAND gate.

This NAND circuit differs from the DTL NAND circuit in figure 6.5(a) as follows:

1. There is multi-emitter transistor at the input in place of *p-n* diode at the inputs *A* and *B*. (Figure 6.5(a)). Four inputs multi emitter–base junction is like a collection of four *p-n* diodes. Multi-emitter junction at transistor *T* forms a circuit equivalent to a diode based AND gate circuit at input stages in Figure 6.1(d).
2. The base-collector junction at *T* and base-emitter junction at *T*' are common, which performs as the replacement of 2-diodes before *T* in circuit of Figure 6.5(a). The output at *T*' emitter is therefore input to the base of the output stage transistor *T*'.
3. In case of a TTL with a totempole output (Section 6.4.3.4) there is active pull up of the collector of *T*' and output *F* in place of passive pull up by 2200 Ohm in Figure 6.5(a) DTL NAND circuit. Advantage of active pull up is to reduce the Joule loss (heat loss) within the resistance in passive case. Active pull up increases the loading fan-out (Note: When the active pull up circuit is externally connected the TTL circuit will be called open-collector TTL gate (Section 8.1)).

6.4.3.2 TTL Circuit Working

The TTL circuit working can be explained as follows:

1. **The both inputs 1:** Two or more inputs to two or more *n*-ends (multi-junction emitter) form multi *n-p* junctions with the base. When all inputs are 1, all the junctions are reversed biased. The transistor *T* base-collector is forward biased. The collector + 5 V voltage makes the base-emitter of *T* also forward biased. Since *T*' emitter connects to *T*'' base, *T*'' is also forward biased. Both *T*' and *T*'' works in saturation mode. *VBE* (sat) at *T*'' ≈ 0.7 V and *VBE* (sat) at *T*' is ≈ (0.7 V + 0.7 V). Voltage at *T* base is ≈ (0.7 V + 0.7 V + 0.5 V). The input voltages at the TTL gates if remain > 2.4 V, the multi-emitter junctions will remain reversed biased. *T*'' collector output at *F* is 0 in saturation mode of T''. The output is ~0.2 V. The collector of *T*'' connects to

supply through the active pull up. This is expected from a NAND gate that the output is low when all inputs are 1.

2. **Any or both inputs 0:** Two or more inputs to two or more *n*-ends (multi-junction emitter) at *T* form the multi *n-p* junctions with the base. When any of the input is low, the emitter-base junctions will be at threshold voltage and base voltage V_{BE} is ~(0.2 V + 0.7 V), transistor *T* goes to saturation and the output at collector of *T* is low enough to make base-emitter of *T*' and base emitter of *T*'', which are in series, conduct any significant current. The base currents at *T*' and *T*'' are negligibly small. The transistor *T*'' is in cutoff mode. *T*'' collector output at *F* is 1 in cutoff mode. The output is close to the supply voltage 5 V as the collector of *T*'' connects to supply through an active pull up. This is expected from a NAND gate that the output is high when any input is low.
3. **Output stage for the next input stage:** A common output from the transistor *T*'' at *F* is given to an input *n-p* junction at next TTL stage(s) multi-emitter junction.

6.4.3.3 Wired Logic (Wired AND or Implied AND) in Case of a TTL Gate with Passive Pull Up Connection

If output *F*s at two four-inputs TTL gates can be connected, provided there is passive pull up (~ 4 kΩ) like 2.2 kΩ in Figure 6.5(a) in place of active pull up like in Figure 6.6(a) are connected, the output can be considered as AND operation between the logic outputs. Because when both the outputs correspond to cutoff stages of the transistors, the output will remain unaffected and will be 1. When any of the outputs correspond to saturation condition ~ 0.2 V, the output from common point will become 0.2 V. If *A*, *B*, *C* and *D* are the inputs at one TTL NAND gate and *E*, *F*, *G* and *H* are inputs at another NAND, the output will be as follows:

$$F = (\overline{A.B.C.D}).(\overline{E.F.G.H}) = (\overline{A.B.C.D + E.F.G.H}) \text{ (Using DeMorgan theorem).}$$

Current dissipation is logic 0 state will increase when two TTL gates with passive pull-ups are ANDed by wired logic. The TTL gates with missing pull up circuit at the collector are also called open collector gates. These are more suitable for the wired logic connections.

6.4.3.4 Totem Pole Output in Case of a TTL Gate with Active Pull Up Connection

Figure 6.6 (a) TTL NAND Circuit is a Totem Pole Output circuit when it has active pull up circuit. Figure 6.6 (a) shows active pull up by combination of a circuit with in a dotted square and a resistance R_C (~ 4 kΩ) between the supply and collector of *T*'' (output *F*).

Let *C* is the capacitance between collector of *T*'' (output at *F*) and emitter of *T*'' (supply ground).

1. When *T*'' is in cutoff mode (output high) and changes to output low, the time constant for discharging when *T*'' goes to saturation mode (output ~ 0.2 V) when all logic inputs becomes 1, *the time constant is very small* (= *T*'' output impedance × *C*) as the *T*'' conducting stage impedance is very low (~20 Ω).
2. When *T*'' goes to off mode (output 1) from the saturation (output 0), the time constant for charging when *T*'' goes to 1 when any of the logic input becomes 0, *the time constant is high* (= *R* × *C*) as the *T*'' is non-conducting stage impedance is very high (~ 20 Ω) (charging is through resistance *R* (~ 4 kΩ) when passive pull up exists instead of active pull shown in Figure 6.6(a)).

Output F when there is active pull up circuit between output F and supply (Figure 6.6(a)) is called Totem Pole Output. It gives an advantage of very low time constant in both cases of transitions (cutoff to saturation and saturation to cutoff) 1 to 0 and 0 to 1 A totem-pole circuit cannot used as wired AND logic (Section 6.4.3.3). Totem pole action is as follows:

1. When T'' output is low, the base current is high. The output at collector of T' is at low (~0.2 V + 0.7 V). It is not sufficient to cause large current through T'' because F is at ~0.2 V and diode between emitter of active pull up transistor and F itself needs ~0.7 V. Voltage V_{BE} at active pull transistor T''' is less than the forward bias threshold. Therefore, the time constant is still controlled by (= T'' output impedance × C) which is very low. The discharge on high to low transition takes place through T'' only.
2. When T'' output is high, the base current is very low in cutoff stage. The output at collector of T' is at high. It is not sufficient to cause large current through T'' because F is also at high and the diode between emitter of active pull up transistor and F itself needs ~0.7 V. Voltage V_{BE} at active pull transistor T''' is less than the forward bias threshold. The transition of the diode from above threshold to below threshold state will take time. However the time constant on charging is still small controlled by {(diode impedance in the beginning conduction state + base-emitter saturation state resistance in the beginning) × C,} which is low. These two resistance collectively are ~150 Ω (<< 4 kΩ in case of use of passive pull up). This is because the when there is a transition from low to high at output F, the diode and T''' will take time to change their conducting state to non-conducting states.

6.4.3.5 TTL Circuit Features

The TTL circuits have the following features:

(i) Less area is needed on the silicon wafer during its fabrication, which results in its faster speed of operation.

(ii) In totem pole TTL circuit version, there are effectively low impedances during both the output transitions; 1 to 0 transition and 0 to 1 transition, provides us an ability to connect its output to a capacitor. We have the effectively low time constants when connecting to next stage(s) (a capacitor charges as well as discharges slowly if charging and discharging is through high impedance). Such ability is also called totem-pole output ability.

(iii) In open collector version (Section 8.1), the pull up can be given externally and wired AND logic can be used. The external pull up helps in operation at higher currents or at higher voltages. The external pull up can also be passive or active.

6.4.3.6 Standard TTL Circuit Parameters

TTL circuit parameters in 7400 series are as follows:

Supply V_{CC} = 5.0 ± 0.5 V V_{EE} = 0 V

V_{OL} (Voltage Output at logic 0) = 0.4 V and I_{OL} (Current Sink at logic 0) = 4 mA

V_{OH} (Voltage Output at logic 1) = 2.4 V and I_{OH} (Current Output at logic 1) = 0.04 mA

V_{IL} (Voltage Input at logic 0) = 0.8 V

V_{IH} (Voltage Input at logic 1) = 2 V

V_{TH} (Threshold Input Voltage) = 1.3 V [A voltage where the transistor in the circuit changes the mode of working between saturation and cutoff).] Note: Actually used inputs are as specified above to ensure correct performance.

Noise Margin at 1 = 0.4 V (between 2.8 V and 2.4 V)

Noise Margin at 0 = 0.4 V (between 0.4 V and 0 V)

6.4.3.7 Unconnected Input Case

If any input is not connected, the multi-emitter junction of transistor *T* the base-emitter at that particular junction does not conduct. As the base connects to the supply V_{CC} through 4 K, the corresponding emitter is also at the potential of the base. Hence, the unconnected input in the TTL logic circuit will behave as logic input = 1.

6.4.3.8 TTL Families

TTL circuits are available in different families. Each family has different speeds and power dissipation standards. Six families are Standard, Low power, High speed, Schottky, Low-power Schottky, and advanced low power Schottky. These families have speed corresponding the 10 ns, 33 ns, 6 ns, 3 ns, 10 ns and 5 ns per gate-propagation-delay times, and the power dissipations per gate of 10 mW, 1 mW, 24 mW, 19 mW, 2 mW and 1 mW. The latter five families are abbreviated as *L*, *H*, *S*, *LS*, and *ALS* respectively. A reciprocal of a propagation delay is a measure of the speed. The delay time is defined by time taken in the logic transition at the output after the change at the inputs.

If a Schottky diode connects a base-collector in a transistor, the capacitive effects within the transistor when operating in saturation mode reduces significantly. The Schottky diode has a metal end (acting as *p*-end). It is connected to the base. The *n*-end connects the collector. This diode switches much faster than the base-collector of the *n-p-n* transistor. (Figure 6.6(a)). The Schottky diode starts conducting at 0.3 V threshold. Schottky diode based TTL dissipates more power. Therefore, by increasing *R* from the to the active pull up transistor, we get the low power Schottky version of TTL. We can also have the advanced Schottky (AS) TTL gates to have low power but high speed. We can have advanced low power (ALS) gates to speed but lower power dissipation from AS-TTL.

6.4.4 TTL other than NAND Gate

TTLs other than NAND gates can be made using NAND as a basic building block.

6.5 EMITTER COUPLED LOGIC (ECL)

6.5.1 ECL OR/NOR Gate

A Schottky diode gives high speed of operation by faster switching of 0 to 1 transitions in a TTL circuit. Transistors are prevented from becoming saturated during the transition period, for example, by adding Schottky diode between base and collector. ECL is a fast speed solution, which of course needs greater power dissipation.

ECL circuit gives a *current mode logic*, which is based upon the current switching actions. Figure 6.7(a) shows the ECL OR/NOR gate circuit. Figure 6.7(b) shows ECL circuit logic level voltages. Figure 6.7(c) shows the conditions of T and other transistors in different cases. Figure 6.7(d) shows the symbol of two inputs OR/NOR ECL gate.

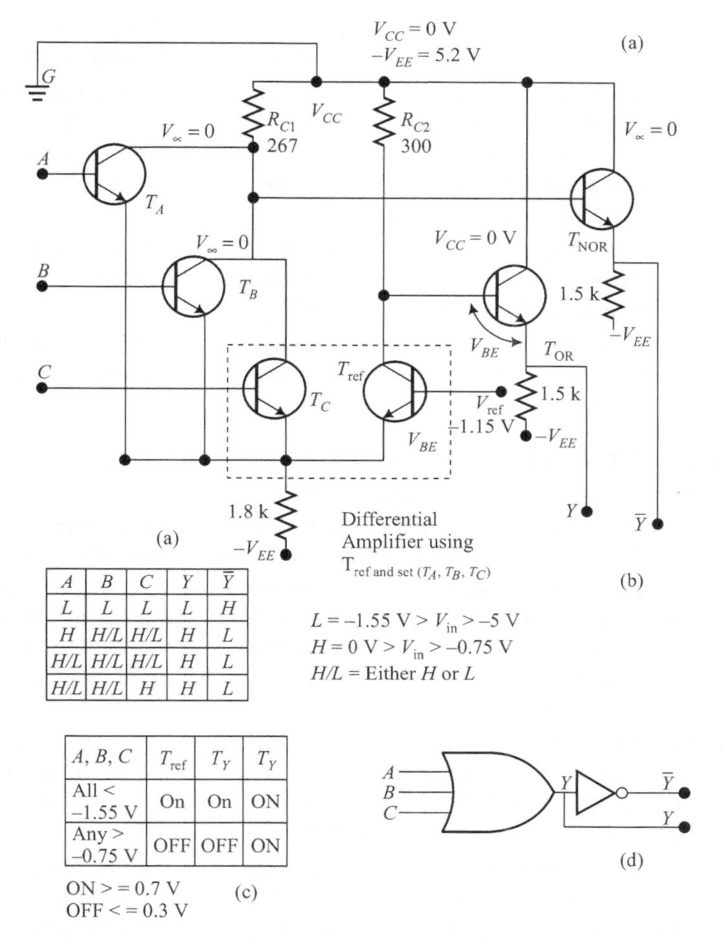

A	B	C	Y	$\overline{Y}$
L	L	L	L	H
H	H/L	H/L	H	L
H/L	H/L	H/L	H	L
H/L	H/L	H	H	L

A, B, C	T_{ref}	T_Y	T_Y
All < −1.55 V	On	On	ON
Any > −0.75 V	OFF	OFF	ON

FIGURE 6.7 (a) ECL OR/NOR gate circuit (b) ECL circuit logic level voltages (c) Conditions of T and other transistors in different cases. (d) Symbol of two-input OR/NOR ECL gate.

6.5.1.1 ECL Circuit Internal Connections

This ECL circuit differs from the RTL/DTL/TTL circuits as follows:

1. Each input of the gate gets the inputs at the base of the individual transistors, T_A, T_B, or T_C.
2. There are the 'OR' output stage transistor (T_{OR}) and 'NOR' output stage transistor (T_{NOR}). The outputs are taken from the emitters of each.
3. There is transistor T_{ref}, which forms a differential amplifier pair with a T in the parallel circuits of T_A, T_B and T_C). The T_{ref} gets the input reference voltage ($V_{REF} = -1.15$ V) from a reference supply circuit. The pairs amplify the difference in the voltages (base currents) between the voltages at the bases of T_A, T_B, T_C and reference T_{ref}.
4. The emitter of T_{ref} and (T_A, T_B, T_C) couples together. (Hence the logic circuit name is emitter-coupled logic, ECL).

5. The emitters of the differential amplifier pair T_{ref} and (T_A, T_B, T_C) connect through a common resistance R_E (≈ 1.18 kΩ) and to the –ve of supply' V_{EE}(≈ –5 V).
6 The collectors of (T_A, T_B, ...) are also common. Common-collectors of the differential amplifier pairs connect through a resistance R_{C1} (~267 Ω) to the GND (which is +ve with respect to the –ve supply). This part of the circuit operates in common collector amplifier (emitter-follower) mode. The collector of T_{ref} connects to GND through a resistance R_{C2} (~300 Ω).
7. Collector of T_{OR} connects to GND in the common collector amplifier mode (also called emitter-follower mode). The emitter gives the output, which also connects to $-V_{EE}$ through a resistance R_{out} (≈ 1.5 kΩ).
8. Collector of T_{NOR} connects to GND in the common collector amplifier mode. The emitter gives the output, which also connects to V_{EE} through a resistance R'_{out} (≈ 1.5 kΩ).

6.5.1.2 ECL Circuit Working

The ECL circuit working can be explained as follows:

1. Consider a differential amplifier pair between T_{ref} and one of the input-stage transistors. If V_{REF} at base of T_{ref} = –1.15 V and input at base of T_i (V_{in}) is –1.6 V (low) T_{ref} emitter has the voltage 0.7 volt less than the V_{REF} = ≈–1.85 Since potential difference between the emitter and base of T_i is only 0.25 V, T_i is in cutoff region, while T is in normal inverting amplifier mode. Hence, the collector of T is at –0.9 V and gives sufficient input base current to T_{OR}. The output from T_{OR} emitter is low (–1.55 V – 5 V).
2. Consider a differential amplifier pair. If V_{in} at base of T_i = – 0.7 V (higher than previous case). The transistor T_i is in normal inverting amplifier mode. The T_i will now be in the cutoff region. Since potential difference between the collector and ground will be 0 V. The output from T_{OR} emitter is high (0 V > 0.75 V).
3. Consider the differential amplifier pairs again. If V_{in} at base of any one or all are at – 0.7 V (high). The transistors with high inputs are in normal mode the currents simply add up. The T_i will again be in the cutoff region. Since potential difference between the collector and ground will be 0 V. The output Y from T_{OR} emitter is high.

The ECL circuit works as OR gate at the output of T_{OR}. Since Step 1 and 2 show that the working of T_{ref} and T_A and T_B are opposite. If T_{ref} is in cutoff, then T_i, is in normal mode and when all inputs are low. If T_i is in cutoff, then T is in normal mode and when any or several or many inputs are high. Therefore, if collector of T_i is used to drive the T_{NOR}, we shall be obtaining the complementary output $\overline{Y}$. ECL circuit of Figure 6.7(a) therefore, gives OR as well as NOR outputs.

6.5.1.3 ECL Circuit Logic Voltage Level Parameters

ECL circuit parameters in 10 K series are as follows:

Supply V_{EE} = – 5.2 V_{CC} = 0 V
V_{OL} (Voltage Output at logic 0) = – 1.7 V
V_{OH} (Voltage Output at logic 1) = –0.9 V
V_{IL} (Voltage Input at logic 0) = – 1.4 V
V_{IH} (Voltage Input at logic 1) = – 1.2 V

V_{TH} (Threshold Voltage) = –1.29 V (a voltage where the transistor in the circuit changes the mode of working between saturation and cutoff). Note: Actually used inputs are as specified above to ensure correct performance.

Noise Margin at 1 and 0 = 0.3 V (difference of – 1.7 V and – 1.4 V)

6.5.1.4 ECL Circuit Wired OR Logic

ECL circuit permits wired OR logic if (i) output of transistor T_{NOR} of one ECL circuit in Figure 6.7(a) connects to another ECL circuit output of transistor T_{OR} and (ii) output of transistor T_{OR} of one ECL circuit in Figure 6.7(a) connects to another ECL circuit output of transistor T_{NOR} gates give output ($Y1 + Y2$).

6.5.1.5 ECL Input Unconnected Input Case

If any input is not connected, the transistor T_i base-emitter will be at cutoff. Therefore, it will be taken as low logic level.

6.6 INTEGRATED INJECTION LOGIC (I^2L)

A resistance needs larger silicon area on a chip than a transistor or diode. Further, there is a Joule heat loss I^2R in a resistance R carrying current I. If base resistance of 450 Ohm in RTL logic circuit of Figure 6.4 (a) is removed, we get a circuit called directly-coupled-transistor-logic (DCTL).

6.6.1 I^2L Circuit Internal Connections

The I^2L circuit (Figure 6.8(a)) differs from the DCTL and RTL circuits as follows:

1. A transistor T or T' base is given the input through a current source within dotted square (Figure 6.8(a)) in place of through the resistor or diode in RTL and DCTL logic circuits.
2. The T and T' emitters are at the GND potential and the collectors connect to a supply potential V_{BB} through an external resistance R_B to each T.
3. The T collector also connects to another transistor T' base.
4. T' base is given another input through another current source.

The transistor T and T' for an I^2L inverter pair.

Figure 6.8(a) shows the basic circuit called Integrated Injection Logic (I^2L).

6.6.2 I^2L Circuit Working

The I^2L circuit working is as follows:

1. When the transistor T base is at logic 0, the injected current i' is 0 and the T is in cutoff mode. Therefore, input current source sinks at the input terminal of next stage T'. Because V_{BE} and thus collector C is at 1 and the current source's current at another input current terminal T' flows (injects) as i'' of second transistor T'. V_{BE} at T' ≈ 0.8 V. V_{CE} at T' ≈ 0.2 V. Therefore, logic states at T' collector = 0 and at $T = 1$.
2. When the V_{BE} transistor T base is at logic 1 ~0.8 V, the i' injects through T. V_{BE} at $T = 0.8$ V and the T is in saturation mode. V_{CE} at $T = 0.2$ V. Therefore, input current source at T collector sinks the current at T and i'' at T' is 0. V_{BE} at T'= 0 V. The collector of T' is at 1 and the T' is in cutoff state.

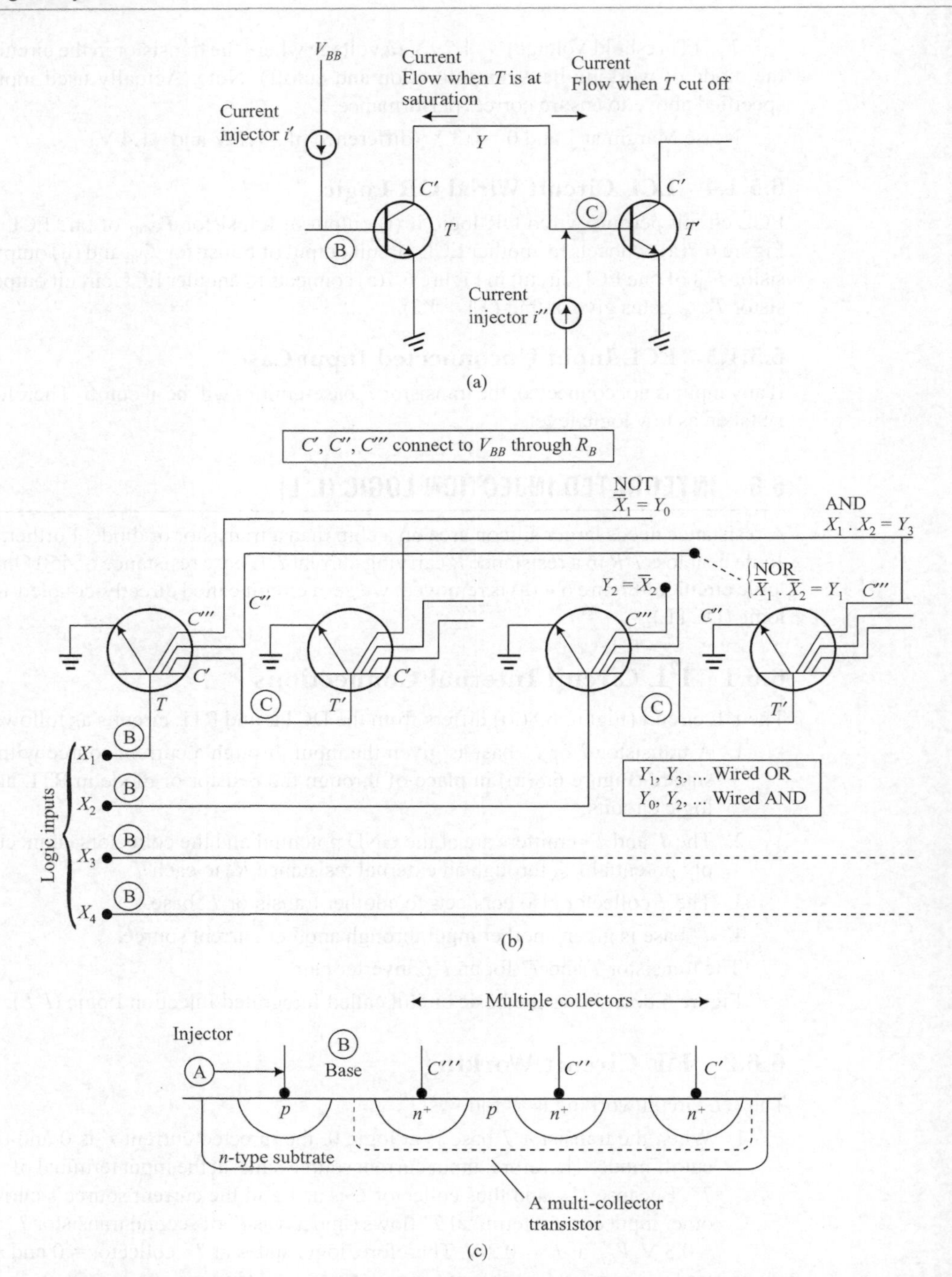

FIGURE 6.8 (a) Basic circuit of Integrated Injection Logic (I^2L) (b) An implementation of AND and NOR using the I^2L circuit by using the multiple collector-junctions in a same transistor (c) The physical placements of the junctions on the silicon for a multicollector transistor.

3. Both transistors T and T' can be made on a silicon as one pair of the multi collector junction transistors. The current sources at the input terminals are made using a *p-n-p* circuit for each. The emitter of a grounded base *p-n-p* transistor act as a current injector or sink (if input terminal is at logic 0). The *p-n-p* collector gets the current through an external resistance connected to V_{BB} + supply. Figure 6.8(b) also shows the input terminals and current source circuits and three collector junctions transistor in place of T and T' and fabrication of the I^2L circuit. Figure 6.8(c) shows the physical placements of the junctions on the silicon.

6.6.3 I^2L Circuit Switching Speed, Delay Times and Power Dissipation

The current source charging current i flows when a base-emitter junction of one of the multi collector junction transistor flow the saturation current. Let base-emitter capacitance is C and saturation resistance is $R_{(sat)}$. (It is very low). The charging time constant is $R_{(sat)}$. Also power dissipation is very small as $i^2R_{(sat)}$ is very small and $R_{(sat)}$ also depends on charging current source. The delay time is inversely proportional to the power dissipation. Smaller delay time means larger power dissipation per gate. Therefore, I^2L gate has high speed high power dissipation.

6.7 HIGH THRESHOLD LOGIC (HTL)

HTL is a high threshold logic based on the modified DTL circuit shown in Figure 6.5(a). The following are the changes done to make an HTL circuit. Figure 6.9 shows a three input HTL NAND circuit obtained after the following modification in the modified DTL.

6.7.1 HTL Connections for the Output at *F*

1. **Operational voltages:** The V_{CC} supply is of 15 V in place of 5 V.
2. **Input stage:** It is based on the application of two or more inputs to two or more *n*-ends of the *p-n* junction diodes in place of passive 450 Ohm resistance in RTL circuit. The *p*-ends are common and connect to 12 kΩ and 3 kΩ resistances R_D and R'_D, which connects to supply voltage of 15 V. The common point of R_D and R'_D, also connects to the collector of a transistor (T). The common point of *p*-ends also connects to the T's base. The emitter of the T connects to one Zener diode of 6.9 V breakdown voltage. The *p*-end of the Zener connects to the transistor (T') base, which also connects to the emitter through 5 kΩ R_B. The emitter of T' is at GND of the supply. The collector of T connects to supply through a resistance R_C of 15 kΩ.
3. **Output stage for the next input stage:** A common output from the transistor T' at F is given to other diodes at the other next stage(s) logic gates (HTL gates), which will get the input from the transistor, T.

Three input stage diodes and transistor form the load of the next input stage circuit(s). The HTL based NAND and its connection to next stages is shown in the Figure 6.9.

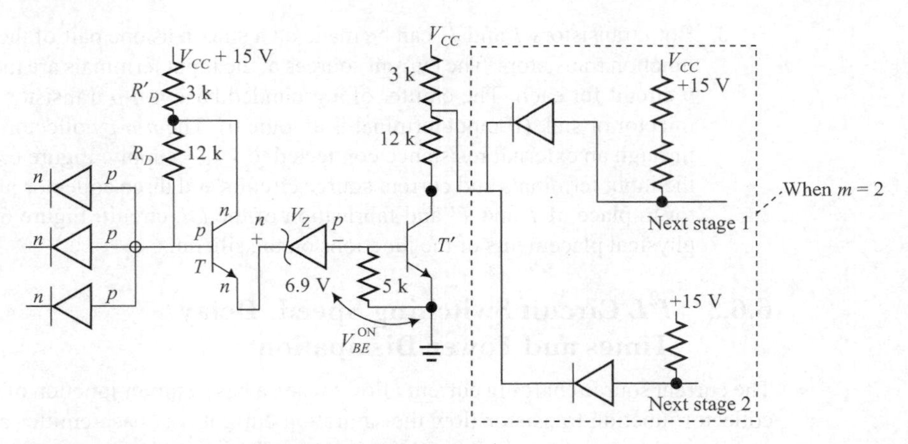

FIGURE 6.9 The HTL based NAND and its connection to next stages.

6.7.2 Logic Operation for the Output at *F*

If input *A* or *B* or *C* is low ~0.2 V (near supply ground), the diode in the path of *A* or *B*, respectively will start conducting. Current through the diode will be approximately equal to (15 V − 0.7 V − 0.2 V)/(15000) ohm = ~0.9 mA. (voltage drop across a conducting *p-n* junction diode is threshold voltage = ~0.7 V.] Voltage drop of 0.9 V connects to the base of *T*' through two diodes, one formed by base-emitter of *T* and other formed by Zener needing threshold voltage of 6.9 V and total 7.6 V to turn ON and conduct. Hence the transistor *T*' base is at ~0 V, below cutoff voltage. When *T*' is not conducting, the output $F = 1$, high (≈ 15 V). *T*' does not conduct when any (or both of the inputs) is low because voltage at the base drops below the cut-off voltage needed at the base.

If all the inputs *A*, *B* and *C* are high (> 0.7 V + 7.6 V), the voltage at common *p*-ends will start exceeding 7.6 V and the Zener diode circuit to the base will start conducting. When input *A*, *B* and *C* exceeds 8.3 V and the voltage at the common *p*-ends exceeds (8.3 V + $V_{BE\,(ON)}$) = 9 V, the base-emitter junction of *T*' starts conducting. When the *A* and *B* inputs exceeds 7.6 V, the diode stops conduction and when exceeds 8.3 V, becomes reverse biased. $V_{BE\,(ON)}$ remains at 0.7 V. If transistor *T*' base-emitter current exceeds a limit, the *T*' goes in saturation mode and it will start conducting current I_C through *R* and V_{CE} = ~0.2 V. Therefore, $F = 0$ when *A*, *B* and $C = 1$.

Property of a NAND is that its output is 0 when all the inputs are 1. Therefore, the HTL circuit in Figure 6.9 works as a NAND gate.

Calculation of Propagation Delay: Let base-emitter capacitance = *C* nF [nF means nanoFarad.] If $m = 4$ stages, which connects to *F*, the total capacitance being all *T* next stages in parallel = 4*C*. Resistance is very small between base and emitter in logic 1 state. Therefore, transistor turn-on delay is small. Resistance in logic 1 state is 15000 Ω, therefore turn-off propagation delay = (15000) *mC* ns. [nF × Ω = ns.] Typically, the turn-on delay is 90 ns and turn-off delay is 240 ns. Circuit temperature sensitivity is small compared to DTL as the Zener has very small temperature coefficient. The high threshold voltages give a higher noise margin a characteristic of HTL gates.

HTL circuit is appropriate suitable for industrial environment.

6.8 NMOS

There are two types of MOSFETs, n-channel and p-channel. Circuits based on n-channel MOSFETS are called NMOS circuits.

6.8.1 NMOS Circuit Connections and Working

Figure 6.10 shows a NMOS inverter. Enhancement mode or deletion mode n-channel MOSFET T' acts as an active pull up load (in place of R, which occupies larger silicon area). An enhancement mode MOSFET T acts as a inverter and gives an output V_o at F.

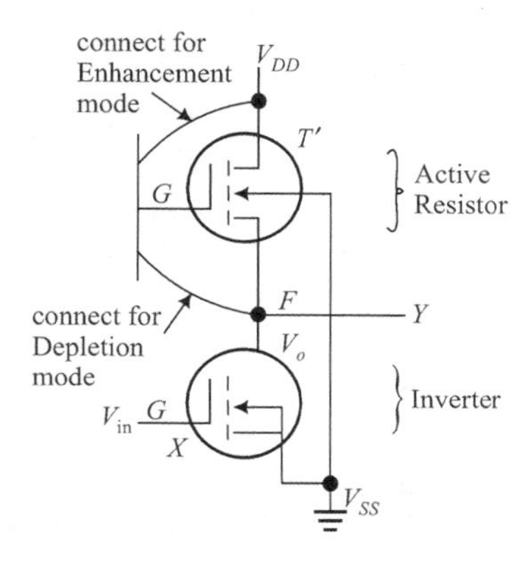

FIGURE 6.10 An NMOS inverter.

Figure 6.11(a) shows a two input NMOS NOR. A depletion mode n-channel MOSFET T'' acts as an active pull up load (in place of R, which occupies larger silicon area). Two enhancement-mode n-channel MOSFETs T_X and $T_{X'}$ are the logic drivers in parallel. These give an output V_o at F through a common point connected to drains of T_X and $T_{X'}$. If both T_X and $T_{X'}$ are off, the output equals, supply voltage V_{DD}. If any one is ON, the output at F equals the V_{SS} (the supply GND).

Figure 6.11(b) shows a two input NMOS NAND A depletion mode n-channel MOSFET T'' acts as an active pull up load (in place of R, which occupies larger silicon area). Two enhancement mode n-channel MOSFETs T_X and $T_{X'}$ are in series. These give an output V_o at F through the upper NMOS drain of $T_{X'}$, which also connects the NMOS MOSFET pull up. If any T_X and $T_{X'}$ are off, the output equals the supply voltage V_{DD}. If both are ON, the output at F equals the V_{SS} (the supply GND).

6.8.2 Calculation of Fan Out

Numbers of logic gates at the next stage(s) that can be loaded are very high due to high input impedance between the gate and channel.

6.8.3 Calculation of Propagation Delay

Let gate-source capacitance = C nF [nF means nanoFarad.] If $m = 40$, the total capacitance being all T_j in parallel = $40C$. Resistance is very high between gate and source in both logic

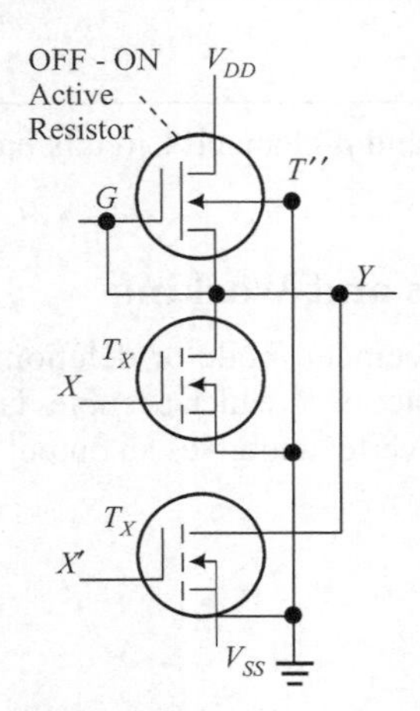

Both X and X' if '0' then only $Y = 1$ as T and T' are in Parrallel in current path from V_{DD}

(a)

V_{DD} G T'' Y T_X X T_X X' V_{SS}

Both X and X' if '1' then only $Y = 0$ as T and T' are in Series in current path from V_{DD}

(b)

FIGURE 6.11 (a) Two-input NMOS based NOR gate. (b) A two-input NAND NMOS gate.

0 and logic 1 states. Therefore, MOSFET turn-on delay is large. Now the technology has been developed to get the very small C to get high speeds compatible with the TTLs.

6.8.4 Calculation of Power Dissipation

NAND gate is ON in one of the 4 conditions of inputs defined in NAND truth table. Therefore like TTL, the NMOS NAND dissipates less average power compared to NMOS NOR.

6.8.5 NMOS Circuit Voltage Levels

NMOS circuit parameters in the 8086 Microprocessor circuit are as follows:

Supply $V_{DD} = 5.0 \pm 0.\ 5\text{V}$ $V_{SS} = 0$ V
V_{OL} (Voltage Output at logic 0) = 0.45 V and $I_{OL} = 2$ mA
V_{OH} (Voltage Output at logic 1) = 2.4 V and $I_{OH} = -400$ µA
V_{IL} (Voltage Input at logic 0) = 0.8 V
V_{IH} (Voltage Input at logic 1) = 2 V
Input Leakage current = $\pm$ 10 µA
Output Leakage current = $\pm$ 10 µA
Noise Margin at 1 = 0.4 V
Noise Margin at 0 = 0.4 V
(Note: Refer Section 6.4.3.6. These parameters are TTL compatible).

6.8.6 Unconnected Input(s) not Permitted

Due to very high gate-source impedance, a small static charge built up can drive a gate voltage higher than the drain. Therefore, the unconnected inputs are not permitted in MOSFET based gates.

6.9 CMOS

6.9.1 Importance and Features of CMOS Logic Circuits

CMOS is most important logic circuit due to the following.

(i) An enhancement mode *p*-channel acts as pull up and *n*-channel enhancement mode MOSFET act as a logic driver (pull down) When *n*-channel is ON, the *p*-channel is OFF and when *n*-channel is OFF, the *p*-channel is ON. It means the D.C. (steady state) current dissipation between the supply ends is very small as the series resistance is always very high. Current will flow only during the transitions from 0 to 1 or 1 to 0. Power will dissipate only during the transitions from 0 to 1 or 1 to 0. This feature makes it possible to fabricate large or very large or very-very large-scale integrated circuits (LSI or VLSI or VVLSI) using the CMOS pairs.

(ii) Since at an instant one of the MOSFET in the pair is ON and has low resistance, the switching speed at charging and discharging is rapid and turn ON delay and turn OFF delay are nearly same.

6.9.2 Operations as Inverter (NOT), NOR and NAND

Figure 6.12(a) shows a CMOS inverter. CMOS (Complementary MOSFET) logic circuit uses an enhancement mode *p*-channel MOSFET *T*' acts as an active pull up load (in place of *R*, which occupies larger silicon area) and an enhancement mode n-channel MOSFET *T* acts as a driver of the output V_o at *F*.

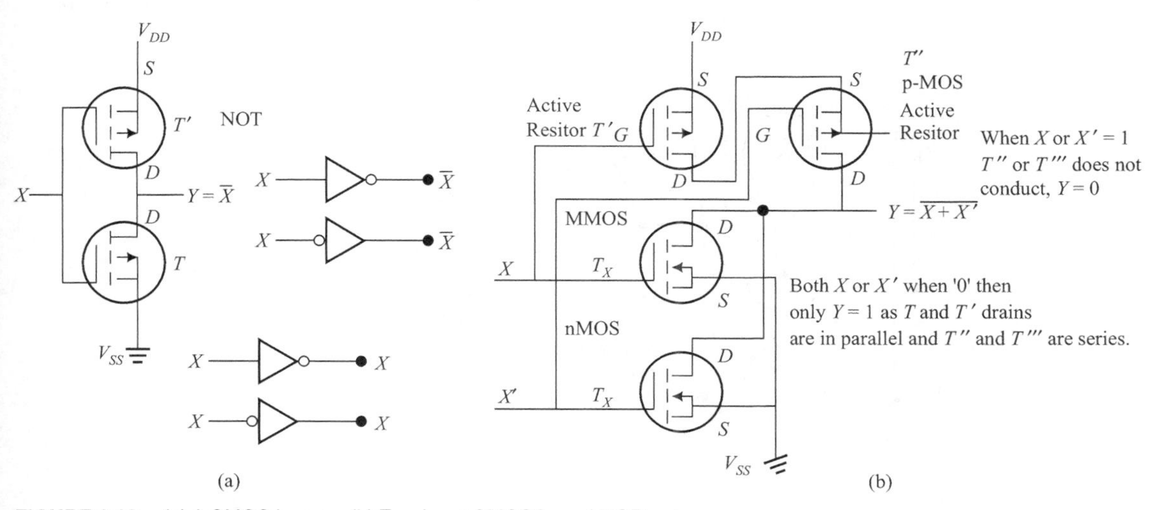

FIGURE 6.12 (a) A CMOS inverter (b) Two input CMOS based NOR gate.

Figure 6.12(b) shows a two input CMOS NOR. Two enhancement mode *p*-channel MOSFETs *T*' and *T*'' give the active pull up loads (in place of *R*, which occupies larger silicon area) to the drains of two *n*-channel logic driver circuits. These two enhancement-mode *n*-channel MOSFETs T_X and $T_{X'}$ are the logic drivers in parallel. These give an output V_o at *F* through a common point connected to drains of T_F and $T_{F'}$. If both T_F and $T_{F'}$ are off, the output = supply voltage V_{DD}. If any one is ON, the output at *F* equals the V_{SS} (the supply GND).

Figure 6.13 shows a two input CMOS NAND. Two enhancement *p*-channel MOSFET *T*' and *T*'' give an active pull up load (in place of *R*, which occupies larger silicon area). Two enhancement mode *n*-channel MOSFETs T_X and $T_{X'}$ are in series. These give an output V_o at *F* through the upper *n*-MOSFET drain of $T_{X'}$, which also connects the *p*-MOSFET pull up. If any T_X and $T_{X'}$ are off, the output equals supply voltage V_{DD}. If both are ON, the output at *F* equals the V_{SS} (the supply GND).

6.9.3 Calculation of Fan out

Numbers of logic gates at the next stage(s) that can be loaded are very high due to high input impedance between the gates and the channels at the logic drivers. There are only capacitive driving loads. Steady state D.C. power dissipation is extremely small. Impedance between gate and source is capacitive. Current flows at the transitions only.

6.9.4 Calculation of Propagation Delay

Let gate-source capacitance = *C* nF (nF means nanoFarad). If $m = 40$, the total capacitance being all T_j in parallel = 40*C*. Resistance is very high between gate and source in both logic 0 and logic 1 states but the charging and discharging occurs from the input logic gate rapidly in both 0 to 1 and 1 to 0 transitions (since at an instant one of the MOSFET in the pair is ON and has low resistance, the switching speed at charging and discharging is rapid and turn ON delay and turn OFF delay are nearly same). Therefore, CMOS MOSFET logic circuit turn-on and turn off delay is little compared to NMOS turn-on delay. Now the technology high speed CMOS (HCMOS) has been developed to get very small *C* to get the large speeds.

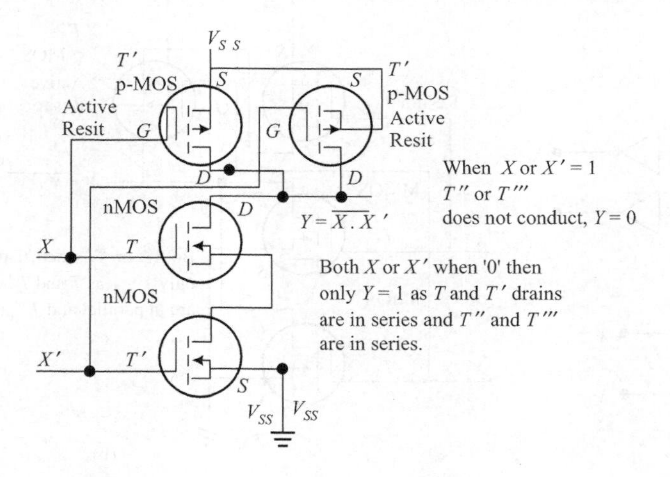

FIGURE 6.13 Two-input NAND CMOS gate.

6.9.5 Calculation of Power Dissipation

When n-channel is ON, the p-channel is OFF and when n-channel is OFF, the p-channel is ON. It means the D.C. (steady state) current dissipation between the supply ends is very small as the series resistance is always very high. Current will flow only during the transitions from 0 to 1 or 1 to 0. Power will dissipate only during the transitions from 0 to 1 or 1 to 0. Therefore, unlike like TTL and NMOS, the CMOS logic gates dissipates very little power compared to NMOSs and TTLs and dissipate power only at the transitions of logic states, not in d.c. state.

6.9.6 CMOS Circuit Voltage Levels

CMOS circuit parameters are as follows:

Supply V_{DD} = 5 V operating version or 3 to 16 V operating version and V_{SS} = 0 V

V_{OL} (Voltage output at logic 0) = (1/3) V_{DD} and I_{OL} = 0

V_{OH} (Voltage output at logic 1) = V_{DD} and I_{OH} = 1 mA (4000B series)

V_{IL} (Voltage input at logic 0) = V_{SS} = 0 V

V_{IH} (Voltage input at logic 1) = (2/3) V_{DD}

V_{TH} (Threshold Voltage) = (1/2) V_{DD} (A voltage where the MOSFET in the circuit change the mode of working between ON and OFF). Note: Actually used inputs are as specified above to ensure correct performance.

Noise Margin at '0' and '1' = ~[(1/3) V_{DD} – (1/2) V_{DD}].

6.9.7 MOS Logic Circuits (CMOSs) and Their Relative Advantages with Respect to TTLs

Table 6.1 gives a comparison of a CMOS with that of a TTL digital circuit.

For a complex integrated circuit (IC) which incorporates millions or more CMOS transistors, the MOSFETs being field effect devices dissipates a total power which is very small compared to the ICs based on the BJTs. This is very significant property of CMOS digital circuits. In place of BJTs based TTLs in an integrated circuit on a silicon chip, the CMOS digital circuits are used in the complex ICs. For this reason including the reasons mentioned in various rows of Table 6.1 the CMOS IC logic gates and digital circuits gained popularity since 1975 at the expense of the TTL circuits. TTL circuits offer the only benefits of (i) higher speed of operation (Tables 6.2 and 6.1) and (ii) the permission to float the inputs (Table 6.1). Very recent advances in the MOSFET fabrication technology (HCMOS) have made it feasible to get the high-speed operations from the CMOS gates also.

6.10 MEANINGS OF SPEED, PROPAGATION DELAY, OPERATING FREQUENCY, POWER DISSIPATED PER GATE, SUPPLY VOLTAGE LEVELS, OPERATIONAL VOLTAGE LEVELS THAT DEFINE LOGIC STATES 1 AND 0

1. **Speed:** It means how many times per second, a logic gate is able to respond to the change in the inputs and give the output as per specified logic.

TABLE 6.1

CMOS	TTL
At an input or output, 1 means a state of a potential difference (p.d.) between ($V_{DD} - V_{SS}$) and 0.66 ($V_{DD} - V_{SS}$).	1 means a p.d. between 5 V and 2.4 V at an output or between 5V and 2.0V at an input.
0 means a state of a p.d. between 0.33 ($V_{DD} - V_{SS}$) and V_{SS}.	0 means a p.d. between 0.8 V and GND at an input or between 0.4V and GND at an output.
($V_{DD} - V_{SS}$) = 3 V to 16 V. Typical values; V_{DD} = 5 V, V_{SS} GND Potential (0 V).	V_{CC} (Figure 6.6) is 5 V ± 0.25 V.
Battery operations are feasible.	Power supply impedance is to be less than one tenth Ohm up to the operating frequencies. Battery operations therefore difficult.
Negligible dissipation at steady state of its inputs. Only about 1 mW per gate per MHz in 4000 series. Power consumption depends upon number of times an input changes i.e. upon the operating frequency.	Power consumption per gate as per Table 6.2.
An enhancement mode p-channel acts as pull up and *n*-channel enhancement mode MOSFET act as a logic driver. When n-channel is ON, the *p*-channel is OFF and when n-channel is OFF, the *p*-channel is ON. It means the d.c. (Steady state) current dissipation between the supply ends is very small as the series resistance is always very high. Current will flow only during the transitions from 0 to 1 or 1 to 0. Power will dissipate only during the transitions from 0 to 1 or 1 to 0. This feature makes it possible to fabricate large or very large or very-very large scale integrated circuits (LSI or VLSI or VVLSI) using the CMOS pairs.	An *n-p-n* transistor and a *p-n* diode form the pull up circuit in totem pole output TTLs. The *n-p-n* BJT act as a logic driver. At each saturation mode operation of the BJT, the current will flow. Significant power will dissipate during the transitions from 0 to 1 or 1 to 0 as well during steady state with output logic state = '0'. This feature inhibits to use the TTLs to fabricate large integrated circuits (LSI) and use is restricted to the interfacing of the logic signals between the processor and memories, and peripherals.
Since at an instant one of the MOSFET in the pair is ON and has low resistance, the switching speed at charging and discharging is rapid and turn ON delay and turn OFF delay are nearly same.	Since in cutoff region the resistance in the charging path is large, the turn ON delay and turns OFF delay are widely different in passive pull up TTL and almost same in totem pole TTL with active pull up circuit at the output.
Gained popularity after 1975 and now almost universally used.	Very popular up to 1985. Loosing popularity exponentially after 1975.
Inputs are not permitted to float (disconnected), and must be connected to state 1 or 0, or to V_{DD} or to V_{SS}.	Inputs are permitted to float (disconnected) as the transistor *T'* (figure 1a) always has a complete circuit (internally) for the currents. Unconnected input logic state is 1.
Output can be as per p.d. ($V_{DD} - V_{SS}$) (Figure 6.12(a)) i.e. fixed at state 0 or 1 as per ($V_{DD} - V_{SS}$). Any Voltage either up to one third of ($V_{DD} - V_{SS}$) or above two third of it can be there.	Output can be below V_{CC} of (5V ± 0.25V) at state 1 and can be down to 2.4 V. For state '0',the output does not exceed 0.8 V.

Table 6.1 Contd.

Inputs and outputs are permitted to be kept at state 1 or 0 as per user wish as steady state current is negligibly small. Direct connections to V_{DD} or V_{SS} at input permitted.	It is better to keep a state 1 because in that case the margin of Voltage fluctuation is (5 V – 2.8 V) i.e. 2.2 V at an output, and is 2.6 V at an input. At a state 1, the steady state current drawn by a transistor at the output is nominal around 50 mA, while at '0' it is much higher. Direct connection is not done to V_{CC} but only through a resistance above 1.1 kΩ.
IC logic gate chips start with 40,41 or 45 with no following nonnumeric character or characters. IC logic gates also start with prefix 74HC in a high speed CMOS version that is followed by numeric numbers analogous to that of 74 series of TTL logic circuit. HCT is used when TTL compatibility is desired from the CMOS logic gates.	IC chips based upon TTL logic start with prefix 74 can be followed by an abbreviation as per Table 6.2.
Maximum operating frequency may be often small, for example for a 40...B it is 2.5 MHz. (Very high in HCMOS)	It depends upon TTL family as per Table 1.2. It can be as high as 100 MHz.
Propagation (speed): For 40...B it is 200 ns, for 74C it is 90ns and for 74HC it is 12ns.	It depends upon TTL type as per Table 1.2
Larger Voltage changes permitted, noise immunity very high compared to TTL.	Small Voltage changes permitted i.e. noise immunity is less than CMOS.

2. **Propagation Delay:** Propagation delay for a logic output from a logic gate means the time interval between change in a defined reference point input voltage and reflection of its effect at the output. It can also be defined as the time interval between changes in a defined logic level input and reflection of its effect at the output logic level. [Note: The slightly or significantly different values of the delays may be obtained when the input change from 0 to 1 and change from 1 to 0. We can take average propagation delay. Also a propagation delay is also subject to variations in power supply and temperature. We can then define a statistical deviation and an average.]
3. **Operating Frequency:** It means how many times logic levels changes per second are permitted without affecting the logic gate characteristics outside the limits, which have been set for the propagation delays, voltage levels and power dissipation per gate.
4. **Power Dissipation Per Gate:** It means how much average power dissipates per gate when a logic circuit is operated within the specified operating frequency.
5. **Voltage Levels that Define Logic States 1 and 0:** A logic level 1 at output is defined by voltage levels V_{OH} maximum and V_{OH} minimum. A logic level 1 at input is defined by voltage levels V_{IH} maximum and V_{IH} minimum. A logic level 0 at output is defined by voltage levels V_{OL} maximum and V_{OL} minimum. A logic level 0 at input is defined by voltage levels V_{IL} maximum and V_{IL} minimum.
6. **Threshold Logic Input Voltage:** A voltage where the transistor in the circuit changes the mode of working between saturation and cutoff. *Actually used inputs are as specified above to ensure correct performance.*

6.11 SPEED, PROPAGATION DELAY, OPERATING FREQUENCY, POWER DISSIPATED PER GATE, SUPPLY VOLTAGE LEVELS, OPERATIONAL VOLTAGE LEVELS THAT DEFINE LOGIC STATES '1' AND '0' FOR VARIOUS FAMILIES OF GATES

Table 6.2 gives the abbreviation for a family, measure of speed per gate and power dissipation of per gate for these families of the logic gates.

TABLE 6.2

S. No.	'Family name'	Abbreviation for the family (Other than standard)	Propagation Delay ns	Power dissipated per gate mW	Maximum operating frequency MHz
1	Standard TTL	-	10	10	30
2	Low Power TTL	L	33	1	3
3	High Speed TTL	H	6	24	50
4	Schottky TTL	S	3	19	100
5	Low Power Schottky TTL	LS	10	2	30
6	Advanced Low Power Schottky TTL	ALS	5	1	35

Table 6.3 gives the propagation delay, power dissipated per gate in mW, speed-power dissipation product in pW.s (pJ), fan-out and maximum operating frequency in MHz.

TABLE 6.3

S. No.	Type of gate	Propagation delay ns	Power dissipated per gate mW	Speed Power product pW.s = pJ (Pico Joule)	Fan out	Maximum operating frequency MHz	Noise immunity
1	RTL	12	12	144	5	8	Average
2	DTL	30	10	300	8	72	Good
3	I^2L	25-250	0.006 to 50 mW	≈1pJ or less Current Source Dependent	Injection	Very High	Small
4	HTL	90	50	5000	10	100	Excellent
5	ECL (10K Series)	2	50	100	25	≈75	Small
6	NMOS in 8085 type VLSI	300	0.2 to ~10	60	20	2	Good
7	Standard TTL	10	10	100	10	35	Very good
6	CMOS in 74HC series	18	2.5 mW				
	static*,	20	60	Very Good			
	600 mW/MHz 0.045 static, ~10 at 1 MHz						

* In steady state, only leakage current glows. It is ~10^{-5}mA

EXAMPLES

Example 6.1

In an RTL logic circuit, $V_{BE(ON)}$ is 0.7 V and input Voltage is 2.4 V. What is the base current when the resistance in base-emitter circuit is 1 kΩ?

Solution

We use the equations $I_B = (V_i - V_{BE(ON)})/R_B = (2.4\text{ V} - 0.7\text{ V})/1000\ \Omega = 1.7\text{ mA}$.

Example 6.2

In an RTL logic circuit, $V_{BE(ON)}$ is 0.7 V and β forward current gain is 20. What will be collector current in saturation state if $I_B = 1$ mA?

Solution

We use the equation $I_C = \beta I_B = \beta(V_i - V_{BE(ON)})/R_B = 20 \times 1\text{ mA} = 20\text{ mA}$

Example 6.3

In an RTL logic circuit, find voltage output at the logic gate if $V_{BE(ON)}$ is 0.7 V and β forward current gain is 10 and collector circuit resistance is 0.5 kW. Assume V_{CC} is 5 V and base resistance is 5 kΩ at logic Input = 3.7 V.

Solution

We use the equation

$$\begin{aligned} V_o &= V_{CC} - I_C.R = V_{CC} - R.\beta(V_i - V_{BE(ON)})/R_B) \\ &= 5\text{ V} - (500\text{ Ohm} \times 10 \times (3.7\text{ V} - 0.7\text{ V})/5000\text{ Ohm}) \\ &= 5\text{ V} - 3\text{ V} = 2\text{ V}. \end{aligned}$$

Example 6.4

A triplet of diode has *p*-ends common at *Y′* and the *n*-ends are for the inputs *A*, *B* and *C* (Figure 6.5(a)). The triplet *p*-end connects to 5.0 V supply through 1 kΩ. It also connects to voltage output *Y* through a *p-n* diode in series with *n*-end of it towards *Y*. Find the output at *Y* under different conditions given in the truth table (Table 6.4). *Y* connects to other loading circuits, the effect of which can be assumed to be negligible. Assume 1 => 4.4 V input and 0 = 0.2 V. Assume all diodes as silicon diodes with 0.7 V threshold voltage.

Solution

TABLE 6.4 Table for logic inputs and voltages after solving the problem

A	*B*	*C*	Output *Y′* at common of all 3 diodes Supply current through 1 kΩ	Output at *Y*
0	0	0	(+ .2 V + .7 V)	.2 V
0	0	1	(+ .2 V + .7 V)	.2 V
0	1	0	(+ .2 V + .7 V)	.2 V
0	1	1	(+ .2 V + .7 V)	.2 V
1	0	0	(+ .2 V + .7 V)	.2 V
1	0	1	(+ .2 V + .7 V)	.2 V
1	1	0	(+ .2 V + .7 V)	.2 V
1	1	1	> 4.4 V	4.3 V to 3.7 V

We draw the circuit for the problem and analyze the circuit as under:

When any of the input = 0 [0.4 V] the diode connected to it is above threshold and voltage at the p-end clamps to (0.7 V + 0.2 V) = 0.9 V as 0.7 V is the drop across this diode and 0.2 V is the input. The output Y will be 0.7 V less than the voltage at the common p-ends. Therefore Y' = 0.2 V.

When all of the inputs = 1 [> 4.4 V] the diode connected to it is cut-off and voltage at the p-end clamps to 5.0 V as 0.6 V is less than the threshold voltage 0.7 V for any of the triplet diodes to conduct. The output Y' will be 0.7 V less than the voltage at the common p-ends. Therefore, voltage at Y' = > 4.4 V and at Y is between 1.3 V to 3.7 V.

Current through the resistance is 0 in case of all inputs = 1 as there is no conduction path available. Current is (5 V – 0.9 V)/1 kΩ = 4.1 mA when conduction path available when the inputs = 0.

Using above solution, now we can fill the columns 4 and 5 Table 6.4 as Table 6.5 as the answer (Table 6.4 also shows Y' and Y potentials).

TABLE 6.5

A, *B* and *C* input states	Output at common of all 3 diodes	Supply current in mA through 1 kΩ	Output at *Y* in Volt
Any input = '0' = 0.4 V	0.9 V	4.1	0.2 V
All input = '1' = > 4.4 V	5.0 V to 4.4 V	0	4.3 V to 3.7 V

Example 6.5 Find how the current will distribute in the diodes of the triplet.

Solution

When any of the input is at 0 = 0.4 V and any diode of triplet is conduction and in case more than one input is at 0, the current through R will distribute among those diodes which connect to input 0. Hence Table 6.4 can be redesigned as following Table 6.6.

TABLE 6.6 Table for input stages and voltages

A	*B*	*C*	Output at common of all 4 diodes in Volts	Through *A* current in mA	Through *B* current in mA	Through C current in mA	Supply Current in mA through 1 kΩ	Output at Y in Volt
0	0	0	1.1	1.3	1.3	1.3	3.9	0.4
0	0	1	1.1	1.95	1.95	0	3.9	0.4
0	1	0	1.1	1.95	0	1.95	3.9	0.4
0	1	1	1.1	3.9	0	0	3.9	0.4
1	0	0	1.1	0	1.95	1.95	3.9	0.4
1	0	1	1.1	0	3.9	0	3.9	0.4
1	1	0	1.1	0	0	3.9	3.9	0.4
1	1	1	5.0	0	0	0	0	4.3

Note: 1.1 V because 0.4 V + 0.7 V = 1.1 V. 1.3 mA because (5 V – 1.1 V)/1 kΩ = 3.9 mA and 3.9 mA/3 = 1.3 mA and 3.9 mA/2 = 1.95 mA.

Example 6.6 Find what will be output Y if the next stage circuits do not load the output.

Solution

When Y is not loaded, the diode between Y' and Y will not conduct. Hence output will be same as that at common triplet p-end Y. Table 6.7 gives the output.

TABLE 6.7

A, *B* and *C* state	Output at common of all 4 diodes in Volt	Output at *Y* in Volt
Any input = 0	1.1	1.1
All inputs = 1	5.0	5.0

Example 6.7 If in Example 6.4 another triplet of diodes (A', B', C') connect and give the output Y'', what will the output if we connect Y and Y''.

Solution

This will correspond to wired AND logic. Therefore the output will

$$Y.Y'' = (A + B + C).(A' + B' + C')$$

Example 6.8 Consider DTL logic circuit of Figure 6.5(a) with a fan-out of 8. If effective Capacitance effect of next stage DTL is 0.04 nF, what will be maximum possible propagation delay in the output change at the next stage?

Solution

Base-emitter capacitance = 0.04 nF. Assume that maximum number of next stages permitted is present. Therefore, $m = 8$. When the output = 1, the next stage transistor-ON delay is small and is 0.04 nF $\times$ 8 $\times$ 150 Ω = 48 ns. When logic state output becomes 0, the discharging (due to 1 to 0 transition) resistance is 5000 Ohm, therefore turn OFF propagation delay = (5000 Ohm) $\times$ 8 $\times$ 0.04 nF = 1600 ns. Average delay = (1648/2) = 824 ns. Average delay per gate next stage gate added = (824/8) = 1103 ns.

Example 6.9 Consider DTL logic circuit of Figure 6.5(a) with a fan-out of 8. What is V_{CE} at saturation? What is V_{CB} at saturation?

Solution

In saturation, the output at F is 0.2 V. Hence V_{CE} = 0.2 V. Therefore, $V_{CB} = V_{BE(\text{sat})} = \sim 0.8$ V $- 0.2$ V $= \sim 0.6$ V.

Example 6.10 Consider circuit of Figure 6.5(a)/ Let β (forward common emitter gain) is 20 and I_B = 0.4 mA. Fan out = 8. Whether will the collector current keep the output stage transistor in saturation or not?

Solution

Let us recall current-sink logic condition. T will remain in saturation if collector current is I_c. < (20 $\times$ 0.4)/8 mA = 1 mA. Actual collector current is (5 V – 0.2 V)/5000. This is less than 1 mA. Therefore, fan out of 8 is permitted.

Example 6.11

Consider circuit of Figure 6.5(a). If a circuit sink $I_C = 4$ mA when output is 0 and source 400 μA when '1', what is the average collector current and average power dissipated.

Solution

1. Average Current = (4 mA+ 0.4 mA)/2 = 2.2 mA.
2. Average Power Dissipated = 5 V × (Average current) = 5 V × (4 mA + 0.4 mA)/2 = 11 mW.

Example 6.12

Consider TTL circuit of Figure 6.6(a). What is the value of active pull totem pole circuit resistance between supply and collector, if logic stat 0 sink current is 2 mA. What is the value of active pull totem pole circuit resistance between supply and collector, if logic state 1 total current to next stage is maximum 0.04mA/gate? Assume fan-out = 10.

Solution

1. At logic 0 output, the transistor is in saturation and is at the voltage ~0.2 V. Total Current = 2 mA = (5 V – 0.2 V)/*R*. Therefore, *R* = 4.8 V/2.0 mA= 2.4 kΩ.
2. At logic '1' output, the transistor is cutoff and is at voltage between 5 V and ~2.4 V. Total Current = 10 × 0.04 mA = (5 V – 2.4 V)/*R*.*R* = 3.6 V/0.4 mA= 9 kΩ.

Example 6.13

In the above example, what is turn-on 0 to 1 transition delay time? What is turn-off 1 to 0 transition delay time? What is average delay time? Assume $C = 0.04$ nF.

Solution

For 0 to 1 transition, the *R* is to be taken as 2.4 kΩ only because the active pull up activates only after a turn-on transition to '1. So delay time = 0.04 nF × 10 × 2.4 kΩ = 9.6 ns. For 1 to 0 transition, the *R* is also to be taken as 2.4 kΩ. So delay time when discharging of next stage input terminals is again = 0.04 nF × 10 × 2.4 kΩ = 9.6 ns. Average delay time remains same.

Example 6.14

Consider the ECL logic circuit of Figure 6.7(a). If two input stage transistors are given the p.d of –1.5 V each, and reference to base of other differential pair transistor is – 1.15 V, show that the input stage transistor are in the cutoff region. If one of the input stage now become at – 0.5 V, what will be changes in the outputs.

Solution

Consider a differential amplifier pair between T_{ref}, one of the input-stage transistors, T_i and another input stage transistor T_j. V_{REF} at base of $T_{ref} = -1.15$V. Input at base of T_i and T_j are – 1.5 V. T_{ref} emitter has the voltage 0.7 volt less than the V_{REF}. Hence voltage at the emitter of T_{ref} is – 1.85. Therefore, the potential difference between the emitter and base of T_i is only 0.35 V the T_i is in cutoff region. Similarly, T_j is in cutoff region while *T* is in normal inverting amplifier mode. Hence, the collector of T_{ref} is at – 0.9 V and gives sufficient input base current to T_{OR}. The output from T_{OR} emitter is low (Figure 6.7(b)).

When one of the inputs (to T_j) becomes – 0.5 V, the p.d. between the base and emitter is now greater than 0.7 V, hence the corresponding transistor operates in normal input mode. The voltage at the emitter of T_j will now drive *T* into the cutoff region. Hence, the collector of *T* is at does not give sufficient input base current to T_{OR}. The output from T_{OR} emitter is high (Figure 6.7(b)).

Example 6.15

Consider a circuit of Figure 6.8. (1) What will be injected current if V_{BB} is 2 V, V_{EB} is 0.7 V and R external is 100 kΩ. (2) If logic input to the base of inverting transistor amplifier is 0, what will be collector current in transistor T?

Solution

1. Since p.d. across R external is ($V_{BB} - V_{EB}$), the injected current = (2 V – 0.7 V)/100 kΩ = .0.013 mA.
2. Since logic input is 0, the current of 0.013 mA just sinks and base current at T is 0. Hence, Collector current at T is negligible because T is at cut-off.

Example 6.16

Draw a circuit to implement AND operation using *n*-channel enhancement MOSFETs alone.

Solution

Refer Figure 6.11 (b), which shows an n-MOS NAND. AND operation is complement of NAND and complement is obtained by common input NAND. Therefore if the output F of first NAND in this figure is connected to a NOT or another identical NAND with both the inputs connected to F, the output of that NAND will be AND operation of tow inputs at the first NAND. Figure 6.14 shows a NAND-NOT bassed circuit.

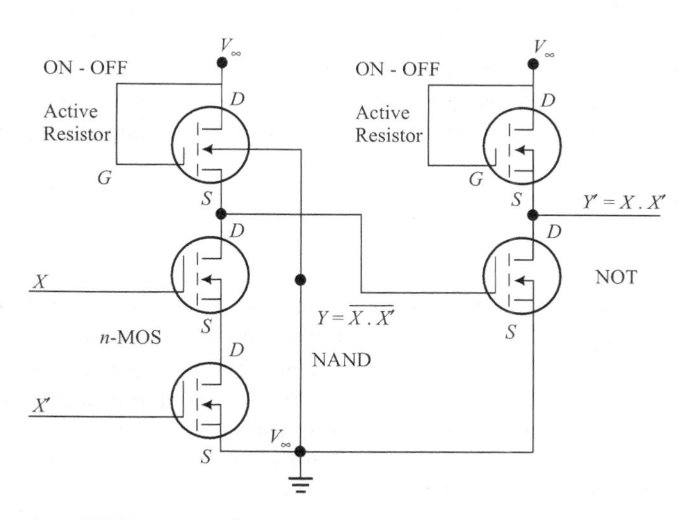

FIGURE 6.14 Two-input NMOS based AND gate using a NAND-NOT.

Example 6.17

Draw a circuit to implement AND operation using CMOS pairs.

Solution

Refer Figure 6.13 (b), which shows a CMOS NAND. AND operation is complement of NAND and complement is obtained by common input NAND. Therefore if the output F of first NAND in this figure is connected to NOT or another identical NAND with both the inputs connected to F, the output of that NAND will be AND operation of tow inputs at the first NAND. Figure 6.15 shows a NAND-NOT based circuit.

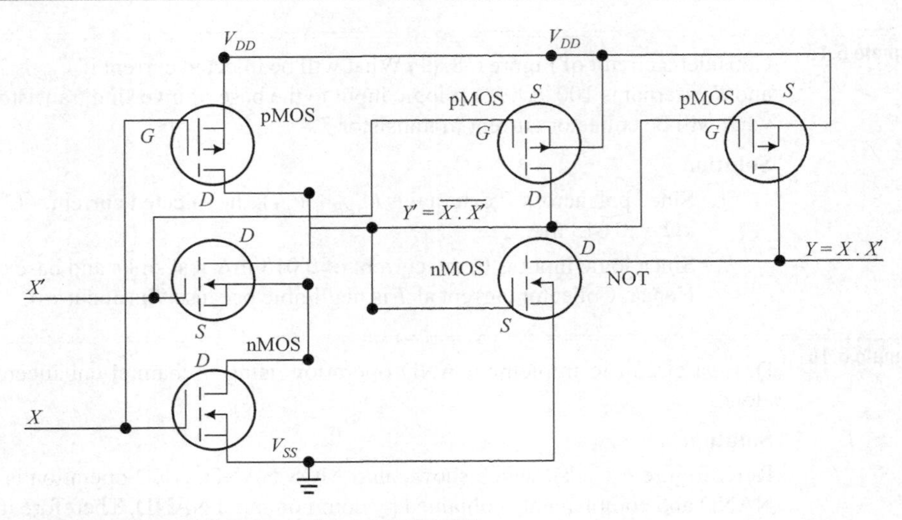

FIGURE 6.15 Two-input CMOS based AND gate using a NAND-NOT.

Example 6.18 A typical CMOS logic circuit dissipates 2.5 μW static, 600 μW/MHz. What is the average power dissipation assuming maximum operational frequency of 10 MHz? What will be the power dissipated at 500 kHz operation?

1. At 100 MHz, the power dissipation will be (2.5 μW + 6000 μW). Average dissipation assuming 50% time static and 50% maximum rate operation = 3001.5 μW.
2. Since power dissipation is 600 μW/MHz, at 500 kHz, it will be half. It means 300 μW.

Example 6.19 What are the correct characteristics among the DTL, ECL, CMOS and TTL gates? (A) or (B) or (C) or (D) [A GATE (2003) Competition Examination Question].

(A) Minimum Fan-out DTL, Minimum Power Dissipation TTL and Minimum Propagation Delay CMOS (B) Minimum Fan-out DTL, Minimum Power Dissipation CMOS and Minimum Propagation Delay ECL (C) Minimum Fan-out TTL, Minimum Power Dissipation ECL and Minimum Propagation Delay TTL (D) Minimum Fan-out CMOS Minimum Power Dissipation DTL and Minimum Propagation Delay TTL

Solution

Using Table 6.3, we find that (B) is correct.

Example 6.20 What are the correct characteristics among the I^2L, ECL, CMOS and TTL gates? (A) or (B) or (C) or (D)

(A) (i) Noise immunity excellent TTL, (ii) Ten MHz operational case minimum power dissipation CMOS and (iii) Minimum propagation delay I^2L (B) (i) Noise immunity excellent I^2L, (ii) Ten MHz operational case minimum power dissipation TTL and (iii) Minimum propagation delay CMOS (C) (i) noise immunity excellent CMOS, (ii) Ten MHz operational case minimum power dissipation ECL and (iii) minimum propagation delay TTL (D) (i) Noise immunity excellent CMOS, (ii) Ten MHz operational case minimum power dissipation I^2L and (iii) Minimum propagation delay ECL

Solution

Using Table 6.3, we find that (D) is correct.

EXERCISES

1. In an RTL logic circuit, if input resistance $R = 5$ kΩ (in place of 450 Ω in Figure 6.4) and collector path resistance = 1 kΩ (in place of 640 Ω. Calculate the noise margins at 1 and at 0, assuming $\beta_F = \sim h_{fe} = 20$, $V_{CE(sat)} = 0.2$, $V_{BE(ON)} = V_{BE(sat)} = 0.7$ V. Also calculate the fan-out of the circuit and propagation delay assuming a next stage effective capacitance = 40 pF.
2. If output stage collector path resistance = 2 kΩ in circuit of exercise 1, what will be the effect on the fan-out.
3. In a DTL logic circuit, (Figure 6.5) if 5 V to the input stage diodes there is an R = 4 kΩ and there are 2 diodes in place of three diodes feeding the next stage base current, what will be collector-emitter currents at T' at logic output of 1 and 0 when $\beta_F = \sim h_{fe} = 20$, $V_{CE\,(sat)} = 0.2$, $V_{BE\,(ON)} = V_{BE\,(sat)} = 0.7$ V with no next stage connection. Also Calculate (i) the fan-out of the circuit, (ii) propagation delay assuming a next stage effective capacitance = 40 pF assuming maximum possible next stages connected, (iii) power dissipated per gate at output = 0 and when output = 1.
4. If output collector path resistance = 2 kΩ in circuit of exercise 3, what will be the effect on the fan-out.
5. (i) What will be logic outputs if the 3 circuits of RTL logic Exercise 1 circuit connect in wired logic configuration (ii) What will be logic outputs if the 3 circuits of DTL logic Exercise 2 circuit connect in wired logic configuration.
6. A triplet of diode p-ends in a multi-emitter input transistor of a TTL circuit (Figure 6.6) are used for the inputs A, B and C. What are the output stage collector currents and voltages under different conditions? What will be the current directions if the output connects to a next stage TTL? Show these in Table 6.8.

TABLE 6.8

A	B	C	Output Voltage	Output Current in supply circuit of the output stage	Current direction when output connects to a next stage TTL
0	0	0			
0	0	1			
0	1	0			
0	1	1			
1	0	0			
1	0	1			
1	1	0			
1	1	1			

7. If reference Voltage in ECL logic circuit (Figure 6.7) changes from –1.15 V to –1.25 V, what will be the changes in the following specifications? V_{EE} = 5.2 V V_{CC} = 0 V, V_{OL} = –1.7 V, V_{OH} = –0.9 V, V_{IL} = –1.4 V and V_{IH} = –1.2 V.
8. What will be output *Y* if another ECL circuit is (i) wired AND and (ii) wired OR?
9. Which are the data not representing the correct situation in Table 6.9?

TABLE 6.9

S. No.	Type of Gate	Propagation Delay ns	Power dissipated per gate mW	Speed *Power Product pW.s = pJ (Pico Joule)	Fan Out	Maximum operating frequency MHz	Noise Immunity
(A)	NMOS operating at 50 MHz	300	0.2 to ≈ 10	60	20	50	Small
(B)	Standard TTL	10	10	100	10	35	Very good
(C)	CMOS in 74HC series static*, 600 mW/MHz 0.045 static, ~10 at 1 MHz	20	18 60	2.5 mW Very Good			
(D)	ECL (10K Series)	2	50	100	25	~75	Small

10. Consider TTL circuit of Figure 6.6(a). (i) What is the value of active pull totem pole circuit resistance between supply and collector, if logic stat 0 sink current is 1 mA. (ii) What is the value of active pull totem pole circuit resistance between supply and collector, if logic state 1 total current to next stage is maximum 0.01 mA/gate. (iii) what is turn-on 0 to 1 transition delay time? What is turn-off 1 to 0 transition delay time? What is average delay time? Assume C = 0.05 nF. Assume fan-out = 20.
11. Consider the ECL logic circuit of Figure 6.7(a). If two input stage transistors are given the p.d of –1.0 V each, and reference to base of other differential pair transistor is –1.15 V, show that the input stage transistor state whether cutoff or saturation or normal region. If one of the input stage now become at –0.9 V, what will be changes in the outputs.
13. Consider a circuit of Figure 6.8. (1) What will be injected current if V_{BB} is 2 V, V_{EB} is 0.7 V and *R* external is 20 kΩ. (2) If logic input to the base of inverting transistor amplifier is 1, what will be collector current in transistor *T*?
14. Consider the circuit of Figure 6.11(a). What will be the circuit and logic function(s) possible if (i) One depletion mode *n*-channel MOSFET *T*' is replaced by two in series (ii) Two enhancement-mode *n*-channel MOSFETs T_X and $T_{X'}$ are replaced by two logic drivers in parallel and one another logic driver in series?
15. Consider the circuit of Figure 6.13(a). What will be the circuit and logic function(s) possible (i) Four enhancement mode *p*-channel MOSFETs *T*' and *T*'' forms a pull up network (ii) There are two enhancement mode n-channel MOSFETs T_X and $T_{X'}$ in series and there is another pair of the logic drivers in parallel at the pull down network?

16. A typical VLSI CMOS logic circuit dissipates 50 nW static, 0.6 nW/MHz. What will be the power dissipated at 500 MHz operation if at a given instant maximum 1000 gate circuit is operational?

QUESTIONS

1. Why is the DTL circuit considered as better logic circuit than RTL?
2. Compare the fan-out, turn-on delay times, turn-off delay times and power dissipated per gate in circuits of RTL, DTL, HTL and TTL?
3. What are the advantages that can be obtained by an open collector TTL?
4. An I^2L circuit needs minimum silicon area in an IC? Why?
5. Compare modified DTL circuit operation with that of DTL.
6. Draw into four circuits a TTL NAND circuit in Figure 6.6(a) that can be used to make AND, OR, XOR and NAND gates.
7. Draw the four circuits for CMOS NOR circuit in Figure 6.13(a) that can be used to make AND, OR, XOR and NAND gates.
8. What are the advantages of using CMOS over NMOS?
9. What are the advantages of using TTL in place of DTL logic?
10. A typical VLSI logic circuit dissipates 1 nW static, 0.06 μW/MHz. What will be the power dissipated at 1MB RAM [1 MB = 1024 × 1024 × 8 bits) if each bit cell has 4 MOSFETs?

CHAPTER 7

Interfacing Circuits between the Logic Gates of Same Family, Different Families and Types

OBJECTIVE

In Chapter 6, we learnt the following topics: RTL, DTL, TTL, ECL, ICL, HTL, NMOS and CMOS, different types of the logic circuits, difference between different families of gates. In this chapter, we shall learn about speed, propagation delay, operating frequency, power dissipated per gate, supply voltage levels, and operational voltage levels that define logic states 1 and 0.

We shall learn in this chapter how to make interconnections between the logic gates to its next stage of same type as well as to the next stages of different families in the same type. We shall also learn interfacing circuits to the net stages of different types.

Let us first revise and learn the following concepts to have ease in learning the topics of this chapter.

7.1 REVISION OF THE IMPORTANT TOPICS

7.1.1 Speed f_o, Propagation Delay, $\bar{t}_p$ and Operating Frequency $\bar{f}_o$

The speed and propagation delays in the various families of the gates are given in Tables 6.2 and 6.3 (see Chapter 6).

Speed of a gate is measured by f_0, the maximum operating frequency. f_o is reciprocal of $\bar{t}_p$, and is usually measured in MHz. f_o is 25 MHz for LSTTL family and 2 MHz for CMOS family of gates. We cannot change an input more number of times than defined by f_o.

$\bar{t}_p$ is measured in nanosecond. The $\bar{t}_p$ is average of two time differences between appearances of 1 after 0 and 0 after 1 at an output of a logic gate of a family from instant when an input or its inputs are changed. $\bar{t}_p$ is found by taking average. $\bar{t}_p$ for a family of the logic gates. $\bar{t}_p$ depends on the family. It is different for each family, 74, 74LS, 74H, 40...B (TTL and CMOS). Each family of the gates will show different $\bar{t}_p$.

A maximum f for TTL gates depends on $\bar{t}_p$. The $\bar{f}_0$ (average value of f_o in a family) for the CMOS family gates or the NMOS family (gates with *n*-channel MOSFETs only to increase speed f_o) depends upon the $\bar{t}_p$ as well as power dissipation per gate (P_g) considerations (refer Chapter 6).

Points to Remember

1. The propagation delay $\bar{t}_p$ (74) is 10 ns and $\bar{t}_p$ (74LS) is also ≈ 10 ns, respectively, assuming 50 pF load at next stages.
2. The $\bar{t}_p$ (40.. B) varies from 75 ns to 40 ns if supply voltage varies from 3.5 V to 7.5 V, respectively.
3. The propagation delays for the CMOS gates for a 0 to 1 and for a 1 to 0 transition at output of an CMOS gate is same.

7.1.2 Power Dissipated per gate P_g

P_g is measured in mW and is nearly frequency—independent for the TTL families. P_g is measured for the MOS families of the gates in μW per kHz (= mW per MHz) change in frequency by which a state at an input changes (P_g ~ 1 μW per kHz change). For GaAs based 10 GHz family gates P_g could be several hundreds of mW.

Points to Remember

1. For TTL gates, power dissipated is almost independent of f_{in} (number of times input states are changed per second).
2. For CMOS gates, power dissipated is almost linearly rises with f_{in}.
3. For TTL gates, in steady state power dissipation differs when the output state 1 and when 0, and is high of the order of –10 mW per gate. (Steady state means input fixed at 1 or 0 for long time).
4. For CMOS 40…B gates, P_g^0 and P_g^1 are negligible in the steady state.
5. Power dissipated from TTL family of gates is like that in a purely resistive load and in the CMOS family of gates is like that in a purely capacitive load up to a limiting f in MHz.
6. For the CMOS 40...B family of the gates, P_g at 1 MHz = 500 μW at 5 V drain-source supply.
7. The variation of propagation delay as a function of V_{DD} and power dissipation as a function of V_{DD} is opposite to each other.

7.1.3 Supply Voltage Levels, V_{CC}, and V_{DD} and V_{SS}

V_{CC} is measured in volt with respect to ground in case of the TTL families. V_{DD} with respect to V_{SS} is important in the MOS families of the gates. Table 7.1 gives the supply voltage levels for the various families of the gates.

TABLE 7.1

Families of the gates	Supply V in volt, higher side (V^+)	Supply V in volt, lower side (V^-)	Remarks
TTL 74, 74LS, 74 HCT (CMOS with TTL buffer)	5.0 ± 0.25 V	Common digital ground	V^+ is also called V_{CC} for a TTL and an NMOS gate.
40.., 74C.. CMOS	+3 V to 18 V above V_{SS}	V_{SS}	V^+ is also called V_{DD} for a CMOS.
74HC (High speed CMOS) 74 AC CMOS	+2 V to +6 V above V_{SS}	V_{SS}	V^- is also called V_{SS} for a CMOS.
NMOS	Usually +5 V	Usually common digital ground	V_{DD} and V_{SS} are usually +ve and 0 V
RS 232C line driver (buffer)	+10 V to +12 V (Normally)	−12 V to −10 V (Normally)	RS 232C Used for the serial communication.

Point to Remember

CMOS family of gates use much smaller currents at steady state, and CMOS family of gates also have equal Fan-in values for 1 and 0 inputs to these.

7.1.4 Operational Voltage Levels which Define the States 1 and 0

Table 7.2 gives the parameters defining state 1 and state 0 in the TTL and MOS family of gates. All the potential differences (p.d.) in this table are given with respect to a common ground of the digital circuit(s) (which should be different from a common ground of analog circuit in an electronic circuit, which has the mixed digital and analog circuits).

TABLE 7.2

Parameter	Unit	Definition	Potential difference		
			TTL	CMOS	NMOS
V^{Max}_{OH}	Volt	Maximum p.d. for state 1 in the output	5.0 ± 0.25	V_{DD}	5.0 ± 2.5
V^{Min}_{OH}	Volt	Minimum p.d. for state 1 in the output	2.8	0.66 V_{DD}	2.8
V^{Max}_{IH}	Volt	Maximum p.d. for state 1 at the input	5	V_{DD}	5
V^{Min}_{IH}	Volt	Minimum p.d. for state 1 at the input	2.4	0.66 V_{DD}	2.4
V^{Max}_{OL}	Volt	Maximum p.d. for state 0 at the output	0.4	0.33 V_{DD}	0.4
V^{Min}_{OL}	Volt	Minimum p.d. for state 0 at the output	0	0	0
V^{Max}_{IL}	Volt	Maximum p.d. for state 0 at the input	0.8	0.33 V_{DD}	0.8
V^{Min}_{IL}	Volt	Minimum p.d. for the state 0 at the input	0	0	0

7.2 CONSIDERATIONS FOR INTERCONNECTIONS AND INTERFACING

7.2.1 Magnitude and Direction of Source and Sink Currents

Let us consider interconnection of an output of a gate with an input of another gate. A current flow takes place (as in any electronic circuit) howsoever small it may be and the amount of current flow will depend upon output at the gate and the input of the second gate. Current always takes the easiest path (least resistance). Direction of this flow will depend on output logic state 1 or 0. For state 1, the transistor at the output stage is OFF (does not conduct). Therefore, the current i_{source} is sourced towards the input of the interconnecting gate. Figure 7.1(a) shows the current direction in this situation and internal connection to V_{CC} at the output end and to ground at input end. The i_{source} depends upon internal input impedance offered and internal resistance between V_{CC} and output.

For the output state 0, the transistor at the output is ON (saturation state and does conduct). Therefore, the i_{sink} is sinking through it. Figure 7.1(b) shows the direction of it as well as internal connection to AND connected to an output and V_{CC} at the input end.

7.2.2 Considerations of Fan-In and Fan-Out

The i_{source} from a standard TTL is twice as large than LS TTL, and 10 times as large than from a CMOS 40... family gates. i_{sink} in a standard TTL gate is four times than in a LS TTL gate (refer Table 7.3.). The reciprocal of it can be called fan-in. Fan-in is a measure of i_{source} and i_{sink} with respect to a standard TTL family gate at the input. Fan in is different in the states 1 and in state 0 for the TTL family gates. Fan in is same for 1 and 0 in CMOS family gates.

TABLE 7.3

Family, *F*	Effect of Loading by *F* due to the state at output				Number of next stage inputs connectable		Output current in μA available during a state	
	'1'		'0'				'1'	'0'
	i_{source} μA@	Fan in with respect to 74	I_{sink} μA *+	Fan-in with respect to 74	Fan-out 74	Fan-out to 74LS	max. I_{OH}	max. I_{OL}
74	40	1	−1600	1	10	20	400	−1600
74LS	20	0.5	−400	0.25	5	20	400	−8000
40..	1	0.025	−40	0.025	0.5*	1**	10†	−500†
NMOS	1	0.025	−40	1	4	200	200	−2000

max. is an abbreviation for maximum.
@denotes as I_{IH}.
*+ denotes as I_{IL}
* means this value is 2 for 40...B family (with TTL driving buffers at the outputs).
** means this value is 6 for 40... B family with TTL driving buffers at the output.
A minus sign in columns 4 and 9 indicates a sinking current towards the output connecting stage transistor. Refer Figures 7.1(b), (d) and (f).
A plus sign means a source. Refer Figures 7.1(a), (c) and (e).
† Assumed +5 V potential difference between V_{DD} and V_{SS}. The current increases upon the operating frequency. (We take 50 pF input capacitance for calculating the currents at a given *f*).

Table 7.3 columns 3 and 5 gives *fan-in* for various families of gates. Fan-in for 74LS family at 1 is 0.5 as i_{source} (20 μA) is half of the value (40 μA) when a standard TTL is at the input. Fan-in for LS family at 0 is 0.25 as i_{sink} is one fourth (400 μA) from it compared to the value (1600 μA) from a > 4 gate at the input. For CMOS 40... series gate fan-in at 1 as well as 0 is only 0.025 as i_{source} is (40 μA multiplied by 0.025 i.e. 1 μA), and i_{sink} is (1600 μA multiplied by 0.025 i.e. 40 μA).

We have learnt the concept of fan-out as the number of input next input stages, which can be connected to a given type of a logic gate. Fan-out is defined for an output logic gate in terms of the identical gates or in terms of standard TTL gates (column 6) or in terms of LSTTL gates (column 7) as the next stage(s) inputs.

Figure 7.1(c) shows that a LS TTL family gate at 1 is interconnected to ten other input terminals in the six logic gates of various types of same LS family. Figure 7.1(d) shows the same family interconnections at output 0. How many is the number of such input terminals sourced and how many are such inputs, which can sink the current from a LS TTL gate? Let us answer by a look at the columns 8 and 9 of Table 7.3. The answer is twenty (ratio of 400 μA and 20 μA) in ratio of I_{OH} max and I_{IH}. Suppose the LS output 1 is connected to a standard TTL input. Then only 10 of the inputs can be sourced at 1 output from it. Only five standard TTLs are allowed to sink current into it. We say that fan-out of a LS TTL gate is ten at 0 and five at 1. It means that, we should never connect an LS gate output to more than five standard TTL inputs. Table 7.3 in the middle part (columns 6 and 7) gives fan-out of various families of gates. Fan-out in columns 6 and 7 can be considered as currents sourcing or sinking (which ever is less) strength of an output state in a family to the number of inputs.

Now, let us consider a CMOS gate receiving input from a standard TTL gate at 1 (refer Figure 7.1(e) for other family). This does not pose any problem because i^{cmos}_{source} is forty times less than the i_{source} in case of standard TTL. However, if a TTL gate is at output state 0 [Figure 7.1(f))] then its output stage transistor will be needed to sink current i_{sink}, which is not possible. Therefore, a TTL gate cannot be interfaced to a CMOS gate without a pull-up resistance (for example, 2200 Ω) at the input to the V_{CC}.

What about a CMOS gate fan-out without any buffer? It can source only half of the i_{source} needed by a standard TTL input, and can source only sufficient current for one LS TTL family input terminal. Therefore, a CMOS gate without buffer at its output has fan-out of 0.5 (Table 7.3).

Table 7.4 gives permissible interconnections.

7.3 INTERFACING CMOS AND TTL GATES

Figures 7.2(a) to (e) shows the interfacing circuits between the CMOS and TTL gates.

1. A pull-up resistance is shown in Figure 7.2(a). It is to keep the voltage levels above 3.3 V as CMOS gate at state 1 for $(V_{DD} - V_{SS}) = 5$ V, and in between 3.3 V and 5 V, while a TTL at 1 the output can be between 2.8 V and 5 V.
2. A pull down resistance (1 kΩ) is shown in Figure 7.2(b), as a CMOS should have for a 0 the V in between 1.66 V and 0 V, while a TTL for a 0 has V less than 0.8 V. This resistance of 1 kΩ will lower the voltage levels to define proper 0 at a TTL input.
3. Figure 7.2(c) shows a + 5 V operated CMOS interconnection to a LS one number input. This is permitted as LS i_{source} and i_{sink} are such that the loading effect is one

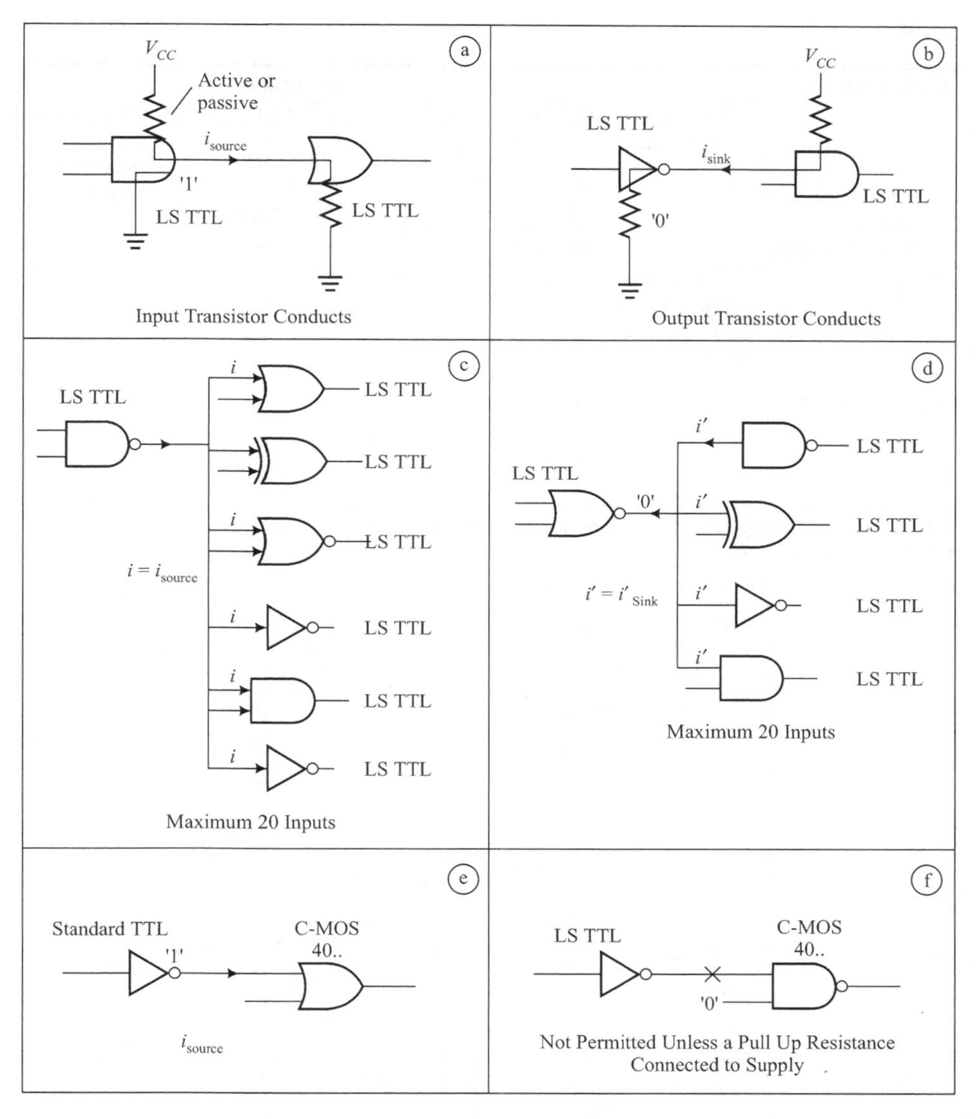

FIGURE 7.1 (a) An interconnection of LSTTL output 1 to an LSTTL input 1 (b) An interconnection of LSTTL output 0 to an LSTTL input 0 (c) An interconnection of LSTTL output 1 to multiple LSTTL gates (d) An interconnection of LSTTL output state 0 to multiple LSTTL gates (e) An interconnection of standard TTL output 1 to a CMOS 40…B gate(s) (f) An interconnection of LSTTL output 0 to a CMOS 40…B gate(s).

half of that in a standard TTL, and the pull-up and pull-down are therefore not needed for one number LS input.

4. Figures 7.2(d) and (e) show the maximum number of interconnections to a standard 74 and 74LS, respectively, in case of a CMOS gate with a buffer to enable higher i_{source} and i_{sink} values (Table 7.3) than possible from the one without buffer.

TABLE 7.4

Output stage	Input stage					
	74, 74LS, 74H etc.	40.. at 5 V	74HC at 3.3 V	74HC at 5 V	LED through a series resistance up to V_{cc}.	16 mA, 300 Ω, relay with one end to V^+
74, 74LS, 74H, 74HCT etc.	✓	Pull up Needed	✓ (HCT no)	Pull up needed	✓	× (except 74 standard only)
40.. at 5 V	✓ LS one (any other × i.e. nil). Pull down therefore needed	✓	×	✓	×	×
74 HC at 5 V	✓	✓	×	✓	✓	×
N-MOS (at 5 V)	✓	Pull up needed	✓ (Needs 5 V CMOS operation)	Pull up needed	✓	×

✓ means an interfacing permissible without a pull up or a pull down or buffer.
× means not permissible.

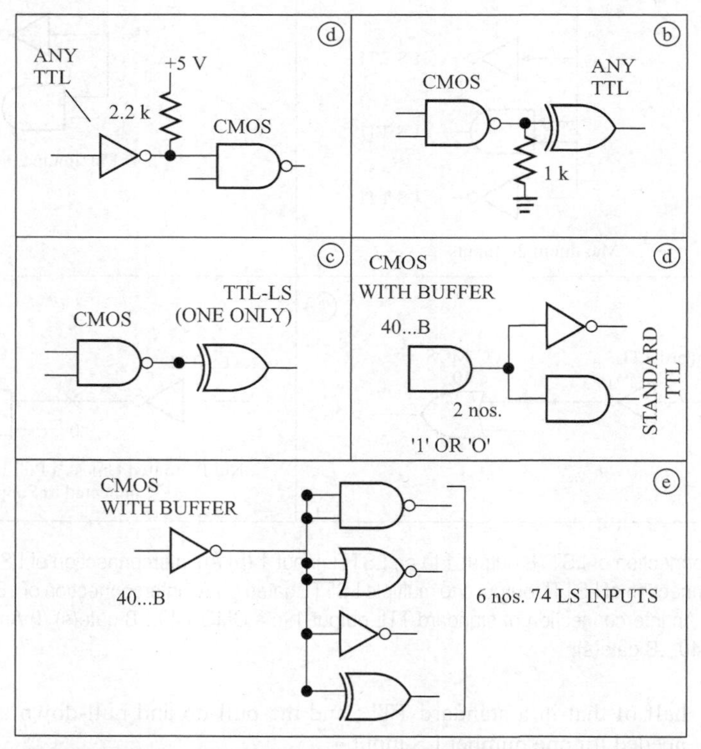

FIGURE 7.2 (a) An interfacing circuit of any family TTL gate with a CMOS using a pull-up resistance (b) An interfacing circuit of a CMOS gate with a TTL using a pull-down resistance (c) An interfacing circuit of a CMOS gate with a single LSTTL without a pull-up or pull down resistance (d) An interfacing circuit of CMOS gate, which includes a buffer for twp standard TTL and interfaces without a pull-up or pull down resistance (e) An interfacing circuit of CMOS gate, which includes a buffer with six LSTTL without a pull-up or pull down resistance.

7.4 INTERFACING CMOS WITH CMOS GATES

What about a CMOS to CMOS interconnection? The current levels are too small in CMOS so that the large number of interconnections does not pose any difficulty unless the operating frequencies become higher than maximum limiting MHz.

Points to Remember when Interfacing

1. A +5 V operated CMOS gate output can always be interfaced to a single LS TTL gate.
2. In CMOS gates, the current ratio i_{sink}/i_{supply} from supply is negligible as the currents are in few μA only if the inputs are not floating but are kept at any instant in logic 0 or at logic 1. This can be done by feeding input either directly to V^- or to V^+ or to an output of a logic gate. Therefore, we cannot float a CMOS.
3. We can float TTL gate inputs as input-current is not rising at the intermediate V_{in} values for these gates, in contrast to the circuit made using CMOS gates.
4. TTL gates sink the current, which is not negligible when the inputs are at logic 0 s.

EXAMPLES

Example 7.1

A circuit is shown in Figure 7.3 with a logic circuit with four number three input LSTTL NANDs with all first stage input unconnected. What will be the output? After how much time last stage output will appear if an input is connected to 0.

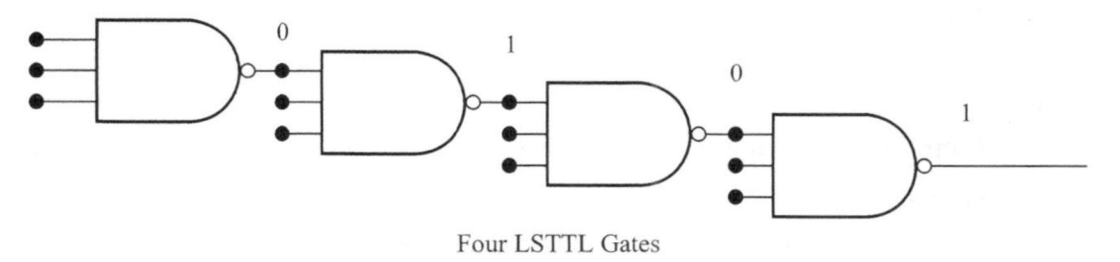

FIGURE 7.3 A logic circuit with four number three input NANDs with all first stage input unconnected.

Solution

1. When an input is not connected in any TTL family gate, it is taken as 1. First stage NAND output will be 0. (NAND give a high output unless any of the input is at 0). Second stage will give output = 1 because at least one of the input = 0. Similarly we can estimate the output from the fourth stage, which will be 1.
2. Propagation delay of LSTTL is 10 ns. Therefore, the last stage will give output after 40 ns.

Example 7.2

A circuit is shown in Figure 7.4(a) for a logic circuit. It has two LSTTL NANDs. An input A connects to a switchs, which when pressed gives the input 1 to the NAND. Input A also connects input B. The connection to B is through another NAND gate whose inputs C and D are common and whose output connects to the input B of the first NAND gate. What will be the NAND gate output after the switch is pressed to make logic input A = 1? Assume that the propagation delay = 10 ns.

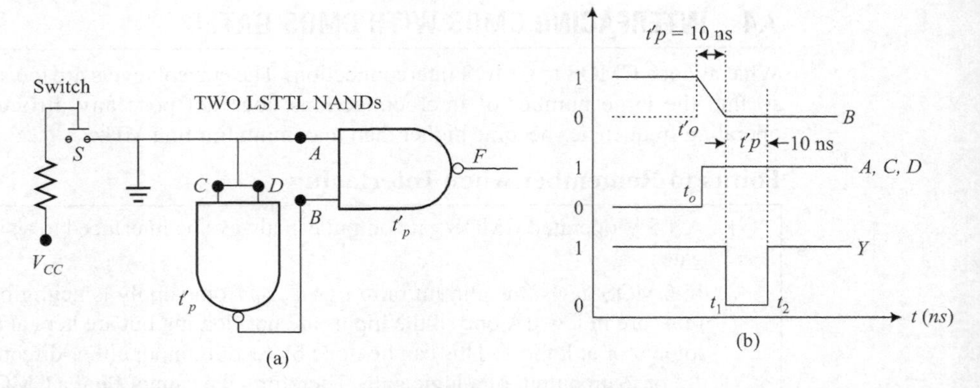

FIGURE 7.4 (a) A logic circuit with two input LSTTL NAND with one of the input connected to another with a NOT operation in-between (b) The timing diagrams of the inputs and outputs at the NAND.

Solution

First NAND has input A connected through second NAND through a NOT gate formed due to common inputs C and D. Therefore, two inputs A and B are always complementary in steady states. The state at NAND at the state is $A = 0$, $B = 1$. However, when the switch at input A is pressed, the input A becomes 1 instantaneously. Both inputs A and B are at 1 at that instant. Therefore, the output F of first NAND will become 0 after a time interval of 10 ns. Now, after 10 ns, the second input becomes at logic 0, after another 10 ns, the first NAND output will again become 1. Therefore, a pulse of 10 ns duration will be the output after a 10 ns delay from the first NAND. Figure 7.4(b) shows the timing diagram of the inputs and outputs at each NAND.

Example 7.3

A circuit is shown in Figure 7.5. Point out the corrections needed in the interfacing circuit.

Solution

Circuit of Figure 7.5 needs the following corrections. (i) A pull up resistance of 2.2 kΩ at each input of CMOS gate. (Figure 7.2(a)) (ii) One buffer is needed as only five inputs (not six) can be connected to 74LS to the standard TTL gates (because fan-out = 5 for 74LS in column 7 Table 7.3).

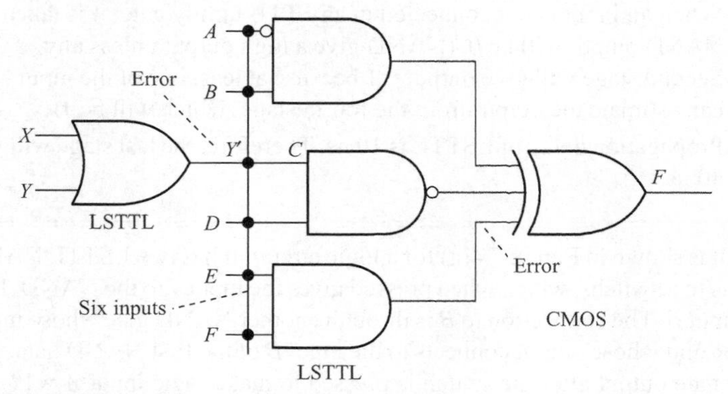

FIGURE 7.5 Logic circuit with LSTTL and CMOS interfaces.

Example 7.4 A circuit is shown in Figure 7.6. Point out the corrections needed if any.

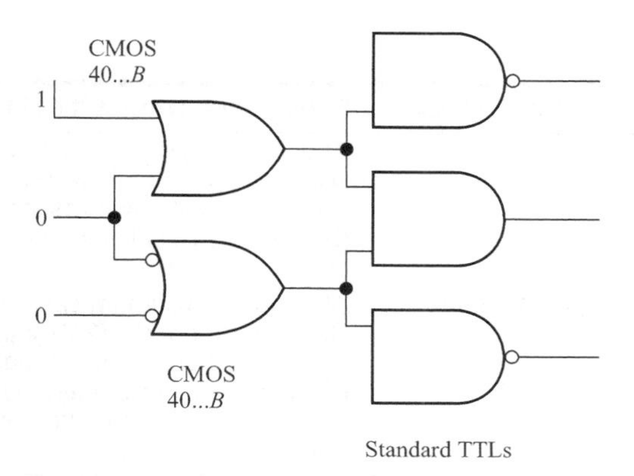

FIGURE 7.6 Logic circuit with all proper matching of fan-in and fan-outs.

Solution

All interfacing logic gates satisfy the fan-in and fan-out requirements (Figure 7.2(d)). Circuit needs no correction as each buffered CMOS connects two TTL inputs only.

Example 7.5 For a CMOS gate operated at 3.3 V, instead of at 5 V potential difference between drain and source, the propagation delay is found to be increased to 1.4 times (by 40%) but the power dissipation is found be decreased to 0.45 times becomes (55%) (Figure 6.12a). Explain why?

Solution

The drain current I_D is smaller at a low voltage operation between V_{DD} and V_{SS}, The power dissipation is given by $(i_D)^2.r_d$. While r_d increases nearly linearly, the current decrease nonlinearly at a faster rate on decrease in V_{DS}. This explains greater decrease in power dissipation. Propagation delay increases due to reduced V_{DG}. The carriers accelerating-potential decreases, hence charging current now builds slowly due to increase in r_d. (r_d is reciprocal of the rate of change of I_D with V_{DS}.).

Example 7.6 CMOS family gate output voltage levels become compatible to TTL output voltage levels when the CMOS is operated at 3.3 V. Why?

Solution

As V_{OH} (minimum) is 2.4 V for TTL gate, the V_{OH} (minimum) is (2/3)V_{DD} for a CMOS gate. Therefore, if V_{DD} is 3.3 V, the V_{OH} (minimum) are of same order of magnitude. Hence the CMOS output can be interfaced with a TTL gate provided that operates at 3.3 V.

Example 7.7 Explain the result of Table 7.3. Give an example each of use of a pull down resistor and of a pull up resistor.

Solution

Table 7.5 below gives the answer.

TABLE 7.5 Use of pull-down and pull-up resistors

Pull down or pull up	Output stage	Input stage	Explanation (why is it so?)
Pull Down	CMOS	LSTTL	At 0 the V_{IH} is 0.8 V, while CMOS gate will give V_{IH} = 1.33 V when operated at 5 V. Therefore, pull-down resistance (Figure 7.2(b)) sinks the current and make the V_{IH} compatible.
Pull up	74 or 74LS or NMOS	CMOS 40x	At 1 the V_{IH} minimum can be 2.4 V, while CMOS gate needs [(2/3) V_{DD}] as V_{IH} minimum, there is pull up passive or active to the 5 V supply is needed to make V_{IH} compatible from the output stage (Figure 7.2(a))

Example 7.8

Power dissipation per gate is say, 2 mW/gate in 74LS family of the gates. On a silicon slice, 1000 of these are stacked together by interconnecting these in an IC. Assume that 20 are active at any instant (i.e. are giving I_{sink}), what is the total power dissipation?

Solution

Since 20 gates are sinking current and current dissipates mostly in current sinking (saturated) transistors, operating 20×2 mW = 0.4 W will be the dissipation.

Example 7.9

Power-dissipation per gate is, say, 1 μW/gate per kHz switching frequency in a CMOS family of the gates. On a silicon slice, 1000000 are stacked together by interconnecting these together in an IC. Assuming on the average 200 are active at any instant, and are getting switched at 2 MHz what is the total power dissipation?

Solution

Power dissipation depends on how many are actually active gates undergoing the transitions. Power dissipation will be (200 gate × 2 MHz × 1 μW/gate/kHz) = 200 × 2000 μW = 400 mW when gates average switching frequency is 2 MHz. (Note: Assume that CMOS gates dissipates equally in logic '1' and '0').

Example 7.10

Let i_{IH}= 1 μA and i_{LL} = −40 μA at 5 V supply Voltage for a 40. B family of the gates on a silicon slice, 1000000 are stacked together by interconnecting these in an IC. Assume 2000 are active at any instant and 1000 are in state 0 at their outputs. What is the steady state current from supply of 5 V (ii) from supply of 3.3 V? (Assuming 40% drop in power consumption in CMOS gates operated at 3.3 V instead of 5 V).

Solution

Power dissipation depends on how many are actually active gates undergoing the transitions. Out of 2000 active gates, 1000 are at 0. Hence we can take average of 1 μA and 40 μA = 20.5 μA. Power dissipation at 5 V will be (2000 gate × 20.5 μA × 5 V). Steady state current at 5 V = 2000 gate × 20.5 μA= 41000 μA. Power dissipation at 3.3 V will be

0.4 × [2000 gate × 20.5 μA × 5 V]. Steady state current at 3.3 V = 0.4 × [2000 gate × 20.5 μA × 5 V]/3.3V = 1.66 × 0.4 × 41 mA = 27.2 mA. (Note: *B* series means buffered gate with TTL. It is due to this i_{IH} and i_{IL} *are differing*).

Example 7.11

Why can't we permit floating of a CMOS logic gate input?

Solution

High impedance between the gate and source layer in unprotected CMOS-device creates a high voltage by the accumulation of static charges. Hence floating inputs are not permitted.

Example 7.12

Why is the floating of a TTL logic gate input permitted?

Solution

Input is given to a multi-emitter junction transistor, which has very low capacitances and resistances. Hence the floating input is permitted. The logic state is 1 in an unconnected input.

Example 7.13

Why does TTL gates sink high current when the output state is 0?

Solution

Output stage transistor is in saturation mode at logic 0 at the output. Therefore, the current sinks from the supply circuit in this state through the output stage transistor.

Example 7.14

Why does we prefer to keep logic 1 at the output followed by input of next stage when in the steady state and only at the time of activating a circuit we force input to next stage 0 in case of the TTL gates? In case CMOS gates, there is no need of such a preference for steady state at 1. Why?

Solution

TTL gate i_{Isource} is in μA while the i_{sink} is in mA. Therefore, there is less power dissipation, for state at 1 when normal steady state is kept at 1. CMOS gates have little dissipation in both the states.

■ EXERCISES

1. A circuit is shown in Figure 7.7 with a logic circuit with four number three input CMOS NORs with all first stage input connected as shown in the figure. What will be the output? After how much time last stage output will appear if an input is connected to 0.
2. What will be the output if all first stage inputs in Figure 7.7 are complemented by the NORs at the inputs? After how much time last stage output will appear if an input is connected to 0.
3. A circuit is shown in Figure 7.8 with a logic circuit. It has the CMOS NORs. Its input A connects to a switch, which when pressed gives the input 0 to the NOR. Input *A* also connects input *B*. The connection to *B* is through another NOR gate whose inputs *C* and *D* are common and whose output connects to the input *B* of the first

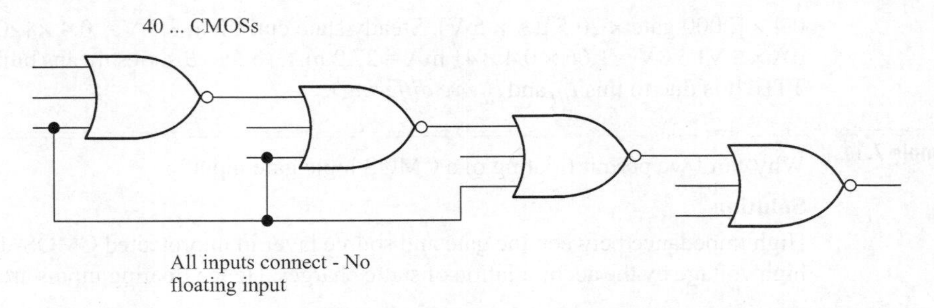

FIGURE 7.7 A logic circuit with four number three input CMOS NORs with all first stage input connected.

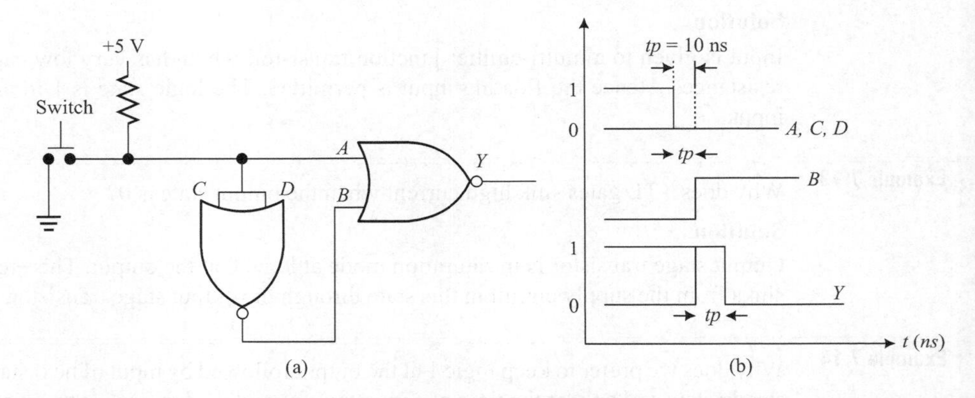

FIGURE 7.8 (a) A logic circuit. It has the CMOS NORs. Its input A connects to a switch (b) Timing diagram for output at Y after switch is pressed.

NOR gate. Figure 7.8(b) shows the NOR gate output after the switch is pressed to make logic input $A = 0$ if propagation delay is taken as 50 ns and output of the NOR is again connected to input A? Show the corrections required in the figure.

4. An interfacing circuit of LSTTL gates is shown in Figure 7.9. Point out the corrections needed in the interfacing circuit.
5. A circuit is shown in Figure 7.10. Point out the ways to connect more LSTTL stages to *A*.
6. For a CMOS gate operated at 12 V, instead of at 5 V potential difference between drain and source. What will be effects on the propagation delay and power dissipation?
7. Power-dissipation per gate is say, 2 μW/gate/kHz in an HCMOS family of the gates. On a silicon slice, 100 of these are stacked together by interconnecting these in an IC. Assume that 10 are active at any instant (i.e. are giving I_{sink}), what is the total power dissipation if ciruit operates at 50 MHz.
8. What should be the power dissipation per gate in nW/gate per kHz if switching frequency in an HCMOS family of the gates is 200 MHz and on the silicon slice 10000000 are stacked together by interconnecting these together in an IC. Find how many of them can be active at any instant if the supply is dissipating 50 mW.

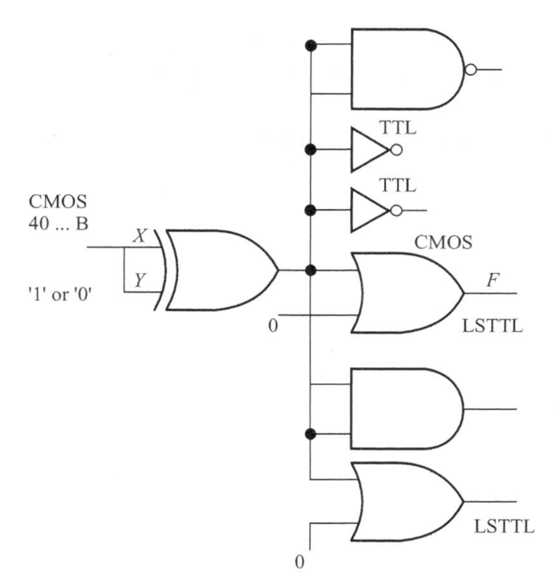

FIGURE 7.9 An interfacing circuit of LSTTL gates.

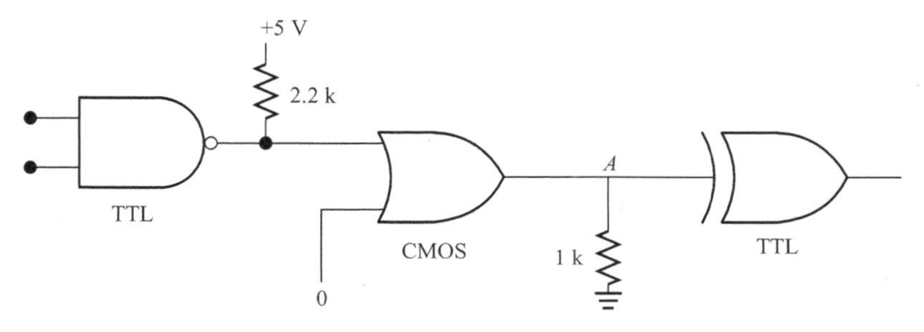

FIGURE 7.10 A logic circuit with short of interfacing LSTTL gates.

QUESTIONS

1. What are the advantages of specifying fan-in and fan-out in terms of standard TTLs?
2. Specify the conditions when a pull-up circuit is needed for interfacing.
3. Specify the conditions when a pull-down circuit is needed for interfacing.
4. How do we calculate the total power needs of a logic circuit consisting of many gates?
5. Describe methods to interface CMOS gates with the TTL families.
6. Describe how do we interface LSTTL gates with standard TTL gates.
7. How will you interface four CMOS gates to the CMOS gates?

8. How will you interface an LSTTL to a CMOS gate?
9. How many will the LSTTL circuits you interface so that at first stage output = 0 state current does not exceed 4 mA?
10. How many will the standard TTL circuits you interface to an LSTTL gate so that at first stage output = 0, the power dissipation is within the limits specified?

CHAPTER 8

Open Collector, Open Drain and Tristate Gates

OBJECTIVE

In this chapter, we shall learn about open collector gates, open drain gates, quasi-bidirectional gates, and the use of an analog/digital switch in analog circuit and in tristate gates.

In Chapter 6, we learnt the different types of the transistor and MOSFET-based logic gates. In Chapter 7, we learnt that when a gate interfaces to the next stage, it has the voltage and current levels as per the output stage pull up circuit and next input stage impedance(s) and these levels change depending upon the number of gates interfaced. Also the next stages driving capabilities (current i_{source} to next stage(s) or current i_{sink} from the next stage(s)) are limited by the pull up circuit of the gate output stage transistor or MOSFET. We shall learn in this chapter how to interface these by the external pull-ups and obtain higher driving capabilities by using the and open collector or open drain gates.

We also learnt about wired AND (Section 6.4.3.3) and wired OR logic (Section 6.5.1.4) interconnections of two logic gate output in Chapter 6. (Note: Such connection is not permitted in TTL gate with totem pole output through internal active pull-up). When a state is wired AND, assume that one output is 0 and other output is 1, the output 1 giving transistor circuit drives the transistor 0 giving circuit and the logic gate 1 state will also be pulled down to 0. Is it possible to enable a wired interconnection of multiple gates on a common rail (line) such that at an instant only a single addressed (enabled) gate gives the output, other gates remain not only in transistor cutoff but also output stage itself in complete cutoff state? An alternative way of putting this question is that can a logic gate have three states; 1, 0 and a tristate. A tristate means high impedance state in which there is no i_{source} or i_{sink} at the output stage transistor or MOSFET. We shall be learning in this chapter use of tristate gates to accomplish the output from different gates on a common wire.

8.1 OPEN COLLECTOR GATE

Recall the NAND TTL shown in Figure 6.6(a). The *T*'' *F* output pulls up to voltage level V_{cc} when the collector of *T*' logic state output is 1. Let us consider an open collector (O.C.) TTL NAND gate. An O.C. gate does not possess the pull-up circuit between output *F* and V_{CC} shown by a dotted square in Figure 6.6(a). It is at *T*'' where we are observing the electronic logic states 1 or 0 at the output *F*. We can use in an O.C. gate circuit [Figure 8.1(a)] the external passive pull up by a resistance [Figures 8.1(b) and (c)] or the active pull up circuit [Figure 8.1(c)] as per needed driving capabilities. The use of an active device is preferred in place of pull up resistances like R_p in order to reduce power dissipation because a resistive element causes a Joule loss.

O.C. gates provides six features as follows:

1. Ability to work at voltage levels of V_p, which is greater than 5 V in the TTL gates. The voltages of operations can be as per need, for example, 12 V. Figure 8.1(b) shows an external pull resistance R_p. The value of R_p is taken much above the V_p/(Maximum I_{OL}) so that *T*' does not draw an excessive current. Maximum I_{OL} is maximum permissible i_{sink} current through *T*' (refer Table 7.3). V_p can be any where from + 5 V to a value which does not cause a Zener breakdown in *T*''. Generally, we limit V_p up to + 12 V and keep R_p above 200 ohm. Let us look at the application of this. If V_p = 5 V and R_p = 250 ohm, then a 20 mA can source or sink from such an arrangement to an O.C. gate. In teletype current loop based logic whereby 16 mA to

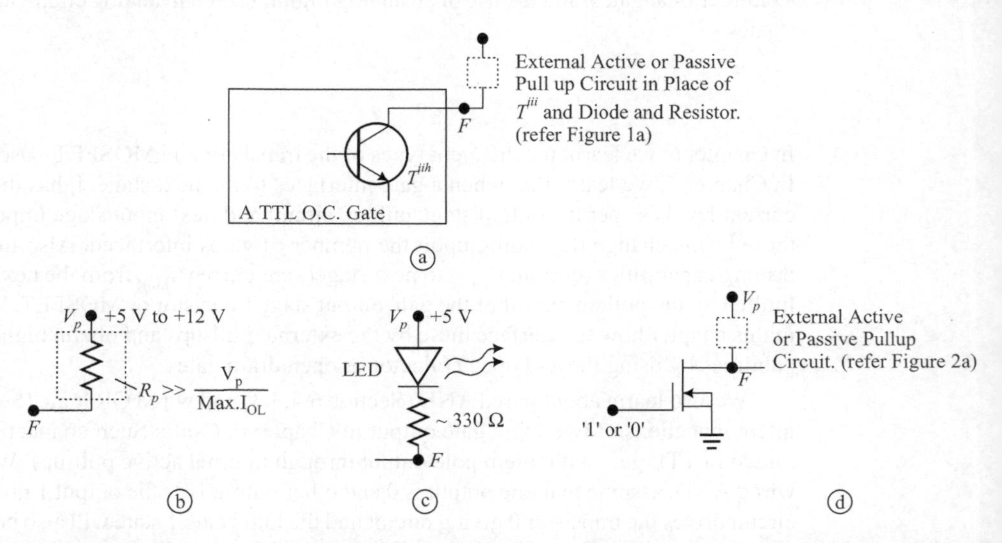

FIGURE 8.1 (a) TTL open collector gate with the external pull up (b) Resistive pull-up (c) Resistive plus LED pull up (d) Open drain gate with external pull up.

20 mA is treated as logic 1, if $V_p = 12$ V, we can then treat state 1 for $V \approx$ of 12 V instead and use the taletype coil in place of R_p.

2. An enhanced current sources from a TTL gate output .
3. Both i_{source} and i_{sink} are now controlled externally. Fan-out to the connected stages can now be as per requirement. We can draw about 10000 μA i.e. 10 mA through an LED whatever may be the TTL family. (We can use a higher power LED also by keeping $V_p = 12$ V and choosing R_L appropriately.) Now, LED glow indicates the logic states.
4. Increased noise margin is present.
5. A higher gate speed is obtained if we work with a reduced R_p compared to one internally used in 74 or 74 LS families.
6. We can interconnect output of the various gates through a common line by a special circuit called wired OR circuit.

However, nowadays, use of the O.C. gates is very much limited. The circuits as per case 1 and 2 above are discouraged nowadays in all designs of the electronics circuits. Further, the noise margins in the CMOS gates are already less and increased speed HCMOS gates are now available.

O.C. gates are used in the inverter gate IC 7406 and 7416, the BUFFER gate IC 7417, the AND O.C. gate IC 7415, the NAND gate IC 7412 and IC 7438 or in binary coded decimal bar displays, and a BCD digit to seven segment numeric display drivers 7445, and 7446 and 7447, respectively, and IC 74159—a 4 to 16-line O.C. decoder.

Nowadays there is large-scale use of the tristate gates and wired 'OR' TTL circuits are row rarely used.

Points to Remember

> If an output stage active pull-up circuit is removed in a TTL logic circuit, we get an open collector TTL gate, which provides (i) the state 1 an higher and/greater current sourcing capability, (ii) the ability to operate at greater voltages and (iii) at the state 0 a greater current sink capability from a TTL output stage.

8.2 OPEN DRAIN GATE

If we have an NMOS logic gate with no internal pull-up to supply by an active resistor like a depletion mode n-MOSFET (Figure 6.11(a)) we call that gate, an open drain (O.D.) gate. The open drain (O.D.) gate is used to source or sink heavier currents than otherwise possible in the NMOS gates (Refer Table 7.3). An O.D. gate circuit output pulls up externally through by a resistor or bipolar junction transistor (BJT) or p MOSFET. We can.

O.D. gate has the four features;

1. Ability to operate at the higher Voltage V_p or currents than with the normal NMOS gates, which are usually, operated at only the 5 V or 3.3 V from V_{supply} (difference of V_{DD} and V_{ss}), the voltages of operations can be as per need, for example, 12 V. An enhanced current sourcing from a TTL gate output.
2. Ability to have desired i_{source} controlled externally as per the interfacing gate type(s), numbers and families.
3. Increased noise margins.

4. Ability to operate at a higher gate speed if we work with a reduced pull-up active resistance or use a BJT for active pull-up compared to one internally used in NMOS gates families.

When there is no pull up at O.D. NMOS, only 0 can be written at an output gate. The previous stage to the output stage MOSFET provides gate voltage to drive it 0. At 0 also there is no i_{source} in steady state. With a pull-up 1 is also written at the O.D. gate.

When we want to use same O.D. gate as an input in bi-directional circuit, we have to first write 1, so that from the previous stage gate voltage level is below threshold and MOSFET drain-source is not conducting. After this if O.D. is used as the input, when input is at 1, it does not face a shorting path.

8.3 QUASI BIDIRECTIONAL GATE

An O.D. gate can be suitably designed as a quasi bi-directional gate. An internal circuit is designed such that momentarily just for about one clock cycle ($\approx 0.1\ \mu s$) it is able to turn-on a single LSTTL gate connected to it. Now, no external pull up is needed if only one LSTTL is to be turned on for a single clock cycle. The gate will work as an O.D. gate for a longer period needed for output = 1 or for a case with more than one LSTTL load, such an O.D. gate is now called quasi-bi-directional buffer.

8.4 TRISTATE GATE

8.4.1 Definition of Tristate

Tristate means a state of logic other than 1 and 0 in which there is a high impedance state and there is no i_{source} or i_{sink} at the output stage transistor (or MOSFET). A gate capable of being in 1, 0 and tristate is known as a tristate gate.

8.4.2 Use of an Analog/Digital Switch in Analog Circuits and in Making the Tristate Gates

A MOSFET analog/digital switch is an extremely useful element in analog-digital mix circuit design. A CMOS analog switch is made from a pair of n-channel and p-channel MOSFET. It provides two features as follows:

(i) It gives a bi-directional current flow and
(ii) Its input and output are symmetrical.

Figure 8.2 shows the CMOS circuit, which acts as an analog switch with above features. CMOS IC chip such as 4016, 4051, 4052, 4053, 4066 use the circuit of Figure 8.2 as one of the basic element for analog switching, analog multiplexing etc.

Since a circuit is also actually an analog circuit but is also a digital one where we consider the signals only in the terms of 1s and 0s. Where 1s and 0s correspond to two discrete regions of Voltages or currents or frequencies (section 1.1 of chapter 1). Therefore, the circuit of Figure 8.2 also works like a digital switch. An analog switch also works as a digital switch provided (i) voltage levels corresponding to the digital logic 1s and 0s are fed and are usable at X end called input/output or at Y end called output/input, and (ii) control voltage can be fed from a digital logic output at both 1 and 0 to set switch in its either ON (close) or OFF (open) (need not be respectively as vice versa is also possible).

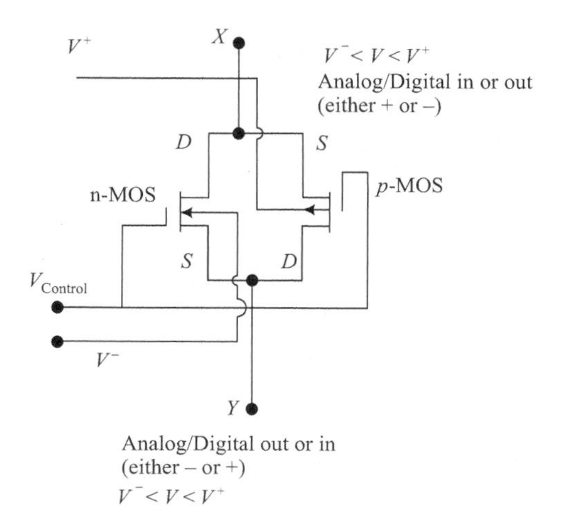

FIGURE 8.2 A CMOS circuit element to work as analog/digital switch.

1. When X is at either state 1 or 0 we shall get state 1 or 0 another end, Y. It when the switch is closed by applying a control voltage ($V_{control}$) of about V^+ (by making the control pin at CMOS logic state 1).
2. No logic output appears at other end if the switch is opened by V^- at the control gate (V^- means logic state 0).

When the switch is opened by applying control voltage of about V^- i.e. control input logic state is 0, we will get no current flow from one end X to another Y (refer Table 7.3) and currents for both the states are nil. Neither state 1 nor state 0 can be transmitted now from one end to another end.

The basic design of Figure 8.2 is now greatly improved, which protects the MOSFETs from the gate over voltage phenomenon (known as SCR *latch-up*).

8.4.3 Tristate Gate

Refer Section 8.4.2. It is possible to have a switching circuit which will neither permit state 1 or state 0 from one end to another unless appropriate control input (enable input) is activated (say, by applying V^+ (logic 1) at it). Enable input if not active then it will give an high impedance state which will neither let a current source nor a current sink (Table 7.3) between X and Y ends. This means enable input inactive then previous stage, whether at 1 or at 0, does not transmit onward.

Figures 8.3(a) and 8.3(b) show two tristate buffer logic gates in the upper portion. Tristate gates have either a control pin $\overline{E}$ (left side circuit) or $\overline{E}$ pin (right side circuit). A tristate gate output 0 or 1 state enables either when $E = 1$(left-side circuit) or when $\overline{E} = 0$ (right side circuit). Figure 8.3(a) lower portion left side gives a truth table in its inset for the tristate gates with $E = 1$ as the gate-output enabling-pin. Figure 8.3(b) lower portion right-side gives a truth table in its inset for the tristate gates with $\overline{E} = 0$ as the gate output enabling pin. A gate output Y can be in three states; state 1, state 0 or a state of high impedance (disconnected output). Third state is denoted by a * in the truth table.

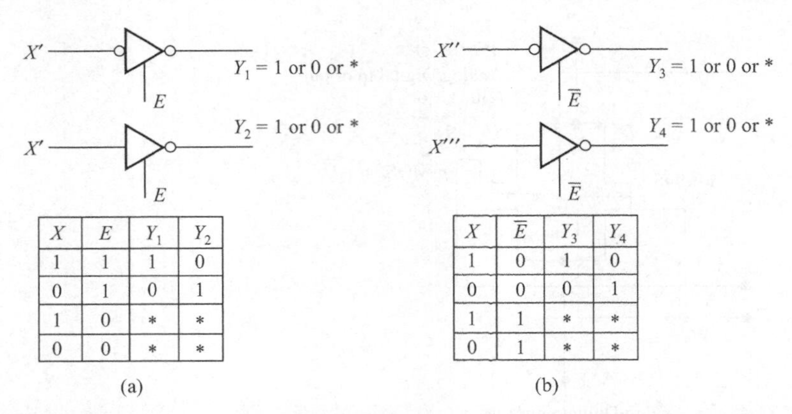

X	E	Y_1	Y_2
1	1	1	0
0	1	0	1
1	0	*	*
0	0	*	*

(a)

X	$\overline{E}$	Y_3	Y_4
1	0	1	0
0	0	0	1
1	1	*	*
0	1	*	*

(b)

FIGURE 8.3 (a) Tristate buffer and NOT gates, which enables 0 and 1 output only when E is at 1 (b) Tristate buffer and NOT gates, which enables 0 and 1 output only when $\overline{E}$ is at 0.

8.4.4 Tristate NOT Gates

Figures 8.3(a) and 8.3(b) also show two tristate NOT logic gates in the middle portion. Enable input (denoted by E) of the left side gate (NOT) is active 1. Enable input (denoted by $\overline{E}$) of the right side NOT gate is active 0. Figure 8.3(a) lower portion at left side give the truth table in its inset for the tristate NOT also with $E = 1$ as the gate output enabling pin. Figure 8.3(b) lower portion right-side gives the truth table in its inset for the tristate NOT also with $\overline{E} = 0$ as the gate output enabling pin. We can also use the notations C or $\overline{C}$, or G or $\overline{G}$ in place E or $\overline{E}$, respectively.

8.4.5 Using Digital Switch in Tristate TTL NAND and NOT Gate Circuits

Figure 8.4(a) shows a tristate circuit for $\overline{A} + \overline{B}$ made with the NAND gates and a digital switch, which enables output 0 and 1 output only when $\overline{E}$ is at 0. Figure 8.4(b) tristate NOT gate and a digital switch, which enables 0 and 1 output only when E is at 1. The switches for the TTL gates are made from the JFETs in place of MOSFETs due to need of fast witching.

Point to Remember

There can be the logic gates with three states 1, 0 and tristate at an output to enable multiple such gates give the output through a common line (wire).

8.4.6 Using a Multi-Emitter Transistor Junction for a Tristate Output for a TTL NOT or NAND

Figure 8.5 shows a tristate gate circuit made with a circuit like a TTL NAND gate in Figure 6.6(a) and two *p-n* junctions at a control input E. One of the input of multi-emitter transistor at the input stage is used a control input E. E input also connects the *n*-end of a *p-n* junction diode, the *p*-end of which connects the next stage transistor T collector. The switching of the TTL data input(s) at the gate is now by the use of two *p-n* junctions in place of MOSFETs based digital switches when there is a need for the fast swtching at the outputs. The output pull up circuit in the Figure 8.5 consists of a pair of transistors T''' and T^{iv}.

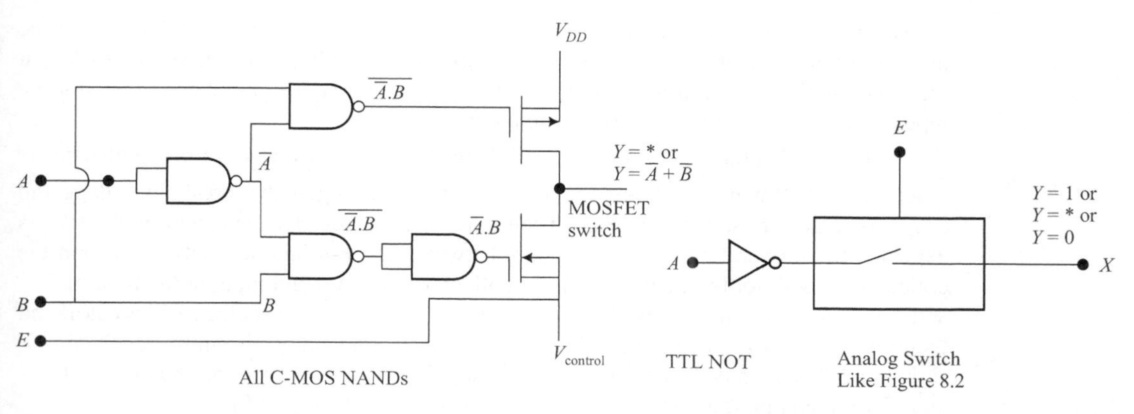

FIGURE 8.4 (a) A Tristate circuit made with the NAND gates and a digital switch, which enables output 0 and 1 output only when E is at 1 (b) Tristate NOT gate and a digital switch, which enables 0 and 1 output only when E is at 1.

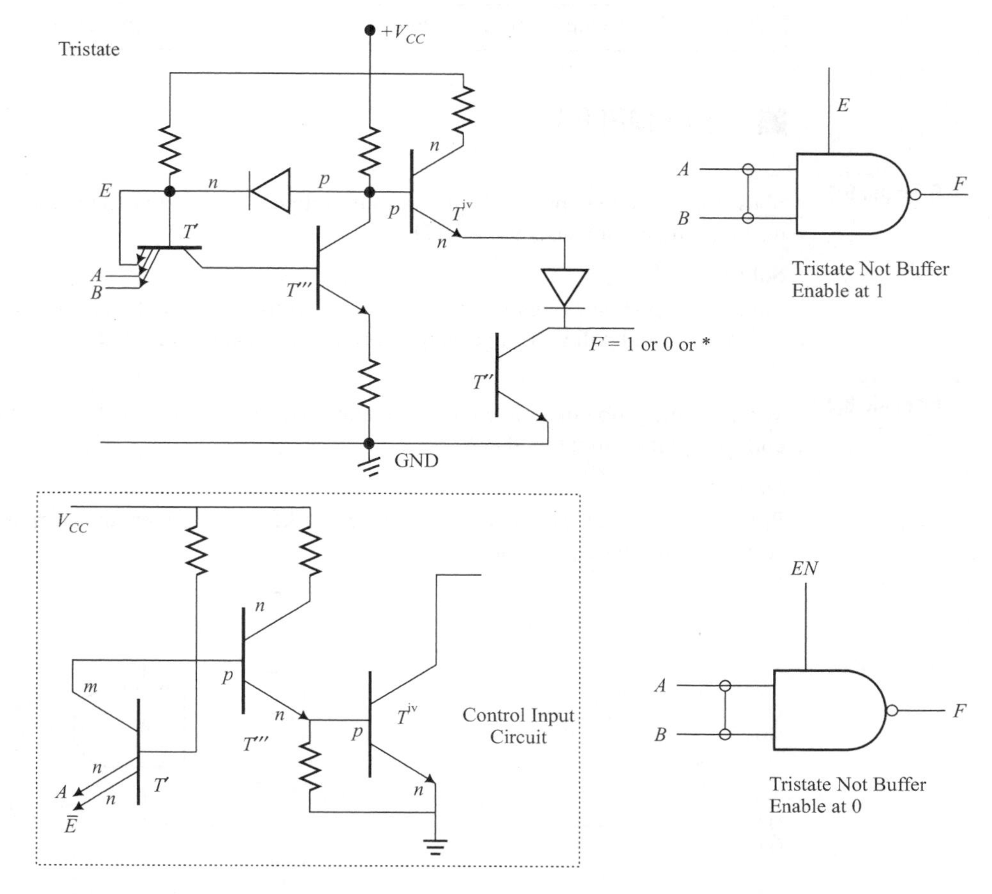

FIGURE 8.5 A Tristate input NOT/ NAND circuit made with a circuit like a TTL NAND gate in Figure 6.6(a) and two p-n junctions at a control input $\overline{E}$ (Rectangular inset shows control input circuit).

When we use two multi-emitter junctions of T, one for control input and others for the input A, the circuit works as tristate input NOT. (Refer inset within the dotted square). When we use more than two multi-emitter junctions of T, one for control input and others for the inputs A, B,.., the circuit works as tristate input NAND.

When $E = 1$, high the output at F will depend on the inputs at the other multi-emitter junctions of T. When $\overline{E} = 0$, low the emitter-base junction gets the threshold voltage and transistor T goes to saturation mode. The output at collector of T' is can go high as T' is expected to be operating in cut-off mode. However the p-n junction between E and the collector T' will not let the voltage at T' collector pulled up and input at the base of T''' active pull up for F remains low. Also T'' emitter will force T'' into cutoff. T''' is cutoff and therefore T''' and T^{iv} as well as T'' are cutoff. It does not let any current source from F or sink into T''. The impedance between F and supply becomes very high. The output at F is cutoff from the inputs at the other multi-emitter junctions of T ($F = *$).

Point to Remember

There are logic gates with three states 1, 0 and tristate at an input also.

EXAMPLES

Example 8.1

Show the use of the open-collector gates for interconnecting two gate outputs. Assume pull up external circuit resistance = 1 kΩ.

Solution

Figure 8.6(a) shows open collector gates the outputs of which are connected together as wired AND and pull up from a supply through a resistance of 1 kΩ.

Example 8.2

What are the problems related to the operational speed and next stage loading in the open collector gates wired together?

Solution

Refer Figure 8.6(b). The pull up resistance of 1 kΩ now delivers the current to all the open collector gates wired together.

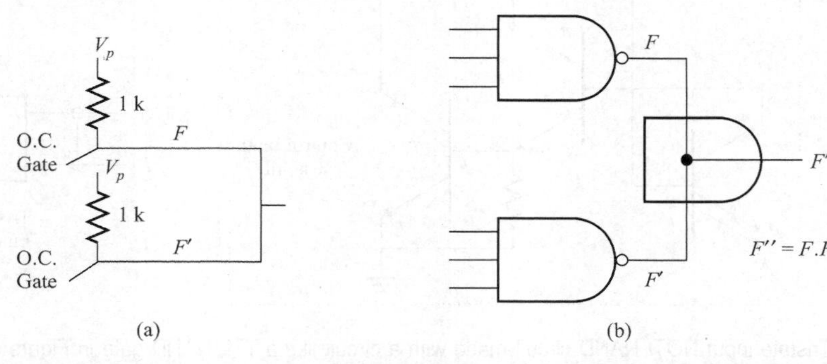

FIGURE 8.6 (a) The open collector gates the outputs of which are connected together as 'wired AND and external pull up from a supply through a resistance of 1 kΩ (b) Wired AND symbol and output.

Therefore the charging current through each gate reduces. This reduces the speed of the gates. The next stage load will be reduced due to reduced supply of current. The open-collector wired together therefore modifies the operational speeds and fan-out.

Example 8.3

Show the use of the tristate gates for interconnecting three totem pole gate outputs (A), (B) and (C) and give the enabling conditions.

Solution

Figure 8.7 shows three totem pole TTL gates the outputs of which are connected together by using a tristate buffer at the each of the output and making common the enable pins of each buffer. When $E = 1$, $E' = 1$ the (B) gate output is enhaled. When $E = 0$, $E' = 1$, (A) enables and when $E = 1$, $E' = 0$ the (C) output enables.

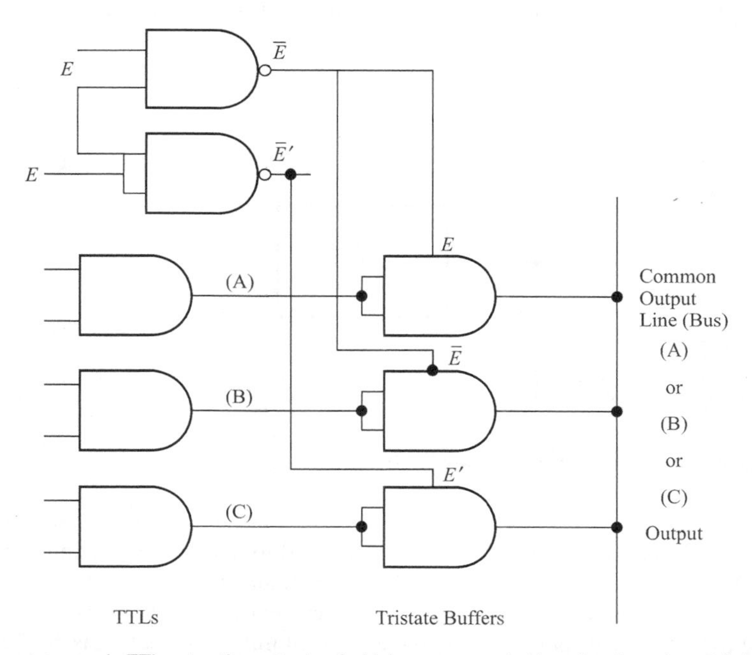

FIGURE 8.7 Three totem pole TTL gates the outputs of which are connected together by using a tristate buffer at the each of the output and making common the enable pins of each buffer.

Example 8.4

Show that the there are no problems related to the operational speed and next stage loading when tristate gates are used.

Solution

Refer Figure 8.7. Only one gate among three totem gates is enabled at a time, other gate outputs are in tristate (means high impedance state). Therefore, the circuit operational speed and next stage fan-out is same as that of one gate.

Example 8.5

A tristate CMOS two input NOR has a control input pin such that enable control $\overline{G} = 0$, the current is 2 nA in tristate. In state 1, the maximum current source to next stage can be 4 μA.

How many tristate CMOS NORs can be connected to a common wire (line) assuming that (i) at any instance all can be enabled and (ii) at any instance only one can be enabled by control input = 0?

Solution

The each gate circuit of this problem will be as shown in Figure 8.4(b) and interfacing will be as shown in Figure 8.8. It is given that only 2 nA can be delivered from an output pin of any of the tristate NOR which has $\overline{G} = 0$. Now, 4 μA is needed by the output stage MOSFET of this gate and rest of the current of (40 μA – 2 nA) from this gate can distribute the other input stages connected to a common line (called bus). Therefore, number of next stage inputs can be (4 μA /2 nA – 1) = 1999. Since the NOR has two inputs each, we can connect ≈ 1000 NORs on a common line. If we assume only one gate is enabled (having $\overline{G} = 0$ at an instance) then any number of gates can connect the common line.

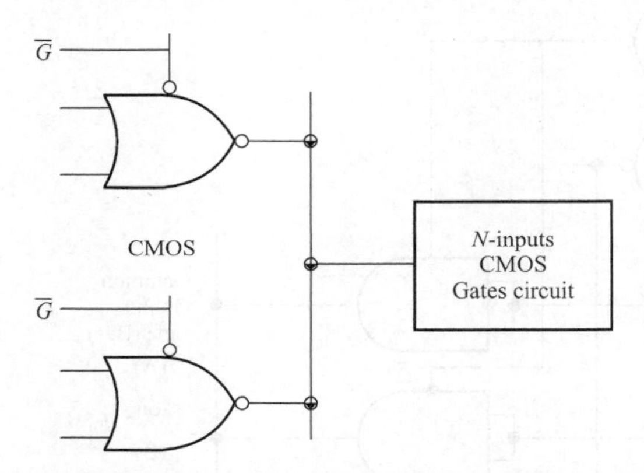

FIGURE 8.8 Each gate circuit interfacing in the problem in Example 8.5.

Example 8.6

A TTL NOT has a control input pin connected to one of the junction of input stage TTL NOT transistor and connected from the collector of next stage transistor through a *p-n* junction. When enable control $E = 0$, the current is 40 μA in tristate. In high state, the maximum current source to next stage can be 4 mA. How many tristate TTL NOTs can be connected to a common wire (line) assuming that at any instance only one can be enabled by control input = 0?

Solution

The each gate circuit of this problem will be as shown in Figure 8.5 and interfacing will be as shown in Figure 8.9. It is given that only 4 mA can be delivered from an output pin of any of the tristate NOT which has $E = 0$. Now, 40 μA is needed by the out stage transistor T'' of this gate and rest of the current of (4 mA – 40 μA) from this gate can distribute the other input stages connected to a common line (called bus). Therefore, number of next stage inputs can be (4 mA/40 μA – 1) = 99. When only one is enabled then there are unlimited connections to the line.

Example 8.7 Why are the properties that enable the circuit of Figure 8.2(a) to work as a better switch compared to *p-n* junction based or transistor-based switch? What will be disadvantages if a BJT is used to make an analog switch in place MOSFETs?

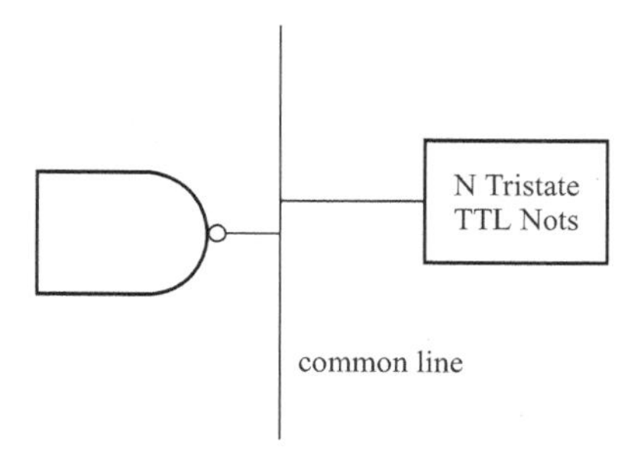

FIGURE 8.9 Each gate circuit interfacing in the problem in Example 8.6.

Solution

1. A CMOS analog switch consumes less power than a bipolar junction transistor based switch. It also shows no D.C. offset voltage. Inputs and outputs can be either up to V_{DD} or down to V_{SS}. When a channel, between the drain and source, is conducting, the resistance is very low (less than 100 Ω). When a channel, between the drain and source, is conducting, the resistance is very low (less than >10 MΩ). A *p-n* junction of BJT based switch has a d.c. offset voltage = threshold voltage.
2. A BJT based switch has disadvantage of lower output impedance compared to a MOSFET in cutoff stage. A BJT based switch has disadvantage of D.C. cutoff voltage offset and therefore unsymmetrical operation in high and low voltage states and therefore is unsuitable for analog switches. Only advantage is fast digital switching compared to CMOS switching.

EXERCISES

1. Calculate the current distribution when the open-collector gates interconnect three gate outputs as shown in Figure 8.10. Assume pull up external circuit resistance = 0.5 kΩ.
2. Calculate the factor by which the operational speeds and next stage loading in the open collector gates wired together modify in circuit of Figure 8.10?
3. List the advantages of the tristate gates when interconnecting six totem-pole NOT-gate outputs as shown Figure 6.11.
4. Show that the operational speed and next stage loading is unaltered in circuit of Figure 8.12.
5. A tristate CMOS two input NOT has a control input pin such that enable control $\overline{G} = 0$, the current is i nA in tristate. In high state, the maximum current source to

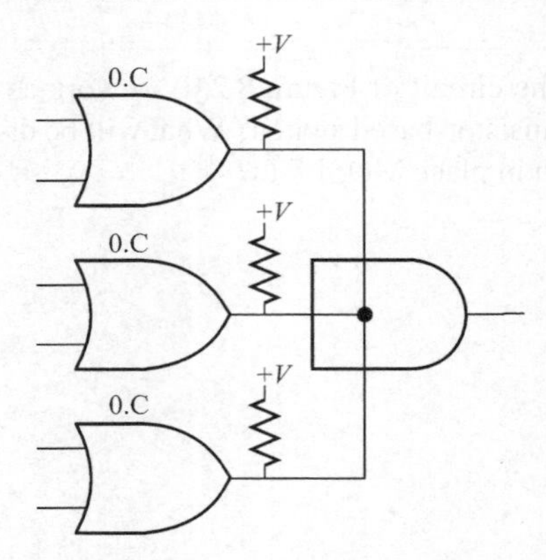

FIGURE 8.10 Three open-collector gates interconnecting three gate outputs.

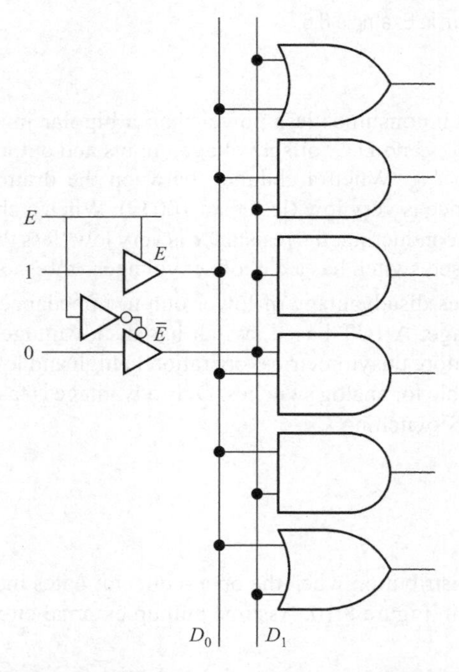

FIGURE 8.11 Tristate gates when interconnecting six totem pole NOT gate outputs in Exercise 8.3.

next stage can be 3 μA. If 1000 tristate CMOS NOTs can be connected to a common wire (line) assuming that at any instance only one can be enabled by control input = 0, the what will be value of *i*.

6. A TTL NOT has a control input pin connected to one of the junction of input stage TTL NOT transistor and connected from the collector of next stage transistor through a *p*-*n* junction. When enable control $E = 0$, the current is 40 μA in tristate. When output = 1, the maximum current source to next stage can be i mA. If 50

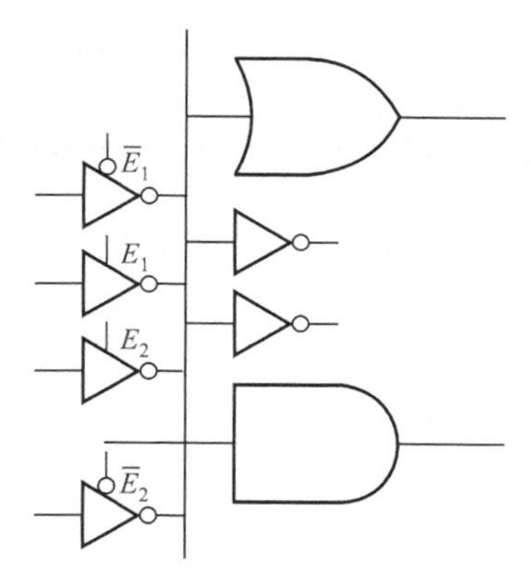

FIGURE 8.12 A gate circuit interfacing circuit in Exercise 8.4.

tristate TTL NOTs can be connected to a common wire (line) assuming that at any instance how many can be enabled by control input then what is the value of i?

QUESTIONS

1. What is an open collector gate?
2. What does O.C. improve upon conventional TTL gate?
3. What is the most inconvenient thing in using the O.C. gates?
4. What is an open drain gate? What does it improve upon?
5. What is the most inconvenient thing in using the open drain O.D. gates? (Hint: external circuit interfacing).
6. Why a bi-directional O.D. gate is first written '1' before using it as input?
7. What is the application of quasi-bi-directional O.D. gate circuit? (Hint: Refer text).
8. Where are the analog switches used?
9. How is an analog switch made from a CMOS pair?
10. C-MOS switches without special protection circuits and latch proofing can be destroyed either due to noise or due to electrostatic charges or when the gate voltages exceed V_{DD}. Why?
11. Two circuits are close by to each other. What do we mean by cross talk between two channels? (Hint: Noise in one effecting the working of a nearby channel due to capacitive effects).
12. Why is an analog switch workable, as a digital switch also but vice versa is not true?
13. What is a tristate BUFFER?
14. Why are the tristate gates extremely beneficial in designing large-scale integrated circuits?
15. What is a tristate? How does this third state differ from a logic state 1 or 0?
16. Why was the concept of tristate introduced?

17. What is difference between a tristate- BUFFER and a BUFFER?
18. What are difference between a tristate NOT and a NOT?
19. How does an output of a tristate BUFFER differ from an open collector or open-drain gate output?
20. How does a tristate BUFFER differ from an analog switch?

CHAPTER 9

Problem Formulation and Design of the Combinational Circuits

OBJECTIVE

We studied Boolean algebra and theorems in Chapter 4. We learnt how to convert a Boolean expression into SOP or POS format. In Chapter 5, we learnt how to build a Karnaugh map and then simplify and minimize it. In Chapter 6, we studied the different types of gates and their families and their characteristics, propagation delay, fan-out, operating voltage levels and noise margins. In Chapter 7, we learnt how to interconnect and interface various types of gate. In Chapter 8, we learnt how to use the open collector, open drain and tristate gates to connect the logic gate outputs to next input stages in the complex logic circuits.

How do we formulate a problem of designing a combinational circuit? How do we then design the combinational circuit? We shall learn the answers of these questions in this chapter.

9.1 COMBINATIONAL CIRCUIT

Figure 9.1(a) shows a combinational circuit representation by a block diagram for n inputs and m outputs, a truth table of 2^n rows, for each output an SOP or POS of 2^n terms and Karnaugh map of n variables and 2^n cells in all layers of 4×4 table when $n > 4$.

Figure 9.1(b) shows the seven primitive building blocks for a combinational circuit; NANDs, NORs, AND-OR, AND-OR-NOT, OR-AND, XOR and XNOR. [Note: Important

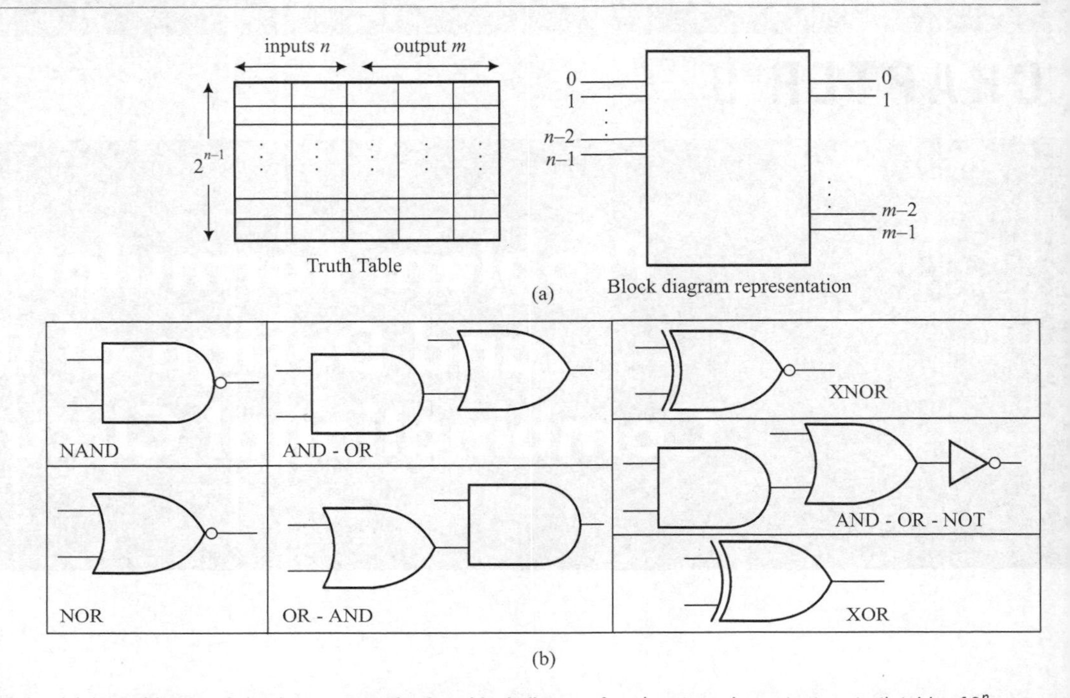

FIGURE 9.1 (a) A combinational circuit representation by a block diagram for n inputs and m outputs, a truth table of 2^n rows, for each output an SOP or POS of 2^n terms and Karnaugh map of n variables and 2^n cells in all layers of 4×4 table when $n > 4$ (b) Seven building blocks; NANDs, NORs, AND-OR, AND-OR-NOT, OR-AND, XOR and XNOR.

new generation building blocks like EPLDs (Electrically Programmable Logic Devices) and FPGAs, (Field Programmable Gate Arrays), we shall be learning later Chapter 21].

Remember the Definition of a Combinational Circuit

A combinational circuit is a circuit made up by combining the logic gates such that the required logic at the output(s) depends only on the input logic conditions, both completely specified by either a truth table or by a Boolean expression or by an SDP or POS. Also (i) An output(s) remains constant, as long as input conditions do not require the change in the output(s), (ii) An output depends solely on the current input condition and not on any past input condition or past output condition, (iii) A combinational circuit has no feedback of the output from a stage to the input of either that stage or any previous stage, and (iv) An output(s) at each stage appears after a delay in few tens or hundred ns depending upon the type or family of the gate used to implement the circuit.

Table 9.1 defines in a truth table the input conditions and the corresponding outputs X and X' in the XOR and XOR-NOT operation, respectively. Figure 9.2(a) gives the corresponding combinational circuit implemented by using four and five NANDs, respectively. Figure 9.2(b) gives the corresponding combinational circuit implemented by using AND-OR gates after defining an SOP Boolean expression. Figure 9.2(c) gives the corresponding combinational circuit implemented by a MSI (medium scale integrated) chip for two XORs. Note: An XOR with one input always = 1 works as a NOT circuit for the other input (refer

TABLE 9.1 Truth Table for the combinational logic circuits with two inputs A and B and two outputs X and X'

Inputs		Outputs	
A	B	X	X'
0	0	0	1
0	1	1	0
1	0	1	0
1	1	0	1

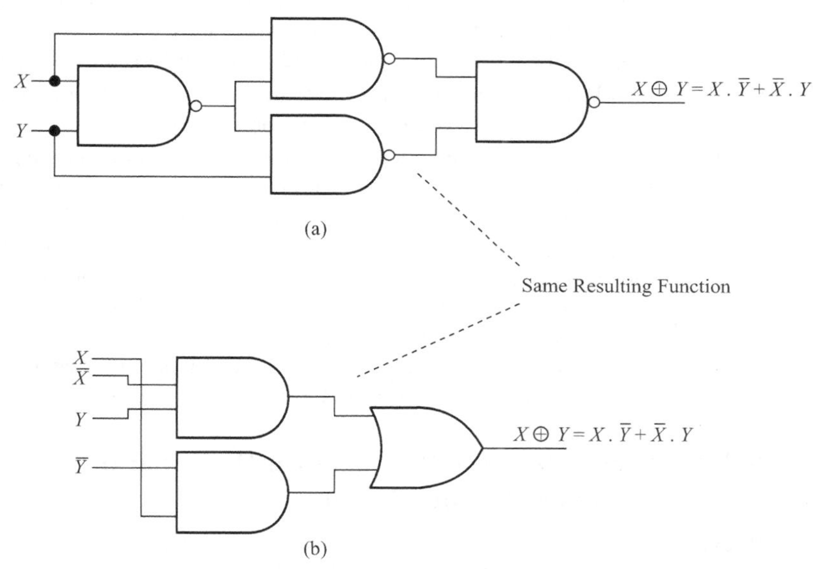

FIGURE 9.2 (a) A combinational circuit implementation by using four NANDs for outputs X and X' for the conditions completely specified by Table 9.1 (b) Corresponding combinational circuit implementation by using AND-OR gates after defining the SOP Boolean expression.

Table 9.1). An XOR with one input always = 0 works as a BUFFER circuit for the other input (refer Table 9.1).

9.2 PROBLEM FORMULATION IN A COMBINATIONAL CIRCUIT

First step is to select the combinational circuit(s) in a logic network for which the problem of designing as per specifications is to be solved. Criteria for whether a problem or its part is solvable by a combination circuit or not, is as follows: (i) check whether the required logic at the output(s) depends only on the input logic conditions, both completely specified by either a truth table or by a Boolean expression, (ii) check whether an output(s) remains same, as long present input condition does not require the change in the output(s), and (iii) check whether an output depends solely on the current input condition and not on any past input condition or past output condition. All three conditions must satisfy for the logic circuit, which can solve the problem by a combinational circuit of the gates.

Now the problem is formulated as follows:

9.2.1 Specification of Each Output as a Function of the Input Conditions

1. Specify the number of inputs, n. The n is also the number of literals in a Boolean expression for an output.
2. Specify the number of outputs, m.
3. Specify the delays permitted at the outputs.
4. Specify the fan-ins permitted at the inputs.
5. Specify fan-outs permitted from the targets gates and building blocks.
6. Design a truth table for n inputs and m outputs. Each output corresponds to each possible combination of input conditions. Make a truth table consisting of 2^n rows, n columns on the left for the n inputs and m columns for the m outputs.
7. Write a Boolean expression for the logic circuit for each output: The n is also the number of literals in a Boolean expression for an output.
8. Specify as SOP or POS standard format: Let us assume that the inputs as well as their complements. It is possible to represent by each Boolean expression of output for the circuits at the two levels. The representation is either for circuit by (i) AND at first level and OR at the second level or (ii) by OR at first level and AND at the second level. The representation is done by writing an SOP or a POS format Boolean operation taking all implicants for which outputs is 1 or for which outputs is 0, respectively. Implicant means a product term (minterm) of a SOP function if product term implies the function (refer section 4.7 of chapter 4 for details of writing a SOP Boolean expression from the truth table).

9.2.2 Accounting for Propagation Delays and Fan-ins at Various Levels

Propagation delays and fan-ins permitted are specified in the problem.

9.2.3 Accounting for Don't Care Conditions

If a possible combination of the input condition is unspecified or is don't care, specify it by 'x'. The Boolean expression for the output is an incomplete Boolean function.

9.2.4 Accounting for Tristate Conditions

If a possible condition is high impedance output tristate, specify it by '*'.

9.3 DESIGN OF A COMBINATIONAL CIRCUIT

A design approach called waterfall model (and also liner incremental model). This model defines the following steps for a design implementation.

(a) Problem formulation (Requirement Specifications)

(b) Design

(c) Implementation of design

(d) Verification test of internal logic and external functions by analysis

Design with a specified target: Use a specific building block for making a combination circuit is preferred due to ease of design consideration. A building bock is selected. For example, among the gates NAND, NOR, AND-OR, AND-OR-NOT, OR-AND, XOR or XNOR and the blocks made from these.

Design with cost minimization: Minimize the number of blocks of AND, OR and AND-OR-NOT gates in the Boolean expression for an output: Using an approach as follows minimizes the terms; (i) Karnaugh map building from the SOP or POS and (ii) minimizing the number of sums or products terms by finding pairs, quads and octets. Finding prime implicants minimizes the number of gates needed. Another approach is computer design tool based minimization (prime implicant means an implement not containing any other implicant with fewer literals than it or else that will not be the part of the Boolean function). It reduces the cost of the circuit measured in terms of number of gates needed to implement the circuit. Prime implicants give an irredundant-disjunctive Boolean function. Irredundant means no term is redundant (means all are necessary) because each term is prime implicant. Disjunctive means no term is removable; else it will modify the function.

Design with delay minimization: Minimize the number of stages (levels) in the Boolean expression for each output. The number of stages (or levels) through which the input(s) passes through up to the final output needs to be minimized by using SOP or POS standard format. Alternatively use a computer-based minimization technique. When we minimize this way the number of levels reduce to 2, it reduces the propagation delay up to two units for a unit delay at each level.

Design with function decomposition using Boolean algebraic simplification: Function decomposition into sub-functions simplifies the circuit. [Note: Factor by using Boolean algebra, for example, use distributive law; $(X.Y + Y.Z) = Y.(X + Z)$. Determine the sub-functions that are sharing the circuits. A sub-function means a partial sum in a Boolean function. The partial sum is for a set of the product-terms [implicants (minterms) or prime implicants]. [Note: An advanced method is to determine the look up tables (LUTs), which are sharing the circuit. It is useful when using FPGA. FPGAs are explained later in chapter 21. FPGAs have the logic cells arranged as the arrays and each cell has the LUTs instead of AND-OR blocks at the programmable logic devices. Each LUT has the inputs and corresponding output like a look-up table, which has a key on the left side and value or answer on the right side.]

Design with Boolean algebraic simplification in terms of target building gate or building: A combination circuit Boolean expression for each output can be simplified by the use of the Boolean rule, laws and theorems minimizes the numbers of gates. For example, use of AND-OR blocks and NOT can be simplified to use of only NANDs or NORs by applying DeMorgan theorems.

(i) Two ANDs and one OR can be made from three NANDs. Figure 9.3 shows how an exemplary circuit of AND-ORs is made from the NANDs. NANDs are easier to implement at the resistor-transistor levels (refer Figure 6.6(a)).

(ii) Two ORs and one AND can be made from three NORs. NORs are easier to implement lesser delay CMOSs and NMOSs. Why? Use of parallel MOSFETs in the NORs increases the fan-in (number of inputs to a gate) for same propagation delay.

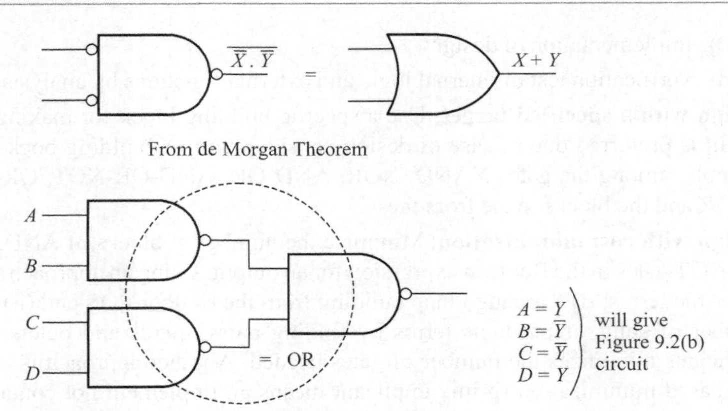

FIGURE 9.3 (a) A combinational circuit implemented by AND-OR circuit from prime implicants (in SOP format) converted to implementation by NANDs alone.

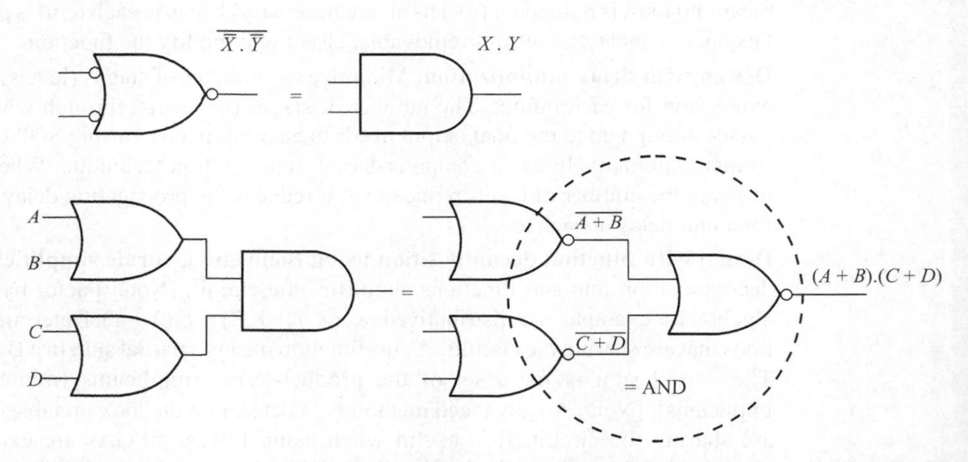

FIGURE 9.4 A combinational circuit implemented by OR-AND circuit from the POS format simplification converted to implementation by NORs alone.

[Refer Figures 6.13(a) and 6.11(a), respectively]. Figure 9.4 shows how an exemplary circuit of OR-ANDs is made from the NORs.

Design with sharing of the gates *in the multi output cases:* For example, two XOR and XOR-NOT outputs can be minimized by first implementing the XOR to get first output and then use NOT to get a second output. The numbers of functions are assumed to be a part of a bigger circuit by sharing the gates within them. Determine the prime implicants, which are sharing the Boolean functions and the circuits for the shared ones form the parts of the bigger circuit. These prime implicants are called tagged prime implicants.

Implementation: Implement the design with the target gates and block or cells of the logic gates.

Point to Remember

> A plus sign in a Boolean expression means use of an OR gate and a implicant or prime implicant product term having dot operations alone means use of the AND gates with the literals or the complements of the literals as the inputs.

Figure 9.5 shows the implementation of post-simplification Boolean expression for $S = \Sigma m(3, 5, 7, 9, 13)]$. The simplification is by finding prime implicants.

$$S = A.\overline{C}.D + \overline{A}.B.D + \overline{A}.C.D$$

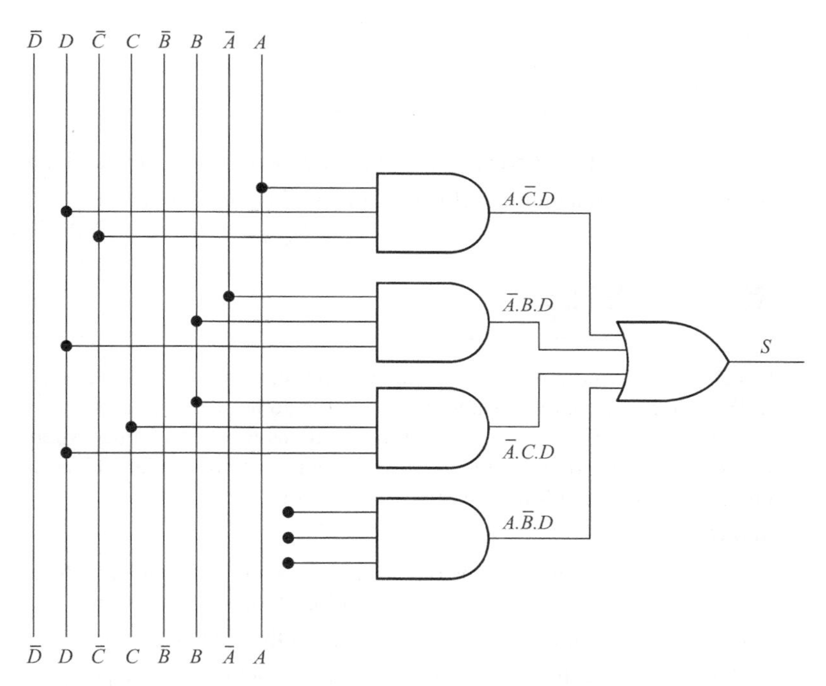

FIGURE 9.5 Implementation of post-simplification by obtaining prime implicants Boolean expression for $S = \Sigma m$ (3, 5, 7, 9, 13)] from a truth-table

Analyse and verify the design using fixed inputs: Any design is incomplete unless it is analysed and its performance verified. Analysis using a probe is done by giving the fixed sets of inputs and verification is done whether all the inputs leads to a reliable output after the specified delays. Table 9.2 shows the verification method. In case F and F' in columns 4 and 5 are identically equal to columns 6 and 7, respectively, except in don't care unspecified conditions of the inputs, the circuit stands verified.

Analyse and verify the design using pulsed waveforms as the inputs: Analysis is done by giving the pulsed waveforms as the sets of inputs and verification is done whether all the inputs leads to a reliable output as a function of time. A graph showing the multiple inputs and multiple outputs as function time is called a timing diagram. Figure 9.6(a) shows a

TABLE 9.2

Inputs			Design outputs		Observed outputs by a probe		Verification
A	*B*	*C*	F	F'	F	F'	
0	0	0	?	?	?	?	
0	0	1	?	?	?	?	
0	1	0	?	?	?	?	
0	1	1	?	?	?	?	
1	0	0	?	?	?	?	
1	0	1	?	?	?	?	
1	1	0	?	?	?	?	
1	1	1	?	?	?	?	

timing diagram of a NAND circuit with a NOT at one of its input. Figure 9.6(b) shows the timing diagram of a multi level AND-OR based circuit $X+\overline{X}.\overline{Y}$.

EXAMPLES

Example 9.1

Check which logic circuit is a combinational circuit. (A) Two outputs are as per the ten truth-table input-conditions and these outputs do not care for the remaining six conditions and thus unspecified. (B) An output that changes from 1 to 0 only when the input changes first from 1 to 0 and then 0 to 1. (C) An output 1 which is independent of any present combination of input conditions, which are possible.

Solution

(A) Both the outputs depend on the current inputs alone and therefore the circuit is combinational circuit of partially specified Boolean expression.

(B) It is not a combinational circuit, as its output depends not on the current input alone.

(C) It is not a combinational circuit, as its output does not depend on any of the present input conditions.

Example 9.2

Check which logic circuit is (i) an output for a given input condition is 0 if previous output was 1 and is 1 if previous output was 0 (ii) two outputs that show the outputs as per Boolean expressions for them (iii) an output for a given current input condition is 0 if previous inputs *A* and *B* were as 1 and 0 and is 1 if previous inputs were 0 and 1, respectively. (iv) two outputs that show the outputs as per truth table input conditions after a time delay at the implementing gates.

Solution

(A) It is not a combinational circuit, as its output does depend on the previous output condition.

(B) It is a combinational circuit, as its outputs depend on the specified Boolean expressions for them.

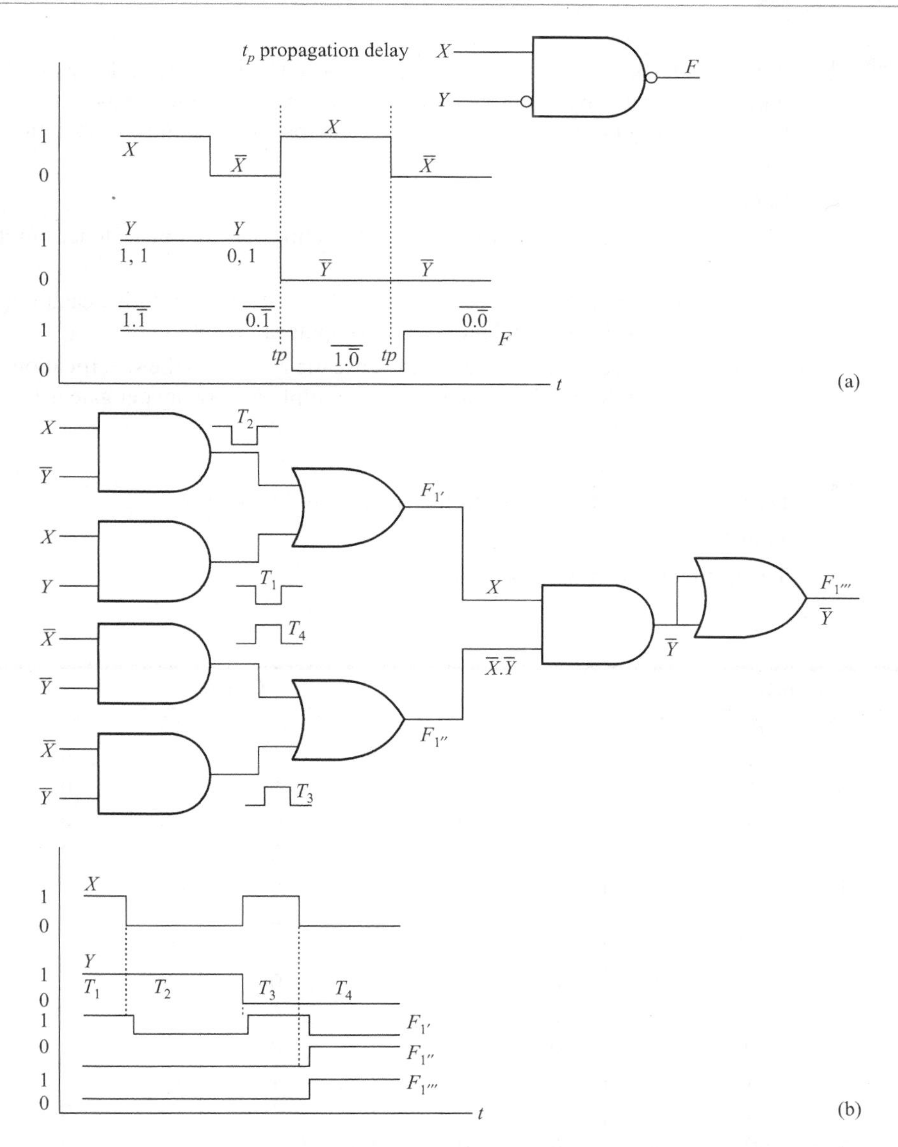

FIGURE 9.6 (a) Timing diagram of a NAND circuit with a NOT at one of its input (b) Timing diagram of a multi level AND-OR based circuit.

(C) It is not a combinational circuit, as its output is depends not on the current input alone, it also depends on the previous inputs.

(D) It is a combinational circuit as output is as per the truth table. The delays are natural at any given gate.

Example 9.3 A four bit binary number between 0 and 15 is converted after a delay of 40 ns into a two-digit BCD number by a combinational circuit. (i) Specify the number of inputs. (ii) Specify the number of outputs. (iii) Specify the delay permitted at the combinational circuit each gate.

Solution

1. A binary number between 0 and 15 decimal has four bits. Hence number of inputs = 4.
2. A BCD number between 0 and 15 decimal has two digits, each digit specified by four bits. Hence total numbers of outputs are eight.
3. Delay given is 40ns. We have the use the gates with the specification that number of levels at the combinational circuit multiplied by delay per gate is less or equal to 40 ns per gate.

Example 9.4 Design the truth table for the problem solved in Example 9.3.

Solution

The truth table will be as per Table 9.3.

TABLE 9.3

Inputs				Outputs							
A3	*A2*	*A1*	*A0*	*Y7*	*Y6*	*Y5*	*Y4*	*Y3*	*Y2*	*Y1*	*Y0*
0	0	0	0	0	0	0	0	0	0	0	0
0	0	0	1	0	0	0	0	0	0	0	1
0	0	1	0	0	0	0	0	0	0	1	0
0	0	1	1	0	0	0	0	0	0	1	1
0	1	0	0	0	0	0	0	0	1	0	0
0	1	0	1	0	0	0	0	0	1	0	1
0	1	1	0	0	0	0	0	0	1	1	0
0	1	1	1	0	0	0	0	0	1	1	1
1	0	0	0	0	0	0	0	1	0	0	0
1	0	0	1	0	0	0	0	1	0	0	1
1	0	1	0	0	0	0	1	0	0	0	0
1	0	1	1	0	0	0	1	0	0	0	1
1	1	0	0	0	0	0	1	0	0	1	0
1	1	0	1	0	0	0	1	0	0	1	1
1	1	1	0	0	0	0	1	0	1	0	0
1	1	1	1	0	0	0	1	0	1	0	1

Example 9.5 Design the Boolean functions for the truth table at Table 9.3 for the excersie in Example 9.3.

Solution

BCD number shows upper digit at place $p = 1$ only when number is more than 9, else upper digit at place $p = 1$ is 0. We can have two parts in the circuit. One part (A) when upper bits = 0s and next part (B) when output bit at place 4 (from right) = 1. Part (B) circuit drives from one of input taken from the output of previous part (A) stage.

(A) We note the followings from truth table at the Table 9.3. $Y7 = 0$, $Y6 = 0$, $Y5 = 0$

Truth table for $Y4$ is shown in part A of Table 9.4.

SOP format for $Y4$ is as follows:

$Y4 = A3.\overline{A}2.A1.\overline{A}0 + A3.\overline{A}2.A1.A0 + A3.A2.\overline{A}1.\overline{A}0 + A3.A2.\overline{A}1.A0 + A3.A2.A1.\overline{A}0 + A3.A2.A1.A0 = \Sigma m\ (10, 11, 12, 13, 14, 15) = A3.A1 + A3.A2$ after Karnaugh map simplification by forming one quad for $m(10, 11, 14, 15)$ and one quad for $m(9, 11, 13, 15)$ (Refer Table 5.9).

We can verify the result by the following: $Y4 = 1$ if $A3.(A2 + A1) = 1$.[when $A3 = 1$ and either $A2$ or $A1$ or both = 1 and $A0 = x$ (don't care) else $Y4 = 0$ when both $A2$ and $A1$ are 0].

Boolean expression for $Y4 = A3.(A1 + A2) = A3.A1 + A3.A2$

(B) We note the followings from truth table (Table 9.4). $Y3$, $Y2$, $Y1$ and $Y0$ differs from $A3$, $A2$, $A1$ and $A0$, respectively, only when $Y4 = 1$. Truth table for $Y3$, $Y2$, $Y1$ and $Y0$ is shown in part B when input $A7 = Y7 = 0$, $A6 = Y6 = 0$, $A5 = Y5 = 0$ and $A4 = Y4$, and the inputs $A3$, $A2$, $A1$ and $A0$ are same as input in part A.

TABLE 9.4

Part A Inputs				Output	Part B Five inputs		Four outputs			
A3	A2	A1	A0	Y4	A4	A0 TO A3	Y3′	Y2′	Y1′	Y0′
0	0	0	0	0	#	$	*0*	0	0	0
0	0	0	1	0	#	$	*0*	0	0	1
0	0	1	0	0	#	$	*0*	0	1	0
0	0	1	1	0	#	$	*0*	0	1	1
0	1	0	0	0	#	$	*0*	1	0	0
0	1	0	1	0	#	$	*0*	1	0	1
0	1	1	0	0	#	$	*0*	1	1	0
0	1	1	1	0	#	$	*0*	1	1	1
1	0	0	0	0	#	$	*1*	0	0	0
1	0	0	1	0	#	$	*1*	0	0	1
1	0	1	0	1	#	$	*0*	0	0	0
1	0	1	1	1	#	$	0	0	0	1
1	1	0	0	1	#	$	0	0	1	0
1	1	0	1	1	#	$	0	0	1	1
1	1	1	0	1	#	$	0	1	0	0
1	1	1	1	1	#	$	*0*	1	0	1

Input to part (B) of the circuit is same as in column 5, which is $A_3.(A_1 + A_2)$. $ four inputs to (B) are same as inputs in columns 4 down to 1 for (A).

SOP format of five variable Karnaugh map will have following result: for $Y3'$, $Y2'$, $Y1'$ and $Y0'$ will be as follows.

$Y3' = \Sigma m(8, 9) = \overline{A}4.A3.\overline{A}2.\overline{A}1.\overline{A}0 + \overline{A}4.A3.\overline{A}2.\overline{A}1.A0 = \overline{A}4.A3.\overline{A}2.\overline{A}1$

$Y2' = \Sigma m(4, 5, 6, 7, 30, 31) = \overline{A}4.\overline{A}3.A2. + A4.A3.A2.A1$

$Y1' = \Sigma m(2, 3, 6, 7, 28, 29) = \overline{A}4.\overline{A}3.A1 + A4.A3.A2.A1$

$Y0' = A0$. [Note: $Y0$ depends only on $A0$. It is always independent of $A4$, $A3$, $A2$ and $A1$. By simplifying $Y0 = \Sigma m(1, 3, 5, 7 \ldots 29, 31)$ also we will get same result].

Final result for four lower BCD output is $Y3 = Y3'.\overline{A}7.\overline{A}6.\overline{A}5$, $Y2 = Y2'.\overline{A}7.\overline{A}6.\overline{A}5$, $Y1 = Y1'.\overline{A}7.\overline{A}6.\overline{A}5$ and $Y0 = Y0'.\overline{A}7.\overline{A}6.\overline{A}5$

Example 9.6 Implement the circuit for the designed Boolean expression with prime implicants for the problem of a 4 bit binary to 2-digit BCD converter in use AND-OR gates. Assume inputs for each literal $A3$, $A2$, $A1$, and $A0$ as inputs and the NOTs for their complements and for $Y4$ complement are also available.

Solution

Figure 9.7 shows the combinational circuit for 4-bit binary to 2 digits BCD converter.

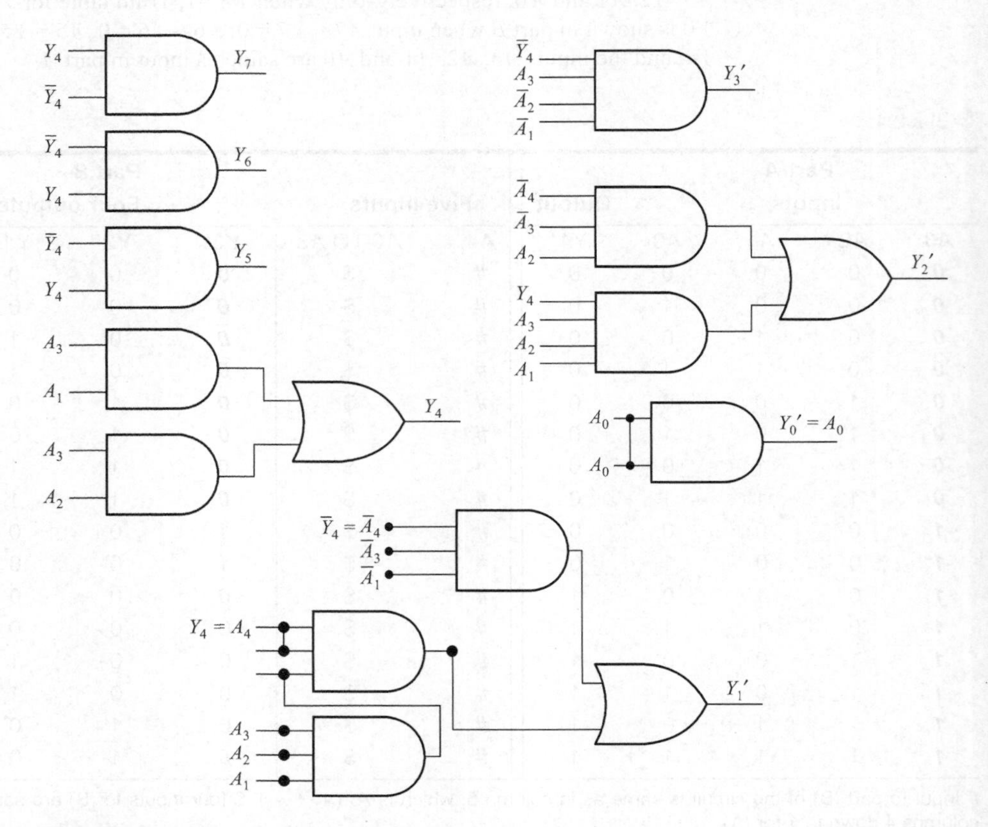

FIGURE 9.7 Combinational circuit for 4 bit binary to 2 digits BCD converter.

(i) $Y3$, $Y2$, $Y1$, and $Y0$ implement by the equations in the Example 9.5 after taking $A4 = Y4$.

(ii) $Y4$ implements by the implementing $A3.A1 + A3.A2$.

(ii) $Y0$, $Y7$, $Y6$ and $Y5$ are obtained from four circuits as follows:

$Y7 = Y4.\overline{Y}4$,

$Y6 = Y4.\overline{Y}4$ and

$Y5 = Y4.\overline{Y}4$

$Y0 = A0.A0$

Example 9.7 Calculate the number of four literal AND-OR gates used for the circuits in Figure 9.7.

Solution

Table 9.5 gives the number of AND-OR gates used. Neglecting ANDing with $\overline{A}7$. $\overline{A}6$. $\overline{A}5$

TABLE 9.5 Number of AND-OR gates used for the BCD converter in Example 9.3

Outputs	Gates used	
	ANDs	ORs
Y7	1	0
Y6	1	0
Y5	1	0
Y4	2	1
Y3	1	0
Y2	2	1
Y1	3	1
Y0	1	0
Total	12	3

Example 9.8 Calculate the number of three literal inputs AND and three input OR gates used. AND gates have complementary inputs also (total 6). Show the combinational circuit also. Use three-input ANDs.

Solution

Table 9.6 gives the number of three literals input AND-OR gates used with each AND having three inputs or their complements and each OR having three input. $Y3$ has four literals in the AND. Therefore, we need two ANDs with three literal inputs each. One of two prime implicants in $Y2$ has four literals in the AND. Therefore we need three ANDs with three literal inputs each for $Y2$. All of three prime implicants in $Y1$ has three literals in the AND. Therefore, we need two stages of ANDs with three literal inputs each for $Y1$. Figure 9.8 shows the combinational circuit implemented by AND-ORs.

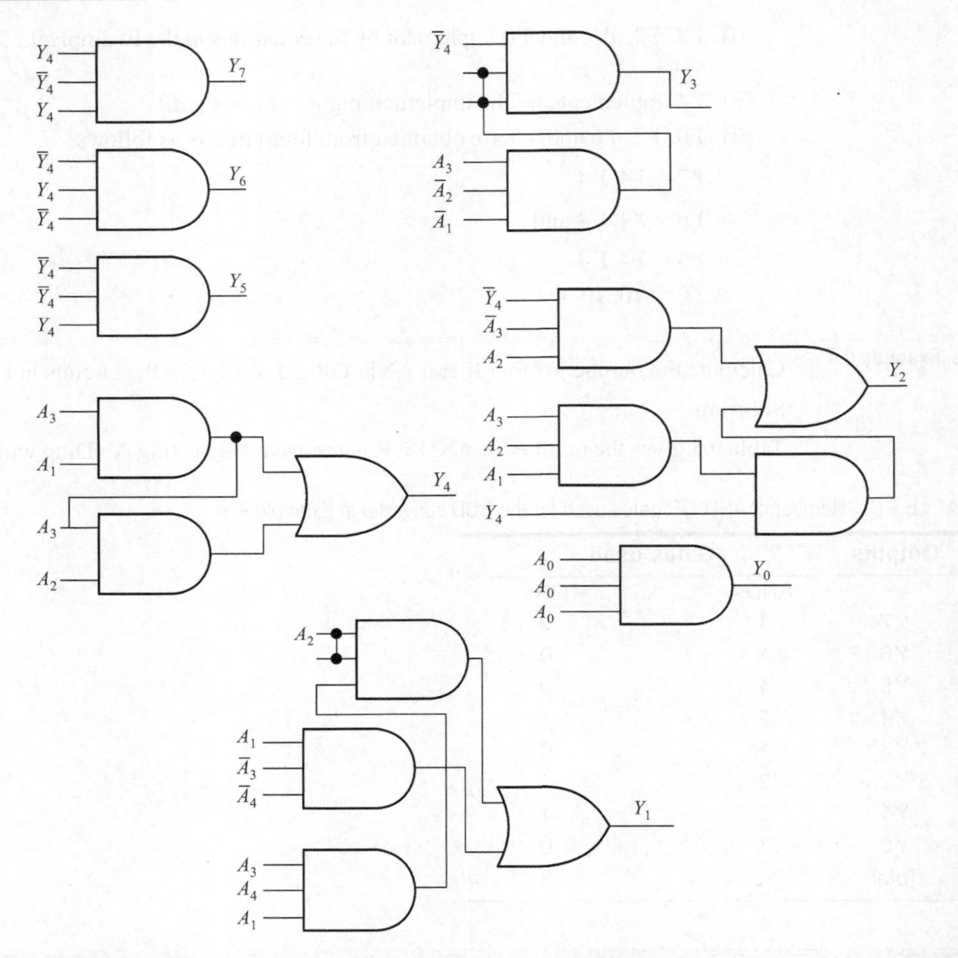

FIGURE 9.8 Combinational circuit for 4-bit binary to 2-digit BCD converter implemented by three literal input ANDs (maximum fan-in = 6) and ORs.

TABLE 9.6 Number of three literal inputs AND-OR gates used for the BCD converter in Example 9.3

Outputs	Gates used	
	ANDs	ORs
Y7	1	0
Y6	1	0
Y5	1	0
Y4	2	1
Y3	2	0
Y2	3	1
Y1	3	1
Y0	1	0
Total	14	3

Example 9.9 Calculate the time delay for each outputs in circuit of Figure 9.8. Assume per gate delay = 10 ns. Neglect the delay at the NOTs.

Solution

Table 9.7 gives the calculations of number of AND gates at level 1 and number of OR gates at level 2 and delays at each level.

TABLE 9.7 Number of AND gates at level 1 and number of OR gates at level 2 and delays at each level in combinational circuit of Figure 9.8

Outputs	ANDs used			ORs used		Total
	Number of ANDs	ANDs placed in series due to more literals	Delay in ns at ANDs at level 1	ORs	Delay in ns at ORs at level 1	Delay in ns at AND-ORs at levels 1 and 2
Y7	1	0	10	0	0	10
Y6	1	0	10	0	0	10
Y5	1	0	10	0	0	10
Y4	2	0	10	1	10	20
Y3	2	2	20	0	0	20
Y2	3	2	20	1	10	30
Y1	3	2	20	1	10	30
Y0	1	0	10	0	0	10
Sum	14	6	20	3	10	30

Note: Delays are summed for the stages, not for ANDs at a stage.

Total delay in conversion is 30 ns, less than the specification in Example 9.3.

Example 9.10 Formulate the problem by building a truth table and the Boolean expressions that detects whether an 8-bit BCD number is having illegal outputs $Y0$ to $Y7$. Output = 0, if any illegal combination of BCD outputs appears else the output is 1 after a delay of 40 ns. (i) Specify the number of inputs. (ii) Specify the number of outputs. (iii) Specify the delay permitted at the combinational circuit each gate. (iv) Formulate truth table and formulate Boolean expressions in terms of implicants (minterms) and prime implicants.

Solution

1. A binary number between 0 and 15 decimal has eight bits in BCD format. Hence number of inputs = 8.
2. Total number of outputs is 1. It is to test whether the inputs are not having any illegal combination.
3. Delay given is 40 ns. We have the use the gates with the specification that number of levels at the combinational circuit multiplied by delay per gate is less or equal to 40 ns per gate.

The truth table will be as per Table 9.8. Y' is an intermediate output taking into accounts inputs from $A4$ down to $A0$ only. Y is the final result

$$Y = Y'.\overline{A}7.\overline{A}6.\overline{A}5$$

TABLE 9.8

Inputs								Output	
A7	A6	A5	A4	A3	A2	A1	A0	Y'	Y
0	0	0	0	0	0	0	0	1	1
0	0	0	0	0	0	0	1	1	1
0	0	0	0	0	0	1	0	1	1
0	0	0	0	0	0	1	1	1	1
0	0	0	0	0	1	0	0	1	1
0	0	0	0	0	1	0	1	1	1
0	0	0	0	0	1	1	0	1	1
0	0	0	0	0	1	1	1	1	1
0	0	0	0	1	0	0	0	1	1
0	0	0	0	1	0	0	1	1	1
0	0	0	1	0	0	0	0	1	1
0	0	0	1	0	0	0	1	1	1
0	0	0	1	0	0	1	0	1	1
0	0	0	1	0	0	1	1	1	1
0	0	0	1	0	1	0	0	1	1
0	0	0	1	0	1	0	1	1	1
Any combination other than the above 16								0	0

Design of the Boolean functions for the truth table at Table 9.8 for the problem is as follows:

$$Y' = \Sigma m(0, 1, 2, 3, 4, 5, 6, 7, 8, 9, 16, 17, 18, 19, 20, 21)$$

Five-variable Karnaugh-map using Table 5.13 is given in Table 9.9.

TABLE 9.9 Five variable Karnaugh Map set with 32 minterms of the SOP expression and upper layer for $A4 = 0$ on the left side and lower layer for $A4 = 1$ on the right side

$A4 = 0$					$A4 = 1$				
A1A0 A3A2	$\overline{A}1.\overline{A}0$	$\overline{A}1.A0$	$\overline{A}1A0$	$A1\overline{A}0$	A1A0 A3A2	$\overline{A}1\overline{A}0$	$\overline{A}1A0$	$A1A0$	$A1\overline{A}0$
	00	01	11	10		00	01	11	10
$\overline{A}3\overline{A}2$ 00	1	1	1	1	$\overline{A}3\overline{A}2$ 00	1	1	1	1
$\overline{A}3A2$ 01	1	1	1	1	$\overline{A}3A2$ 01	1	1		
$A3A2$ 11					$A3A2$ 11				
$A3\overline{A}2$ 10	1	1			$A3\overline{A}2$ 10				

The map shows on the left side one octet with $\overline{A}4.\overline{A}3$ common. Left hand side also shows one pair $\overline{A}4.A3.\overline{A}2.\overline{A}1$. On right hand side, there are two quads for $A4.A3.A2$ and $A4.A3.A1$. [$Y4 = A4$]

Result is $Y' = \overline{A}4.\overline{A}3 + \overline{A}4.A3.\overline{A}2.\overline{A}1 + A4.\overline{A}3.\overline{A}2 + A4.\overline{A}3.\overline{A}1$

Problem is now formulated for implementation of Boolean expression for Y from four prime implicants.

Example 9.11

Implement the circuit for the designed Boolean expression with prime implicants for the problem formulated in Example 9.10. Use two literal inputs to AND-OR gates and to an OR = 2. The NOTs for $A0$ to $A7$ complements and for Y complement are also available for the ANDs and ORs, but assume that delays at the NOTs are not accounted.

Solution

Figure 9.9 shows an implementation for Y in Example 9.10. 1st term needs one AND. 2nd terms have 3 ANDs, each pair is in series. 3rd and 4th terms need 2 ANDs each with one AND common. Total ANDs (1 + 3 + 3) = 7. Delay at the ANDs level 1 to 3 = 30 ms. Assume each gate delay = 10 ns. Expression has 3 OR operations. Therefore, 2 ORs are needed in series. Delay at the OR level 2 = 20 ns. Total delay for the BCD code verifier is 50 ms when implemented using AND-ORs.

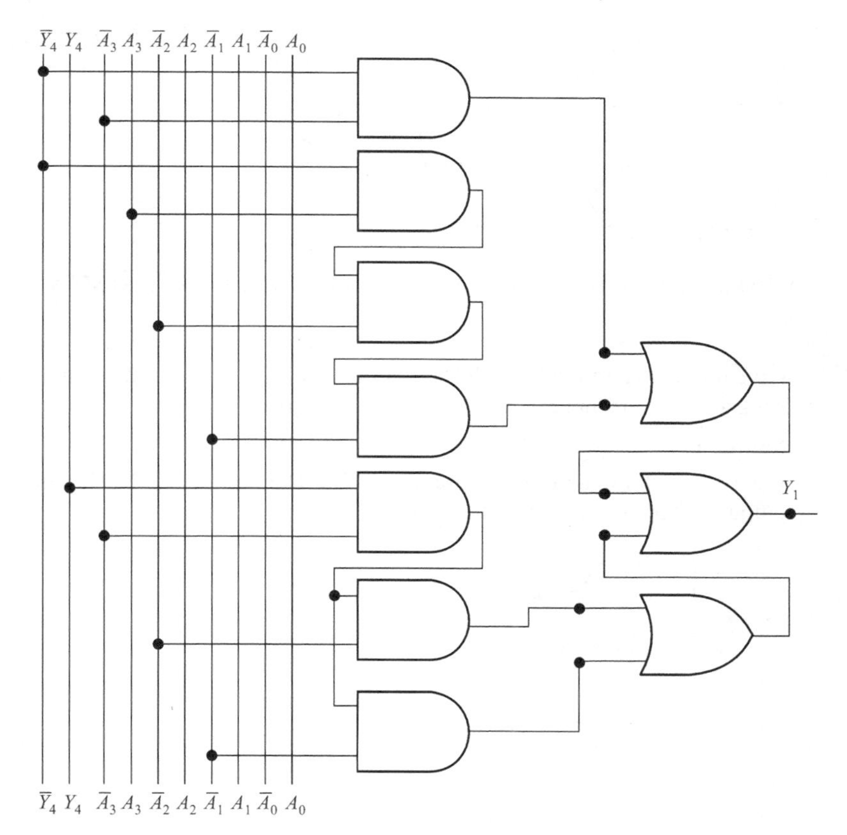

FIGURE 9.9 Combinational circuit for checking any illegal BCD input implemented by two literal input ANDs (maximum fan-in = 4) and ORs (maximum fan-in = 2).

Example 9.12 Implement the combinational circuit in Figure 9.9 by three input NORs alone.

Solution

Max terms $M10$ to 15 and 22 to 31 are present.

Figure 9.10 shows an implementation with three-input NORs.

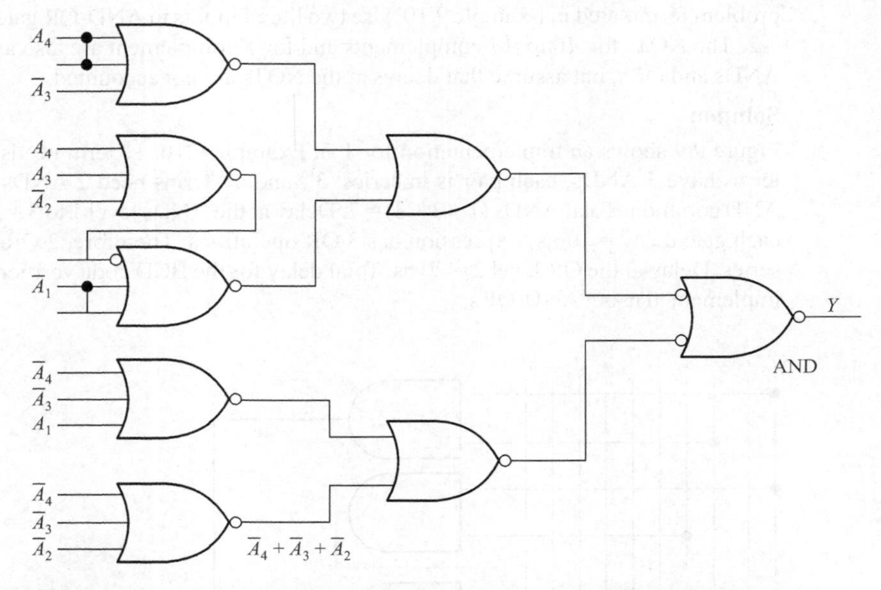

FIGURE 9.10 Combinational circuit for BCD illegal code detector implemented with three-input NORs.

Example 9.13 Give timing diagram of a NAND circuit with a NOT at one of its input and output is feedback at one of its input.

Solution

Figure 9.11 shows a timing diagram of a NAND circuit with a NOT at one of its input and output is feedback at one of its input.

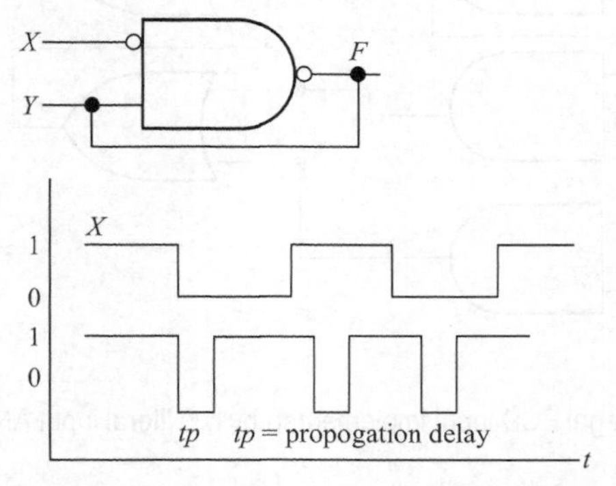

FIGURE 9.11 Timing diagram of a NAND circuit with a NOT at one of its input and output is feedback at one of its the input.

EXERCISES

1. Check which logic circuit is a combinational circuit. (A) A motor is rotating and associated circuit is giving digital outputs for its velocity. After it there is a circuit, which connects to LCD drivers. (B) A circuit for counting the motor rotations. (C) A circuit, which gives an output that toggles on any input change. (D) A circuit for detecting an illegal combination of inputs (E) A circuit in which when numbers of 1s at the inputs are odd, it gives output = 1 else 0.
2. Formulate the problem for a circuit that finds one's complements for the first and third input and buffers the second and fourth input. ($Y0 = \overline{A}0$, $Y1 = A1$, $Y2 = \overline{A}2$ and $Y3 = A3$). Give implementation by XORs.
3. Give timing diagram of a NOR circuit with a NOT at one of its input and output is feedback at one of its input.
4. A four bit binary number between 0 and 15 is converted after a delay of 40 ns into another binary number four more than the original number by a combinational circuit. (i) Specify the number of inputs. (ii) Specify the number of outputs. (iii) Specify the delay permitted at the combinational circuit each gate.
5. Formulate the problem by a truth table of four inputs and five outputs in Exercise 4. Then design the Boolean expressions with prime implicants.
6. Implement the circuit for the designed Boolean expressions with prime implicants using AND-OR gates for the problem in Exercise 4. Assume inputs for each literal as inputs and the NOTs for their complements and for complements are also available. Assume maximum fan-in = 4 for the two inputs ANDs and four input ORs.
7. Calculate the numbers of ANDs and ORs in a circuit for XOR-NOT that is used to check the number of 1s at the 8 inputs, whether these are odd or even.
8. Calculate the time delay for each of the output in circuit of Exercise 6. Assume per gate delay = 10 ns. Neglect the delay at the NOTs.
9. Formulate the problem by building a truth table and then the Boolean expressions that detect the divisibility of a number by four using a four bit binary number. Output = 0, if any number between 0 and 15 can't be divided by four else, the output =1. (i) Specify the number of inputs. (ii) Specify the number of outputs. (iii) Specify the delay permitted at the combinational circuit each gate. (iv) Formulate the truth table and formulate Boolean expressions in terms of implicants (minterms) and prime implicants.
10. Implement a circuit with OR at first level and ANDs at second level for $Y = \Sigma m(0, 1, 2, 3, 4, 5, 6, 7, 8, 9, 16, 17, 18, 19, 20, 21)$. (Hint: First convert SOP form into POS form as that will have OR terms at first stage)
11. Implement the circuit for the designed Boolean expression with SOP form prime implicants for the problem formulated in Exercise 10. Use two literal inputs AND-OR gates. Assume maximum literals to an AND = 4 and to an OR = 2. The NOTs for $A0$ to $A5$ complements and for Y complement are also available for the ANDs and ORs, but assume that delays at the NOTs are not accounted.
12. Implement the combinational circuit for the following Boolean expression

$$S = A.B.C.\overline{D} + A.B.\overline{C}.D + A.\overline{B}.C.D + \overline{A}.B.C.D + A.B.C.D.$$

QUESTIONS

1. What are the steps followed when to formulate the problem when designing a combinational circuit?
2. Given two Boolean expressions, what are the steps followed to implement it with minimum number of NOR gates.
3. Given two Boolean expressions what are the steps followed to implement it with minimum number of NAND gates.
4. Given two Boolean expressions what are the steps followed to implement it with minimum delays at each level.
5. How do you equalize the delays due to the different number of literal terms in a Boolean expression?
6. Why do we design TTL circuits by the NANDs as building block and CMOS circuits with NORs as the building blocks?
7. How do we convert AND-OR blocks based combinational circuit to the NANDs based circuit?
8. How do we convert an OR-AND blocks based combinational circuit to the NORs based circuit?
9. What is cost minimization? What is delay minimization?
10. When do we get the glitches in the output of a combinational circuit? (Glitch means 0 to 1 or 1 to 0 transition momentarily. For example, in circuit of Figure 9.6. One reason is unequal delays at gates through which the different inputs pass before showing a steady-state output).

CHAPTER 10

Binary Arithmetic and Decoding and Mux Logic Units

OBJECTIVE

We learnt binary arithmetic addition, subtraction, multiplication and division in Chapter 3. How do we formulate the problem and implement the design of combinational circuits for the 4-bit and 8-bit arithmetic functions? We shall learn the answers of these questions in this chapter.

In Chapter 9, we learnt that AND-OR arrays or OR-AND arrays or NANDs or NORs could be used as basic building blocks for designing a combinational logic circuit. We shall study decoders, encoders, demultiplexers and multiplexers in this chapter.

We shall learn that the binary arithmetic circuits fast addition by carry lookahead circuit; decoders and multiplexers can also be used as the bigger building blocks for the logic design.

10.1 BINARY ARITHMETIC UNITS

10.1.1 Binary Addition of Two Bits

10.1.1.1 Half Adder

Let us assume that Cy'_0 is a carry bit obtained at an output in an adder circuit, and S is a sum bit obtained in another output of the adder. Then, the circuit is called a half adder (H.A.) if we have

$$A \text{ add } B = S \text{ plus } Cy'_0, \quad ...(10.1)$$

where the left hand side and right hand side values are as per Table 10.1. This equation is in fact representing four sub-equations, one for each row of the table.

TABLE 10.1 Truth table for half adder

Inputs Left Hand Side		H.A. Outputs Right Hand Side		Equation Number
A	B	S	Cy'_0	
0	0	0	0	...(10.1a)
0	1	1	0	...(10.1b)
1	0	1	0	...(10.1c)
1	1	0	1	...(10.1d)

A circuit of the logic gates as shown in Figure 10.1(a) implements it. Whenever an XOR is required in a circuit the adder can be used as a fuilding block. Examples 11.5 and 11.6 in next chapter will show use of the adder as building block.

10.1.1.2 Full Adder

When we add two digits in decimal system, we also add any previous carry. For example, let us consider a decimal addition of 59 + 07. When lower digits 9 + 7 are first added, we get 6 and a carry 1 to the left. We do not take into account any previous carry for the lower digits. We add just like shown in equations (10.1a to 10.1d) above for a half adder. When the upper digits 5 and 0 are added, we get 5 only. Result of the decimal addition will be 56, which is wrong. This is due to an error. We are not adding previous lower digit addition's carry while adding 5 and 0. Actually, we will get the correct answer 66 the previous stage carry also adds. This example shows that any carry from previous stage must also be added. A full adder takes care of it. Its working can be understood by truth table in Table 10.2.

Figure 10.1(b) shows the logic circuit of a full adder and also shows a shorter representation of full adder as *FA*. If *Cy*, a carry at an input of a *FA*, is 0 then the *FA* is equivalent to a half adder. *Cy'* is a new carry which is taken as previous (input) carry, if we use it in the next stage. *FA* is represented by equation

$$Cy \text{ add } A \text{ add } B = \text{S plus } Cy', \quad ...(10.2)$$

where the left hand side and right hand side values are as per Table 4.2. This equation is in fact representing six sub-equations, one for each row of the Table 10.2.

10.1.2 Addition of Two Arithmetic Numbers Each of 4 Bits

Let us assume that two arithmetic number **A** and **B** each of 4 bits, which are to be added. Let us assume bits in **A** are A_3, A_2, A_1 and A_0 and bits in **B** are B_3, B_2, B_1 and B_0. Figure 10.2(a) shows four *FA*s processing the addition operation the four bits of **A** and four of **B**. Previous stage carry at right most stage is reset at 0 so that at least significant bit position *FA* works as a half adder. Figure 10.3 shows how to a subtrator fuilds from *FA*.

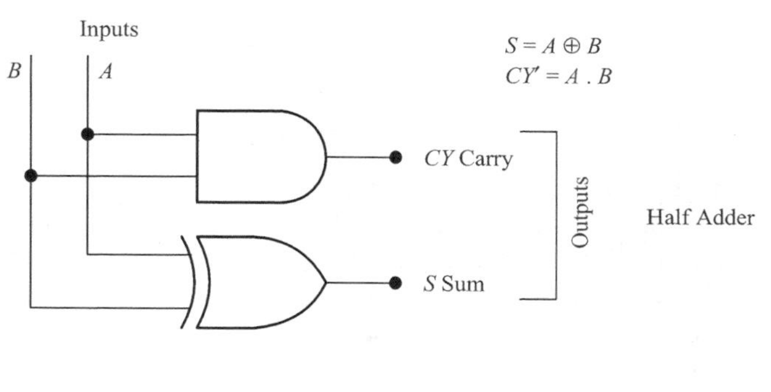

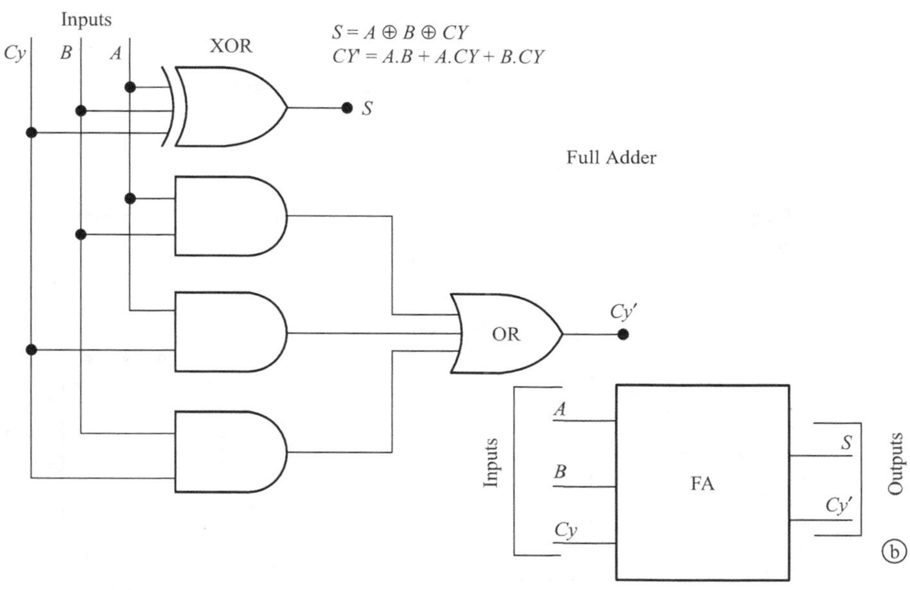

FIGURE 10.1 (a) Half Adder (addition of two bits without previous stage carry) (b) Full Adder (addition of two bits with previous stage carry) logic circuit and its simpler representation by an *FA* block.

TABLE 10.2 Truth table for full adder

Inputs Left Hand Side			FA Outputs Right Hand Side		Equation Number
A	*B*	*Cy*	*S*	*Cy'*	
0	0	0	0	0	...(10.2a)
0	0	1	1	0	...(10.2b)
0	1	0	1	0	...(10.2c)
0	1	1	0	1	...(10.2d)
1	0	0	1	0	...(10.2e)
1	0	1	0	1	...(10.2f)
1	1	0	0	1	...(10.2g)
1	1	1	1	1	...(10.2h)

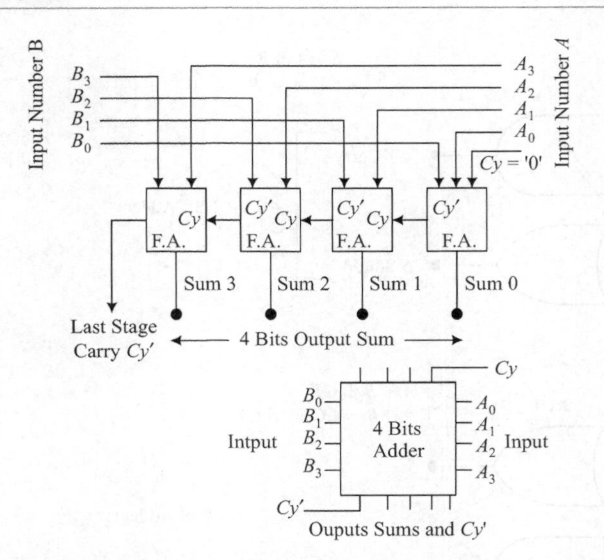

FIGURE 10.2 Four FAs processing the addition operation on four bits of *A* and four bits of *B* (Previous stage carry at right most stage is reset at 0).

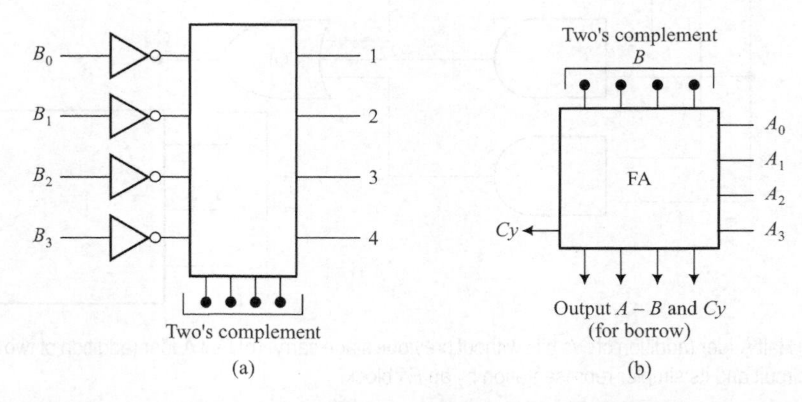

FIGURE 10.3 (a) Circuit for two's complement of *B*. (b) Implementation method subtraction by the addition of *A* with two's complement of *B*.

Karnaugh map from the truth table for an implementation of logic circuit for *S* (Sum) operations of Table 10.2 is as follows.

TABLE 10.3 Three variable Karnaugh Map for *S*

B.Cy \ A		$\overline{A}$	A
$\overline{B}.\overline{Cy}$	00	0	1
$\overline{B}.Cy$	01	1	0
$B.Cy$	11	0	1
$B.\overline{Cy}$	10	1	0

Recall section 5.5.4 for a method to minimize the logic circuit from the Karnaugh map offset and diagonal adjacencies. Therefore, $S = Cy$.XOR.(A.XOR.B). (There are two diagonal adjacencies with offset to each other.)

Karnaugh map from the truth table for a logic circuit implementation for Cy' (next stage carry) operations of Table 10.2 is as follows:

TABLE 10.4 Three variable Karnaugh Map for Cy'

B.Cy		$\overline{A}$ 0	A 1
$\overline{B}.\overline{Cy}$	00	0	0
$\overline{B}.Cy$	01	0	1
$B.Cy$	11	1	1
$B.\overline{Cy}$	10	0	1

Minimizing the logic circuit from the Karnaugh map showing three quads, we get $Cy' = A.B + A.Cy + B.Cy = A.B + (A + B).Cy$.

Point to Remember

> Full adder circuit is a basic building block in the adders and subtractors. Next stage carry $Cy' = A.B + A.Cy + B.Cy$ and sum $= A \oplus B \oplus Cy$ in a full adder. It finds wide application in the arithmetic operation circuits.

10.1.2.2 Adder and Subtractor Implementation Using MSI ICs

For laboratory experiments, we can use an MSI IC of CMOS family 74HC82 or of TTL family 7482. This IC is a two-bit *FA*. Other ICs, which are usable, are CMOS based 74HC 83 or 74 HC 283 or TTL based 7483 OR 74283. These ICs are the four bits *FA*s. TTL based 74283 has a faster processing of the Cy at the higher stages than the 7483 TTL. (It also differs in pin numbers). In a computer, there is no single 4 bits adder unit, as in Seventies, but nowadays the 32 bit and 64 bit adder units are there. In the computer circuits, the *FA*s are implemented with NANDs or NORs only.

10.1.3 Subtraction of Two Arithmetic Numbers Each of 4 Bits

Let us assume two arithmetic numbers of **A** of 4 bits A_3, A_2, A_1 and A_0 and **B** of 4 bits B_3, B_2, B_1 and B_0. Figure 10.3(a) showed implementation of two's complement of **B**. Figure 10.3(b) showed an implementation method for subtraction by the addition of **A** with two's complement of **B.**

Figure 10.4 shows a detailed circuit made for subtracting using the *FA*s as well as the XORs. Figure shows an adder cum subtractor circuit and implements two's complement of **B** as well as subtraction by an adder circuit. It uses a property of XOR gate that a two-input XOR gate if given input = 1, then its output is complement of the other input and if given input = 0, then its output is same as that of complement of the other input. (XOR is working

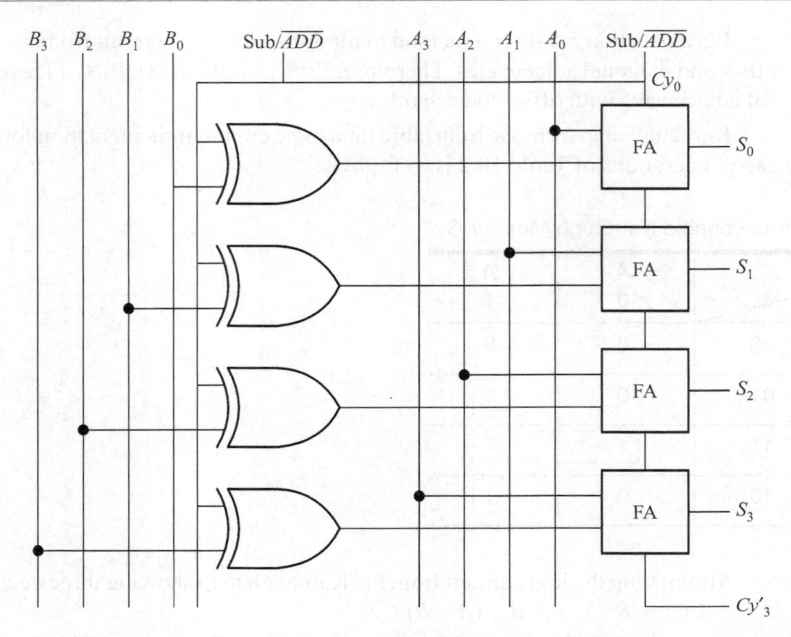

FIGURE 10.4 Adder cum subtractor circuit and implement two's complement of **B**. as well as subtraction by an adder circuit using the property of XOR gate that a two-input XOR gate if given input = 1, then its output is complement of the other input.

like a controlled inverter. We get NOT operation if control input = 1 else output = input when control input = 0).

Case 1: For active 0 at *Cy* input (Sub/$\overline{ADD}$ line = 0) of *FA*, which is also one of two inputs of all the four XORs, obtain an arithmetic addition only by this circuit. Case 2: For active 1 at *Cy* input (Sub/$\overline{ADD}$ line = 1) of first *FA*, we get an increment by one circuit and we simultaneously get the one's complement of a bit of **B** at one of three inputs of each *FA*s. In this way, an active 1 generates 2's complement of **B**. The 2's complement of **B** is like a negative number corresponding to **B**. Therefore; by adding it to **A** we get an arithmetic subtraction.

Figure 10.4 circuit is useful both as a subtractor and an adder. It also work as a two's complement generator. It works as a two's complement generator when we connect all the four inputs of number **A** to A_3, A_2, A_1 and A_0 to the logic state 0s as the two's complementation is simply a subtraction from (called negation 0).

Point to Remember

A two-input XOR gate if given input = 1, then its output is complement of the other input. Two's complement of a binary number is first generated using an XOR left side circuit. Subtraction is then done with full adder circuit as a basic building block in the subtractor. A two-input XOR gate if given input = 0, then its output is same as the other input. The same circuit will then work as an adder.

10.2 DECODER

A decoder is a circuit that converts the binary information from one form to another. A decoder selects a unique combination of inputs and according to that information generates a unique output(s) at one-line (or at multiple lines).

10.2.1 Decoder (Line Decoder)

Let us assume that we have eight circuits to implement the different functions–One is for addition, other is for subtraction, other for incrementing, and so on. We have to select only one by giving appropriate instruction input. A decoder will let us select only one.

For example, we want to decode and get a selected output when the input is 000 (like an instruction) and activate the addition circuit, when 001 activate the subtraction, when 010 activate the increment circuit, and so on. Decoder can generate an output $Y0 = 0$ or $Y1 = 0$, or $Y2 = 0$ or $\ldots Y6 = 0$ or $Y7 = 0$, respectively, for the next stage circuit so that either addition or subtraction or increment circuit or any one of the eight circuits can activate as per the inputs; 000, 001, 010, ..., respectively. Such a decoder circuit is called 'line decoder'. It is also called 3-line to 8-line decoder or 1 of 8 decoder. Its truth table will be as per Table 10.5. Example 10.5 will describe an exemplary combination circuit for the decoder for implementing Table 10.5.

TABLE 10.5 A 3 to 8 line decoder for selecting 1 of the circuit out of 8 from the given 3 binary inputs

Inputs			Outputs							
A_2	A_1	A_0	$\overline{Y}_0$	$\overline{Y}_1$	$\overline{Y}_2$	$\overline{Y}_3$	$\overline{Y}_4$	$\overline{Y}_5$	$\overline{Y}_6$	$\overline{Y}_7$
0	0	0	0	z	z	z	z	z	z	z
0	0	1	z	0	z	z	z	z	z	z
0	1	0	z	z	0	z	z	z	z	z
0	1	1	z	z	z	0	z	z	z	z
1	0	0	z	z	z	z	0	z	z	z
1	0	1	z	z	z	z	z	0	z	z
1	1	0	z	z	z	z	z	z	0	z
1	1	1	z	z	z	z	z	z	z	0

Note: The z is either 1 or tristate. An appropriate next stage circuit enables by appropriate Y becoming = 0. Bar over Ys shows that line output is active 0. It represents active state for the next stage when at 0. When $z = 1$, an output Y reflects a maxterm. We can implement the Boolean function(s) by ANDing(s) the various maxterm outputs.

10.2.1.1 Decoder with Outputs Enabling Control (Gate) Input(s)

Certain decoders ICs (MSIs-Medium Scale Integrated Circuits) have in addition either a single control gate input or two or three control gate inputs. All control gate pins are activated then only the any of the output of the decoder can be activated else all the outputs will be in the tristate. Figure 10.5(a) shows a 3-line to 8-line (1 of 8) decoder with three control inputs $\overline{G}0$, $\overline{G}1$, and $G2$. $\overline{G}0$, and $\overline{G}1$ control input are activated by 0 and $G2$ input is activated by 1. Table 10.6 gives the truth table of 3-line to 8-line decoder with 3 control gate input. The sign * means an output condition *tristate* instead of 1 or 0. A MSI chip 74138 has the pin corresponding to the decoder shown in the Figure 10.5(a).

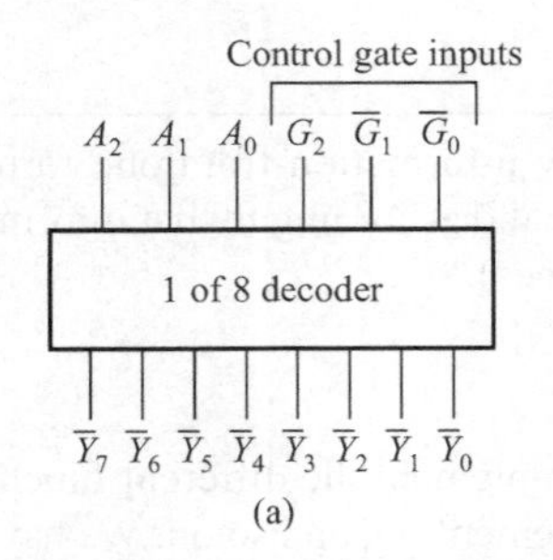

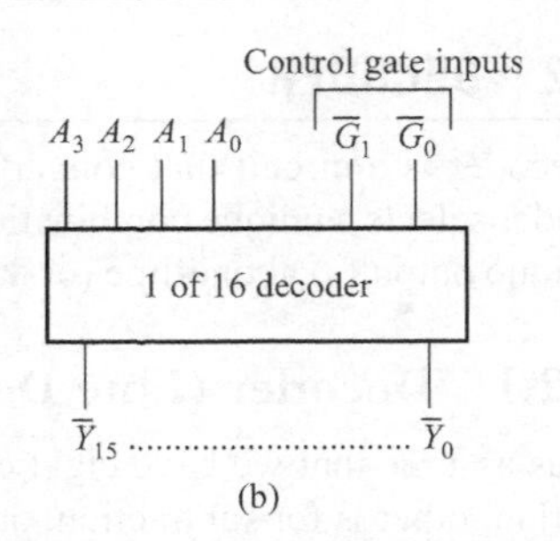

FIGURE 10.5 (a) 3-line (*A*, *B*, *C*) to 8-line decoder for selecting 1 of 8 with two active 0 and one active 1 control gate inputs $\overline{G}_0$, $\overline{G}_1$, $\overline{G}_2$ (b) 4-line to 16-line(A_0 to A_3) decoder for selecting 1 of 16 with two active 0 control gate inputs G_0, G_1.

TABLE 10.6 Truth table of 3-line to 8-line decoder with 3 control gate inputs

Control (Gate) Inputs			A_0	A_1	A_2	$\overline{F}_0$	$\overline{F}_1$	$\overline{F}_2$	$\overline{F}_3$	$\overline{F}_4$	$\overline{F}_5$	$\overline{F}_6$	$\overline{F}_7$
'$\overline{G}_1$	'$\overline{G}_2$	$\overline{G}_3$	0	0	0	0	z	z	z	z	z	z	z
0	0	1	1	0	0	z	0	z	z	z	z	z	z
0	0	1	0	1	0	z	z	0	z	z	z	z	z
0	0	1	1	1	0	z	z	z	0	z	z	z	z
0	0	1	0	0	1	z	z	z	z	0	z	z	z
0	0	1	1	0	1	z	z	z	z	z	0	z	z
0	0	1	0	1	1	z	z	z	z	z	z	0	z
0	0	1	1	1	1	z	z	z	z	z	z	z	0
Any combination other than above			*	*	*	*	*	*	*	*	*	*	*

* All *F* outputs = *z* tristate or 1.

Points to Remember

1. Decoder(s) finds wide application in selecting a memory chip or port in a computer systems from a given address of it at the input lines and/or at control lines.
2. A number of decoders can be arranged in a parallel or a tree topology to obtain a bigger number of input-bits decoder by using the control gate (enable) input(s) defining the specific decoder and applying the inputs to the enable inputs of all the decoders at the tree also in different combinations so that only one decoder activates at an instant.

An output-enabling pin (input) helps in placing more decoders in parallel. For example, suppose, we need (4 of 16) decoder from two 3 to 8 decoders. (1) A_0 A_1 and A_2 are given to first (3 to 8) decoder at A_0, A_1, A_2 pins. A_3 is the input at G_2. (2) A_0, A_1 and A_2 are given to second (3 to 8) decoder at A_0, A_1, A_2 pins. A_3 is the input at G_2 after a NOT operation. (3) $\overline{G}_0$ and $\overline{G}_1$ are made common in both the decoders, and are enabled by giving 0s as inputs to them.

10.2.2 The 1 of 2 and 1 of 4 Line Decoders

Figures 10.6(a) and (b) show the 1-line to 2-line and 1 of 4 decoders. Figure shows both active 1 and active 0 decoders. Truth tables are at the insets.

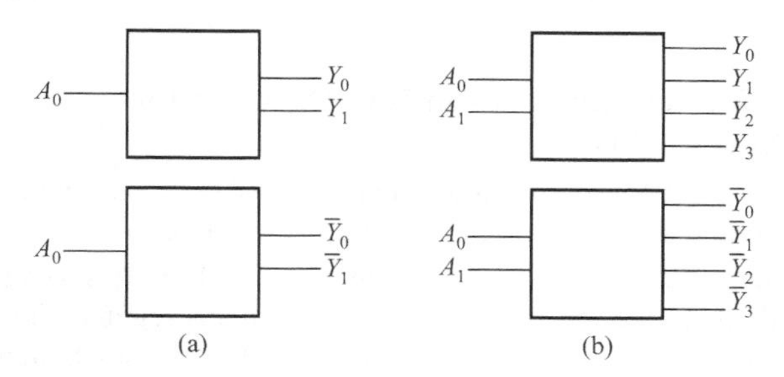

FIGURE 10.6 (a) 1-line to 2-line decoder (b) 1 of 4 decoders (truth tables at the inset).

10.2.3 The Four-line to 16-line Decoder

Figure 10.5(b) showed a 4 to 16 decoder with two active 0 control gate inputs. A MSI chip 74154 has the pins corresponding to the decoder shown in the figure. It has two gate enabling pins; both are active 0. When are then decoding action (active 0) rakes place, else all the output remains 1. (MSI IC 74159 is open collector outputs version of 74154).

10.2.4 Function Specific Decoders

There can be function specific decoder. Consider the following type of decoder.

10.2.4.1 BCD to Decimal Line Decoder

Let a BCD to decimal decoder has 12 outputs, each one activating corresponding to the given BCD input. Truth table is per Table 10.7. Example 10.6 will describe an exemplary combination circuit for the decoder for implementing Table 10.7. MSI IO 7442 is a BCD to decimal decoder. It has 11 output pins, Y_0 to Y_{10}. It has four pins A_0, A_1, A_2 and A_3 for the BCD inputs.

TABLE 10.7 A BCD to decimal line decoder for selecting 1 of the circuit out of 10 from the given BCD inputs

Inputs					Outputs											
A_4	A_3	A_2	A_1	A_0	Y_0	Y_1	Y_2	Y_3	Y_4	Y_5	Y_6	Y_7	Y_8	Y_9	Y_{10}	Y_{11}
0	***0***	0	0	0	1	0	0	0	0	0	0	0	0	0	0	0
0	***0***	0	0	1	0	1	0	0	0	0	0	0	0	0	0	0
0	***0***	0	1	0	0	0	1	0	0	0	0	0	0	0	0	0
0	***0***	0	1	1	0	0	0	1	0	0	0	0	0	0	0	0
0	***0***	1	0	0	0	0	0	0	1	0	0	0	0	0	0	0
0	***0***	1	0	1	0	0	0	0	0	1	0	0	0	0	0	0
0	***0***	1	1	0	0	0	0	0	0	0	1	0	0	0	0	0
0	***0***	1	1	1	0	0	0	0	0	0	0	1	0	0	0	0
0	***1***	0	0	0	0	0	0	0	0	0	0	0	1	0	0	0
0	***1***	0	0	1	0	0	0	0	0	0	0	0	0	1	0	0
1	***0***	0	0	0	0	0	0	0	0	0	0	0	0	0	1	0
1	***0***	0	0	1	0	0	0	0	0	0	0	0	0	0	0	1

10.2.4.2 Multi-line Decoder

The line decoders described above when selected activated only one output. It is possible to activate multi-line outputs for a specific combination of inputs. For example, gives three line outputs as 011 when input is 0000 and as 100 when input is 0001 (three more than the input).

10.2.4.3 Four Binary Input Seven line-Decoder for the Seven LED Segments

The line decoders described above when selected activated only one output. It is possible to activate multi-line outputs for a specific combination of inputs.

For example, consider the truth table of a binary to LED one to seven segment display outputs. An LED digit has the seven segments. Assume that a segment lights up if its input logic is 1 (active 1) and light off if 0. When inputs are 0000, the outputs should be *a*, *b*, *c*, *d*, *e* and $f = 1$ and $g = 0$. Table 10.8 shows the outputs needed by different segments of LED. MSI IC 7447 is a BCD to 7-segment decoder cum driver. All outputs are active 0 ($\overline{y}_0$, ...,). It has a control gate pin for blanking input and ripple blanking output RBI/RBO when 0. It has a control gate pin for ripple blanking input $\overline{\text{RBI}}$ when 0. It has a control gate pin for LED test LT, which makes all outputs 0, when 0. Ripple means a ($\overline{y}_6$) carry from a previous stage (carry means blank the previous stage digit and light up the new one). These three pins enable all the multiple 7447s connected together to display digits with a set of multiple seven segments.

Example 10.7 will show an exemplary combination circuit with truth table as per Table 10.8. The circuit gives the seven line decoded outputs for the LED segments for displaying the as per the BCD input. The example will also show the requirements of an eight segment LED with a figure there.

TABLE 10.8 Seven segment (7-line) LED decoder for selecting one set of the circuit out of 10 different BCD inputs

Inputs				Outputs						
A_3	A_2	A_1	A_0	Y_0 *a*	Y_1 *b*	Y_2 *c*	Y_3 *d*	Y_4 *e*	Y_5 *f*	Y_6 *g*
0	0	0	0	1	1	1	1	1	1	0
0	0	0	1	0	1	1	0	0	0	0
0	0	1	0	1	1	0	1	1	0	1
0	0	1	1	1	1	1	1	0	0	1
0	1	0	0	0	1	1	0	0	1	1
0	1	0	1	1	0	1	1	0	1	1
0	1	1	0	1	0	1	1	1	1	1
0	1	1	1	1	1	1	0	0	0	0
1	0	0	0	1	1	1	1	1	1	1
1	0	0	1	1	1	1	1	0	1	1

10.2.4.4 Logic Design and Boolean Function Implementation Using the Decoders

Assume a Boolean function F is $\Sigma m(3, 7, 9, 10)$. Consider a decoder outputs with active 1 output and the n binary inputs (number of binary input combinations $= 2^n$, where n is the number of literals in F. Its outputs reflect the minterms with each minterm at each of the output. Therefore, if the OR operations are done on $Y3$, $Y7$, $Y9$ and $Y10$ outputs of the four active 1 decoders with four binary inputs and 16 line outputs from $Y0$ to $Y15$, then the F implements by $Y3 + Y7 + Y9 + Y10$. A decoder circuit can be used to implement AND-OR circuit SOP Boolean expression when active output is 1. Example 10.8 will give a logic design and implementation using a decoder for $F_1 = \Sigma m(3, 7, 9, 10)$ and F_2 is $\Sigma m(2, 7, 12, 15)$.

Assume a Boolean function F' is $\Pi M(1, 7, 9, 13)$. Consider a decoder outputs with active 0 output and the n binary inputs (number of binary input possible combinations = 2^n, where n is the number of literals in F. Its outputs reflect the maxterms with one term each at each of the output. Therefore, if the AND operations are done on $Y1$, $Y7$, $Y9$ and $Y13$ outputs on the active 0 decoders with four binary inputs and 16 line outputs from $Y0$ to $Y15$, then the F implements by $F = Y1.Y7.Y9.Y17$. A decoder circuit can be used to implement OR-AND arrays based circuit for the POS Boolean expression when active output is 0. Example 10.9 will give a logic design and implementation using a decoder for $F' = \Pi M(1, 7, 9, 13)$.

Point to Remember

A decoder with one-line output decoder is also a minterm or maxterm generator. Therefore, a decoder can be used to implement a Boolean expression(s).

10.3 ENCODER

1. An encoder is a circuit that converts the binary information from one form to another.
2. An encoder gives a unique combination of outputs according to the information at a unique input at one-line (or at multiple lines).
3. Action of a one active line input encoder is opposite of that of a one active line output decoder.
4. An encoder, which has multi-lines as the active inputs, is also called 'priority encoder'.
5. Encoder can be differentiated from a decoder by greater number of inputs than the outputs when compared to the decoder.

A widely used application of an encoder is keypad (or keyboard) encoder. (A keypad has limited number of keys as in a telephone or a mobile phone. Keyboard has many more keys). At an instant, when a key presses, the input after an appropriate de-bouncing circuit applies the input to an encoder. The encoder generates an 8-bit code for a given active input corresponding to the pressed-key.

10.3.1 Encoder (Line Encoder)

Let us assume that we have eight inputs from the different functions. One is from completion of addition, other is for completion of subtraction, other on completing increment, and so on. We have to find which of the operation has been completed. An encoder will let us find one.

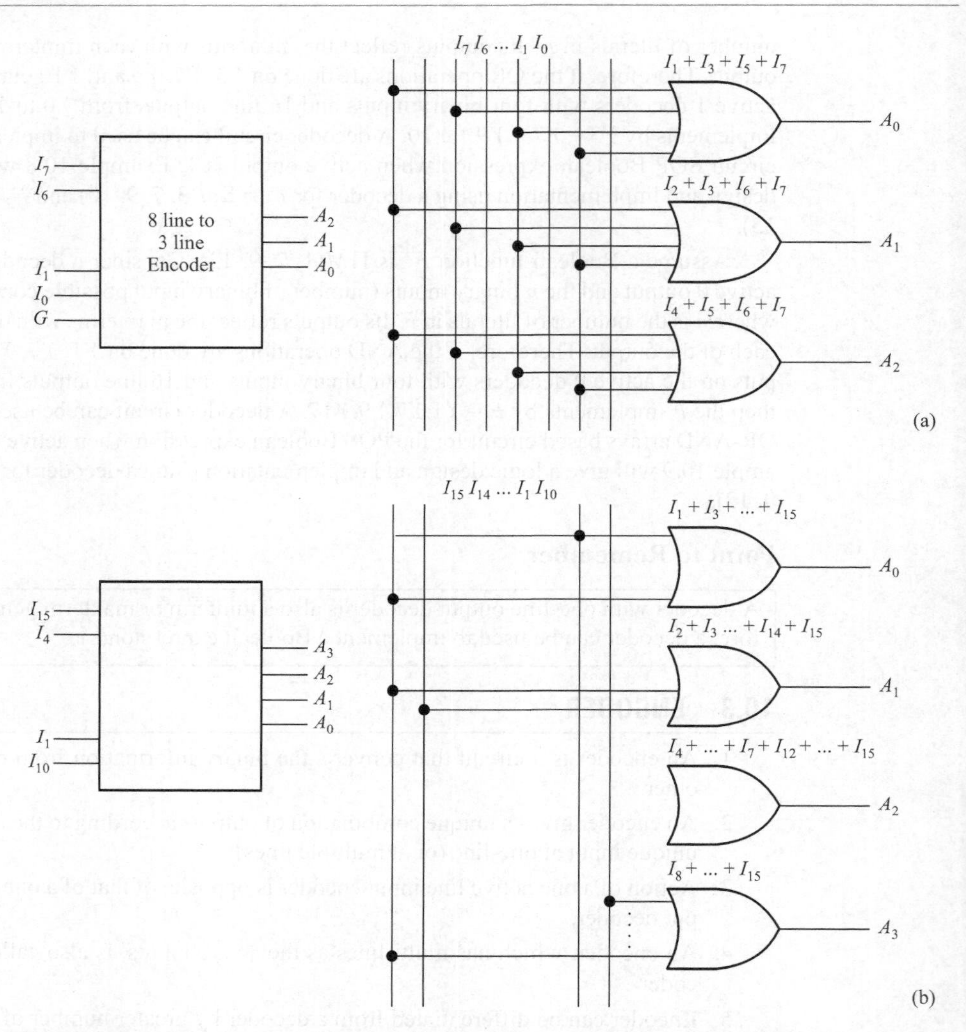

FIGURE 10.7 (a) 8-line to 3-line encoder for finding which 8 out of 1 (b) 16-line to 4-line encoder for finding 16 of 1.

For example, we want to encode and get an output 000 when addition completes and activates an input $A0$, get an output 001; when a subtraction completes it activates another input $A1$, get an output 010, and so on. Encoder can generate an outputs $Y2Y1Y0 = 000$ or 001 or 010 or 011 and so on depending on whether $A0 = 1$ or $A1 = 1$ or $A2 = 1$ or $A3 = 1$ and so on, respectively. Such an encoder circuit is called '2^n to n line encoder' also. It is also called 8-line to 3-line encoder or 8 of 1 encoder when $n = 3$. Table 10.9 gives a part of the truth table as a functional table showing nine rows only in place of 256 rows needed for the 8-input case. Example 10.13 will describe an exemplary combination circuit for the encoder for implementing Table 10.9.

MSI IC74148 is an 8 to 3 encoder with an active 0 gate-enable pin for the enabling the inputs and another active 0 gate-enable pin for enabling the outputs.

Points to Remember

> An application of an encoder is to generate at an instant set of outputs, which can be a code or an address corresponding to a given input which is active at that instant. If there are N inputs and only one of it is active at a time, then the encoder outputs at m pins where $2^m = N$ will provide us the means to find which of the N inputs is active at present with the remaining of the inputs (z and z' in Table 10.9) are in either the tristate or inactive states.

10.3.2 Encoder (Priority Encoder)

Suppose multiple inputs can activate simultaneously. Assume that $A6$ and $A1$ activate simultaneously. When input $A1 = 1$, the output = 001 and $A6 = 1$ then output = 110. A priority encoder will resolve the simultaneous activations of multiple inputs, by giving the output as per the highest priority one. It will give the output = 110 for $A6$. Such an encoder is called a priority encoder. Table 10.9 is also a truth table for a priority encoder when an encoder design is implemented for $z = X$ (don't care) and only z' = either 1 or 0 depending upon whether the line encoder design is for input active 1 or active 0, respectively.

TABLE 10.9 Functional Table 8 to 3 line encoder for finding I at the input circuit out of 8 inputs and give the 3-bit binary output

Inputs								Output			
*A*0	*A*1	*A*2	*A*3	*A*4	*A*5	*A*6	*A*7	*S*	*Y*2	*Y*1	*Y*0
0	0	0	0	0	0	0	0	0	0	0	0
I	z'	z'	z'	z'	z'	z'	z'	1	0	0	0
z	I	z'	z'	z'	z'	z'	z'	1	0	0	1
z	z	I	z'	z'	z'	z'	z'	1	0	1	0
z	z	z	I	z'	z'	z'	z'	1	0	1	1
z	z	z	z	I	z'	z'	z'	1	1	0	0
z	z	z	z	z	I	z'	z'	1	1	0	1
z	z	z	z	z	z	I	z'	1	1	1	0
z	z	z	z	z	z	z	I	1	1	1	1

Note: (1) $S = 0$ when none of the input is active. $S = 1$ when at least one input is active. (2) Both z and z' are either 0 or 1 depending upon whether the line encoder design is for I active 1 or active 0, respectively. (3) When an encoder design is for a priority encoder, then the design is implemented for $z = X$ (don't care) and only z' either 0 or 1 depending upon whether the line encoder design is for input active 1 or active 0, respectively.

10.3.3 BCD 10 of 1 four-bit Encoder

Table 10.10 gives the truth table of the combinational circuit for BCD priority encoder. X means don't care. p means active. $q = \overline{p}$. The input G means gate pin to control the output. If G is active then only all the outputs S, B_5, B_4, B_3, B_2 and B_1 are not in idle state (not in tristate). Figure 10.8(a) shows a decimal to BCD output encoder [Also called (10 of 1) four bits encoder (BCD priority encoder)] MSI IC 74147 is a decimal (9-active 0 inputs) to BCD (four outputs active 0) encoder. BCD output = 1111 when none of the input is 0.

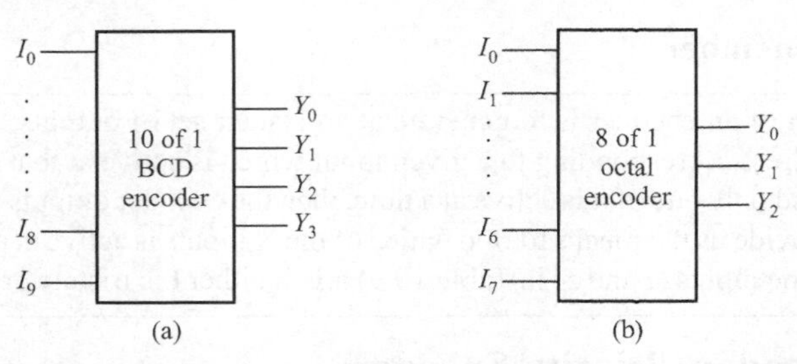

FIGURE 10.8 (a) A decimal to (BCD) output encoder [Also called (10 of 1) four bits encoder (BCD priority encoder)] (b) an octal to binary (8 of 1) three bits encoder.

TABLE 10.10 Decimal to BCD Encoder

Gate	Only One out of 10 is an Active input											Encoded BCD outputs					
	I_0	I_1	I_2	I_3	I_4	I_5	I_6	I_7	I_8	I_9	I_{10}	S	B_4	B_3	B_2	B_1	B_0
G	p	q	q	q	q	q	q	q	q	q	q	1	0	0	0	0	0
G	X	p	q	q	q	q	q	q	q	q	q	1	0	0	0	0	1
G	X	X	p	q	q	q	q	q	q	q	q	1	0	0	0	1	0
G	X	X	X	p	q	q	q	q	q	q	q	1	0	0	0	1	1
G	X	X	X	X	p	q	q	q	q	q	q	1	0	0	1	0	0
G	X	X	X	X	X	p	q	q	q	q	q	1	0	0	1	0	1
G	X	X	X	X	X	X	p	q	q	q	q	1	0	0	1	1	0
G	X	X	X	X	X	X	X	p	q	q	q	1	0	0	1	1	1
G	X	X	X	X	X	X	X	X	p	q	q	1	0	1	0	0	0
G	X	X	X	X	X	X	X	X	X	p	q	1	0	1	0	0	1
G	X	X	X	X	X	X	X	X	X	X	p	1	1	0	0	0	0
G	0	0	0	0	0	0	0	0	0	0	0	0	0	0	0	0	0

Note: X don't care in priority encoder, else equals q.

10.3.4 Octal 8 of 1 three-bit Encoder and Hexadecimal Encoder

Figure 10.8(b) shows an octal to binary (8 of 1) three-bit encoder. Table 10.11 gives the truth table of the combinational circuit of a hexadecimal to binary (14 of 1) four-bit encoder.

TABLE 10.11 Hexadecimal (16 bits) to Nibble (4 bits) Encoder

(Input) Control Pin	One out 16 active inputs active Input	Address outputs if an (output) control pin active			
		B_3	B_2	B_1	B_0
Active	I_0, active, rest inactive	0	0	0	0
Active	I_1 active, rest inactive	0	0	0	1
Active	I_2 active, rest inactive	0	0	1	0
....	....	...	...	...	...
....	....	...	...	...	...
Active	I_{13} active, rest inactive	1	1	0	1
Active	I_{14} active, rest inactive	1	1	1	0
Active	I_{15} active, rest inactive	1	1	1	1
Inactive	I_0 .... $I_{15} = x$	*			

The x means don't care. The sign * means tristate.

10.4 MULTIPLEXER

A multiplexer is a circuit that selects a input line among the input lines as per the channel-selector logic-inputs and gives that at the output. A multiplexer selects a unique input line according to the channel selector inputs to it.

10.4.1 Multiplexer (Line Selector)

Let us assume that we have two logic circuits that provide the outputs. One I_1 is for a logic function F_1 and other I_2 is for F_2. We have to select only one by giving appropriate instruction. A multiplexer will select for the output only one. Figure 10.9(a) shows a 2-channel input-selector 2 to 1 multiplexer with one channel selector pin A (0 for channel F_0 and 1 for channel F_1). Its functional table is given in the inset at the figure.

Figure 10.9(b) shows 4-channels input selector; 4 to 1 multiplexer with two channel-selector pins A_0 and A_1 (when 00 the channel F_0; when 01 the channel F_1; when 10 the channel X_2 and when 11 the channel X_3). Its functional table is given in the inset at the figure. MSI IC74156 is a four-channel multiplexer. It selects the inputs I_0, I_1, I_2 and I_3 as per the channel selector lines A_0 and A_1. When A_0 and A_1 are 00, $F = I_0$, 10 then $F = X_1$, and 01 then $F = X_2$, 11 then $F = I_3$, respectively.

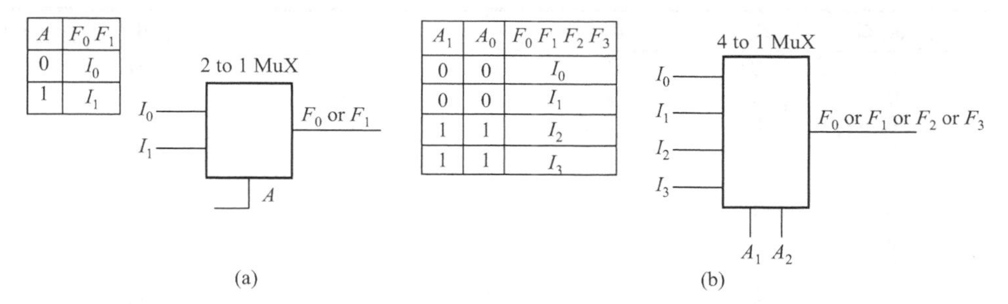

A	$F_0\ F_1$
0	I_0
1	I_1

A_1	A_0	$F_0\ F_1\ F_2\ F_3$
0	0	I_0
0	0	I_1
1	1	I_2
1	1	I_3

FIGURE 10.9 (a) A 2-channel input-selector using a 2 to 1 multiplexer with one channel selector pin A = (0 for channel F_0 and 1 for channel F_1) (b) 4-channel input selector (4 to 1 multiplexer) with two channel selector pins A_0 and A_1, when 00 the channel F_0, when 01 the channel F_1, when 10 the channel F_2 when 11 channel F_3 is selected.

Points to Remember

> A multiplexer provides one of the output path (channel) for the one of the channel data from a number of channels' data present at a given instant. Its important application is in sharing the circuits, ports, devices and resources.

Examples 10.15 and 10.16 will describe an exemplary combination circuits for the multiplexers for implementing tables in insets of Figures 10.9(a) and 10.9(b), respectively.

10.4.2 Multiplexer with Outputs Enabling Control (gate) Pin(s)

Certain multiplexers ICs [MSIs (Medium Scale Integrated circuits)] have in addition either a single control gate pin or two or three control gate pins. A control gate pin(s) when activated then only the output of the multiplexer activate else all the outputs will be in the tristate. Figure 10.10 shows a 16-line to 1-line (16 of 1) multiplexer with one control pins $\overline{G}$

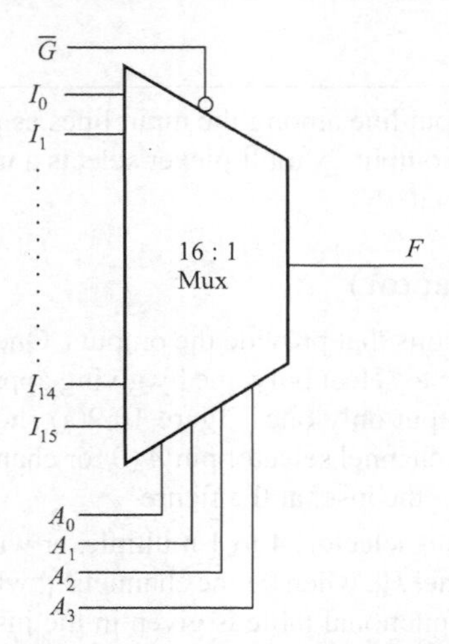

FIGURE 10.10 A 16-line to1-line (16 of 1) multiplexer with one control pins '$\overline{G}$ control pin activates the output when it is at 0.

TABLE 10.12 Truth table of 4-line to 26-line multiplexer with one control gate pin

Gate '$\overline{G}_1$	A_0	A_1	A_2	A_3	Input I	Output F
0	0	0	0	0	I_0 to I_{15}	I_0
0	1	0	0	0	I_0 to I_{15}	I_1
0	0	1	0	0	I_0 to I_{15}	I_2
0	1	1	0	0	I_0 to I_{15}	I_3
0	0	0	1	0	I_0 to I_{15}	I_4
0	1	0	1	0	I_0 to I_{15}	I_5
0	0	1	1	0	I_0 to I_{15}	I_6
0	1	1	1	0	I_0 to /15	I_7
0	0	0	0	1	I_0 to I_{15}	I_8
0	1	0	0	1	I_0 to I_{15}	I_9
0	0	1	0	1	I_0 to I_{15}	I_{10}
0	1	1	0	1	I_0 to I_{15}	I_{11}
0	0	0	1	1	I_0 to I_{15}	I_{12}
0	1	0	1	1	I_0 to I_{15}	I_{13}
0	0	1	1	1	I_0 to I_{15}	I_{14}
0	1	1	1	1	I_0 to I_{15}	I_{15}
1	X	X	X	X	X	*

X means 0 or 1 (don't care condition). I_0 to I_{15} means 16 input lines. * means tristate.

control pin activates the output when it is at 0. A 16 to 1 line multiplexer for selecting 1 of the circuit out of 16 from the given 4 binary channel selection inputs. There is a control gate pin with active 0. The sign * means an output condition *tristate* instead of 1 or 0.

10.4.2.1 Multiplexers Arranged as a Tree

How do we get the (m of 1) multiplexing from i numbers of the (m' of 1) multiplexers? Here $m = i.m'$ where i is an integer and $m' = 2^n$ where n is the number of channel selector lines at each of the i multiplexers. Multiplexers arranged as a tree facilitates this. Figure 10.11 shows an exemplary tree type arrangement of the five multiplexers. Four are at the leaves

TABLE 10.13 An exemplary tree type arrangement for (16 of 1) multiplexing from our multiplexers at the leaves

Inputs at Leaf *A*				Inputs at Leaf *B*				Inputs at Leaf *C*				Inputs at Leaf *d*				Inputs at Root *R*			*Y*
GA	*I*	*A*0	*A*1	*GB*	*I*	*A*0	*A*1	*GC*	*I*	*A*0	*A*1	*GD*	*I*	*A*0	*A*1	*GR*	*A*2	*A*3	*F*
1	*IA*	0	0	1	*IA*	0	0	1	*IA*	0	0	1	*IA*	0	0	1	0	0	*I0A*
1	*IA*	1	0	1	*IA*	1	0	1	*IA*	1	0	1	*IA*	1	0	1	1	0	*I1A*
1	*IA*	0	1	1	*IA*	0	1	1	*IA*	0	1	1	*IA*	0	1	1	0	1	*I2A*
1	*IA*	1	1	1	*IA*	1	1	1	*IA*	1	1	1	*IA*	1	1	1	1	1	*I3A*
1	*IB*	0	0	1	*IB*	0	0	1	*IB*	0	0	1	*IB*	0	0	1	0	0	*I0B*
1	*IB*	1	0	1	*IB*	1	0	1	*IB*	1	0	1	*IB*	1	0	1	1	0	*I1B*
1	*IB*	0	1	1	*IB*	0	1	1	*IB*	0	1	1	*IB*	0	1	1	0	1	*I2B*
1	*IB*	1	1	1	*IB*	1	1	1	*IB*	1	1	1	*IB*	1	1	1	1	1	*I3B*
1	*IC*	0	0	1	*IC*	0	0	1	*IC*	0	0	1	*IC*	0	0	1	0	0	*I0C*
1	*IC*	1	0	1	*IC*	1	0	1	*IC*	1	0	1	*IC*	1	0	1	1	0	*I1C*
1	*IC*	0	1	1	*IC*	0	1	1	*IC*	0	1	1	*IC*	0	1	1	0	1	*I2C*
1	*IC*	1	1	1	*IC*	1	1	1	*IC*	1	1	1	*IC*	1	1	1	1	1	*I3C*
1	*ID*	0	0	1	*ID*	0	0	1	*ID*	0	0	1	*ID*	0	0	1	0	0	*I0D*
1	*ID*	1	0	1	*ID*	1	0	1	*ID*	1	0	1	*ID*	1	0	1	1	0	*I1D*
1	*ID*	0	1	1	*ID*	0	1	1	*ID*	0	1	1	*ID*	0	1	1	0	1	*I2D*
1	*ID*	1	1	1	*ID*	1	1	1	*ID*	1	1	1	*ID*	1	1	1	1	1	*I3D*
0	*X*	*X*	*X*	0	*X*	*X*	*X*	0	*X*	*X*	*X*	0	*X*	*X*	*X*	0	*X*	*X*	*

IA means *I*0 to *I*3 at leaf *A*, *IB* means at *B*, *IC* means at *C* and *ID* means at *D*. * means tristate. *I0A* means *F* = Input *I*0 at leaf *A* multiplexer. *Y* is the output

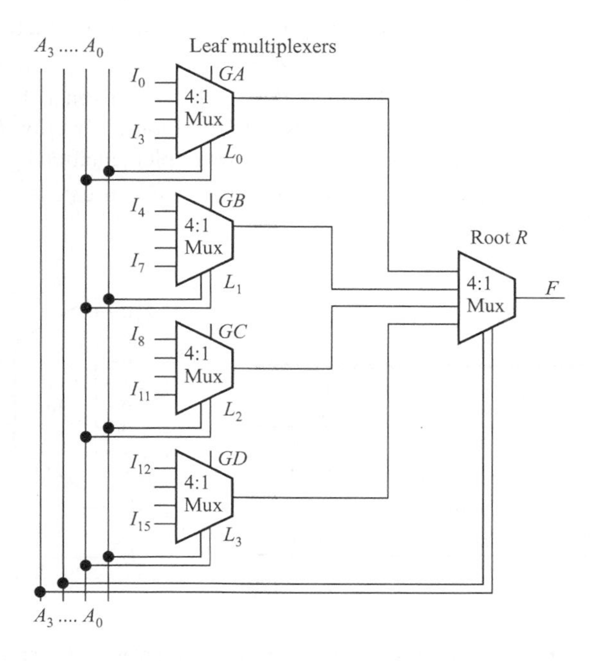

FIGURE 10.11 An exemplary tree type arrangement of the five multiplexers: four at the leaves and one at the root. We use the multiplexers of truth table shown in Figure 10.9 (b) and $m' = 4$, $n = 2$, $m = 16$ and $i = 4$.

and one at the root. We use the multiplexers of truth table shown in Figure 10.9(b). $m' = 4$, $n = 2$ and design circuit for $m = 16$ and $i = 4$ in this example. We get the (16 of 1) multiplexing from five numbers (4 of 1) multiplexers. Function table is as per Table 10.13. (*GS* are control bits for the MUXs.)

Channel selector inputs at the root multiplexer *R* is selecting a leaf multiplexer (L_0 or L_1 or L_2 or L_3) among the four multiplexers at leaves L_0, L_1, L_2 and L_3. The channel selector inputs A_0 and A_1 select the one out of the four inputs among I_0, I_1, I_2 and I_3 at each of the multiplexer.

Point to Remember

A number of multiplexers can be arranged in a tree topology to obtain a bigger numbers of channels multiplexer.

10.4.2.2 Logic Design and Boolean Function Implementation Using a Multiplexer

Example 10.18 will show a logic design and implementation using a multiplexer for $F_1 = \Sigma m(3, 7, 9, 10)$ and F_2 is $\Sigma m(2, 7, 12, 15)$. Example 12.5 will show a use of mux as building block for Gray code converter. Assume a Boolean function *F* is $\Sigma m(3, 7)$ for a three variable expression. Assume channel selector inputs are A_0, A_1 and A_2. Let the inputs to an 8 of 1 multiplexer are X_0, X_1, …., X_6 and X_7. If a Karnaugh map is drawn for a three variable input, then at $m(3)$ position and at $m(7)$ position there is 1. (Refer Figure 10.12a).

The output F' is as follows:

$$F' = X_0.A_0.A_1.A_2. + X_1.A_0,A_1.A_2 + X_2.A_0,A_1.A_2 + X_3.A_0,A_1.A_2 + X_4.A_0,A_1.A_2 + X_5.A_0,A_1.A_2 + + X_6.A_0,A_1.A_2 + X_7.A_0,A_1.A_2.$$

Now if $F = \Sigma m(3, 7)$ is to be implemented, if X_3 and X_7 are given as 1 (at the positions of 1s at the Karnaugh map) and remaining inputs = 0 then Boolean function *F* is $\Sigma m(3, 7)$ is implemented by the multiplexer. Figure 10.12(b) shows implementation by 8 of 1 Mux.

Suppose we are using a gated output multiplexer. $F' = F'^*$ if gate is disabled and $F = F'.G$ if gate is enabled.

Three variable Karnaugh Map

AB \ C	C	$\overline{C}$ 0	C 1
$\overline{A}\,\overline{B}$	00	$m(0) = 0$	$m(1) = 0$
$\overline{A}\,B$	01	$m(2) = 0$	$m(3) = 1$
$A\,B$	11	$m(6) = 0$	$m(7) = 1$
$A\,\overline{B}$	10	$m(4) = 0$	$m(5) = 0$

(a)

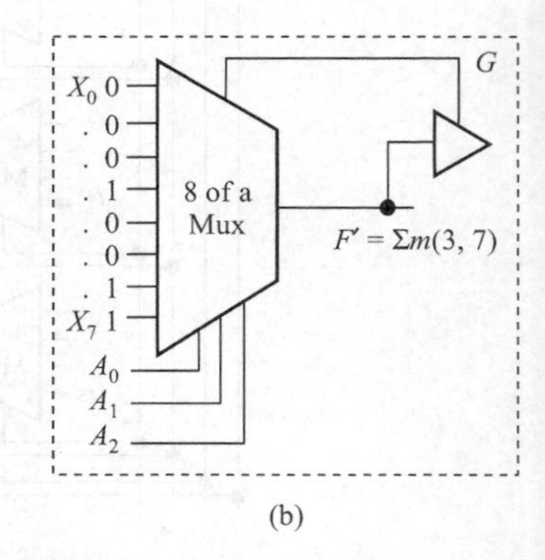

(b)

FIGURE 10.12 (a) Karnaugh map for *F* is $\Sigma m(3, 7)$ for a three variable expression with 1 at the cells corresponding to $m(3)$ and $m(7)$. Therefore the *X*3 and *X*7 are given as 1 (at the positions of 1s at the Karnaugh map) and remaining inputs = 0. (b) Boolean variables are given as the inputs at *A*0, A_1 and A_2 when implementing the circuit with the help of a multiplexer.

Example 10.19 will give a logic design of 8 of 1 Mux using three 4 of 1 Muxs. Example 10.20 will give a logic design and implementation using a multiplexer for $F' = \Pi M(1, 7, 9, 13)$ (Implementation of POS Boolen expression).

Point to Remember

> A multiplexer with one-line output is also a Boolean expression implementer and implements of a SOP (or POS cells) Karnaugh map. The literal variables are the inputs to the channel selector lines and minterm (or mux-term) selectors are given at the input lines of a multiplexer.

A three-variable expression and its Karnaugh map can also be implemented by a 4 to 1 multiplexer with two variables at the channel selector lines and third variable, its complement or 1 or 0 at the channel input lines (Example 10.19).

10.5 DEMULTIPLEXER

10.5.1 Demultiplexer Definition

A demultiplexer is a circuit that takes the binary information and sends it to select appropriate channel. A demultiplexer gives an output at a unique channel (one-line or multiple line) among according to a unique combination of the channel selector inputs at the input at one-line (or at multiple lines).

MSI IC74155 is a dual four-channel demultiplexer. It gives two output Y as well as its complement at the selected channel. It gives output $Y0$, $Y1$ $Y2$ and $Y3$ as per the channel selector lines A_0 and A_1. When $A0$ and A_1 are 00, $Y0 = X$, 10 then $Y1 = X$, and 01 then $Y2 = X$, 11 then $Y3 = X$, respectively. X is the input to be sent to an addressed channel.

MSI IC 74156 has open collector outputs. It has dual inputs, input and its complement.

Point to Remember

> Action of a one active line input demultiplexer is opposite of that of a one active line output multiplexer. Demultiplexer can be differentiated from multiplexer by a greater number of outputs than inputs compared to a multiplexer.

10.5.1.1 Demultiplexer (Line Demultiplexer)

Let us assume that we have eight line-outputs for activating the different functions at the different channels. One is for initiating an addition; other is for initiating a subtraction; other initiating an increment and so on. We select one operation from the channel selection inputs. A demultiplexer will let us find one. An input 1 is sent to the addressed channel by the demultiplexer.

For example, we want to get output at address 0, $Y0 = 1$ when the three channel selector (address selector) pins are at 000, output at address 1, $Y1 = 1$ when selector pins are at 001, and so on. Output at address 7, $Y7 = 1$ when selector pins are at 001. Remaining outputs remain at 0s. (Active 1 output case).

TABLE 10.14 Truth table of a demultiplexer for sending I at output using 8-line F_0 to F_7 selector from 3 channel select bits

Channel (address) select inputs			Data		Outputs at the 8 channels							
A_0	A_1	A_2	I	G	$\overline{F}_0$	$\overline{F}_1$	$\overline{F}_2$	$\overline{F}_3$	$\overline{F}_4$	$\overline{F}_5$	$\overline{F}_6$	$\overline{F}_7$
0	0	0	*I*	1	I	z	z	z	z	z	z	z
1	0	0	*I*	1	z	I	z	z	z	z	z	z
0	1	0	*I*	1	z	z	I	z	z	z	z	z
1	1	0	*I*	1	z	z	z	I	z	z	z	z
0	0	1	*I*	1	z	z	z	z	I	z	z	z
1	0	1	*I*	1	z	z	z	z	z	I	z	z
0	1	1	*I*	1	z	z	z	z	z	z	I	z
1	1	1	*I*	1	z	z	z	z	z	z	z	I
Any combination of inputs			x	0	*	*	*	*	*	*	*	*

'*I* is the input at pin *I* to be sent to the addressed channel. The z = 0 or 1 or * as per the demultiplexer design. *x* means don't care condition.

Alternatively, $Y0 = 0$ at one line when the three channel selector (address selector) pins are at 000, output $Y1 = 0$ when selector pins are at 001, and so on. Output $Y7 = 0$ when selector pins are at 001. Remaining outputs remain at 0s. (Active 0 output case).

Example 10.21 will describe an exemplary combination circuit for the demultiplexer for implementing Table 10.14.

10.5.1.2 Demultiplexer as a Decoder

Let us compare the functional truth tables in Table 10.6 and 10.14. We note that in a demultiplexer if the encoded inputs are given at the channel address lines and a data is applied at control enable pin, then demultiplexer also functions as a line decoder. Therefore MSI IC74154 is a four-channel demultiplexer, which also functions as 2 to 4 decoder. (MSI IC 74159 is open collector outputs version of 74154).

Point to Remember

> Like a decoder, a demultiplexer can also be used as a minterm or maxterm generator. Therefore, a demultiplexer can be also used to implement a Boolean expression(s). It is when the channel select pins are used to give Boolean inputs. If I = 0 and z = 1, the circuit is maxterm selector. If I = 1 and z = 0, the circuit is minterm selector.

10.5.1.4 Demultiplexers arranged in Tree Topology

A set of output-enabling pin helps in placing more demultiplexers in parallel. For example, consider a circuit for four pins for the addressed selector for the 16 outputs (channels). [1] A_0 A_1 and A_2 are given to first 3 to 8 demultiplexer and data to *I* pin. Enable pin $E = 1$ (active 1 case). [2] A_0, A_1 and A_2 are given to second demultiplexer also. *I'* is the input at *I* pin. *E* is connected to *E* of the first demultiplexer after a NOT operation. [3] I and *I'* are made common in both the demultiplexers. Now when $E = 1$, the first demultiplexer gives the output, else the second.

Point to Remember

> A number of demultiplexers can be arranged in tree topology to obtain a bigger number of input-bits decoder by using the control gate (enable) pin(s) and applying inputs to the enable pins also in different combinations.

EXAMPLES

Example 10.1

Write a general formula for an i-th stage carry to next stage using AND operation as a new term and OR operation as a propagating term at the i-th stage.

Solution

Using the formula; $Cy = A.B + A.Cy' + B.Cy'$, the carry at stages 0 to $(m - 1)$th for an m-bit adder can be written as follows:

$Cy'_0 = 0 = Cy_{-1}$; Cy_{-1} = Previous stage carry at the 0th stage = 0.

$Cy_0 = A_0.B_0 = n0 + Cy_{-1}$; (Let $n0$ is a new term product of the adding elements at stage 0).

$Cy_1 = A_1.B_1 + (A_1 + B_1), n0 = n1 + p1.n0$; (Let $p1$ is the propagating term, which is equal to OR operation of the adding elements at stage 1. The $n1$ is the new term at the stage 1).

$Cy_2 = A_2.B_2 + (A_2 + B_2)(n1 + p1.n0) = n2 + p2.(n1 + p1.n0)$; (Let $p2$ is the propagating term, which is equal to OR operation of the adding elements at stage 2. The $n2$ is the new term at the stage 2, equal of product of the adding elements).

Therefore, a general formula for next stage carry from an i-th stage is as follows:

$Cy_i = n_i + p_i.n_{i-1} + p_i.p_{i-1}.n_{i-2} + p_i.p_{i-1}.p_{i-2}.n_{i-3} + \ldots + p_i.p_{i-1}.p_{i-2} \ldots\ldots p_0.n_0 + p_i.p_{i-1}.p_{i-2} \ldots\ldots p_{-1}.\ Cy_{-1}$, where $p_{-1} = 1$, Cy_{-1} is the input carry to the 0th stage, $p0$ is 0th stage OR operation of adding elements, …, pi is the i-th stage OR operation and n_i is the i-th stage product (AND) operation between the adding elements.

An adder circuit of m bits and 0th to $(m - 1)^{th}$ stages is based upon the four stage exempling circuit of Figure 10.2 and Cy of last stage is given by the above equation. Each calculation of ORs among the propagating terms, second, third, fourth takes longer time than the previous stage.

Example 10.2

Calculate the delay at an i-th stage in finding Cy_i assuming that each stage of FA in Figure 10.2 takes propagation time = t_s.

Solution

Every stage gives output carry after t_s to the next stage. Note that inputs A and B are readily available but the carry is available after the previous stage result is obtained. Time taken at the i-th stage = $i.t_s$.

Example 10.3

Implement a carry lookahead circuit for each stage using the general formula derived in Example 10.1 for an i-th stage carry to next stage and generate the next stage carry using AND operation as a new term and OR operation as a propagating term at the i-th stage.

Solution

Look ahead carry generator is a logic circuit to implement at each *i*-th stage, n_i, p_i and s_i. It calculates the sum term in advance so that carry input at each stage is readily available. This creates a fast additer circuit. Figure 10.13(a) shows a summation element at each stage, which has two outputs n_i and p_i. Figure 10.13(b) shows a fast addition circuit using the carry lookahead general formula derived in Example 10.1.

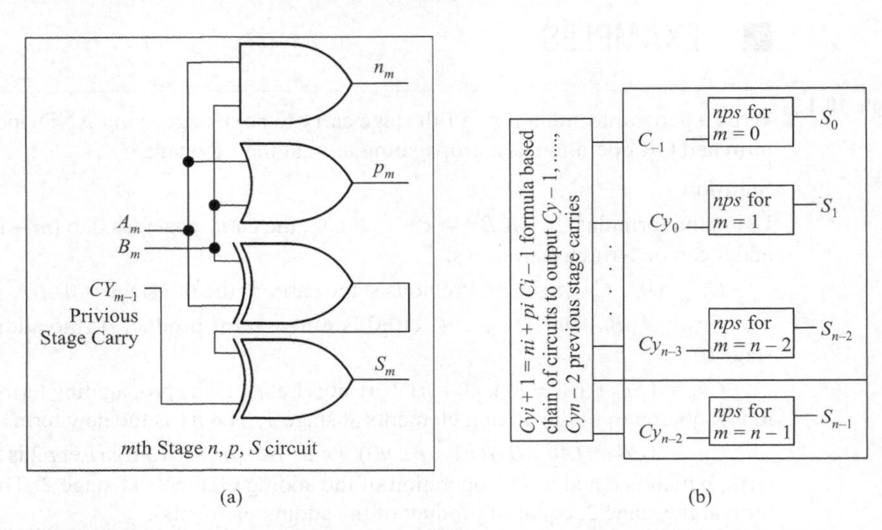

FIGURE 10.13 (a) Summation element at each stage, which has two outputs n_i and p_i. (b) Fast addition circuit using the carry lookahead general formula derived in Example 10.1.

Example 10.4 Calculate the delay at an *i*-th stage using the lookahead carry generator circuit. Assume that summation unit in Figure 10.3 takes time = t_s. Time taken by lookahead carry generator = t_c.

Solution

Time taken at the *i*-th stage = $t_c + t_s$. (Every stage gives output carry after t_s to the next stage).

Example 10.5 Formulate the problem and how will you implement a 3-line to 8-line decoder of truth table as per Table 10.5. Output is active at 0 and inactive at = 1 (output when line not selected).

Solution

Assume that outputs are $Y0$ to $Y7$. Since 0 occurs only at one output and remaining outputs remains 1, therefore, it will easier to implement the logic design by POS form of three variable (literals). From the Table 10.5, the outputs in terms of the maxterms are as follows: $\overline{Y}0 = M(0)$; $\overline{Y}1 = M(1)$; $\overline{Y}2 = M(2)$; $\overline{Y}3 = M(3)$; $\overline{Y}4 = M(4)$; $\overline{Y}5 = M(5)$; $\overline{Y}6 = M(6)$; $\overline{Y}7 = M(7)$.

$\overline{Y}0 = (A + B + C)$; $\overline{Y}1 = (A + B + \overline{C})$; $\overline{Y}2 = (A + \overline{B} + C)$; $\overline{Y}3 = (A + \overline{B} + \overline{C})$; $\overline{Y}4 = (\overline{A} + B + C)$; $\overline{Y}5 = (\overline{A} + B + \overline{C})$; $\overline{Y}6 = (\overline{A} + \overline{B} + C)$; $\overline{Y}7 = (\overline{A} + \overline{B} + \overline{C})$; $A = A_2$, $B = A_1$, and $C = A_0$.

Figures 10.14(a) and (b) show the logic circuit using OR-AND gates and NAND gates.

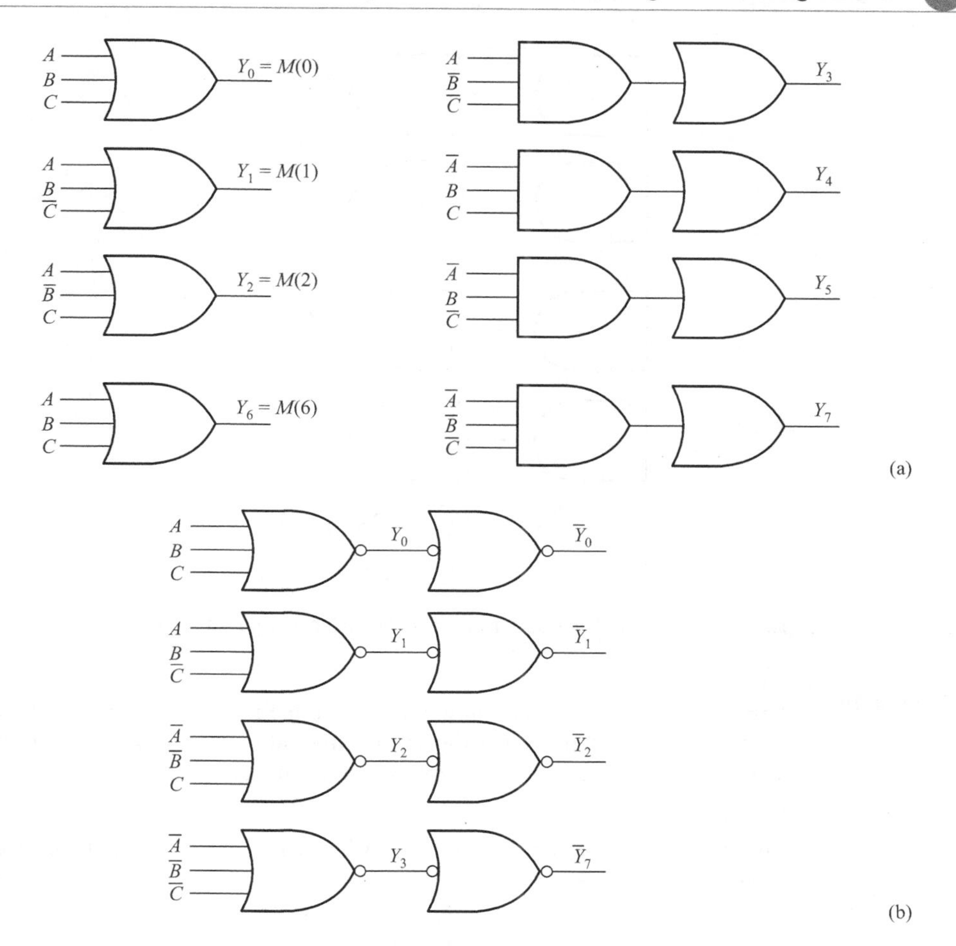

FIGURE 10.14 (a) Logic circuit using OR-AND gray of gates (b) Logic circuit using NOR gates for a decoder truth table as per Table 10.5.

Example 10.6 Formulate the problem and how will you implement a BCD- to 12 -line decimal decoder on defined in truth table (Table 10.7). Output is active at 1 and inactive at = 0 (output when line not selected).

Solution

From the Table 10.7, the outputs in terms of the minterms are as follows: $Y0 = m(0)$; $Y1 = m(1)$; $Y2 = m(2)$; $Y3 = m(3)$; $Y4 = m(4)$; $Y5 = m(5)$' $Y6 = m(6)$; $Y7 = m(7)$; $Y8 = m(8)$; $Y9 = m(9)$; $Y10 = m(16)$; $Y11 = m(17)$

The combinational circuit is as per Figure 10.15.

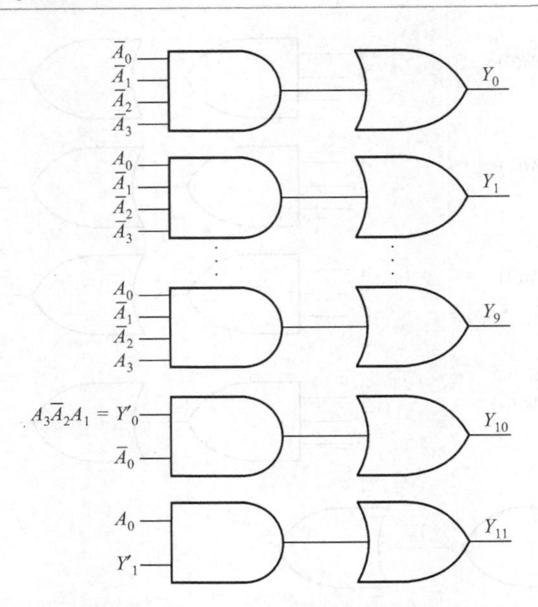

FIGURE 10.15 Implement a BCD to 12-line Decimal decoder of truth table as per Table 10.7.

Example 10.7 Formulate the problem and how will you implement a BCD- to seven segments LED decoder with segments as per Figure10.16(a) and truth table as per Table 10.8. Output is active at 1 and inactive at = 0 (output when line not selected).

Solution

Recall columns of Table 10.8 for *a* to *g*. We observe that the set of outputs for the segments should be as follows.

$$a = \Sigma m\,(0, 2, 3, 5, 6, 7, 8, 9);$$
$$b = \Sigma m\,(0, 1, 2, 3, 4, 7, 8, 9);$$
$$c = \Sigma m\,(0, 1, 3, 4\ 5, 6, 7, 8, 9);$$
$$d = \Sigma m\,(0, 2, 3, 5, 6, 8, 9);$$
$$e = \Sigma m\,(0, 2, 6, 8);$$
$$f = \Sigma m\,(0, 4, 5, 6, 8, 9);$$
$$g = \Sigma m\,(2, 3, 4, 5, 6, 8, 9).$$

One of the way of implementing above functions are using OR gates to the outputs of a decoder *Y*0 to *Y*9 as inputs and get the outputs *a*, *b*, *c*, *d*, *e*, *f* and *g*. Figure 10.16(b) shows the combinational circuit made from a decoder.

Example 10.8 Logic design and implement the Boolean functions $F_1 = \Sigma m(3, 7, 9, 10)$ and F_2 is $\Sigma m(2, 7, 12, 15)$ by using a decoder.

Solution

Since highest minterm is $m(15)$, a four bit input based decoder is needed. Since the minterms implement both the Boolean functions, we need a decoder with active 1 (inactive state = 0) output. Figure 10.17 shows an implementation.

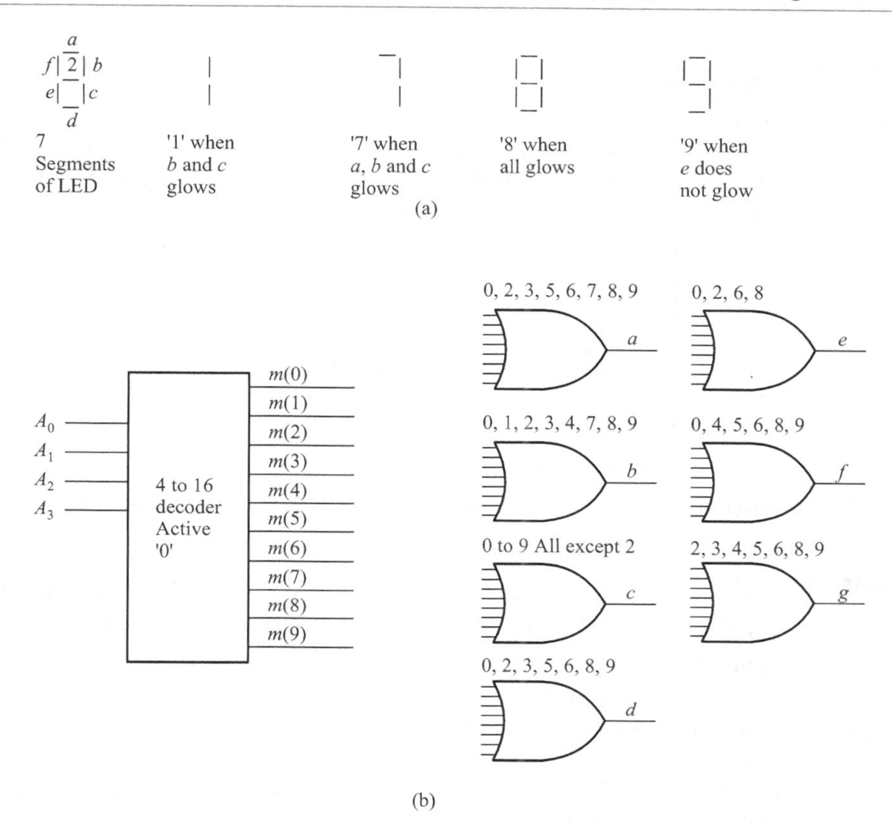

FIGURE 10.16 (a) Seven segments of an LED and how these display 0, 1, 2, up to 9 decimal digits. (b) Combinational circuit made from a decoder and the OR at the outputs of the decoder.

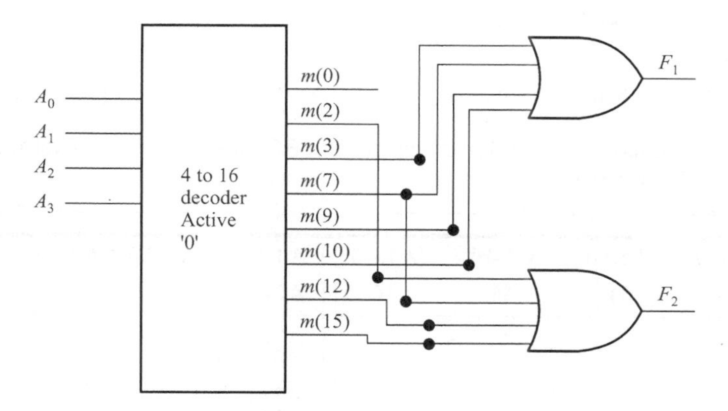

FIGURE 10.17 Implement the Boolean functions $F_1 = \Sigma m(3, 7, 9, 10)$ and F_2 is $\Sigma m(2, 7, 12, 15)$ by using a decoder.

Example 10.9 Logic design and implement a Boolean function $F' = \Pi M(1, 7, 9, 13)$ by using a decoder.

Solution

Since highest minterm is $M(13)$, a four bit input based decoder is needed. Since the maxterms implement the Boolean function, we need a decoder with active 0 (inactive state = 1) output. Figure 10.18 shows the implementation.

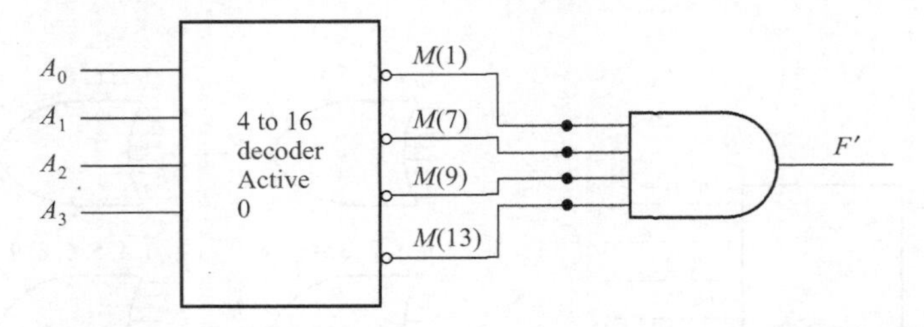

FIGURE 10.18 An implementation of a Boolean function $F' = \Pi M(1, 7, 9, 13)$ by using a decoder.

Example 10.10 Design a circuit for a decoder with three inputs and active 0 and 1 outputs and implement the decoder circuit by NANDs alone. How will you extend the circuit for four inputs.

Solution

Since a decoder with three inputs and active 0 gives the outputs and after NOTs gives the minterms, we can implement the decoder circuit by AND-OR arrays first. Figure 10.19(a) shows this circuit. We convert the circuit into NANDs based circuit using DeMorgan theorem. Figure 10.19(b) shows the circuit. We can generate in identical manner the terms $m(0)$ to $m(15)$ in case of 4 inputs.

Example 10.11 Logic design and implement a Boolean function, which Karnaugh map in Table 10.14 represents. Use a decoder.

Solution

A five-variable Karnaugh-map is given in Table 10.14

TABLE 10.14 Five variable Karnaugh Map set with 32 minterms of the SOP Expression and upper layer for $A4 = 0$ on the left side and lower layer for $4 = 1$ on the right side

$A3\overline{A2}$ \ $A1A0$		$\overline{A1}\,\overline{A0}$ 00	$\overline{A1}A0$ 01	$A1A0$ 11	$A1\overline{A0}$ 10	$\overline{A1}\,\overline{A0}$ $A3A2$	$\overline{A1}\,\overline{A0}$ 00	$\overline{A1}A0$ 01	$A1A0$ 11	$A1\overline{A0}$ 10
		$A4 = 0$				$A4 = 1$				
$\overline{A3}\,\overline{A2}$	00	1.	1	1	1	$\overline{A3}\,\overline{A2}$ 00	1	1	1	1
$\overline{A3}A2$	01	1	1	1	1	$\overline{A3}A2$ 01	1	1		
$A3A2$	11					$A3A2$ 11				
$A3\overline{A2}$	10	1	1			$A3\overline{A2}$ 10	1	1	1	1

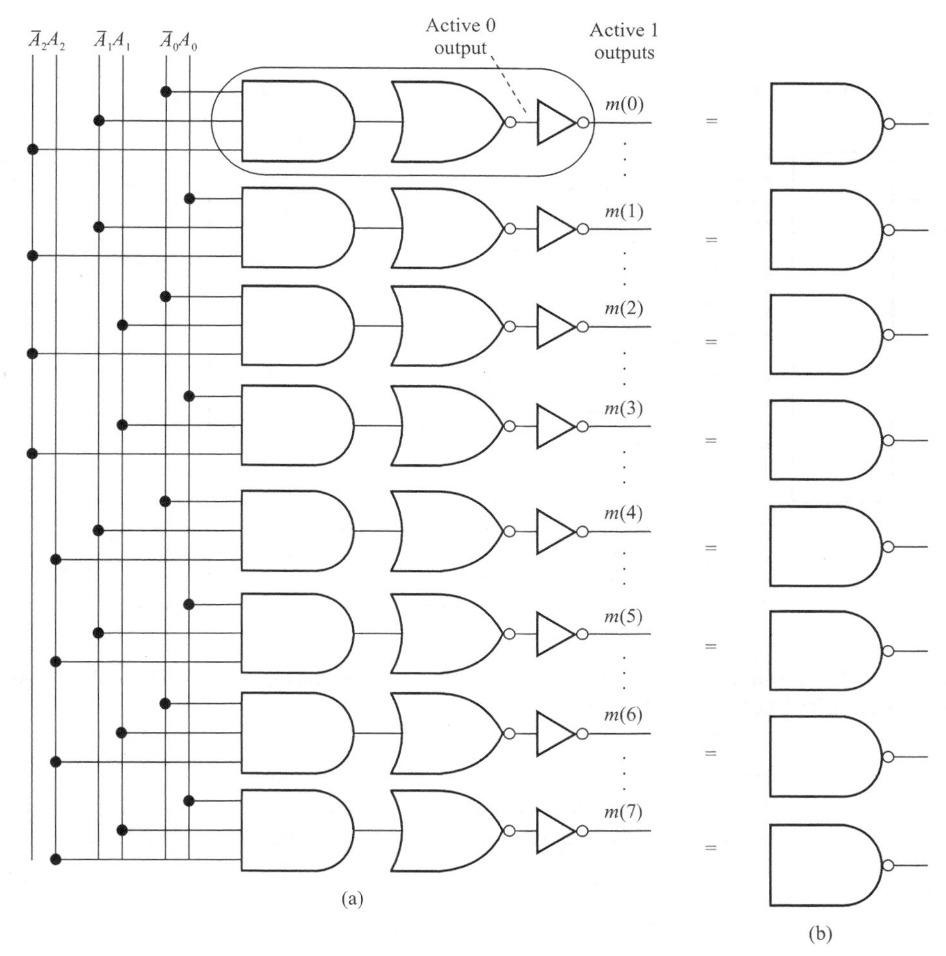

FIGURE 10.19 (a) A decoder-circuit, which implements the minterms at the outputs using AND-OR arrays (b) Circuit with NANDs alone.

The map shows on the left side one octet with $\overline{A}4.\overline{A}3$ common. Left hand side also shows one quad $\overline{A}4.\overline{A}2.\overline{A}1$. On right hand side, there are, we can redesign the map as four variable map with 2 minterms a quad for $A4.\overline{A}3.\overline{A}1$ and and one octet (wrapping adjacencies).

Result is $Y = \overline{A}4.\overline{A}3 + \overline{A}4.\overline{A}2.\overline{A}1 + A4.\overline{A}3.\overline{A}1 + A4.\overline{A}2$.

Problem is now formulated for implementation of Boolean expression for Y from prime implicants. Since a decoder with five binary inputs and output active 1 gives the outputs, which are also the implementation of the minterms, we can design logic circuit using the minterm outputs only by placing the OR gate. Figure 10.20 shows the implementation.

Example 10.12 Formulate the problem that how will you implement an 8-line to 3-line encoder of truth table as per Table 10.9. Input is active at 1 when that input line selected and is to be encoded) and inactive at = 0 (input when that line not selected).

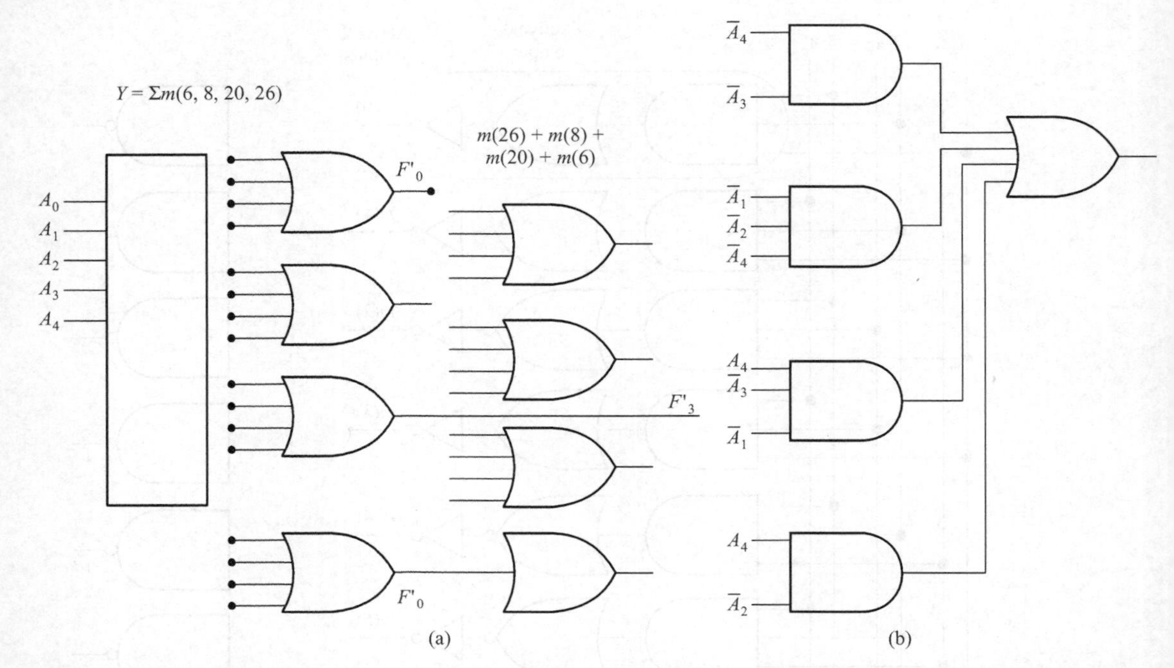

FIGURE 10.20 (a) An implementation of a Boolean function, which Karnaugh map in Table 10.14 represents using a decoder as a basic building block (b) Using three input AND-OR array.

Solution

Assume that output is 000 when input is $A0 = 1$ and $A1$ to $A7 = 0$. Since there are three outputs, three Boolean Expressions $Y0'$, $Y1'$ and $Y2'$ are as follows.

$$Y0' = A1.\overline{A}3.\overline{A}5.\overline{A}7 + \overline{A}1.A3.\overline{A}5.\overline{A}7 + \overline{A}1.\overline{A}3.A5.\overline{A}7 + (\overline{A}1.\overline{A}3.\overline{A}5.A7)$$

$$Y1' = A2.\overline{A}3.\overline{A}6.\overline{A}7 + \overline{A}2.A3.\overline{A}6.\overline{A}7 + \overline{A}2.\overline{A}3.A6.\overline{A}7 + (\overline{A}2.\overline{A}3.\overline{A}6.A7)$$

$$Y2' = A4.\overline{A}5.\overline{A}6.\overline{A}7 + \overline{A}4.A5.\overline{A}6.\overline{A}7 + \overline{A}4.\overline{A}5.A6.\overline{A}7 + (\overline{A}4.\overline{A}5.\overline{A}6.A7)$$

Let S is 1 when at least one input is 1. It means all the inputs are not 0 or input is not invalid, it is a valid condition. Therefore,

$$S = A0 + A1 + A2 + A3 + A4 + A5 + A6 + A7 \text{ and } Y0 = (S.Y0').\overline{A}_0$$

$$Y1 = (S.Y1').\overline{A}_0 \text{ and } Y2 = (S.Y2').\overline{A}_0.$$

Figure 10.21 shows the logic circuit for $Y0$ and S.

Example 10.13

Formulate the problem that how will you implement a 16-line to 4-line encoder of truth table as per Table 10.11. Input is active at 1 Input when that line selected and is to be encoded and active at = 0 (Input when that line not selected).

Solution

Assume that output is 0000 when input is $A0 = 1$ and $A1$ to $A15 = 0$. Since there are four outputs, three Boolean expressions $Y0'$, $Y1'$, $Y2'$ and $Y3'$are as follows.

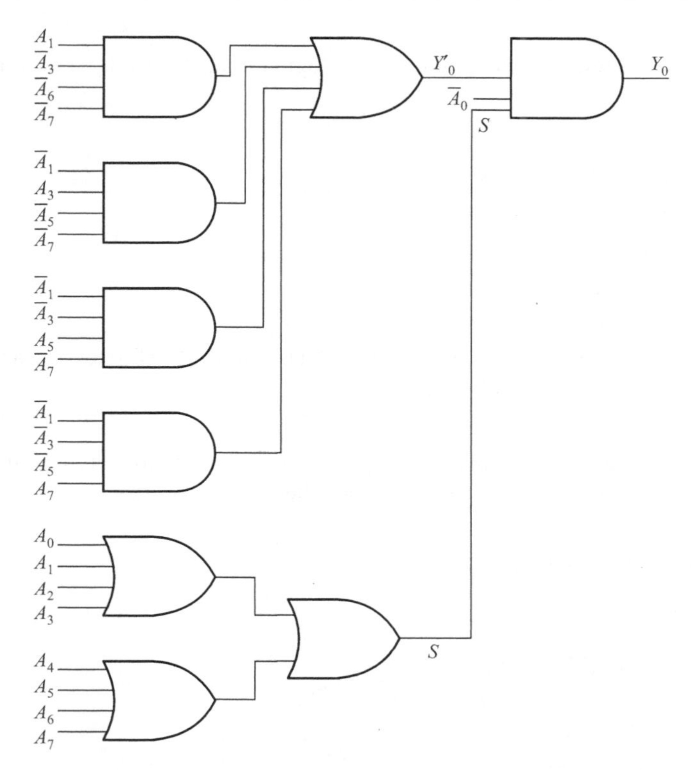

FIGURE 10.21 Implementation of an 8-line to 3-line encoder of truth table as per Table 10.9 (Only Y_0 output shown. We can draw in similar way for Y_1 and Y_2).

$$Y0' = \Sigma m(1) + m(3) + m(5) + m(7) + m(9) + m(11) + m(13) + m(15)$$

$$Y1' = \Sigma m(2) + m(3) + m(6) + m(7) + m(10) + m(11) + m(14) + m(15)$$

$$Y2' = \Sigma m(4) + m(5) + m(6) + m(7) + m(12) + m(13) + m(14) + m(15)$$

$$Y3' = \Sigma m(8) + m(9) + m(10) + m(11) + m(12) + m(13) + m(14) + m(15)$$

Let S is 1 when at least one input is 1. It means all the inputs are not 0 or input is not invalid, it is valid condition. Therefore,

$$S = A0 + A1 + \ldots + A14 + A15 \text{ and } Y0 = (S.Y0').\overline{A}_0;$$

$$Y1 = (S.Y1').\overline{A}_0;\ Y2 = (S.Y2').\overline{A}_0 \text{ and } Y3 = (S.Y3').\overline{A}_0;$$

Example 10.14 Implement by logic design combinational logic circuit for truth table in inset of Figure 10.9(a).

Solution

Boolean expressions are as follows:

$F0 = A.I_0; F1 = A.I_1$. This circuit easily implements by two ANDs and one NOT gate.

Example 10.15

Implement by logic design combinational logic circuit for truth table in inset of Figure 10.9(b).

Solution

Boolean expressions are as follows:

$$F_0 = \overline{A}_1.\overline{A}_2.\,I_0;$$
$$F_1 = A_1.\overline{A}_2.\,I_1;$$
$$F_2 = \overline{A}_1.A_2.\,I_2;$$
$$F_3 = A_1.A_2.\,I_3;$$

This circuit easily implements by four ANDs and two NOT gates. F is a common line in case only one is active at an instant.

Example 10.16

Draw logic design for combinational logic circuit for truth table in inset of Figure 10.9(b) with two control pins one with active 1 and other with active 0.

Solution

Figure 10.22 shows a logic design.

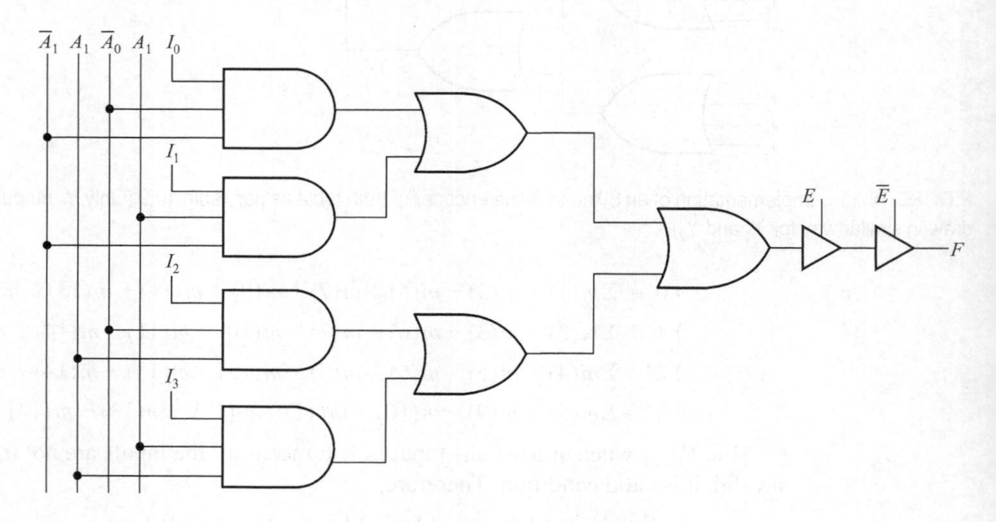

FIGURE 10.22

Example 10.17

How will we arrange in tree topology for (32 of 1) multiplexing using the nine numbers of (4 of 1) multiplexers?

Solution

Refer to the Figure 10.23, which shows the tree topology. Eight Muxs are given inputs between $I_{31} - I_0$, four each one. Then their output are given as inputs to two Muxs. At third stage, the 2 Muxs output are given to a Mux as inputs. The output of this Mux is one among $I_0 - I_{31}$ de pending on $A_4A_3A_2A_1A_0$.

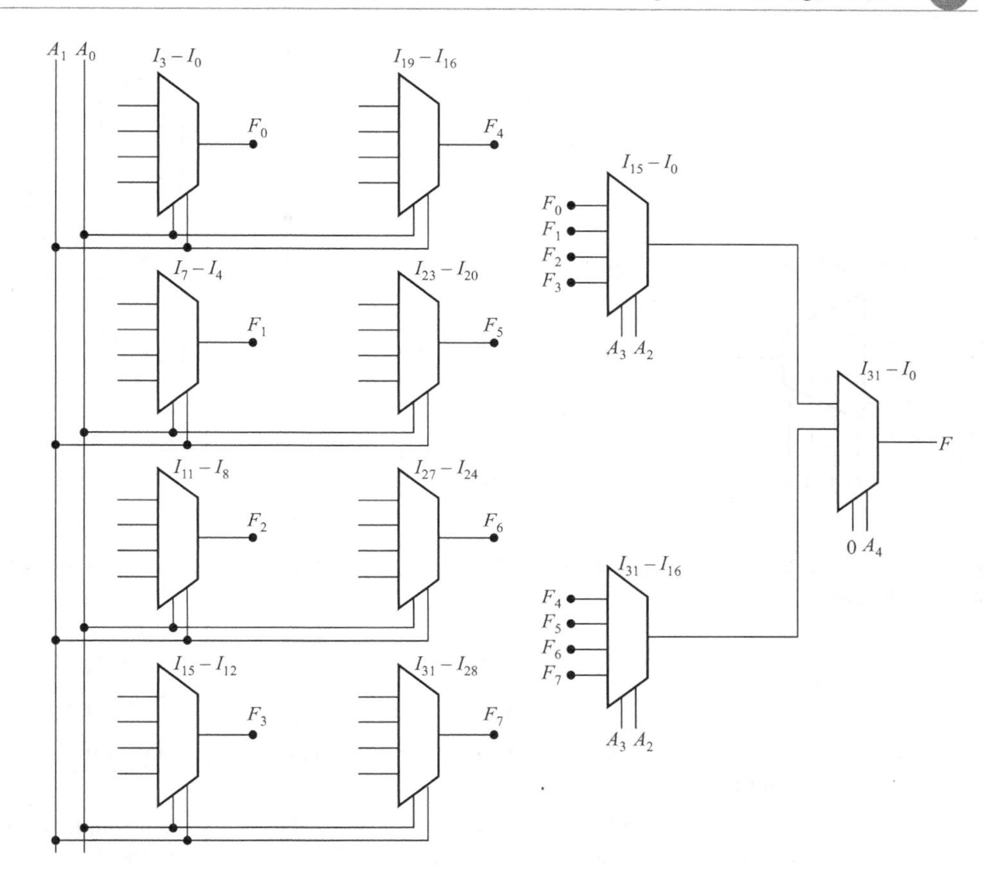

FIGURE 10.23

Example 10.18 Give a logic design and implementation using the multiplexers for $F_1 = \Sigma m(3, 7, 9, 10)$ and F_2 is $\Sigma m(2, 7, 12, 15)$.

Solution

Multiplexer is Boolean expression implementer when the inputs are given at the channel selector pins and the inputs are used also as a minterm generator for F_1 give inputs $X3$, $X7$, $X9$ and $X10 = 1$, other inputs as 0s. Apply the four inputs at the channel select input pins. For F_2, make $X2$, $X7$, $X12$ and $X15$ as 1 and apply inputs at the channel select pins. Figure 10.24(a) shows an implementation for F_1 and F_2.

Example 10.19 Give a logic design and implementation using the multiplexers for $F_1 = \Sigma m(3, 7)$ using a 4 of 1 multiplexer.

Solution

The output F' is as follows:

$$F' = X0.\overline{A}_0.\overline{A}_1.\overline{A}_2. + X1.A_0.\overline{A}_1.\overline{A}_2 + X2.\overline{A}_0.A_1.\overline{A}_2 + X3.A_0.A_1.\overline{A}_2$$
$$+ X4.\overline{A}_0.\overline{A}_1.A_2 + X5.A_0.\overline{A}_1.A_2 + X6.\overline{A}_0.A_1.A_2 + + X7.A_0.A_1.A_2;$$

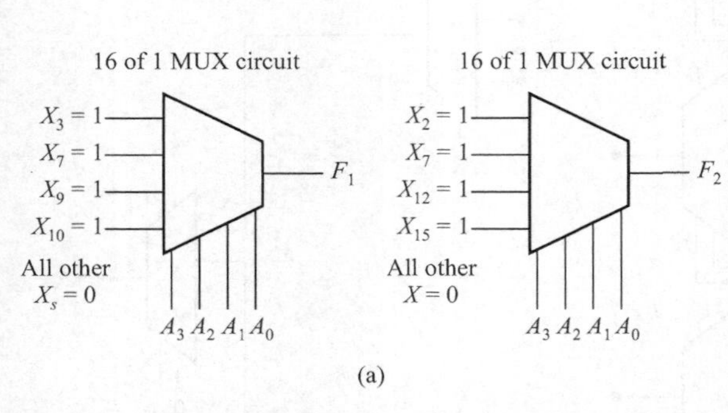

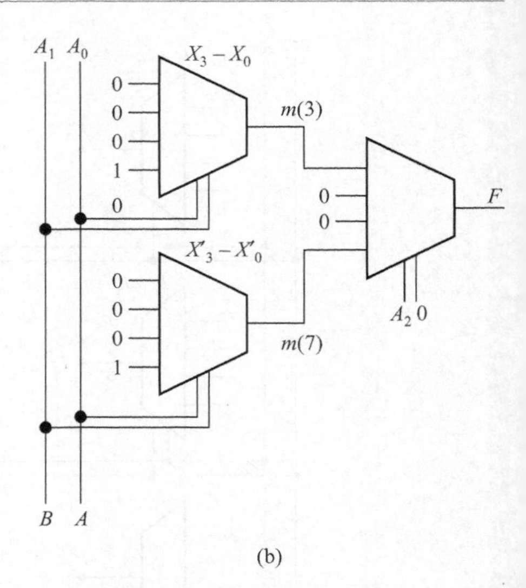

FIGURE 10.24 (a) Implementation of $F = \Sigma m(3, 7, 9, 10)$ and $\Sigma m(2, 7, 12, 15)$ (b) Implementation of $F_1 = \Sigma m(3, 7)$ using a 4 of 1 multiplexer.

$$F' = \overline{A}_2.\overline{A}_1\,(X0.\overline{A}_0. + X1.A_0) + \overline{A}_2.A_1\,(X2.\overline{A}_0. + X3.A_0)$$

$$+ A_2.\overline{A}_1\,(X4.\overline{A}_0. + X5.A_0) + A_2.A_1\,(X6.\overline{A}_0. + X7.A_0)$$

Let $F' = A_0.A_1.(I0) + A_0.A_1.(I1) + A_0.A_1.(I2) + A_0.A_1.(I3)$ in terms of inputs to a 4 of 1 multiplexer.

I0, I1, I2 and I3 are either 0 or 1 for A_2 or $\overline{A}_2 = 1$. F' can be implemented by using A_0 and A_1 channel selector inputs and result of the sum terms in the above four terms as the inputs X_0', X_1', X_2' and X_3'. When $F_1 = \Sigma m(3, 7)$, we have $X0 = X1 = 0$ and therefore I0 = 0.I1 and I2 = 0 and I3 =1, and for other mux also I3 = 1. Figure 10.24(b) gives the implementation circuit. I3 and I′3 are given inputs = $\overline{A}_2$.I3 and A2.I′3.

Example 10.20

Give a logic design and implementation using the multiplexers for $\overline{F}_1 = \Pi M(1, 7, 9, 15)$ using a 4 of 1 multiplexer

Solution

Assume a Boolean function $\overline{F}$' is $\Pi M(1, 7, 9, 13)$. It means there are four literals in Boolean expression. We can implement this by a (16 of 1) multiplexer. It equivalent to $F' = \Sigma\, m(0, 2, 3, 4, 5, 6, 8, 10, 11, 12, 13, 14)$. We give complements of the four variables as the inputs at the four channel selector lines, A_0, A_1 and A_2. Also give inputs 1 at the $X0, X2, X3, X4, X5, X6, X8, X11, X12, X13$, and $X14$. Figure 10.25 gives the implementation circuit.

Example 10.21

How will you implement the design of the circuit as for logic as per truth table of a demultiplexer for sending output 1 using a 8-line F_0 to F_7 selector from three channel select bits, Assume inactive output state $z = 0$.

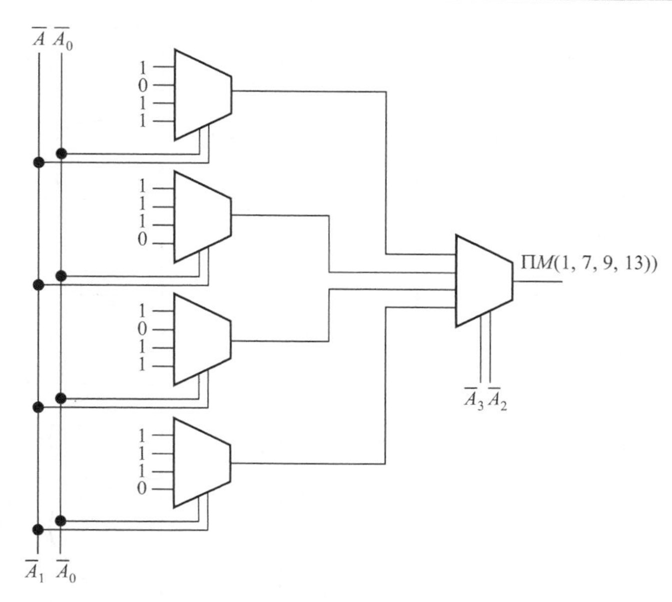

FIGURE 10.25 Implementation of Boolean expression $\Pi M(1, 7, 9, 13)$ using multiplexers.

Solution

Consider Table 10.14. First let us ignore I (4th column). The outputs in terms of the minterms are as follows: $Y0' = m(0)$; $Y1' = m(1)$; $Y2' = m(2)$; $Y3' = m(3)$; $Y4' = m(4)$; $Y5' = m(5)$; $Y6' = m(6)$; $Y7' = m(7)$ if we assume that A_0, A_1 and A_1 are the inputs (in place of the channel or address select bits).

Now consider I. $Y0 = Y0'.I$; $Y1 = Y1'.I$; $Y2 = Y2'.I$; $Y3 = Y3'.I$; $Y4 = Y4'.I$; $Y5 = Y5'.I$; $Y6 = Y6'.I$. The combinational circuit is as per Figure 10.26.

EXERCISES

1. Draw on a drawing sheet an 8-bit adder circuit using full adders as the building blocks.
2. Make the circuits for the following Boolean Expressions for a 4-stage *FA* circuit shown in Figure 10.2.

   $Cyi = n_i + p_i.n_{i-1} + p_i.p_{i-1}.n_{i-2} + p_i.p_{i-1}.p_{i-2}.n_{i-3} + \ldots + p_i.p_{i-1}.p_{i-2}\ldots\ldots p_0.n_0 + p_i.p_{i-1}.p_{i-2}\ldots\ldots p_{-1}.\ Cy_{-1}$, Assume $i = 4$.
3. Formulate the problem and how will you implement a 3-line to 8-line decoder of truth table as per Table 10.6. Output is active at 0 and inactive state is at * (tristate output when line not selected). The outputs are available only when two control gate inputs are at 0 (low) and one control gate pin is at 1 (high).
4. How will you implement (6 of 64) decoder from 74138, a 3 to 8 decoder with three control gate pins? [Hint: Use eight number (3 to 8) decoders. Give three inputs $A0$, $A1$ and $A2$ to $A0$, $A1$, $A2$ of each decoder. Give remaining three inputs $A3$, $A4$ and

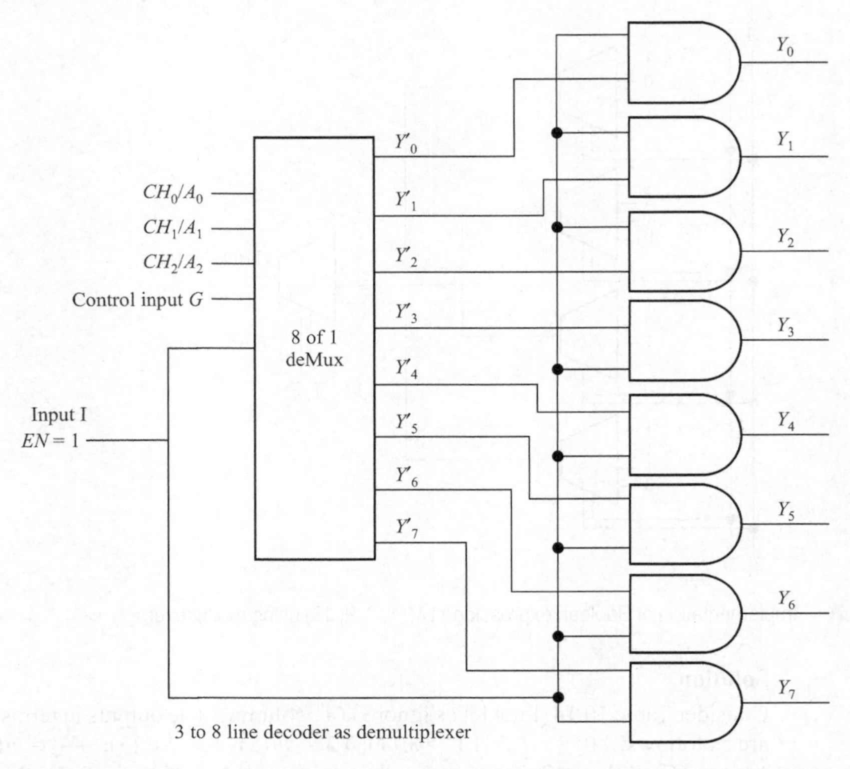

FIGURE 10.26 Combinational circuit for an 8 output demultiplexer with addressed output = input as per the channel selection bits A_0, A_1 and A_2. Except the selected output channel, all outputs $z = 0$. The addressed output = 1.

$A5$ at $G0$, $G1$, and $G2$ to each of the 8 decoders in 8 possible combinations. Only one decoder activates at a given inputs $A0$, $A1$, $A2$, $A3$, $A4$ and $A5$.]

5. How will you implement (6 of 64) decoder from 74154, a 4 to 16 decoder with two control gate pins, both active 0 inputs? [Hint: Use four number (4 to 16) decoders and control gate pins also for the additional two inputs $A4$ and $A5$.]
6. Formulate the problem and implement an 11-line to 5-line encoder (decimal to BCD priority encoder truth table as per Table 10.10. Input is active at 1 Input when that line selected and is to be encoded) and inactive = 0 (input when that line not selected).
7. Formulate a problem for implementing a 16 key keypad encoder.
8. Formulate the problem that how will you implement an 8-line to 3-line encoder of truth table as per Figure 10.8(b). Input is active at 1 input when that line selected and is to be encoded) and inactive = 0 (input when that line not selected).
   [Hint: Use maxterms in Example 10.14.]
9. Using a multiplexer implement a four variable Karnaugh map of Table 10.15 circuit.
10. Using a decoder tree implement (8 to 64) decoder.

TABLE 10.15 Four variable Karnaugh Map

AB \ CD	00	01	11	10
00			1	
01		1	1	
11		1		
10		1	x	

11. Design a multiplexer tree circuit with seven (4 to 1) multiplexers. How many channels are multiplexed at the root multiplexer?
12. Design a demultiplexer tree circuit with seven (1 to 4) demultiplexers. How many channels are demultiplexed at the root demultiplexer output?
13. Show how will you use a 4 to 16 channel multiplexer for a four-variable Boolean-expression.
14. Design a decoder based logic circuit for a Karnaugh map of three variables in which the even minterms are 0s.
15. Design a multiplexer based logic circuit for a Karnaugh map of three variables in which the even maxterms are 0s.

QUESTIONS

1. Explain a four-bit binary adder circuit.
2. Explain a 4-bit adder cum subtractor circuit, which uses the XORs as a controlled inverter.
3. Draw a circuit for a two's complement implementer using the 4-bit adder cum subtractor circuit.
4. Why is a four-bit adder circuit implemented with full adders like Figure 10.2 also called a ripple adder?
5. Why does a carry look-a-head generator give a fast adder? How much is the speed up for an 8-stage circuit?
6. Give four exemplary applications of a decoder.
7. What is a difference between a decoder and a digital demultiplexer? Explain their truth table differences by taking an example.
8. Show that decoder circuit in Figure 10.8 is also useful for generating minterms and maxterms.
9. What is the purpose of a control gate pin in a decoder?
10. What is the purpose of multiple control gate pins in a decoder?
11. What is a difference between an encoder and a decoder? Explain with an example.
12. A digital multiplexer (MUX) cannot be used as DMUX (digital demultiplexer). Why? (Hint: A digital gate is not bilateral; an analog gate can be).

13. What is a difference between an encoder and a digital multiplexer?
14. Give four exemplary applications of a multiplexer.
15. Explain the truth table differences of an encoder and a DMUX by taking an example of each.
16. What is the difference between a digital multiplexer and a digital demultiplexer? Explain with an example?
17. Implement a 4 to 1 digital multiplexer using a decoder and four tristate buffers.
18. Show a decoder circuit can be is a minterms generator.
19. Show a decoder circuit can be is a maxterms generator
20. Show a multiplexer is also a Boolean expression implementer.

CHAPTER 11

Code Converters, Comparators and Other Logic Processing Circuits

OBJECTIVE

In Chapter 9, we learnt that AND-OR arrays, OR-AND arrays, NANDs and NORs are the building blocks of combinational logic circuits. In Chapter 10, we learnt that full adder circuits, decoders, encoders and multiplexers can also be used as the bigger building blocks for the logic design.

We will learn some important logic processing circuits, namely, code converters, digital comparator for magnitude and equality, parity generators and checkers, and bit-wise AND, OR, NOT logic processing circuits.

11.1 CODE CONVERTERS

11.1.1 Codes for Decimal Numbers

11.1.1.1 Common BCD Code

Recall the Equation (2.6) of Chapter 2.

A general formula to get the total N is again as under with only w_0 changed

$$N = y_{p\max-1} \times w_0^{p\max-1} + y_{p\max-2} \times w_0^{p\max-2} + y_{p\max-3} \times w_0^{p\max-3} + \cdots + y_2 \times w_0^2 + y_1 \times w_0^1 + y_0 \times w_0^0, \qquad ...(11.1)$$

where $p = p\max - 1, p\max - 2, p\max - 3, ..., 2, 1$ and 0, and $p\max$ is maximum, the number of places used in the representation and $w_0 = 2$ for the binary numbers. The $y_0, y_1, y_2, ...$ are the digit = 0 or 1 at the right-most, left first from that, left second from that and so on.

A general formula to get the total N for the common BCD code (called 8421 code) is again as under:

$$N = y_{p\max-1} \times w'_7 + y_{p\max-2} \times w'_6 + y_{p\max-3} \times w'_5 + \cdots$$
$$+ y_2 \times w'_2 + y_1 \times w'_1 + y_0 \times w'_0, \qquad ...(11.2)$$

where $p = p\max - 1, p\max - 2, p\max - 3, ..., 2, 1$ and 0, and $p\max$ is 8 (maximum) the number of places used in the BCD) representation and $w_0 = 1; w_1 = 2; w_2 = 4, w_3 = 8, w_4 = 10\, w_0, w_5 = 10\, w_1, w_6 = 10\, w_2, w_7 = 10\, w_3$ for the binary numbers when representing BCD code. The $y_0, y_1, y_2, ...$ are the bit = 0 or 1 at the right-most, left first from that, left second from that and so on. The BCD code represented by above equation is also called 8421-code. At lowest nibble weights are 1, 2, 4 and 8. At next to lowest nibble weights are 10, 20, 40 and 80, and so on. Example 11.17 will show a design of 8421 circuit.

11.1.1.2 Excess-3 (XS-3) BCD Code

If instead of stating from 0000, we start from 0011, then the XS-3 code will be 0011, 0100, 0101, 0110, 0111, 1000, 1001, 1010, 1011 and 1100 for 0, 1, 2, 3, 4, 5, 6, 7, 8 and 9 for ten decimal numbers, respectively. A specialty of this code is that for each decimal number is self-complementing. Complement of 0 is 9. Similarly XS-3 code for 0 is 0011 and for 9 it is 1100. Complement of 1 is 8. Similarly XS-3 code for 0 is 0100 and for 8 it is 1011.

An adder for adding the BCD code and 0011 implements a logic circuit for the excess-3 code. Example 11.16 will show the design.

11.1.1.3 7536, 2421, and 5421 Format BCD Codes

(1) Formula to get the total N for the 7536 BCD code is as under:

$$N = y_{p\max-1} \times w'_7 + y_{p\max-2} \times w'_6 + y_{p\max-3} \times w'_5 + \cdots$$
$$+ y_2 \times w'_2 + y_1 \times w'_1 + y_0 \times w'_0, \qquad ...(11.3)$$

where $p = p\max - 1, p\max - 2, p\max - 3, ..., 2, 1$ and 0, and pmax is 8 (maximum) the number of places used in the BCD) representation but $w_0 = -6; w_1 = 3; w_2 = 5, w_3 = 7, w_4 = 10w_0, w_5 = 10w_1, w_6 = 10w_2, w_7 = 10w_3$ for the binary numbers when representing 7436 BCD code. The $y_0, y_1, y_2, ...$ are the bit = 0 or 1 at the right-most, left first from that, left second from that and so on. At lowest nibble weights are –6, 3, 5 and 7. At next to lowest nibble weights are –60, 30, 50 and 70, and so on.

(2) Formula to get the total N for the 2421 BCD code is again as under:

$$N = y_{p\max-1} \times w'_7 + y_{p\max-2} \times w'_6 + y_{p\max-3} \times w'_5 + \cdots$$
$$+ y_2 \times w'_2 + y_1 \times w'_1 + y_0 \times w'_0, \qquad ...(11.4)$$

where $p = p\max - 1, p\max - 2, p\max - 3, ..., 2, 1$ and 0, and $p\max$ is 8 (maximum) the number of places used in the BCD) representation but $w_0 = 1; w_1 = 2; w_2 = 4, w_3 = 2, w_4 = 10\, w_0, w_5 = 10\, w_1, w_6 = 10\, w_2, w_7 = 10\, w_3$ for the binary numbers when representing 2421 BCD code. The $y_0, y_1, y_2, ...$ are the bit = 0 or 1 at the right-most, left first from that, left second from that and so on. The 2421-BCD code represented by above equation. At lowest nibble weights are 1, 2, 4 and 2. At next to lowest nibble weights are 10, 20, 40 and 20, and so on.

(3) Formula to get the total N for the 5421 BCD code is again as under:

$$N = y_{p\max-1} \times w'_7 + y_{p\max-2} \times w'_6 + y_{p\max-3} \times w'_5 + \cdots$$
$$+ y_2 \times w'_2 + y_1 \times w'_1 + y_0 \times w'_0 \quad \text{...(11.5)}$$

where $p = p\max - 1, p\max - 2, p\max - 3, ..., 2, 1$ and 0, and $p\max$ is 8 (maximum) the number of places used in the BCD) representation and $w_0 = 1$; $w_1 = 2$; $w_2 = 4$, $w_3 = 5$, $w_4 = 10\ w_0$, $w_5 = 10\ w_1$, $w_6 = 10\ w_2$, $w_7 = 10\ w_3$ for the binary numbers when representing 5421 BCD code. The $y_0, y_1, y_2, ...$ are the bit = 0 or 1 at the right-most, left first from that, left second from that and so on. At lowest nibble weights are 1, 2, 4 and 5. At next to lowest nibble weights are 10, 20, 40 and 50, and so on.

11.1.2 Unit Distance Code Converter

In the Equation (11.1) described above, suppose 3 increments to 4, then 8421 BCD code changes will be from 0011 to 0100. The digits are changing at $p = 0, 1$ as well as 3 (0 to 1, 1 to 0 and 0 to 1).

Assume that there are circular slots in a shaft at $0^0, 3^0, 6^0, 9^0, 12^0$,..., up to 357^0. When shaft is rotating, its 8-bit position will be found from the 00000000, 00000001, 00000010, 00000011, and so on. The numbers of places where the changes occur either from 0 to 1 or from 1 to 0 are variable; 1, 2, 1 and so on.

When we convert analog signal to the digital bits, it is more reliable if for each increment or decrement step, the bit change is only at one place. Effect of some misalignment in the sensors can then be detected from the sequence of signals that is received from the rotating shaft and accounted for.

Unit distance code is a code in which next increment or decrement causes the bit-transition only at one place.

11.1.2.1 Gray Code Converter

Gray code is a unit distance code. The code is given in Table 11.1.

TABLE 11.1

Decimal value representation	Four bit binary representation variables				Four bit Gray code	Cell and minterm at the Karnaugh map of four	
	A	*B*	*C*	*D*	Y3Y2Y1Y0	Corresponding Cell number	Minterm at the Karnaugh
0	0	0	0	0	0000_{gray}	0	*m*(0)
1	0	0	0	1	0001_{gray}	1	*m*(1)
2	0	0	1	0	0011_{gray}	2	*m*(3)
3	0	0	1	1	0010_{gray}	3	*m*(2)
4	0	1	0	0	0110_{gray}	4	*m*(6)
5	0	1	0	1	0111_{gray}	5	*m*(7)
6	0	1	1	0	0101_{gray}	6	*m*(5)
7	0	1	1	1	0100_{gray}	7	*m*(4)
8	1	0	0	0	1100_{gray}	8	*m*(12)

Table 11.1 Contd.

9	1	0	0	1	1101_{gray}	9	m(13)
10	1	0	1	0	1111_{gray}	10	m(15)
11	1	0	1	1	1110_{gray}	11	m(14)
12	1	1	0	0	1010_{gray}	12	m(10)
13	1	1	0	1	1011_{gray}	13	m(11)
14	1	1	1	0	1001_{gray}	14	m(9)
15	1	1	1	1	1000_{gray}	15	m(8)

Four-variable Karnaugh Map with Gray codes and cell numbers shown inside the cells is as follows:

AB \ CD		$\overline{C}\,\overline{D}$ 00	$\overline{C}D$ 01	CD 11	$C\overline{D}$ 10
$\overline{AB}$	00	0000_{Gray} 1	0001_{Gray} 2	0011_{Gray} 3	0010_{Gray} 4
$\overline{A}B$	01	0100_{Gray} 7	0101_{Gray} 6	0111_{Gray} 5	0110_{Gray} 5
AB	11	1100_{Gray} 8	1101_{Gray} 9	1111_{Gray} 10	11110_{Gray} 11
$A\overline{B}$	10	1000_{Gray} 15	1001_{Gray} 14	1011_{Gray} 13	1010_{Gray} 12

Examples 11.4 to 11.8 will explain the gray code logic circuits.

11.1.3 ASCII (American Standard Code for Information Interchange) for the Alphanumeric Characters

Examine your computer keyboard. Alphanumeric characters (Table 11.2) are *a* to *z*, *A* to *Z*, 0 to 9 and signs *, !, @, #, $, % and so on. Examine your telephone keypad. Alphanumeric characters are 0 to 9 and signs * and #. ASCII codes are now universally used. Maximum significant bit (msb) Bit 7 is used as parity bit to check transmission error at the receiver when an ASCII code transfers as a byte on a line or network. (For international characters representation, an extension of ASCII code is nowadays used. It is called 16-bit unicode.) Example 11.18 explain the code with an example.

TABLE 11.2 Seven bits of ASCII Codes for the alphanumeric characters

ASCII Code lower nibble				Character for ASCII Code upper nibble lower 3 bits							
Y3	Y2	Y1	Y0	000	001	010	011	100	101	110	111
0	0	0	0	z	z	SP	0	@	P	‘	p
0	0	0	1	z	z	!	1	A	Q	a	q
0	0	1	0	z	z	“	2	B	R	b	r
0	0	1	1	z	z	#	3	C	S	c	s
0	1	0	0	z	z	$	4	D	T	d	t

Table 11.2 Contd.

0	1	0	1	z	z	%	5	E	U	e	u
0	1	1	0	z	z	&	6	F	V	f	v
0	1	1	1	z	z	'	7	G	W	g	w
1	0	0	0	z	z	(	8	H	X	h	x
1	0	0	1	z	z	)	9	I	Y	i	y
1	0	1	0	z	z	*	:	J	Z	j	z
1	0	1	1	z	z	+	;	K	[	k	{
1	1	0	0	z	z	,	<	L	\	l	|
1	1	0	1	z	z	-	=	M	]	m	}
1	1	1	0	z	z	.	>	N	^	n	~
1	1	1	1	z	z	/	?	O	_	o	DEL

Note: *z* means that it is one of the keyboard specific special codes—for example, 0001101 is ASCII code for CR (carriage return). SP in third ASCII code column means code generated when the space key of keyboard presses. ~ sign mean keyboard specific Escape character. DEL means Delete character.

11.2 EQUALITY AND MAGNITUDE COMPARATORS BETWEEN TWO FOUR-BIT NUMBERS

An important operation in the computations requires a comparison of two binary words of 8 or 16 bits and to find whether these are equal or one greater than other or vice versa. A digital comparator has many applications like during executions of while … do... repeat ... until if ... then ... else type of computer statements. It is one of the important logical units.

A digital comparator (Examples 11.9 and 11.10) differs from a analog voltage comparator in the sense that (i) the former compare only the logic levels of one number's binary bits with that of another number's binary bits while the later compares the two potential differences each with respect to a common ground potential, and (ii) the former is made from the digital logic gates while the latter is made from an operation amplifier.

A comparator, which is not a magnitude comparator and just an equality comparator has only output terminal for **A** = **B.** It will be set to, say 1 if condition is satisfied and show complementary output if the equality condition is not satisfied. (Alternatively, it will be set to, say, 0 if condition is satisfied and show complementary output 1 if the equality condition is not satisfied). A circuit for the comparator of **A** and **B** four bit digital words has only one terminal to show equality condition. Figure 11.1 shows such a circuit of a digital (equality) comparator.

Another four bit digital comparator, called a magnitude comparator, means as following: Let A is binary word 1101of four bits; A_3, A_2, A_1 and A_0, i.e. $A_0 = 1$, $A1 = 0$, $A2 = 1$, and $A3 = 1$, the corresponding decimal number for is 13 (thirteen). Let **B** the another binary word of four bits B_3, B_2, B_1 and B_0. Let $B_0 = 0$, $B_1 = 1$, $B_2 = 1$ and $B_3 = 0$—the corresponding decimal number for 0110, 6 (six). The comparator will compare and find if **A** > **B**. It then exhibits a logic state 1 (or alternatively, a state 0) at first output terminal designated as **A** > **B** output. This output is complement of other two outputs.

At second output terminal designed as **A** = **B**, if **A** = **B** then it exhibits a state 1 (or alternatively, a state 0) which is also the zero flag or equality bit.

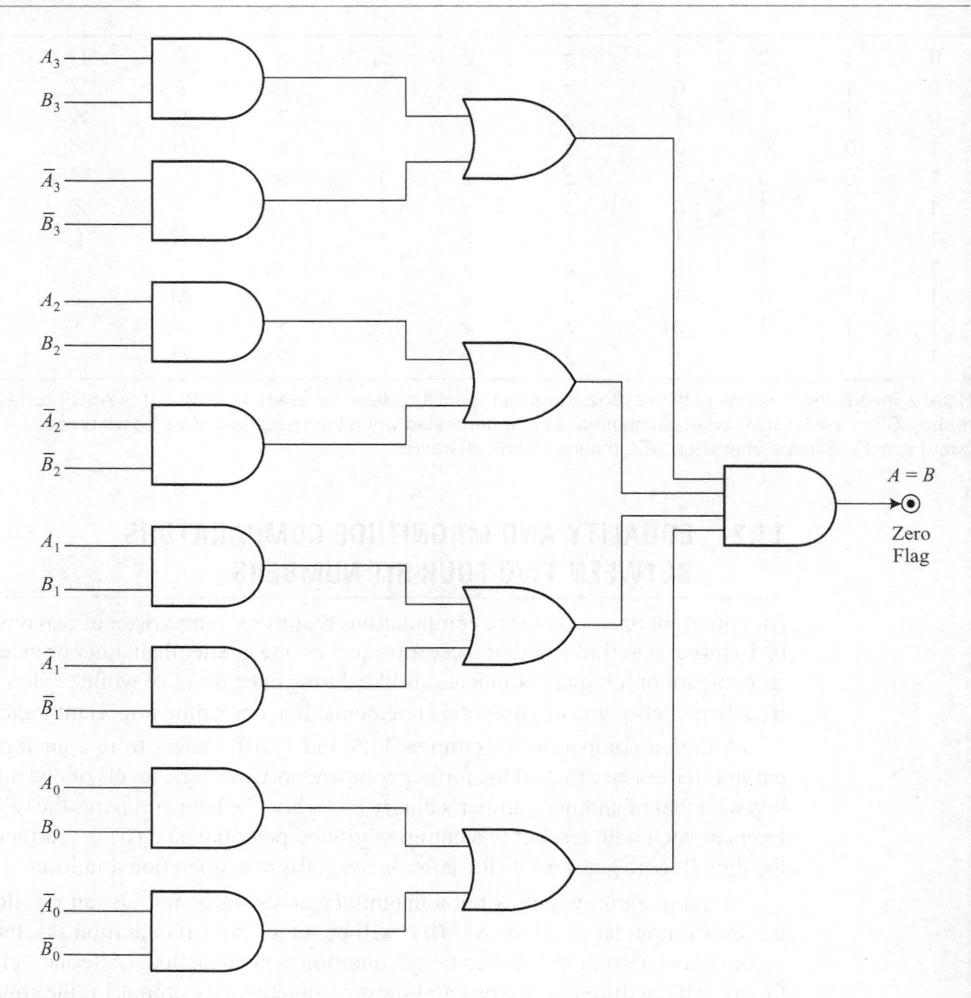

FIGURE 11.1 Digital (equality) comparator.

Complement of the first as well as third output terminals: At third **A** < **B** output terminal, if **A** < **B** then it exhibits a logic state 1 (or alternatively, a state 0) which is also the complement to that at the other two terminals.

MSI IC 7485 (or its HCMOS version 74HC85) is a 4-bit magnitude comparator. It also as well incorporates all the requirements shown in Figure 11.1. Its block diagram is described in Figure 11.2. An implementation example is given for the comparator in Example 11.10. It shows how to implement a four bit digital comparator by AND-OR arrays.

Let us consider a bit as a flag. If we assume that a flag sets if a condition is satisfied then we can define three flags from a digital magnitude comparator. Three results of an imaginary subtraction of **B** from **A** can be positive or negative or zero. One flag is for the positive sign and one flag is for the negative sign of the result. These 2 flags can be denoted by *PSF*.

A MSF or SF flag indicates minus sign when $A < B$. However, some times two flags, ZF and SF, suffice to exhibit at the outputs the four conditions, which are possible after comparison, (the imaginary subtraction). These four conditions are zero, nonzero, positive and negative.

Point to Remember

> We get a one-bit result of certain arithmetic or logic operation or comparison operation. The result indicates by a bit, which is called flag. Carry, zero, sign and parity flags are the examples of the one-bit results. Comparison is a hypothetical subtraction.

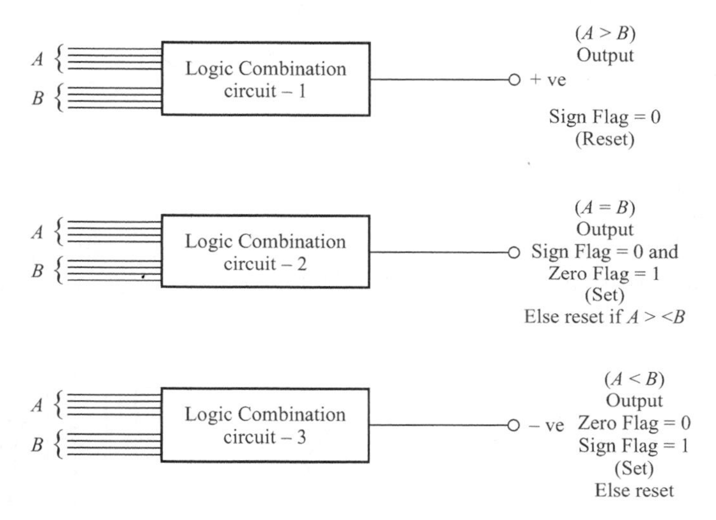

FIGURE 11.2 Block diagram of a four-bit digital magnitude comparator and results at $A < B$, $A = B$ and $A < B$ pins and at ZF and SF.

11.3 ODD PARITY AND EVEN PARITY GENERATORS

Some times the parity of a set of bit for a number is also important. Figures 11.3(a) and 11.3(b) show the even and odd parity flag (OPF and EPF) bit circuits (Refer also to Example 11.11). When the bits transmit from one source to another, the generated parity is compared at destination with the expected parity from the bits received from the source. If both are same, then it is presumed that there is no error. Parity check is successful when only one bit is in error. When two bits have error, the check is not successful. However, the chances of two errors are much smaller than the chance of a single error.

11.4 THE 4-BIT AND, OR, XOR BETWEEN TWO WORDS

11.4.1 AND

Figures 11.4(a) shows bit wise AND between two word of four bits. $Y0' = A0.B0$; $Y1' = A1.B1$; $Y2' = A2.B2$; $Y3' = A3.B3$; $ZF = Y0'.Y1'.Y2'.Y3'$ (If bit wise AND results in all output bits = 0s, then ZF sets to 1). (Example 11.13)

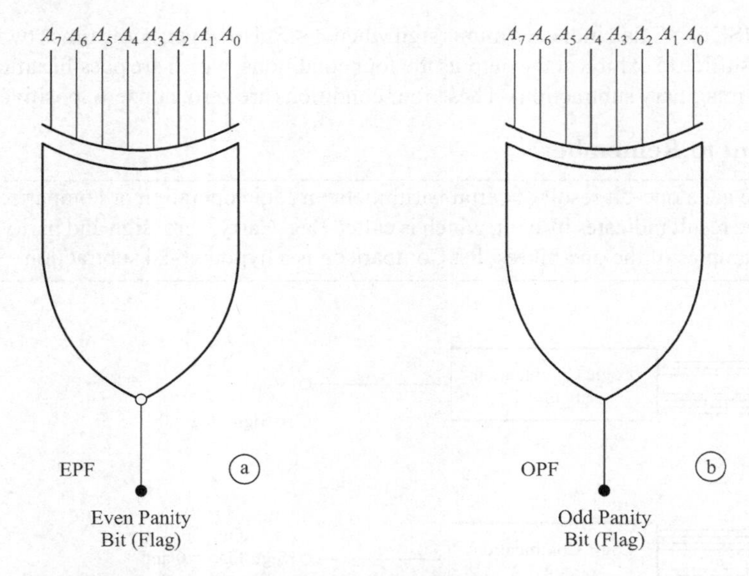

FIGURES 11.3 (a) and (b) Even and odd parity flag (OPF and EPF) bit generator.

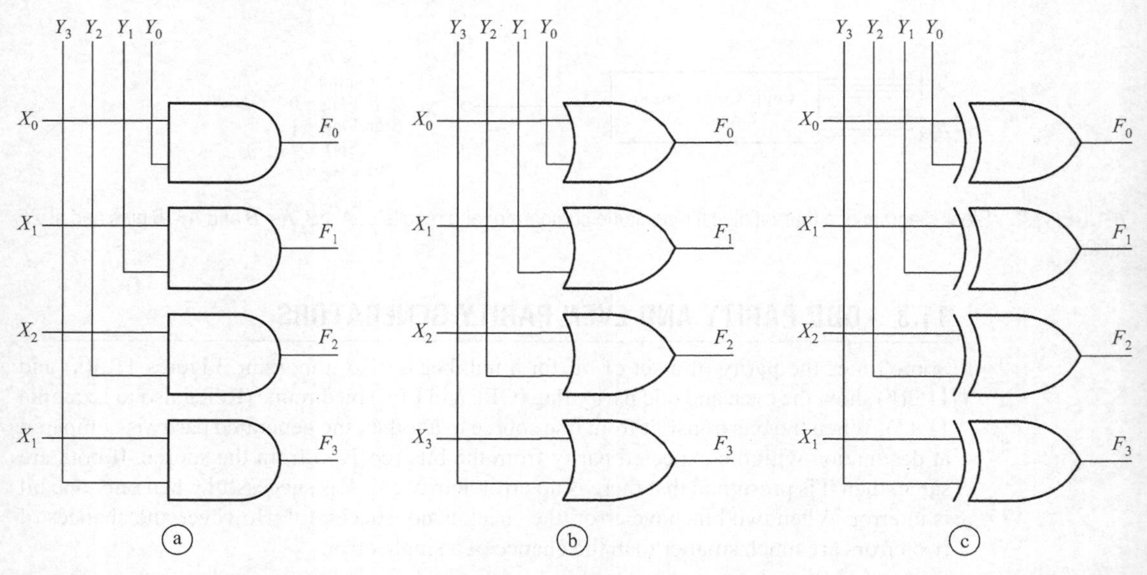

FIGURES 11.4 (a) Bit wise AND between two word of four bits (b) Bit wise OR between two word of four bits (c) Bit wise XOR between two words of four bits.

11.4.2 OR

Figure 11.4(b) shows bit wise OR between two word of four bits. $Y0' = A0 + B0$; $Y1' = A1 + B1$; $Y2' = A2 + B2$; $Y3' = A3 + B3$; $ZF = Y0'.Y1'.Y2'.Y3'$ (If bit wise OR results in all output bits = 0s, then ZF sets to 1). (Example 11.12)

11.4.3 XOR

Figure 11.4(c) shows bit wise XOR between two words of four bits. $Y0 = A0$.XOR.$B0$; $Y1 = A1$.XOR.$B1$; $Y2 = A2$.XOR.$B2$; $Y3 = A3$.XOR.$B3$; $ZF = \overline{Y}0.\overline{Y}1.\overline{Y}2.\overline{Y}3$ [If bit wise XOR results in all output bits = 0s, then ZF sets to 1. It also means that all the bits are equal $A = B$.] (Example 11.14)

11.4.4 Test

Just is a comparison is hypothetical subtraction, test is a hypothetical AND operation. Both give only flags at the outputs. Y = test flag = $Y0'.Y1'.Y2'.Y3'$, where $Y0' = A0.B0$; $Y1' = A1.B1$; $Y2' = A2.B2$; $Y3' = A3.B3$; the test is to find whether A and B.

We same use the logic processing circuits for a word of 32 bits the AND, OR and XOR with the other 32 bits word in a computer system.

EXAMPLES

Example 11.1 Using Equations (11.2), (11.3), (11.4) and (11.5), give the BCD format codes 8421, 7536, 2421 and 5421, respectively for the following decimal numbers in column Table 11.3.

TABLE 11.3 8421, 7536, 2421 and 5421 Codes for decimal numbers

Decimal	8421 Standard BCD	7536	2421	5421
13	0001 0011	1001 0010	0001 0011	0001 0011
5	0101	0100	0101	1000
7	0111	100	0111	1010

We verify each code and find it as per Equations (11.2) to (11.5). For example, in 7536 code, $1 \times (-6) + 0 \times 3 + 0 \times 5 + 1 \times 7 = 1$ and $0 \times (-6) + 1 \times 3 + 0 \times 5 + 0 \times 7 = 3$ in first row for first and second decimal digits.

Example 11.2 Write a general formula for 5043210- code for BCD representation called biquinary code.

Solution

Formula to get the total N for the 5421 BCD code is again as under:

$$N = y_{p\max-1} \times w'_7 + y_{p\max-2} \times w'_6 + y_{p\max-3} \times w'_5 + \cdots$$
$$+ y_2 \times w'_2 + y_1 \times w'_1 + y_0 \times w'_0, \quad ...(11.6)$$

where $p = p\max - 1, p\max - 2, p\max - 3, ..., 2, 1$ and 0, and pmax is 8 (maximum the number of places used in the BCD) representation and $w_0 = 0$; $w_1 = 1$; $w_2 = 2$ $w_3 = 3$, $w_4 = 4$ w_0, $w_5 = 0$ $w_6 = 5$ for the binary numbers when representing 5043210 BCD code. The $y_0, y_1, y_2, ...$ are the bit = 0 or 1 at the right-most, left first from that, left second from that, and so on. At lowest level weights are 0, 1, 2, 3, 4, 0 and 5. From the Equation (11.6) decimal digit 9 is 1010000 and decimal 0 is 01000001.

Example 11.3 Write Excess–3 (XS-3) code for 8, 9, 2 and 6.

Solution

Excess-3 code is obtained by adding 0011 binary into the standard binary form. Binary codes for 8, 9, 2 and 6 are 1000, 1001, 0010 and 0110. Therefore XS-3 codes are 1011, 1100, 0101 and 1001 after performing binary addition with 0011.

Example 11.4 Design a four bit binary (ABCD) number to Gray code $Y3Y2Y1Y0$ converter.

Solution

A binary to Gray code converter design using a Karnaugh map (Table 11.4) is as follows:

Y3: 4-bit Gray code $Y3$ and A in binary number are equal in the truth table. However, this can also be proven by Karnaugh map for the output $Y3$.

TABLE 11.4 Map to Y3 in gray code converter

AB \ CD		$\overline{C}\,\overline{D}$ 00	$\overline{C}D$ 01	CD 11	$C\overline{D}$ 10
$\overline{A}\,\overline{B}$	00				
$\overline{A}B$	01				
$\overline{A}\overline{B}$	11	1	1	1	1
$\overline{A}B$	10	1	1	1	1

An octet at the map forms from the adjacent cells. Therefore

$$Y3 = A \qquad ...(11.7)$$

Y2: Boolean expression for code can be defined by first filling the Karnaugh map (Table 11.5) cells with 1s from the truth table in Table 11.1.

TABLE 11.5 Map for Y2 in gray code inverter

AB \ CD		$\overline{C}\,\overline{D}$ 00	$\overline{C}D$ 01	CD 11	$C\overline{D}$ 10
$\overline{A}\,\overline{B}$	00				
$\overline{A}B$	01	1	1	1	1
AB	11				
$A\overline{B}$	10	1	1	1	1

There are two quads between which there is offset adjacency. Therefore

$$Y2 = \overline{A}.B + A.\overline{B} = A.\text{XOR}.B \qquad ...(11.8)$$

Y1: Boolean expression for code can be defined by first filling the Karnaugh map (Table 11.6) cells with 1s from the Table 11.1.

TABLE 11.6 Map for Y1 in gray code converter

AB \ CD		$\bar{C}\bar{D}$ 00	$\bar{C}D$ 01	CD 11	$C\bar{D}$ 10
$\bar{A}\bar{B}$	00			1	1
$\bar{A}B$	01	1	1		
AB	11	1	1		
$A\bar{B}$	10			1	1

There are two quads between which there is offset adjacency. Therefore

$$Y1 = \bar{B}.C + \bar{C}.B = B.\text{XOR}.C \qquad ...(11.9)$$

Y0: Boolean expression for a 4-bit Gray code can be defined by first filling the Karnaugh map (Table 11.7) cells with 1s from the Table 11.1.

TABLE 11.7 Map for Y0 in gray code converter

AB \ CD		$\bar{C}\bar{D}$ 00	$\bar{C}D$ 01	CD 11	$C\bar{D}$ 10
$\bar{A}\bar{B}$	00		1		1
$\bar{A}B$	01		1		1
AB	11		1		1
$A\bar{B}$	10		1		1

There are two quads between which there is offset adjacency. Therefore,

$$Y0 = \bar{D}.C + \bar{C}.D = C.\text{XOR}.D \qquad ...(11.10)$$

Figure 11.5 gives the implementation of binary to Gray code converter using AND-OR array (equivalent to use of NANDs).

Example 11.5 Define a simple method to convert a binary code to Gray code and show that Gray code converter can be made from adder circuit as a building block.

Solution

From the Boolean expressions in Equations (11.7) to (11.10), we note a simple method as under:

1. Upper most bit-3 of Gray code is taken same as binary number upper-most bit.
2. Lowest Gray code bit is addition without considering any carry of bit 0 and bit 1 of the binary number. (Figure 10.1)

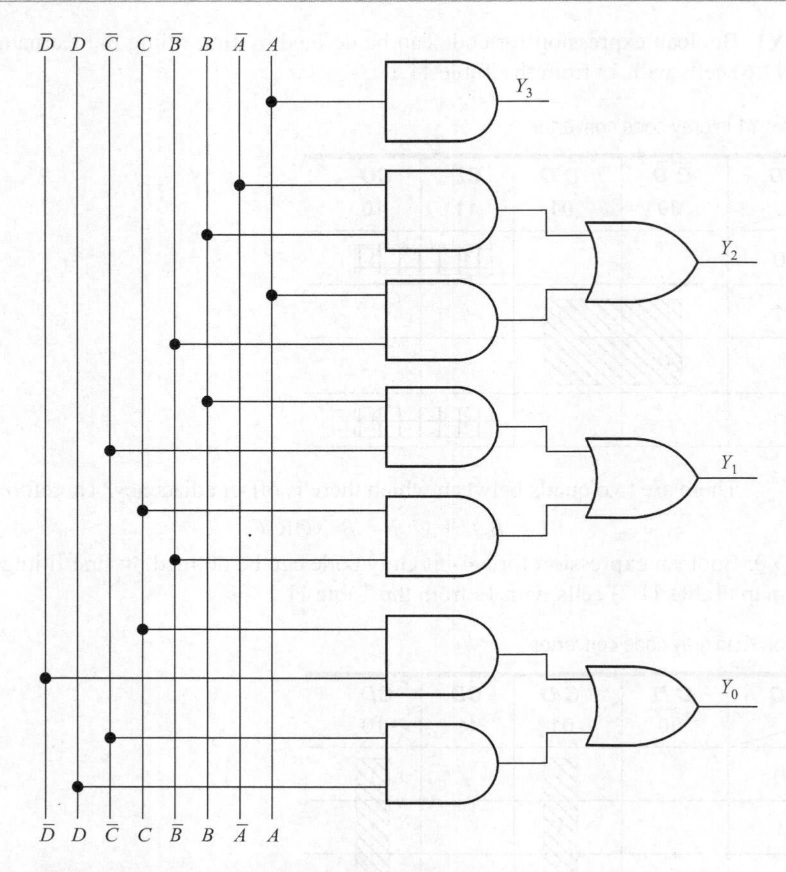

FIGURES 11.5 Implementation of the binary 4-bit number to the Gray code converter circuit.

3. Bit 1 (left to the lowest) of Gray code is addition without considering any carry of bit 1 and bit 2 of the binary number. (Figure 10.1)
4. Bit 2 (right to the uppermost) of Gray code is addition without considering any carry of bit 2 and bit 3 of the binary number. (Figure 10.1)

Example 11.6

Convert 1101 binary into Gray code using a simple method of bit addition of each bit by 1 without carry considerations (except *Y*3).

Solution

1. Gray code MSB $Y3 = 1$, because it does not change on conversion.
2. $Y0 = \text{bit } 0 + \text{bit } 1 = 0 + 1 = 1$.
3. $Y1 = \text{bit } 1 + \text{bit } 2 = 0 + 1 = 1$.
4. $Y2 = \text{bit } 2 + \text{bit } 3 = 1 + 1 = 0$.

Gray code is 1011. (Answer is as per Table 11.1). We can use adder circuit in Figure 10.1 as building block.

Example 11.7 Give a simple method to convert Gray code 0011 to binary code converter.

Solution

A careful look at the Table 11.1 reveals a simple method as under: from the truth table in Table 11.1.

1. Upper most bit-3 of binary code be taken as 0 as Gray code upper most bit.
2. Lowest binary code bit 1 is added without considering any carry to 1, 0 and (bits 1, 2, 3).
3. Bit 1 (left to the lowest) of binary code is addition without considering any carry to 1 + 0 + 0 to get b1 = 1.
4. Bit 2 (right the uppermost) of binary code is addition without considering any carry of bit 2 and bit 3 0 + 0 to get b2 = 0.

Karnaugh map based verification of the above method is left as an exercise to the reader (Exercise 7). We get the answer = 0010 as code.

Example 11.8 Convert 0111 Gray code into binary 4-bit code using the simple method of bit addition without carry considerations.

Solution

1. Binary code MSB $A = 0$, because it does not change on conversion.
2. $D =$ bit 0 + bit 1 + bit 2 + bit 3 = 0 + 1 + 1 + 1= 1. (Add all 4 bits of Gray code).
3. $B =$ bit 2 + bit 3 = 1 + 0 = 1. (Add upper 2 bits of Gray code).
4. $C =$ bit 3 + bit 2 + bit 1 = 0 + 1 + 1 = 0. (Add upper 3 bits of Gray code).

Binary code ABCD is 0100. (Answer is as per Table 11.1).

Example 11.9 Give truth table of a 3-bit digital comparator for six sample bit combinations for binary numbers **A** and **B**.

Solution

Result is in Table 11.8.

TABLE 11.8 A Three Digital comparator for sample bits for binary numbers **A** and **B**

Inputs						Outputs		
	Input A			Input B		ZF	PSF	MSF
A_0	A_1	A_3	B_0	B_1	B_2	=	>	<
0	0	0	0	0	0	1	0	0
1	1	1	1	1	1	1	0	0
1	1	1	1	1	0	0	1	0
1	1	0	1	0	0	0	1	0
1	1	0	1	1	1	0	0	1
1	0	0	1	1	0	0	0	1

PSF bit means $A > B$ bit. MSF bit means $A < B$ bit. ZF means $A = B$ bit.

Example 11.10 Formulate the problem for implementing an 8-bit digital comparator.

Solution

Equality:

$$ZF = (\overline{A_0 \text{ XOR } B_0}).(\overline{A_1 \text{ XOR } B_1})\ldots(\overline{A_7 \text{XOR } B_7})$$

An XOR circuit when both inputs are 0s or 1s hives output = 0. Hence complement after an XOR operation between each bit will give 1s only when each bit $A_0 \ldots A_7$ of **A** equals $B_0 \ldots B_7$ of **B**. As XOR between X and Y is $X.Y + X.Y$, we get the following Boolean expression in terms of AND-OR arrays for zero sign flag output.

$ZF = (\overline{A}_0.B_0 + \overline{A}_0.B_0) + (\overline{A}_1.B_1 + A_1.\overline{B}_1) \ldots + (\overline{A}_7.B_7 + A_7.\overline{B}_7)$. Figure 11.6(a) shows the logic circuit from AND-OR array.

Greater $A > B$:

0th stage $P_0 = A_0.\overline{B}_0$; (find whether $A_0 > B_0$) (shown at Figure 11.6(b) top).

1st stage $P_1 = A_1.\overline{B}_1 + A_1.P_0 + \overline{B}_1.P_0$; (find whether $A_1 > B_1$ or A_1 and P_0 both 1 or B_1 = 0 and P_0 =1) (shown at Figure 11.6(b) middle).

2nd stage $P_2 = A_2.\overline{B}_2 + A_2.P_1 + \overline{B}_2.P_1$; (find whether $A_2 > B_2$ or A_2 and P_1 both 1 or B_2 = 0 and P_1 =1).

Last stage $P_7 = PSF = A_7.B_7 + A_7.P_6 + B_7.P_6$. (Find whether $A_7 > B_7$ or A_7 and P_6 both 1 or B_7 = 0 and P_6 =1). Figure 11.6(b) shows the logic circuit from AND-OR array (shown at Figure 11.6(b) bottom).

Lesser $A < B$:

0th stage $M_0 = \overline{A}_0.B_0$; (find whether $A_0 < B_0$).

1st stage $M_1 = \overline{A}_1.B_1 + \overline{A}_1.M_0 + B_1.M_0$; (find whether $A_1 < B_1$ or B_1 and P_0 both 1 or A_1 = 0 and M_0 =1):

2nd stage $M_1 = \overline{A}_2.B_2 + \overline{A}_2.M_1 + B_2.M_1$; (find whether $A_2 < B_2$ or B_2 and M_1 both 1 or A_2 = 0 and M_1 =1).

Last stage $M_7 = MSF = \overline{A}_7.B_7 + \overline{A}_7.M_6 + B_7.M_6$. (Find whether $A_7 < B_7$ or B_7 and M_6 both 1 or A_7 = 0 and M_6 =1). Figure 11.6(c) shows the logic circuit from AND-OR array (using the circuit of Figure 11.6(b) and complements of As and Bs.

Example 11.11 Implement the parity generator circuits of Figure 11.3(a) and (b) by AND arrays.

Solution

Circuits of Figures 11.3(a) and (b) are same except that there is a NOT gate at the last stage. Figure 11.7(a) implements by AND-OR arrays as follows:

0th stage $PF_1 = (A_0.\overline{A}_1 + \overline{A}_0.A_1)$;

1st stage $PF_2 = (\overline{A}_1.PF_1 + A_2.\overline{PF}_1)$;

7th stage $PF_7 = (\overline{A}_7.PF_6 + A_7.\overline{PF}_6)$

Figure 11.7 gives the circuit using AND OR array for a parity generator.

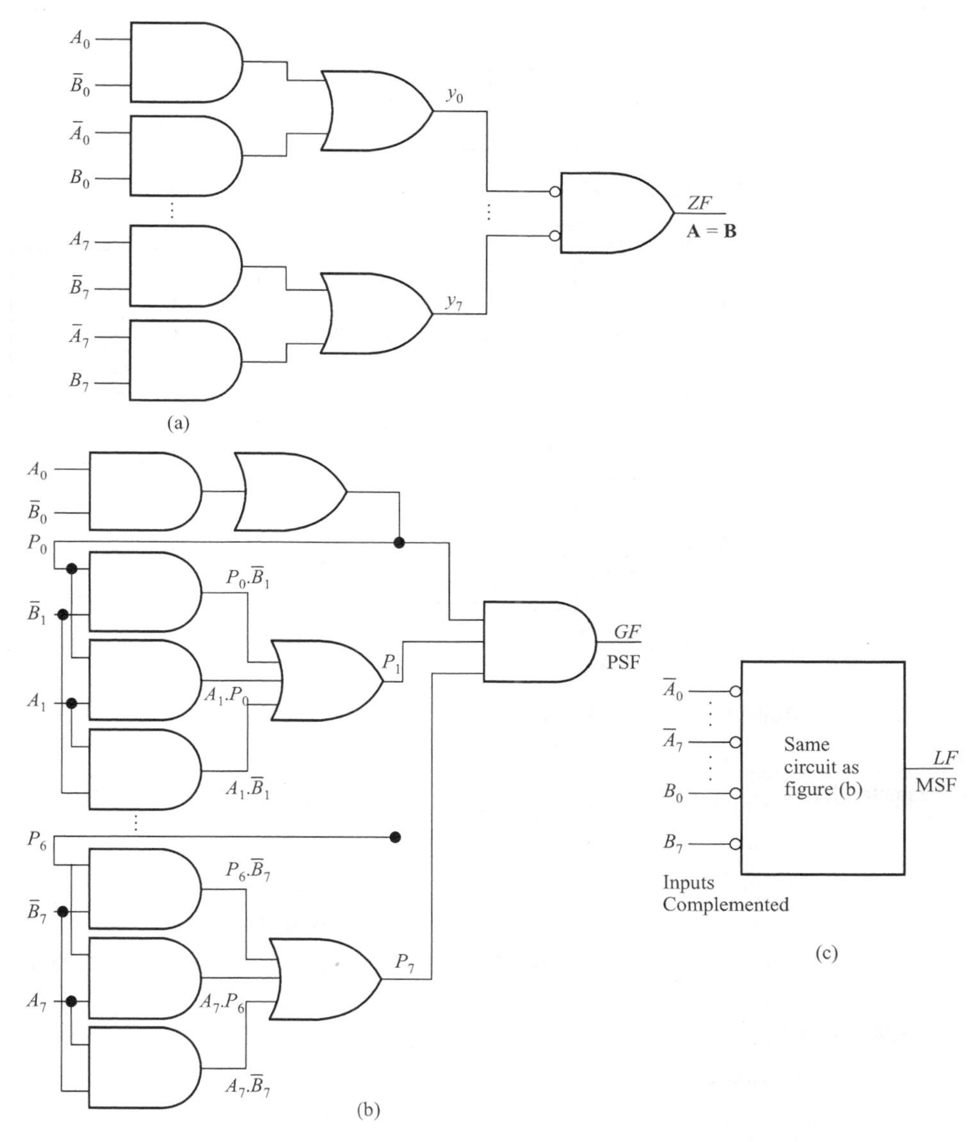

FIGURES 11.6 (a) AND-OR Array for equality test (b) AND-OR Array network for positive test $A > B$ (c) AND-OR Array network for negative test $A < B$ using the $\overline{A}$ s and Bs.

Example 11.12 Find bit wise OR of 10101010 and 01010101.

Solution

10101010

01010101

Bit wise OR 'operation gives all bits $Y0$' to $Y7$' = 1. Result is 11111111 and $ZF = 0$.

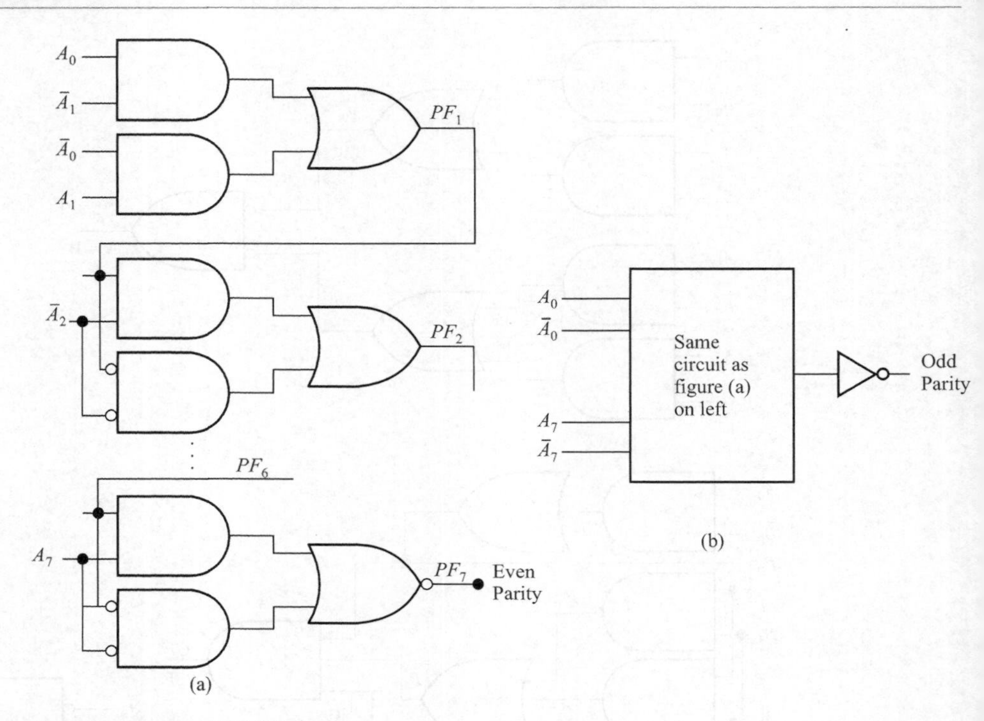

FIGURES 11.7 Parity generator using AND-OR arrays.

Example 11.13 Find bit wise AND of 10101010 and 01010101

Solution

10101010

01010101

Bit wise AND operation gives all bits $Y0$' to $Y7$' = 0s. Result is 00000000 and $ZF = 1$.

Example 11.14 Find bit wise XOR of 10101010 and 01010101.

Solution

10101010

01010101

Bit wise XOR 'operation gives all bits $Y0$' to $Y7$' = 1s. Result is 11111111 and $ZF = 0$.

Example 11.15 Find bit wise NOT of 10101010.

Solution

10101010

Bit wise NOT 'operation gives bits $Y7$' down to $Y0$' = 01010101.

Example 11.16 Design an Excess-3 (XS-3) code generator.

Solution

We take a 4-bit binary adders. It is given inputs **A** and **B**. **B** input pins are given 0011 as input. The output of the adder gives the XS-3 code.

Example 11.17 Design an adder for the numbers in BCD standard (8421) format.

Solution

Figure 11.8 shows the logic circuit of the BCD adder. We take two 4-bit binary adders. Both adders have 0th stage carry input = 0. One is given BCD inputs **A** and **B**. Outputs of this $S3$, $S2$, $S1$ and $S0$ are given as an input to **B** input pins of another. A_0 and A_3 pins of second adder are given 0000 as input. The output of second adder gives the lower nibble result. Least significant bit (lsb) of upper nibble (number of 10s) is obtained by following Boolean operation using four NAND gates. Cy_3 is output last stage carry of the first adder. It is independently given to one of the NAND. S_2 and S_3 are inputs to second NAND and S_1 and S_3 are inputs to third NAND. Outputs of these three NANDs are the inputs to the fourth NAND. Fourth NAND output is the lsb of the upper BCD nibble in the result.

$$\text{lsb of Upper nibble} = \overline{(\overline{S_1 . S_3} . \overline{S_2 . S_3} . \overline{Cy_3})} = S_1.S_3 + S_2.S_3 + Cy_3$$

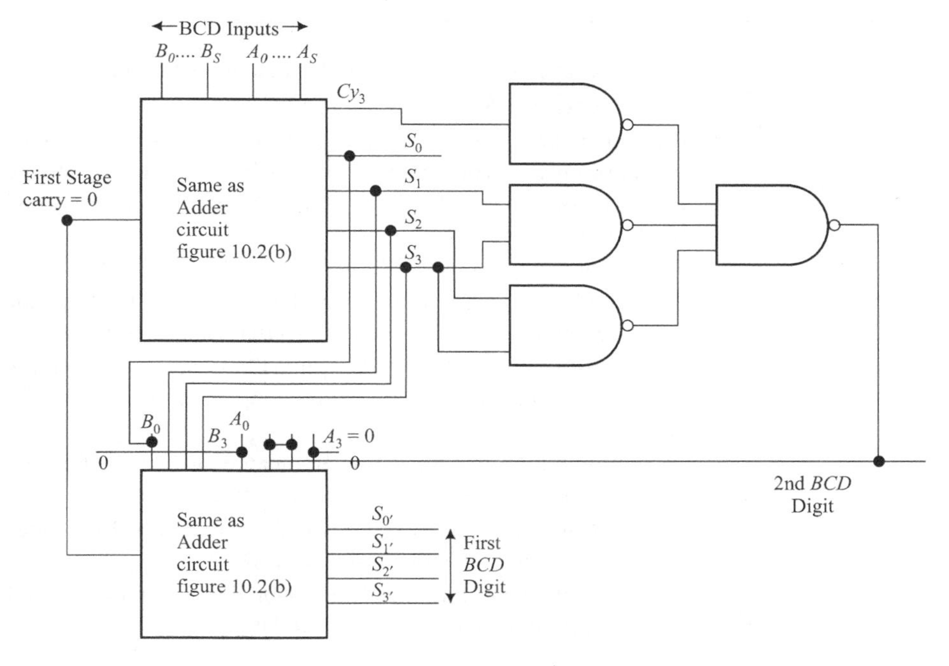

FIGURE 11.8 Logic circuit of the BCD adder using two four-bit binary adders.

Example 11.18 Write ASCII codes for nineteen characters in 'Digital Electronics'.

Solution

Using Table 11.2, we get the following answer:

100 0100; 110 1001; 110 0111; 110 1001; 111 0100; 110 0001; 110 1100; 010 0000; 100 0101; 110 1100; 110 0101; 110 0011; 111 0100; 111 0010; 110 1111; 110 1110; 110 1001; 110 0011; 111 0011.

EXERCISES

1. Using Equations (11.2), (11,3), (11.4) and (11.5), give the BCD format codes 8421, 7536, 2421 and 5421, respectively for the following decimal numbers in Table 11.9, Verify your answers for each column.

TABLE 11.9 8421, 7536, 2421 and 5421 Codes for decimal numbers

Decimal	8421 Standard BCD	7536	2421	5421
9				
12				
6				

2. Write 5043210 biquinary code for decimal digits 6, 9 and 12.
3. Write Excess–3 (XS-3) codes for 3 and 7.
4. The outputs from the sensors at the slots when a shaft moves in steps of 30^0 clockwise are Gray codes 0000; 0001; 0011; 0010; ... If the shaft moves from 0^0 towards anticlockwise in steps of 60^0 each, then what shall be the Gray code outputs during one rotation of the shaft.

   (Hint: Write Gray codes for 0^0, 300^0, 240^0, 180^0, 120^0, 60^0, 0^0).
5. Using Boolean expressions to convert the binary code to Gray codes for 1001, 0010 and 1010. Implement a Gray code converter using decoder as building block.
6. Convert 1001, 0010 and 1010 binary into Gray codes using the simple method of bit addition without carry considerations. Implement Gray code converter by using MUX.
7. Design logic circuit and develop Boolean expressions after minimization using Karnaugh map approach for Gray code to binary code converter.
8. Convert 1001, 0010 and 1010 Gray codes into binary 4-bit code using the simple method of bit addition without carry considerations.
9. Give 8 sample rows of truth table of a 4-bit digital comparator (Table 11.10) for six sample bit combinations for binary numbers *A* and *B*.

TABLE 11.10 A Four-bit digital comparator for sample bits for binary numbers *A* and *B*

Inputs								Outputs		
Input *A*				Input *B*				ZF	PSF	MSF
*A*3	*A*2	*A*1	*A*0	*B*3	*B*2	*B*1	*B*0	=	>	<

10. Formulate the problem and design logic circuit for implementing a 3-bit digital comparator.
11. Implement the odd and even parity generator circuits to get eighth bit of 7-bit ASCII code implement by AND-OR arrays.
12. Find bit wise OR of 11101011 and 01010101.
13. Find bit wise AND of 11101011 and 01010101
14. Find bit wise XOR of 10101010 and 10101010.
15. Find bit wise NOT of 10101010 and then XOR the output and input bits. .
16. Design an Excess-3 (XS-3) codes adder.
17. Design an 8421 standard BCD inputs adder/subtractor circuit.
18. Write the ASCII codes for characters in your college or university name.

QUESTIONS

1. Explain standard 8421 BCD, 7536, 2421, and 5421 BCD codes by two examples each.
2. What do you mean by self-complementing code?
3. Why is the ASCII code a widely used code? What is the 16-bit new extension of it?
4. Why do we use Gray code in certain applications?
5. Describe a digital comparator circuit.
6. If bit wise XOR the two equal binary numbers, we get resulting output number as zero. Explain it.
7. Show that bit wise XOR between a binary number and its NOT gives all bits as 1s.
8. If we bit wise NOT a number and then add 1 in it, we get two's complement. Design a circuit for finding two's complement.
9. How can we use a two input XOR as a controlled inverted of a logic state?
10. If we AND binary numbers and get the same result as the input numbers, then both numbers are said to be equal. Why?

CHAPTER 12

Implementation of Combinational Logic by Standard ICs and Programmable ROM Memories

OBJECTIVE

We learnt in earlier chapters that a combinational circuit has following characteristics:

1. These have n inputs and m outputs. For examples, (i) A NOT (inverter) gate, $n = 1$ and $m = 1$. (ii) An XOR gate, $n = 2$ and $m = 2$. (iii) An 8-bit adder, which accepts in its inputs a carry bit, 8 bits of X and 8 bits of Y and results at outputs the 8 bit sum Z and final carry CY, $n = 17$ and $m = 9$.
2. Logic state, at any of the outputs in the combinational circuits, depends only on the inputs at any given instant (not considering always present propagation delay period) and is not correlated at all with any of its previous outputs or outputs.
3. A truth table of a combinational circuit gives the values of all the m outputs for each possible combination of the n inputs and has 2^n rows and $(n + m)$ columns.
4. A combination circuit can be designed for obtaining the m outputs using the m Boolean expressions.
5. Each expression can be represented by SOP or POS format. A Karnaugh map-based technique or computer-based minimization technique is used to get least cost (minimum number of gates) or least delay (minimum number of levels between inputs and the corresponding output).

We learnt that the minimized circuit is then implemented easily using AND-OR arrays or NANDs or OR-AND arrays or NORs.

Alternatively, a logic design implements by decoders, encoders, multiplexers or demultiplexers, or binary arithmetic adders, adder/subtractors, code converters, comparator, bit-wise 8-bit AND, OR, XOR circuits or parity generators. Standard ICs are commercially available. We will learn the standard ICs and PROMS that are used for these in this chapter.

A combinational circuit may be complex enough to assemble using standard ICs. We wish to design a circuit for a character in a line printer, which gives a 64-bit output for each pixel in the character when the ASCII code of that is given as the input (refer section 11.1.3 for ASCII codes). These 64-bits are input to the printer head pins driving circuit. It means that we need a combination circuit for each ASCII character which has $n = 8$ (7-bit ASCII code + one parity bit) and $m = 64$. Another example is of a LCD line display or multi line display circuit, that has further complexity. Another example is of an advertisement displayed by an array or matrix of LEDs.

The implementation of the complex circuits with standard ICs can be extremely bulky in many situations. How do we then design such a complex circuit? One method nowadays for a complex combinational circuit is using a programmed or programmable logic memories, ROM or EPROM or EEPROM. Another is by using programmable logic devices PALs, PLAs, GALs and FPGAs.

We will learn the ROM, EPROM and EEPROM memories and implementation of complex logic circuits using these in this chapter. Other method of using PALs and PLAs will be described in Chapter 13 and FPGAs in Chapter 21.

12.1 STANDARD ICS FOR DESIGN IMPLEMENTATION

We have learnt in chapters 10 and 11 the binary arithmetic circuits, decoders, encoders, multiplexers, code converters, digital comparator for magnitude and equality, parity generators and checkers and bit wise AND, OR, NOT logic processing circuits. Standard ICs are available for these circuits. The combinational circuits are made using these ICs.

12.1.1 Adder/Subtractor IC and Magnitude Comparator

CMOS based 74HC 83 and 7483 family ICs are the four bits full adders. Figure 12.1 shows the IC 7483 connections when using it as an adder/subtractor.

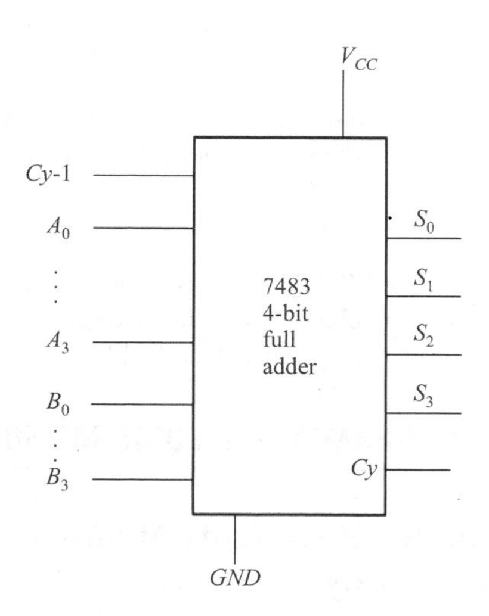

FIGURE 12.1 IC 7483 connections and its use as an adder/subtractor.

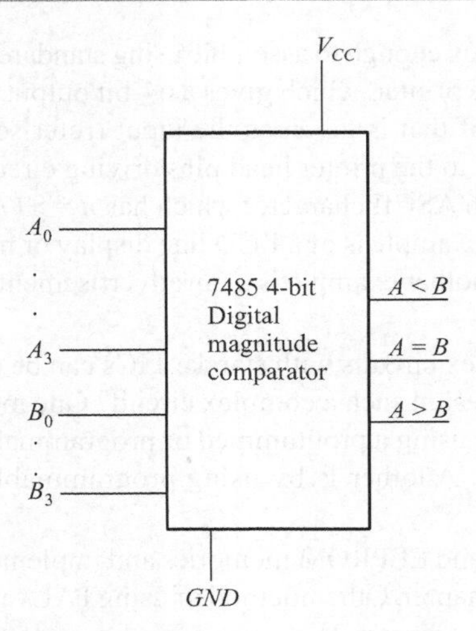

FIGURE 12.2 Digital comparator IC 7485.

MSI IC 7485 (or its HCMOS version 74HC85) is a 4-bit magnitude comparator. It also as well incorporates all the requirements shown in Figure 11.5. Its block diagram was described in Figure 11.6. Figure 12.2 shows the IC 7485.

12.1.2 Decoder IC

Figure 10.5(a) showed 3-line to 8-line (1 of 8) decoder with three control pins $\overline{G}1$, $\overline{G}2$, and $G3$. The $\overline{G}1$, and $\overline{G}2$ control pins are activated by 0 and the $G3$ pin is activated by 1. A MSI chip 74138 has the pin corresponding to the decoder shown in the figure. Figure 12.3(a) shows an exemplary connection in 74138.

12.1.3 Encoder IC

MSI IC74148 is an 8 to 3 encoder with an active 0 gate-enable for the inputs and an active 0 gate-enable pin for the outputs. Figure 12.3(b) shows connections using 74148.

12.1.4 Multiplexer IC

Figure 10.10 showed a multiplexer IC. The MSI IC74156 is a four-channel multiplexer. It selects the inputs I0, I1 I2 and I3 as per the channel selector lines A_0 and A_1. Figure 12.4 shows the connections using 74156. The stroke pin is MUX enable pin.

12.2 PROGRAMMING AND PROGRAMMABLE LOGIC MEMORIES

12.2.1 ROM (Pre-Programmed Read Only Memory) and PROM (Programmable Read Only Memory)

A read only memory, abbreviated as ROM or read only memory, is also a ready to use set of combinational circuits. A ROM is a previously programmed 'Decoder-OR' array based logic device using appropriate masks at the manufacturing stage.

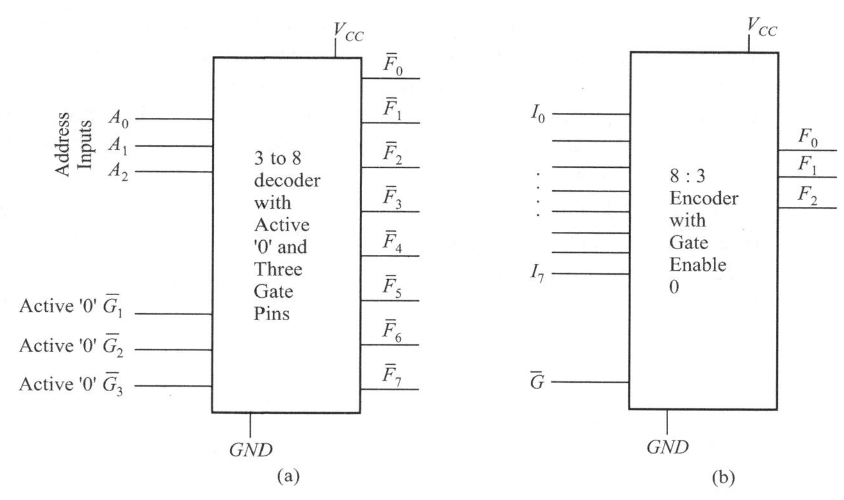

FIGURE 12.3 (a) Connections in using 74138 (b) Connections in using 74148

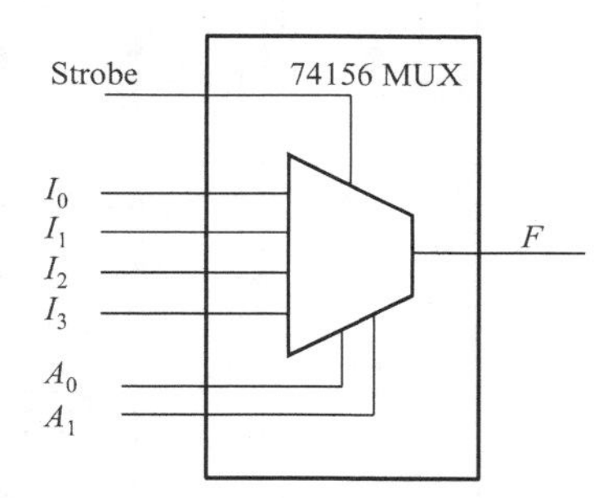

FIGURE 12.4 Connections using a 74156 multiplexer IC.

Points to Remember

(i) A decoder with one-line output decoder is also a minterm or maxterm generator, and (ii) a decoder can be used to implement a Boolean expression each by using an OR gate with those decoder outputs, which happens to be the minterms of the expression. If given Boolean variables at the address input points, then a programmed ROM implements a combinational circuit, each with an OR gate at the output.

ROM special versions programmable logic devices (PLDs) called PROMs (Programmable Read Only Memories) [An EPROM or an E2PROM or flash or OTP ROM] is programmable at the laboratory scale].

The ROMs do not change or lose its programmed data upon an interrupt of power to it. (This feature is also called non-volatility of a ROM). Non-volatility is a most important feature in ROM and PROM that is useful in a computer system, where the ROMs or EPROMs or E^2PROMs store the frequently needed programs and data sets for the system users or the truth tables for implementing the combinational circuits. Let us consider applications, such as a preprogrammed toy circuit, a preprogrammed robot circuit, a standard look up table, or an arithmetic function table generator, a user defined code generator, a character generator, a printable or displayable fonts table, a set of the data or a set of instructions or an arithmetic function table generator. The instructions, table, and /or the data) must remain stored in the memory in these applications, even after a power interruption or power switched OFF. A ROM or PROM is used for it.

Figure 12.5 shows block diagram of a ROM with n address inputs, one select input and m data bit outputs and one read input [control gate (enabling) input]. As diodes between 1 and 2, and 5 and 6 are masked, when the address inputs are all 0s then Y_0 activates and $D_0 = 0$, $D_1 = 1$, $D_2 = 0$, $D_3 = 1$. The output is 1010. The masks are as programmed before manufacturing the chip.

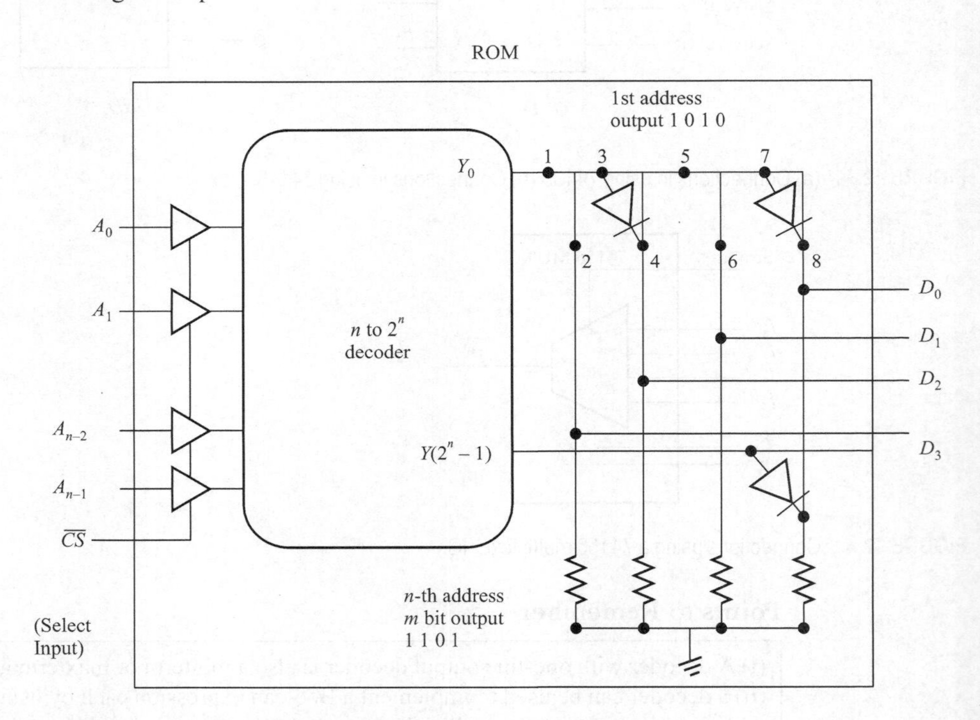

FIGURE 12.5 Block diagram of a ROM with n address inputs, one select input and m outputs.

Points to Remember

Each ROM or PROM has n inputs, called address bits (for one of the 2^n memory locations). This ROM on activating an input, called read input (actually a control gate input), generates for each set of address inputs, a distinct set of m outputs, called data bits (most often $m = 8$) from the addressed memory location. The data bits are as per programmed bits during manufacturing of ROM by silicon masking or any other processing. When we use a ROM for the implementation of the combinational circuits, the Boolean variables are the inputs at the address input pins.

12.2.1.1 An Exemplary 3 × 2 ROM

A ROM has an address decoder (n to 2^n decoder) within it. Figure 12.6(a) shows a 3 × 2 ROM block diagram.

1. The decoder thus consists of $2^3 = 8$ memory locations (outputs) and thus implements 8 combination circuits. A decoder output is activate 1 when a minterm happens to be present in the Boolean expression for the given input address bits, A_2, A_1 and A_0.
2. The 8 outputs of the decoder are given to two number eight-input OR gates to obtain two outputs ($m = 2$), D_0 and D_1.
3. Total 16 fusible links exist with 8 fuses per OR gate. Fusing the links generates the outputs as per Boolean expression and its truth table.

Figure 12.6(b) shows truth table, which the 3 × 2 programmed ROM implements.

Figure 12.6(c) shows a 3 × 2 ROM programmed ROM, which has the fused (snapped) links to implement the truth table of Figure 12.6(b) by appropriate masking at the manufacturing stage.

Characteristic of a ROM circuit of Figure 12.6(c) is that just apply the address bits A_2, A_1 and A_0 and activate $\overline{\text{RD}}$ input and we get the output data bits D_0 and D_1 as per the truth table logic from the row as per the address bits. When $\overline{\text{RD}} = 1$, the D_0 and D_1 are in tristates.

Figure 12.6 shows a simpler representation of fuse links at the multi-input OR gate(s). This representation has become conventional for the fuse links at the multi input AND or OR in case of the AND-OR or OR-AND arrays. In place of many input lines to an AND or OR, a single line is shown with each dot representing a separate line, which is limited.

12.2.1.2 An Exemplary 4k × 8 ROM (4 kB ROM)

A 4k × 8 ROM has $m = 8$, eight data bits at the output. It is also called 4 kB ROM (B is for a byte). Data bits are D_0, D_1, D_2, ... , and D_7.

4k means (4 .1024) addresses. [With reference a memory unit, 1 k is conventionally taken as 1024 ($=2^{10}$)]. $2^n = 4\ .1024$. There $n = 12$ and address bits will be A_0, A_1, A_2, ..., A_{11}. 4k × 8 ROM implements 4096 combinational circuits, each with 8 outputs and 12 Boolean variables as the inputs. The outputs are as per the programming designed according to a truth table used during manufacturing.

> ROM Programming means selecting the appropriate fusible links, which are snapped in the programmed ROM.

12.2.1.3 ROM Special Versions for Laboratory scale Programming [EPROM, E²PROM, flash and OTP ROM]

A PROM [EPROM, an E^2PROM and an OTP ROM] is the user programmable logic device (called EPLD, electrically programmable logic device).

An EPROM is a special ROM. It can be erased as well as programmed in a laboratory. A user can program and thereby store the data at the various referencable address locations. A special unit called an EPROM programmer does the programming. Programming at a selected address is done with the help of a 12.5 V or 25 V supply at one of its input, and then at another input, a programming pulse of duration say 50 ms or less. The EPROM then stores the data bits at the selected address. These bits can be read when this EPROM is used

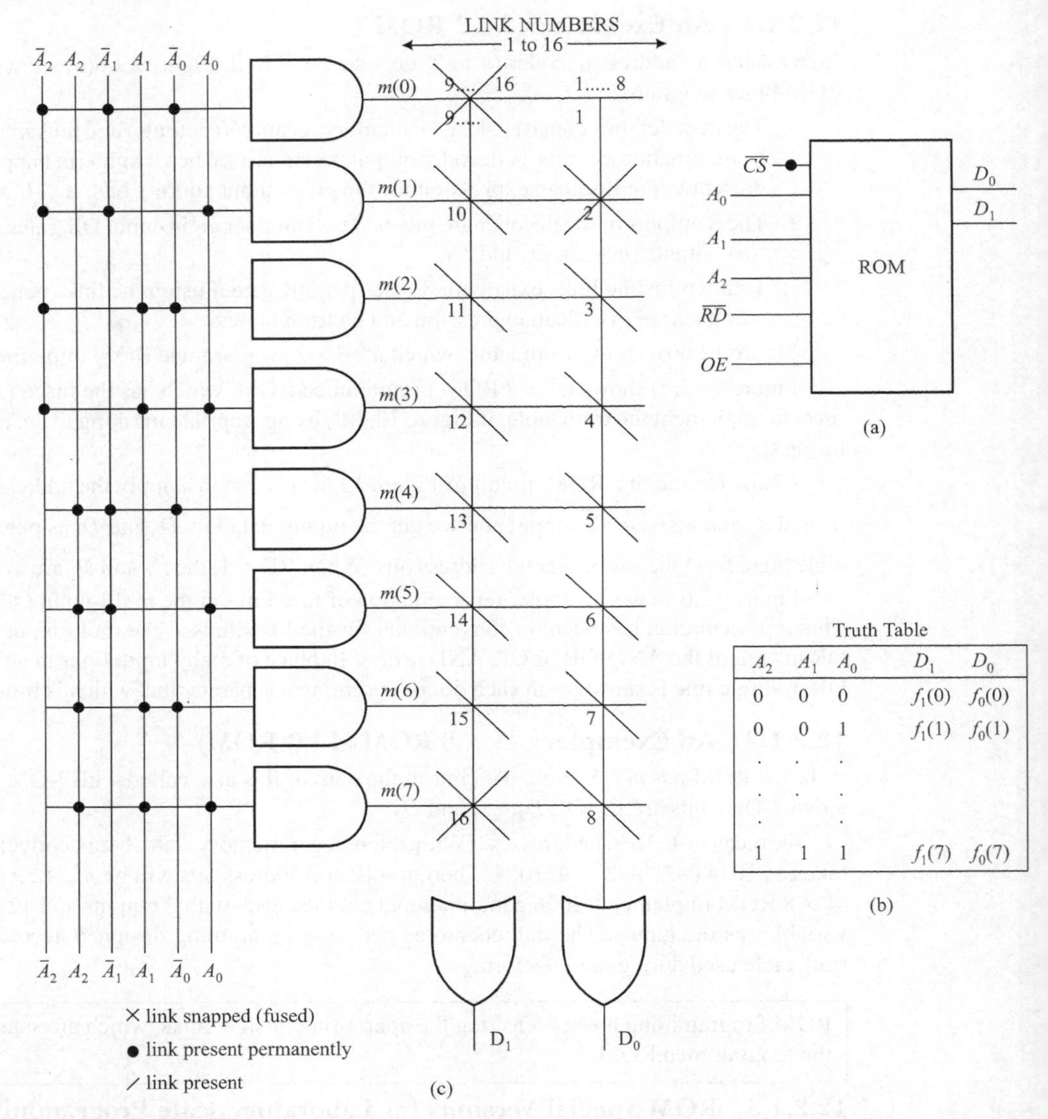

A_2	A_1	A_0	D_1	D_0
0	0	0	$f_1(0)$	$f_0(0)$
0	0	1	$f_1(1)$	$f_0(1)$
.	.	.		
.	.	.		
.	.	.		
1	1	1	$f_1(7)$	$f_0(7)$

FIGURE 12.6 (a) Block diagram of a 3×2 ROM (b) Truth table to be implemented by the 3×2 programmed ROM (c) 3×2 ROM programmed ROM, which has the fused (snapped) junctions to implement the truth table. A simpler representation of multi-input OR gate with fuse links is used here.

in another circuit. *Read* means obtaining the outputs corresponding to a selected address. These are the inputs to other circuits like a processor, which reads these.

The programmer can program all the 2^n addresses. The programmed data bits then have erase immunity from any power interruption or power OFF. The EPROM can now only be erased through a UV light of, say, a 12 mW/cm^2 intense UV source at 2.5 cm above the quartz window on the EPROM IC chip. The IC is exposed for a period of about 20 m.

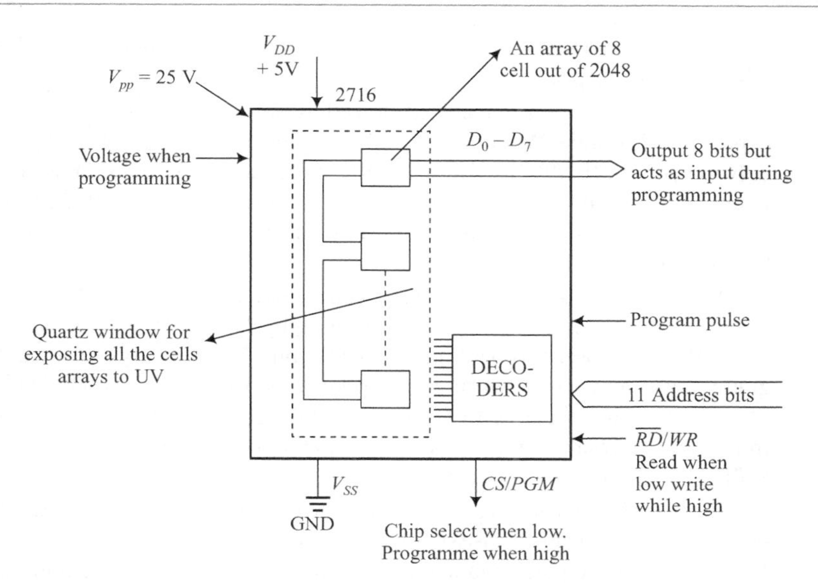

FIGURE 12.7 An EPROM model diagram for an IC chip 2k × 8 2716 showing the address inputs, one select input and *m* outputs and other pins.

The laboratory erase and programming facilities are not available in case of ROM, unlike an EPROM because a ROM has to be programmed at its manufacturing stage itself.

Figure 12.7 shows an EPROM model diagram for a 2k × 8 IC chip 2716.

(1) There is an additional pin in the EPROM that should be applied a specified voltage, Vpp (typical values are +25 V, + 21 V or + 12.5 V) during programming (writing) of data bits (byte) at a selected address location inside it.

(2) A pin, $\overline{\text{CS}}$/PGM, when at 1, enables programming of this EPROM through a pulse for a period of 50 ms to another pin.

(3) The $\overline{\text{CS}}$/PGM pin, when at 0, enables the chip selection and the inputs and outputs, D_0 to $D_{m-1,}$ and disables the programming in order to permit only a reading of the D bits from a select address.

(4) The $\overline{\text{RD}}$ pin, when at 1, makes the outputs, D_0 to $D_{m-1,}$ in tristates and disables the programming in order to permit only a reading of the D bits from a select address.

An EPROM laboratory programmer unit performs the following steps in a sequence:

1. Applies gives the n bits (address bits) to the decoder of the array of cells and applies as inputs the D bits, which are meant for the outputs later on during the read operation corresponding to the selected address.
2. Applies a high voltage to make programming feasible and applies a very short duration (as per EPLD specification) to cause fusing (snapping) of the desired links in the array due to the high voltage.
3. Repeat from the step 1 by applying the next higher address than the previous one.
4. Repeat till all addressed are programmed.
5. Verify the programmed bits.

During read operation, the V_{DD} and V_{SS} are the supply inputs in EPROM chip and are at 5 V and GND, respectively.

An IC E^2PROM can be erased as well as programmed like (2k × 8) 2816 without an exposure to UV light. We call such EEPROM; electrically erasable and programmable read only memory i.e. E^2PROM. These E^2PROMs, in case possess a short access as well as programming time and has full erasibility in one cycle then are called flash memories, for example, an E^2PROM IC 28F256 is a flash memory. (F in an IC number of a ROM denotes a flash memory).

E^2PROM is erasable and programmable above 10000 (in latest E^2PROMs) times by the electrical means. Erase of a byte is by writing by 1111 1111 at D_0 to D_7 inputs from some external circuit. Before programming, the 0s must be erased by replacing them by 1s. Programming of a byte is by writing appropriate bits (for example 1000 1111 for writing 1000 1111) by the successive D_0 to D_7 inputs and the corresponding address inputs one by one from some external circuit (processor or laboratory programmer). Erase is byte by byte in E^2PROM and full sector wise in flash.

Flash is also erasable and programmable above 10000 times (in latest flash memories) by the electrical means. Erase of a whole sector of bytes (in a flash) is by writing by 1111 1111 at D_0 to D_7 inputs from some external circuit. Before programming, the 0s must be erased by replacing them by 1s. Programming of a byte is by writing by appropriate bits (for example 1000 1111 for writing 1000 1111) by successive D_0 to D_7 inputs and the corresponding address inputs one by one from some external circuit (processor or laboratory programmer). Flash can have a sector reserved to work as the OTP.

A certain E^2PROM chip, in case is one time programmable then it is called an OTP. An OTP does not have a quartz window. It stores data, which is unmodifiable by a user except once by an application of a high voltage pulse. (An OTP is like an electronic paper onto which writing is done by permanent ink).

■ EXAMPLES

Example 12.1 Show how will we design an eight-bit adder using a four-bit adder IC 7483.

Solution

We will cascade two adders by an adder IC for lower nibble addition given input carry = 0 and next upper nibble adder given the next stage carry of lower nibble as the input carry for its 0th stage. Figure 12.8 shows the circuit for implementing an eight-bit adder using two numbers four-bit adder IC 7483s.

Example 12.2 Show how will we verify using a digital comparator whether outputs from the adder circuit in Example 12.1 gives all bits 0s.

Solution

We will give the eight bits of Figure 12.8 adder to a comparator IC 7485. The inputs are as shown in the Figure 12.9. Comparison will be done with other eight inputs for B as shown there. The output ZF_1 = 1 shows all 7 bits of A as 0s, else output will be 0.

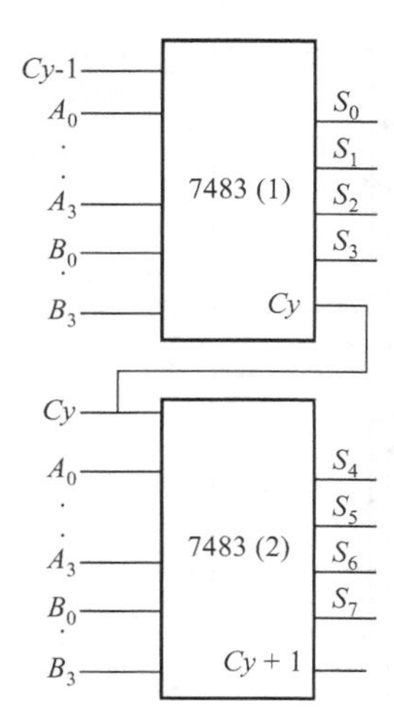

FIGURE 12.8 Circuit for implementing an eight-bit adder using two numbers four-bit adder IC 7483s.

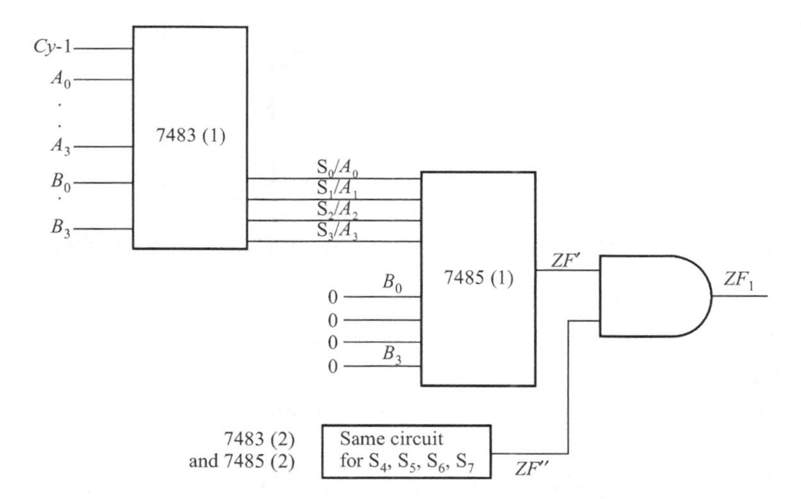

FIGURE 12.9 Two digital comparators to find whether outputs from the adder circuit have all 7 bits as 0s.

Example 12.3 Show how will we design a decoder tree for 6 six encoded inputs A_0 ... A_5 using 74138.

Solution

Figure 12.10 shows a decoder tree for six encoded inputs A_0 ... A_5 using 74138.

Example 12.4 Show how will we design a multiplexer tree for 16 inputs using 74156.

Solution

Figure 12.11 shows a multiplexer tree for 16 inputs using 74156.

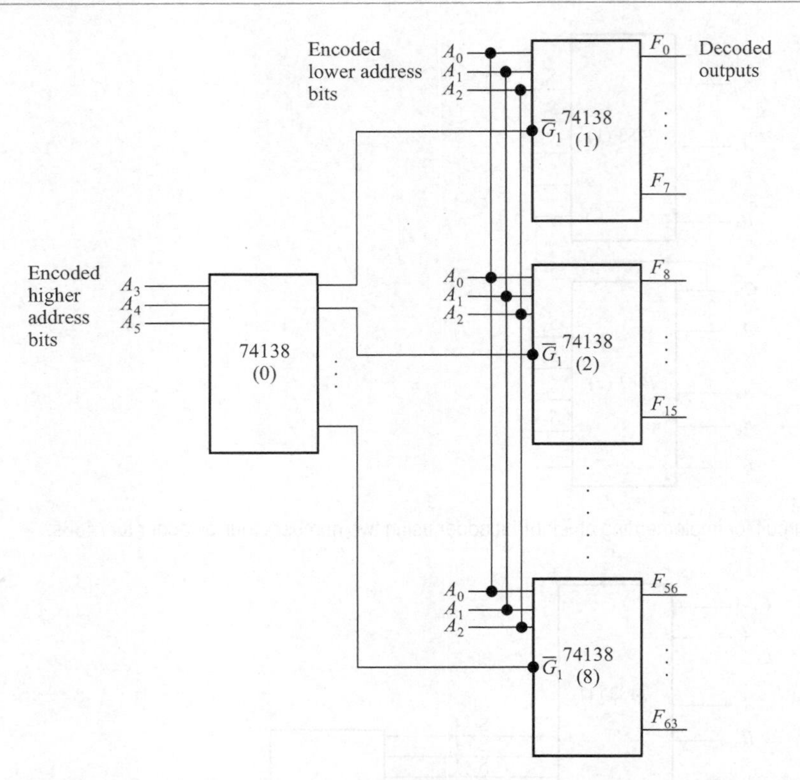

FIGURE 12.10 Decoder tree using 9 ICs 74138 for six encoded inputs using 74138. $\overline{G}_2$ and G_3 are 0 and 1 in each 74138.

Example 12.5

Show how will we design a BCD to Gray code converter using 74156.

Solution

Figure 12.12 shows a BCD to Gray code converter using the 74156.

Example 12.6

Show how to program the fusible links to get a 4-bit Gray code from the binary inputs using a PROM.

Solution

Step 1: Finding the PROM address bits and data bits and number of fused links needed:

Four OR gates are needed to get a four bit Gray code. There are four binary inputs. There will be $2^4 = 16$ AND gates in the decoder. We need 16×4 PROM. There will be 64 fusible links, 1 to 64. (Figure 12.13(a)). Let OR_0, OR_1, OR_2 and OR_3 give the outputs D_0, D_1, D_2 and D_3, respectively.

Step 2: Finding the fused links:

Table 12.1 gives the address inputs and Gray code for each set of inputs. Now, let us number these starting from $OR_0\ m(0)$ taken as 1 to $OR_0\ m(15)$ taken as 16, $OR_1\ m(0)$ taken as 17 to $OR_1\ m(15)$ taken as 32, $OR_2\ m(0)$ taken as 33 to $OR_2\ m(15)$ taken as 48 and OR_3

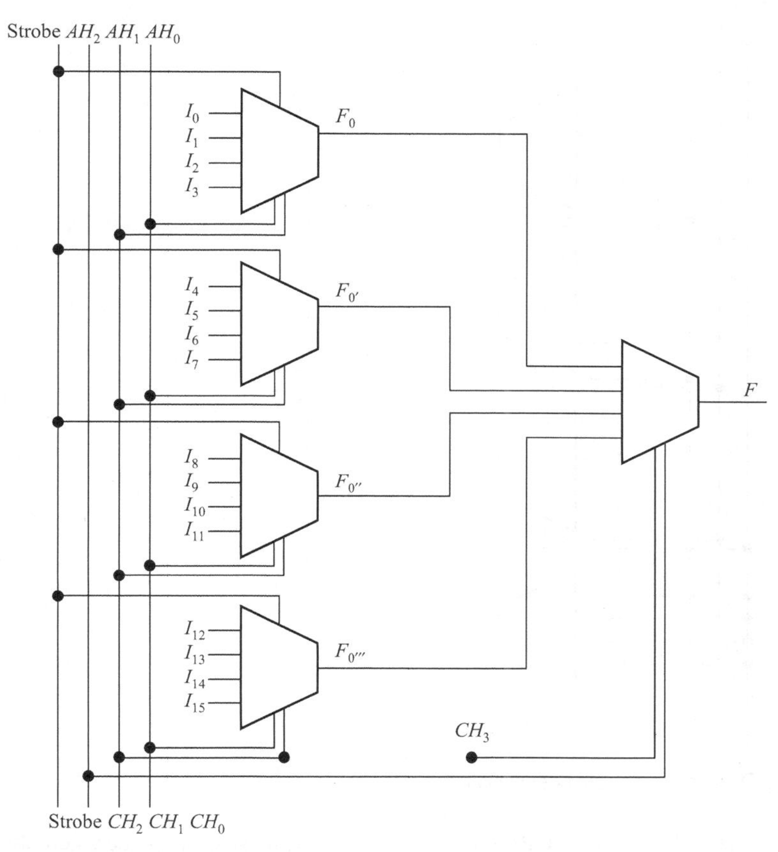

FIGURE 12.11 Multiplexer tree for 16 inputs using 74156D.

TABLE 12.1 Snapped link numbers corresponding to four gray code bits from solution in Example 11.4 for the Karnaugh map

Address inputs				Gray code	Snapped link numbers			
A_3	A_2	A_1	A_0	$D_3D_2D_1D_0$	OR_3	OR_2	OR_1	OR_0
0	0	0	0	0000_{gray}	49	33	17	1
0	0	0	1	0001_{gray}	50	34	18	
0	0	1	0	0011_{gray}	51	35		
0	0	1	1	0010_{gray}	52	36		4
0	1	0	0	0110_{gray}	53			5
0	1	0	1	0111_{gray}	54			
0	1	1	0	0101_{gray}	55		23	
0	1	1	1	0100_{gray}	56		24	8
1	0	0	0	1100_{gray}			25	9
1	0	0	1	1101_{gray}			26	
1	0	1	0	1111_{gray}				
1	0	1	1	1110_{gray}				12
1	1	0	0	1010_{gray}		45		13
1	1	0	1	1011_{gray}		46		
1	1	1	0	1001_{gray}		47	31	
1	1	1	1	1000_{gray}		48	32	16

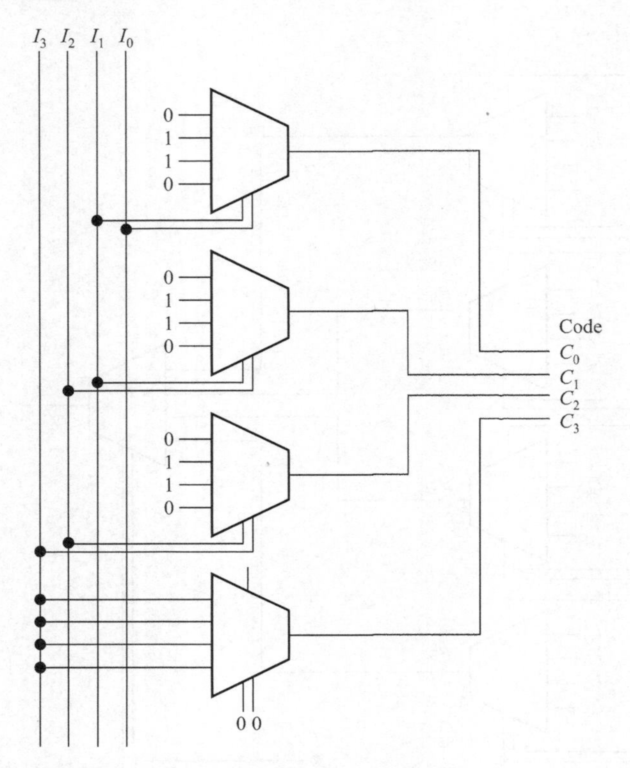

FIGURE 12.12 BCD to Gray code converter using the 74156.

$m(0)$ taken as 49 to OR0 $m(15)$ taken as 64. Link snaps at place where there are 0s in bits in column 5 (Gray codes). Figure 12.13(b) shows the solution for OR_0 and OR_1.

Example 12.7

Show how to program the fusible links to get the four Boolean expressions correspond to the given sets of four combinational circuits implemented using PROM.

Solution

$$F_0 = \Sigma m(0, 1, 4, 5);\ F_1 = \Sigma m(1,3, 5, 7);\ F_2 = \Sigma m(0, 2, 4, 6);\ F_3 = \Sigma m(0)$$

Step 1: Finding the PROM address bits and data bits and number of fused links needed:

Four OR gates are needed to obtain F_0, F_1, F_2 and F_3 outputs. There are three inputs, because maximum minterm given is $m(7)$. There will be eight AND gates in the decoder. We need 8 x 4 PROM. There will be 32 fusible links, 1 to 32 in it. (Figure 12.14).

Step 2: Finding the fused links needed:

Let OR_0, OR_1, OR_2 and OR_3 give the outputs F_0, F_1, F_2 and F_3, respectively (Table 12.2). Now, let us number the links starting from OR_0 $m(0)$ as 1 to OR_0 $m(7)$ as 8, OR_1 $m(0)$ taken as 9 to OR_1 $m(7)$ taken as 16, OR_2 $m(0)$ taken as 17 to OR_2 $m(7)$ taken as 24, and OR_3 $m(0)$ taken as 25 to OR_3 $m(7)$ taken as 32 in Figure 12.14. Link snaps at places corresponding to the 0s (shown by a × sign) in column for the outputs at the Table 12.2. Link is present at places shown by the dot.

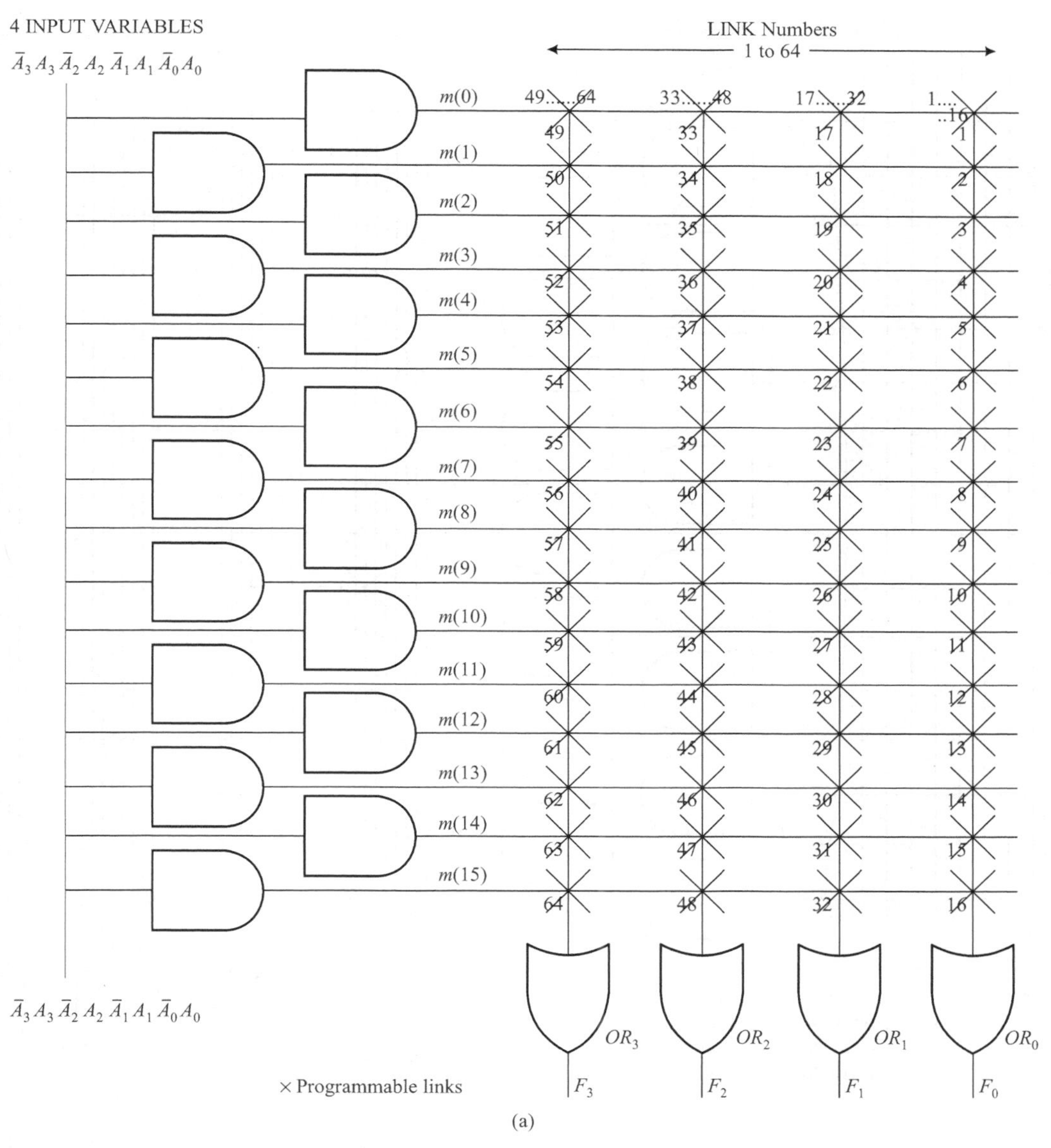

FIGURE 12.13 (a) 64 fusible links, 1 to 64 in a 16 × 4 PROM (b) Snapped links for OR_0 and OR_1 outputs F_0 and F_1 as per D_0 and D_1 in column 5 Table 12.1.

TABLE 12.2

Address inputs			Outputs	Present link numbers			
*A*2	*A*1	*A*0	*F*3*F*2*F*1*F*0	OR3	OR2	OR1	OOR0
0	0	0	1 1 0 1		25	17	1
0	0	1	0 0 1 1			10	2

Table 12.2 Contd.

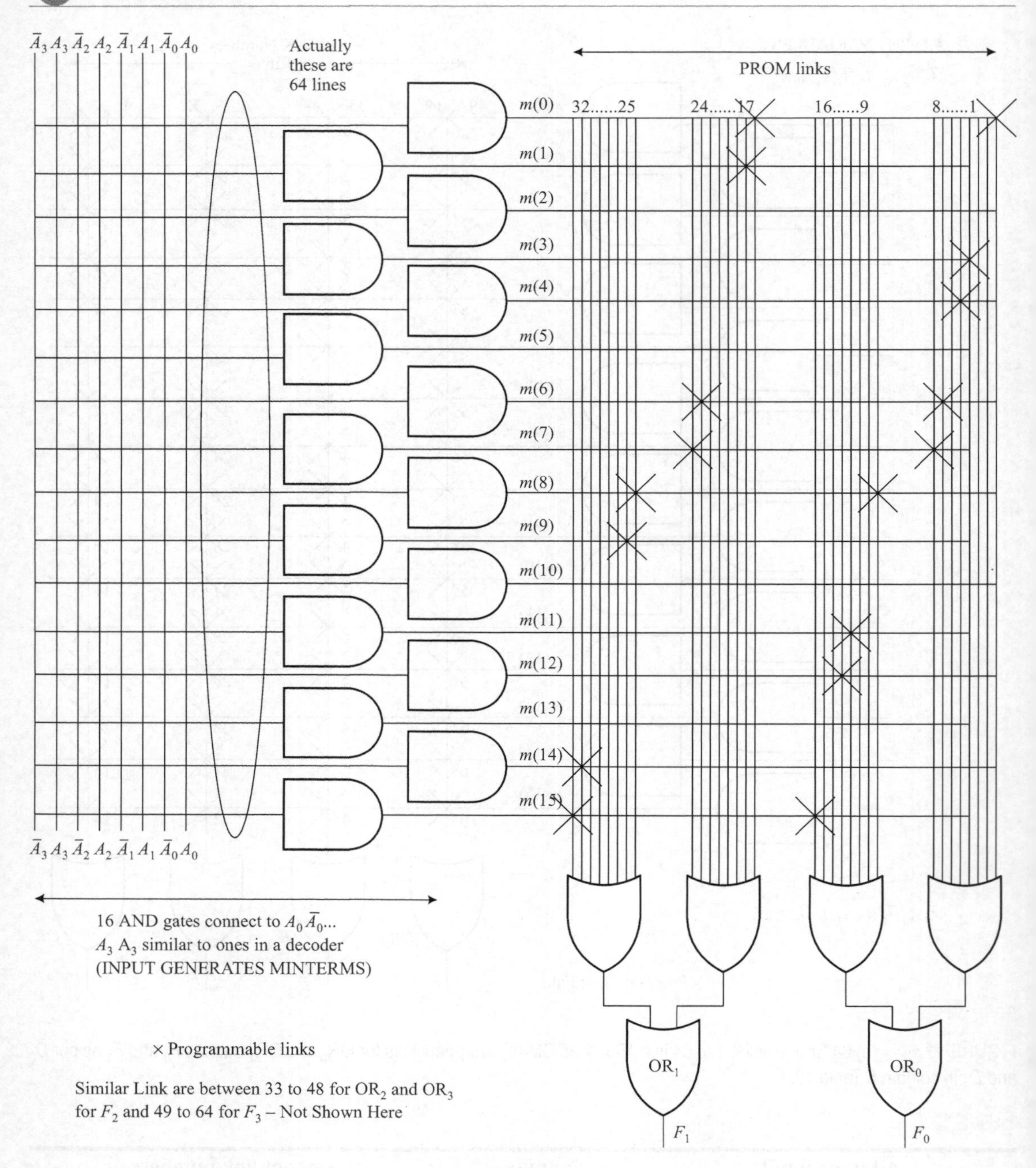
$\overline{A}_3 A_3 \overline{A}_2 A_2 \overline{A}_1 A_1 \overline{A}_0 A_0$
Actually these are 64 lines
PROM links
m(0)
m(1)
m(2)
m(3)
m(4)
m(5)
m(6)
m(7)
m(8)
m(9)
m(10)
m(11)
m(12)
m(13)
m(14)
m(15)
32......25
24......17
16......9
8......1
$\overline{A}_3 A_3 \overline{A}_2 A_2 \overline{A}_1 A_1 \overline{A}_0 A_0$
16 AND gates connect to $A_0 \overline{A}_0$...
A_3 A_3 similar to ones in a decoder
(INPUT GENERATES MINTERMS)
× Programmable links
Similar Link are between 33 to 48 for OR_2 and OR_3
for F_2 and 49 to 64 for F_3 – Not Shown Here
OR_1
OR_0
F_1
F_0

(b)

0	1	0	0 1 1 0	19		
0	1	1	0 0 1 0		12	
1	0	0	0 1 0 1	21		5
1	0	1	0 0 1 1		14	6
1	1	0	0 1 0 0	23		
1	1	1	0 0 1 0		16	

Note: $F3$ is 1 for $m(0)$ only. So fuse-link number 25 remains as such. $F2$ is 1 for $m(0)$, $m(2)$, $m(4)$ and $m(6)$. Therefore, 17, 19, 21 and 23 fuse-links are not snapped. Similar is the method for determining the links for $F2$ and $F1$.

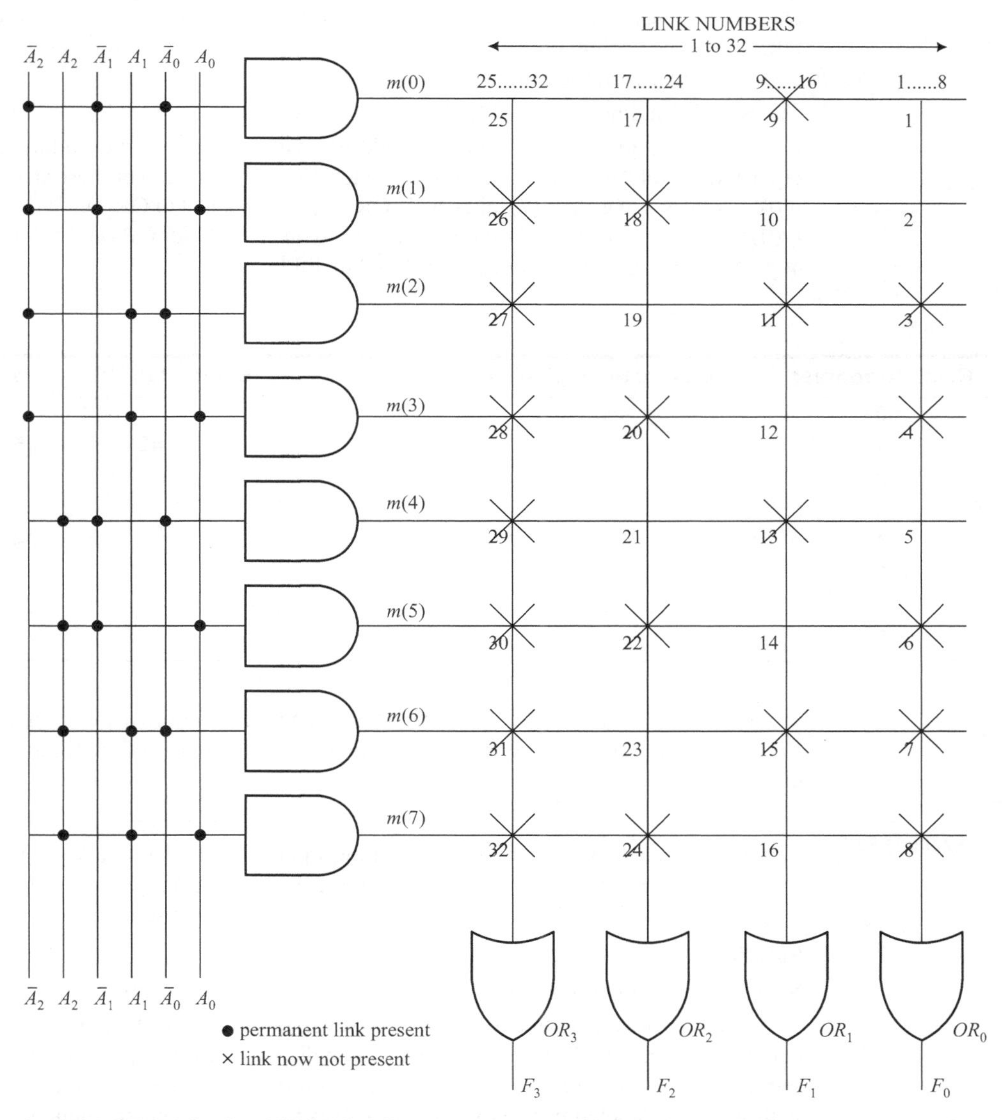

FIGURE 12.14 Implementation using 8 × 4 PROM four Boolean expressions $F0 = \Sigma m(0, 1, 4, 5)$; $F1 = \Sigma m(1,3, 5, 7)$; $F2 = \Sigma m(0, 2, 4, 6)$; $F3 = \Sigma m(0)$ for four combinational circuits.

Example 12.8

Take an 8 mm × 8 mm area on a centimeter graph paper and put points with a pencil so that these points correspond to a character *A*. Show how to design and program the fusible links to get the 8 successive set of outputs to print '*A*' using a PROM.

Solution

Step 1: Finding the PROM address bits and data bits and number of fused links needed:

(i) There are three inputs, because maximum minterm is $m(7)$. There will be eight AND gates in the decoder. (eight rows in the truth table). (ii) Eight OR gates are needed for 8 pixels at a row. We need 8 × 8 PROM. There will be 64 fusible links, numbered 1 to 64 in it. Let OR_0, OR_1, OR_2, ... and OR_7 give the outputs F_0, F_1, F_2, ... and F_7, respectively, to drive the dots in the matrix at the print head.

Step 2: Finding the Fused Links:

Table 12.3 gives relative address (address with respect to a base address) at which *A* eight rows pixel data stores as per column 2 to 9. Now, let us number these starting from OR_0 $m(0)$ taken as 1 to OR_0 $m(7)$ taken as 8, OR_1 $m(0)$ taken as 9 to OR_1 $m(7)$ taken as 16, OR_2 $m(0)$ taken as 17 to OR_3 $m(7)$ taken as 24 and so on up to 64 for OR_7 $m(7)$. A Link is present at place where there is a 0 in column 2 to 9 in Table 12.3.

TABLE 12.3

Relative address	Eight × Eight pixels for '*A*'							Eight bit (8 ORs) links needed						
0				*							33			
1			*		*					42		26		
2		*				*			51				19	
3	*						*	60						12
4	*	*	*	*	*	*	*	61	53	45	37	29	21	13
5	*						*	62						14
6	*						*	63						15
7	*						*	64						16

Figure showing the fused (snapped) links can be easily drawn using Figure 12.13(a) in a identical way shown in Figures 12.13(a) and 12.14. Drawing of the figure is left as exercise to the reader.

Example 12.9

Consider the seven bits of ASCII codes for the alphanumeric characters and assume 8th parity bit = 0. Show how to design and program the fusible links to get the eight successive sets of outputs to display 'Welcome' using a PROM using sixteen segment character displays.

Solution

Step 1: Finding the PROM address bits and data bits requirements:

(i) Given are the number of inputs = 8 (seven bit code + one bit parity). (ii) Since 'Welcome' has characters W and m, which can't be displayed by seven-segment unit, a sixteen-segment unit is needed for each character. [Figure 12.15(a) shows a sixteen segment unit with segments labeled from a to *p* as well as 0 to 15]. For sixteen-segment unit, sixteen out-

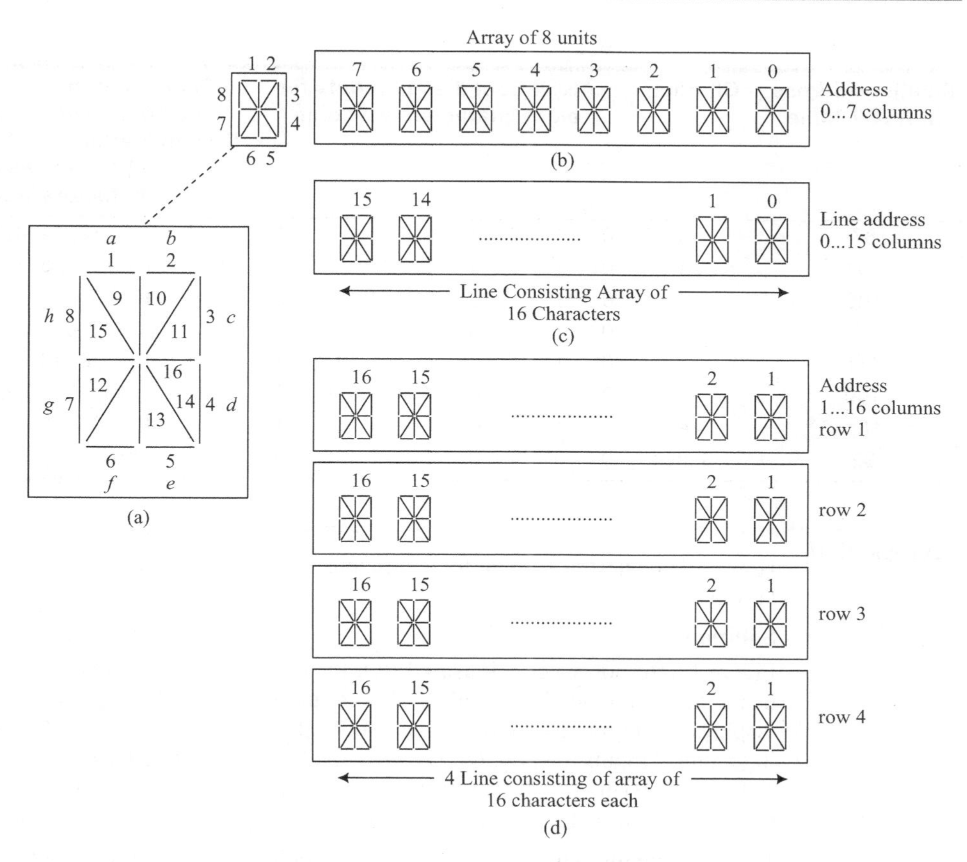

FIGURE 12.15 (a) Sixteen-segment unit with segment numbers labeled from *a* to *p* as well as 0 to 15 (b) An array of 8 units to display 'Welcome' (c) An array of 16 units to display a line (d) A matrix of 64 units to display four lines, 16 characters, each.

puts are needed. Therefore, an 8×16 PROM is required. [Note: Figure 12.15 (b) also shows sixteen-segment units in an array of 8 units to display 'Welcome'. Figure 12.15(c) shows an array of 16 units to display a full line message. Figure 12.15(d) shows a (16×4) 64-unit matrix to display four lines, 16 characters, each using the PROM outputs.

Step 2: Finding the fused links:

Let base address b be the starting address in PROM for the 256 addresses for placing the ASCII table. Table 12.4 gives the relative address r (address with respect to the base address at which the segments for a character in the 'Welcome' selects among the sixteen segments. If links are not fused (snapped) at the places shown then the segments shown in column 10 display. Column 10 gives segment numbers, which equal the [fuse Link numbers with respect to (relative address + base address)*16)] for the eight rows for the characters of 'Welcome'. If segment number = 2 then corresponding fuse link is $(b + r)$, if 3 then $(b + r + 1)$, if 3 then $(b + r + 2)$ and so on up to $(b + r + 15)$. From the segment numbers we find the fuse link numbers for the fuses not to be snapped.

TABLE 12.4

Relative address in ASCII Table	Character	ASCII Coded eight inputs for eight characters welcome								Segment numbers = {fuse link numbers with respect to the (Relative Address + Base address) × 16)} for 8 rows for eight characters welcome
87	*W*	0	1	0	1	0	1	1	1	8, 7, 12, 14, 4, 3
101	*e*	0	1	1	0	0	1	0	1	15, 10, 1, 8, 7, 6
108	*l*	0	1	1	0	1	1	0	0	10, 13
99	*c*	0	1	1	0	0	0	1	1	1, 8, 7, 6
111	*o*	0	1	1	0	1	1	1	1	15, 7, 6, 13
109	*m*	0	1	1	0	1	1	0	1	7, 15, 13, 16, 4
101	*e*	0	1	1	0	0	1	0	1	15, 10, 1, 8, 7, 6
32	Space	0	0	1	0	0	0	0	0	None

Example 12.10

How will you design a circuit for continuously incremental display from 000 to 999 at regular intervals?

Solution

Figure 12.16 shows a circuit using PROM to display continuously 000 to 999 sequentially using three numeric display units, nd_0, nd_1 and nd_2, each of 7 segments. A seven segment display unit (Figure 10.16(a)) displays decimal numbers 0, 1, 2, 3, 4, 5, 6, 7, 8 and 9 depending on the 7 inputs bits, O_0, O_1, O_2, O_3, O_4, O_5, O_6, and O_7 as (i) as 1, 1, 1, 1, 1, 1, 0 (ii) 0, 1, 1, 0, 0, 0, 0 (iii) 1, 1, 0, 1, 1, 0, 1 (iv) 1, 1, 1, 1, 0, 0, 1 (v) 0, 1, 1, 0, 0, 1, 1 (vi) 1, 0, 1, 1, 0, 1, 1 (vii) 1, 0, 1, 1, 1, 1, 1 (viii) 1, 1, 1, 0, 0, 0, 0 (ix) 1, 1, 1, 1, 1, 1, 1 and (x) 1, 1, 1, 1, 0, 1, 1, respectively. The unit has a common cathode interconnected to a bit, called $\overline{\text{EN}}$ bit. This $\overline{\text{EN}}$ bit, if at logic state 0 or 1 a display unit for a digit glows or not, respectively. At three consecutive addresses in the PROM, the details of the three bits corresponding to nd_0, nd_1, and nd_2 are stored. Three sequences are as follows. The first digit glows when $A_0 = 0$ and $D_7 = 1$. Second digit glows when $A_0 = 1$ and $D_7 = 1$. Third digit glows when $A_1 = 1$, $A_1 = 0$ and $D_7 = 1$.

At each address, the $D_0 ... D_7$ bits corresponds to 7 segments O_0, O_1, O_2, O_3, O_4, O_5, O_6, and O_7, respectively, where $i = 1, 2, 3$. The total number of addresses, which are programmed in the sequential pattern, will be 3000 as there are 3 display units, and there are one thousand sequences to be displayed.

EXERCISES

1. Recall the concept of lookahead carry for the fast processing of the carry. IC74283 has a faster processing of the *Cy* at the higher stages. Show the 74283 IC connections.
2. Show how will you design an eight-bit adder cum subtractor using a four-bit adder IC 7483s and XOR ICs.

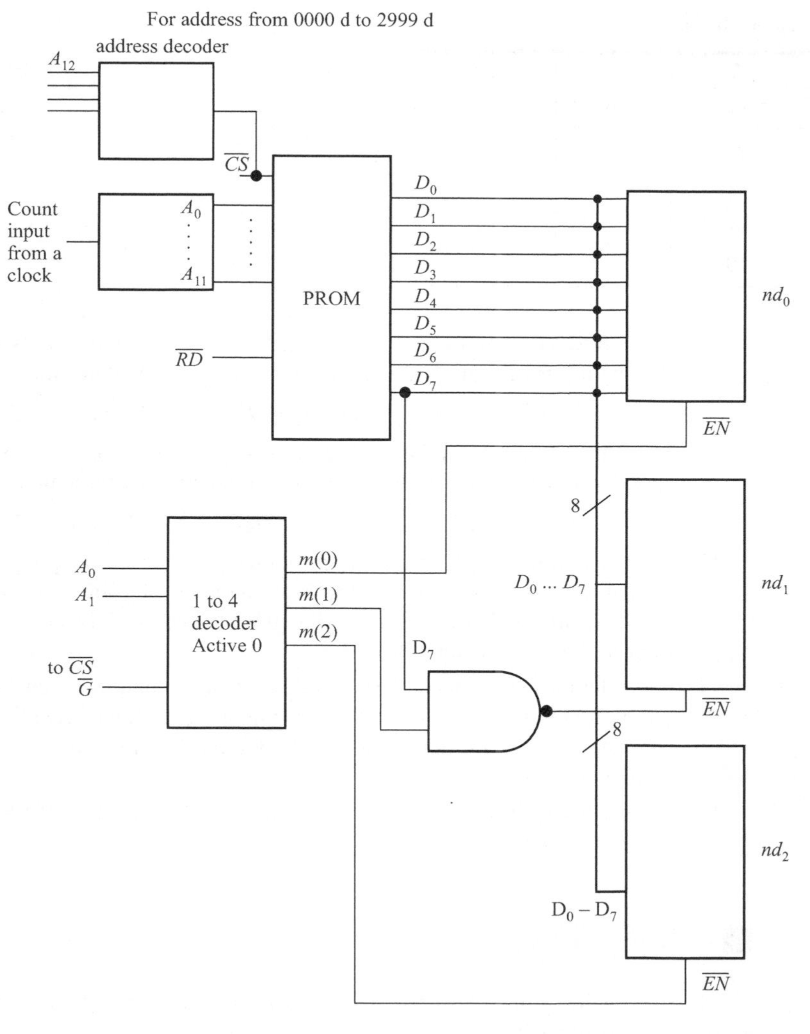

FIGURE 12.16 Circuit using 4k × 8 PROM with 3000 programmed addresses to display continuously 000 to 999 sequentially using three numeric display units, *nd*0, *nd*1 and *nd*2, each of seven segments.

3. Show how will we verify using a digital comparator whether the outputs from a subtractor circuit will generate carry.
4. Show how will we design a decoder to decode eight address lines using 74138.
5. Design using ICs an encoder for encoding inputs from the 16 keys.
6. Show how to program the fusible links to get a 4-bit Excess –3 Gray code from the binary inputs using a PROM.
7. Show how to program the fusible links to get the Boolean expression correspond to the given Karnaugh map implemented using PROM.

Three variable Karnaugh Map

AB \ C		$\overline{C}$ 0	C 1
$\overline{A}\,\overline{B}$	0		
$\overline{A}\,B$	01		1
AB	11		1
$A\overline{B}$	10		1

8. Show how to program the fusible links to get the Boolean expression correspond to the given sets of four combinational circuits implemented using PROM.

$$F_0 = \Sigma m(1, 4);\ F_1 = \Sigma m(3, 5);\ F_2 = \Sigma m(7);\ F3 = \Sigma m(2)$$

9. Show how to program the fusible links to get the four Boolean expressions correspond to the given sets of four combinational circuits implemented using PROM.

$$F_0 = \Sigma m(3, 4, 5);\ F_1 = \Sigma m(1, 3, 5);\ F_2 = \Sigma m(4, 6);\ F_3 = \Sigma m(7)$$

10. Take an 8 mm × 8 mm area on a centimeter graph paper and put points with a pencil so that these points correspond to first character of your university's name. Show how to design and program the fusible links to get the eight successive set of outputs to print the name using a PROM.
11. Consider the seven bits of ASCII codes for the alphanumeric characters and assume 8th parity bit = 0. Show how to design and program the fusible links to get the eight successive sets of outputs to display 'Good Bye!' using a PROM using sixteen segment character displays.
12. How will you design a circuit for continuously decremented display from 00 to 99 at regular intervals?

QUESTIONS

1. List the ICs you will use for the arithmetic and logic operations. Give two exemplary circuits of each.
2. List the ICs you will use for the decoders, encoders, demultiplexers and multiplexers. Give two exemplary circuits of each.
3. What is a PROM? Describe four applications of PROMs as (a) implementer of Boolean expressions and combinational circuit, (b) memory unit for data and instructions of a program, (c) code converter and (d) display units driver.
4. Is a ROM or PROM not accessible randomly?
5. When do we use an EPROM, when a E^2PROM, when a flash, when an OTP and when factory marked ROM?
6. What do we mean by the *erase* in EPROM, in flash and in E^2PROM? (Hint: All bytes erase by UV, all bytes of a sector by writing all 1s, and single byte erase by writing 1s).

7. How many distinct addresses are programmed in a PROM, when we wish to drive the six numeric display units, each unit with 16 segments?
8. Why should it be that an active 0 two inputs 1 out of 4 address decoder is used to connect a common cathode of the numeric display units?

CHAPTER 13

Implementation of Combinational Logic by Programmable Logic Devices

OBJECTIVE

We learnt in Chapter 12 that instead of implementing a minimized circuit using AND-OR arrays or NANDs or OR-AND arrays or NORs or ICs for the decoders, encoders, multiplexers or demultiplexers, or binary arithmetic adders, adder/subtractors, code converters, comparator, bit-wise 8-bit AND, OR, XOR circuits or parity generators, a complex combinational circuit can be easily assembled by programmable logic memories (ROM, EPROM, EEPROM or Flash or OTP).

In this chapter, we shall learn about two other forms of PLDs called PAL and PLA. We shall also study the implementation of the circuits using PAL and PLA.

13.1 BASICS POINTS TO REMEMBER WHEN USING THE PLDS (PROMS, PALs, PLAs)

For any logic function implemented by a combination circuit, there is a truth table. A row in a truth table can be specified as a minterm in a Boolean expression form, called sum of the products (SOPs) expression. There is one SOP expression each for each output column of the table. For example, consider a sum of the products expression for an output, Y.

$$Y = \overline{A}.B.\overline{C} + A.B.C \qquad ...(13.1)$$

This means that we have a truth table at which in an output column for the output Y the logic state = 1 in those two rows where the inputs, (A, B, C), are (0, 1, 0) and (1, 1, 1), respectively. For the remaining 6 rows of the truth table, we have $Y = 0$. A complemented variable at above expression, like $\overline{A}$, means its state is logic 0, and a variable without complement sign like B means state 1. Maximum number of rows in case of the three input variables is 8, and therefore equation (13.1) can have maximum eight terms on the right hand side. There can be maximum eight sums of product (SOP) terms with each product term having 3 variables, either as such or in its complemented form.

Similarly, consider another SOP expression for another truth table.

$$Y_0 = \overline{A}.B.\overline{C}.\overline{D} + \overline{A}.B.C.D + A.\overline{B}.\overline{C}.\overline{D} \quad ...(13.2a)$$

$$Y_1 = \overline{A}.\overline{B}.\overline{C}.D + \overline{A}.B.\overline{C}.\mathrm{D} \quad ....(13.2b)$$

It means that $Y0 = 1$ for the three rows of out of the 16 rows of the corresponding truth table for four inputs A, B, C and D. $Y = 0$ in all rows except the ones in which A, B, C and D are 0, 1, 0 01 and 0; 1, 1, 1 and 1, 0, 0 0, respectively. Maximum number of rows in four input variables are there are 16 and therefore, two expressions (13.2a and b) can have maximum 16 terms each on the right hand side, since there can be maximum 16 sums of product terms with each product term having four variables either as such or in a complemented form. Each output, Y, in its sum of products form can be implemented by a circuit having the AND-OR arrays. This is explained as follows:

Let us consider a logic circuit in Figure 13.1. Figure 13.1 is for n inputs (2^n truth table rows) a matrix of the $2n$-input AND gates and 2^n input m OR gates. ($2n$) input AND is used because of n variables and their n complements). Here, $n = 3$ and $m = 1$. Let us consider the array of six ANDs in the figure. Each AND has a distinct set of three plus three inputs each among A, B and C plus complements $\overline{A}$, $\overline{B}$ and $\overline{C}$. The 6 inputs of the AND are interconnected as per 8 possible logic states of the 3 input variables in the Boolean expression.

A horizontal line known as product line represent a set of inputs for a truth table row or a minterm. Let a connection mean permanent link and intersection mean temporary fusible link. There can thus be 48 points of intersections or connections in the AND array inputs for six ANDs. An AND array output connects to an eight input-line OR (shown by single vertical line) through another eight connections or intersections.

In general for the n inputs, there are maximum 2^n numbers $2n$-input AND gates (called AND plane) and m OR gates (called OR plane). A matrix of these AND-OR arrays have maximum 2^n rows and m columns. (The $n = 3$ and $m = 1$ in Figure 13.1).

At each intersection of the horizontal line (AND plane inputs) and vertical line (AND plane outputs), there can be a fusible link, which may be present or missing. The maximum number of intersections is equal to the number of elements in the matrix of 2^n rows and are in m columns. Maximum number of fusible links can be ($2n.2^n.m$). There can thus be ($2n.2^n.m$) links, which are to be programmed in general for a link present or link (missing, fused and snapped). (Figure 13.2 shows case of $2n.l.m$ links).

Figure 13.3 shows circuit of AND-OR arrays to implement a sum of the products expression for the case of three input variables with one output Y.

Figure 13.3 also shows the intact links (by the crosses), which should be present in order to implement a function defined by SOP expression $C + \overline{B}.\overline{C}$ and implement the logic

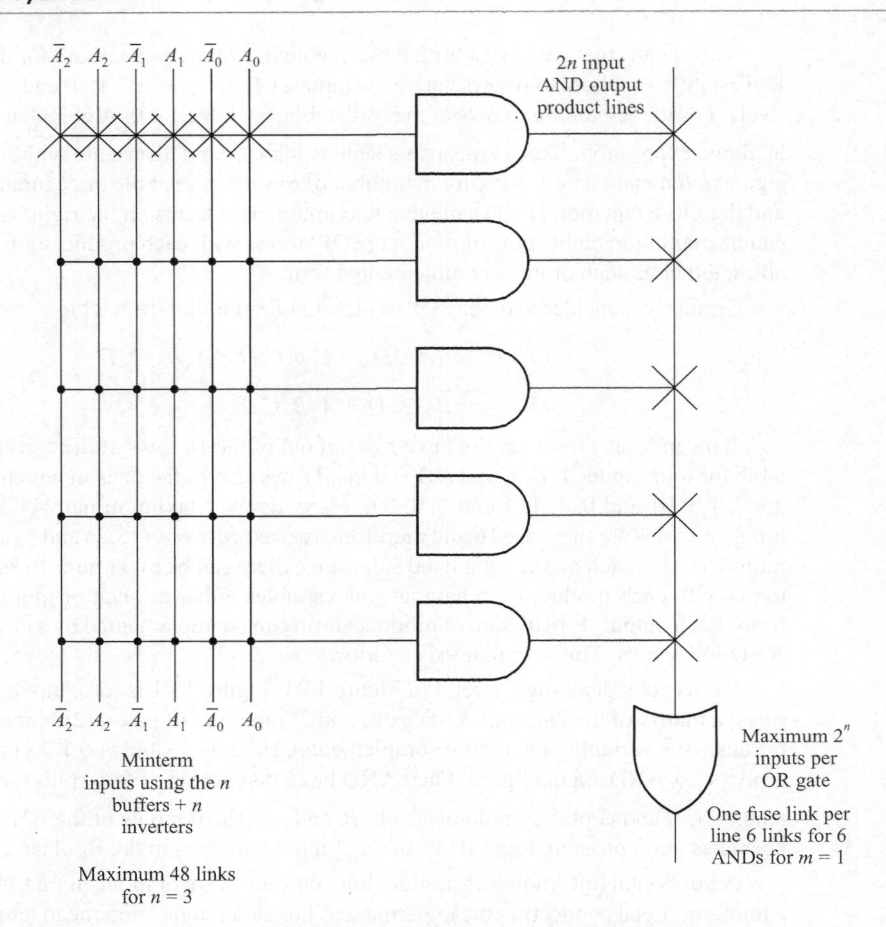

FIGURE 13.1 For the n inputs (2^n truth table rows) a matrix of $2n$-input AND gates and m OR gates, when $n = 3$ and $m = 1$.

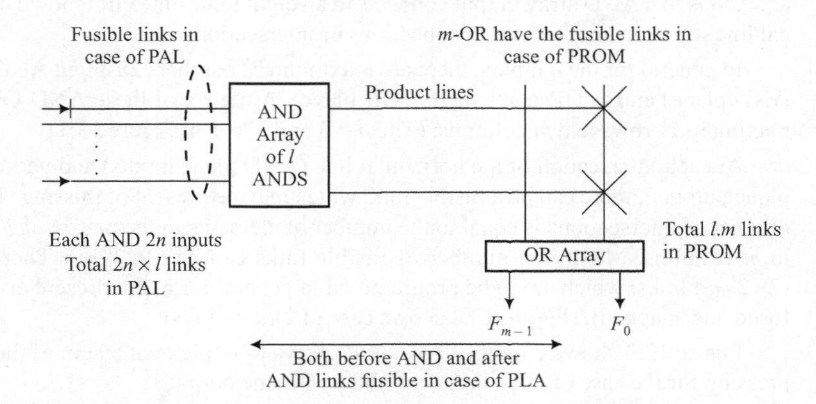

FIGURE 13.2 Circuit of the AND-OR arrays of l ANDs to implement any sum of the products expression for the case of n input variables with one output Y and the fusible links present for obtaining the F outputs as per required $Y_0, \ldots, Y_{m-1}$ when using a PAL or PROM of PLA.

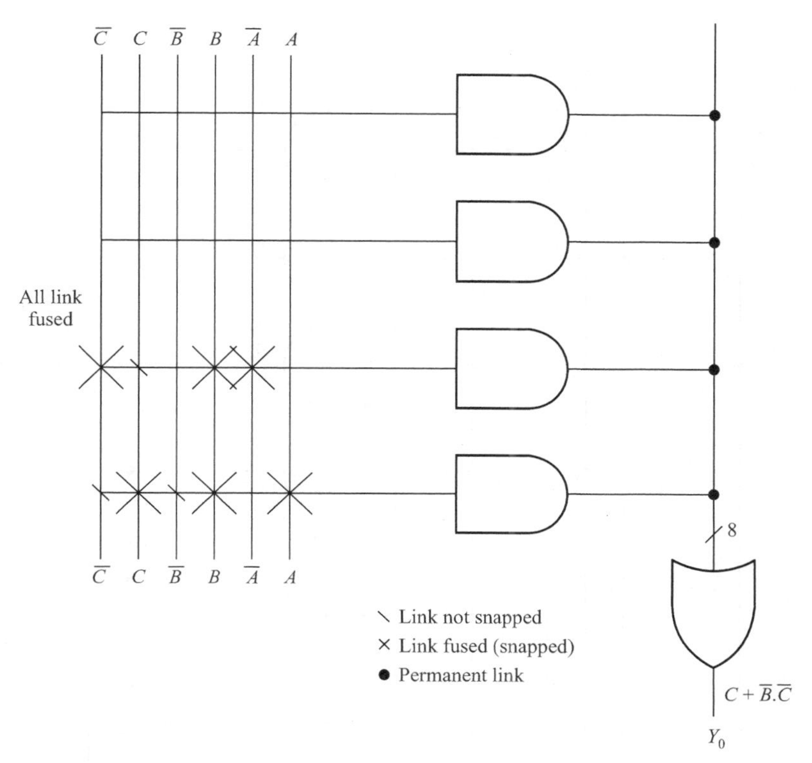

FIGURE 13.3 PAL circuit of AND-OR arrays to implement any sum of the products expression for the case of three input variables with one output Y_0.

states. The fused links for $Y_0 = C + \overline{B}.\overline{C}$ are shown by the cross signs. In this expression, there are only two terms on right hand side. Therefore, there are only two present links to an eight input OR gate.

Figure 13.3 also shows the intact links (by the dots), which should be present in order to implement a function defined by SOP expression. In Figure 13.4 the present links at the AND arrays [in order to implement truth table function corresponding to the SOP expressions (13.2a and 2b)] are shown by the cross signs. In the expressions, there are only two and three terms on the right hand side for Y_1 and Y_0. Therefore, there are only two and three present links to the16 input OR gates are shown.

When n is large, this number becomes too big. Practically, the numbers of links are therefore limited in PROMs, PALs and PALs. PROM has only OR links and PAL only AND links. FIgures 12.6, 12.13 and 12.14 shows applications of PROMs.

Points to Remember for a PROM

1. Programmable ROM is a special programmable logic device (PLD) in which each input of the OR gate in the OR-arrays have the fusible links, which are fused as per the truth table or the Boolean expressions or Karnaugh maps requirements for the given set of outputs. (AND array inputs are not programmable in a PROM).

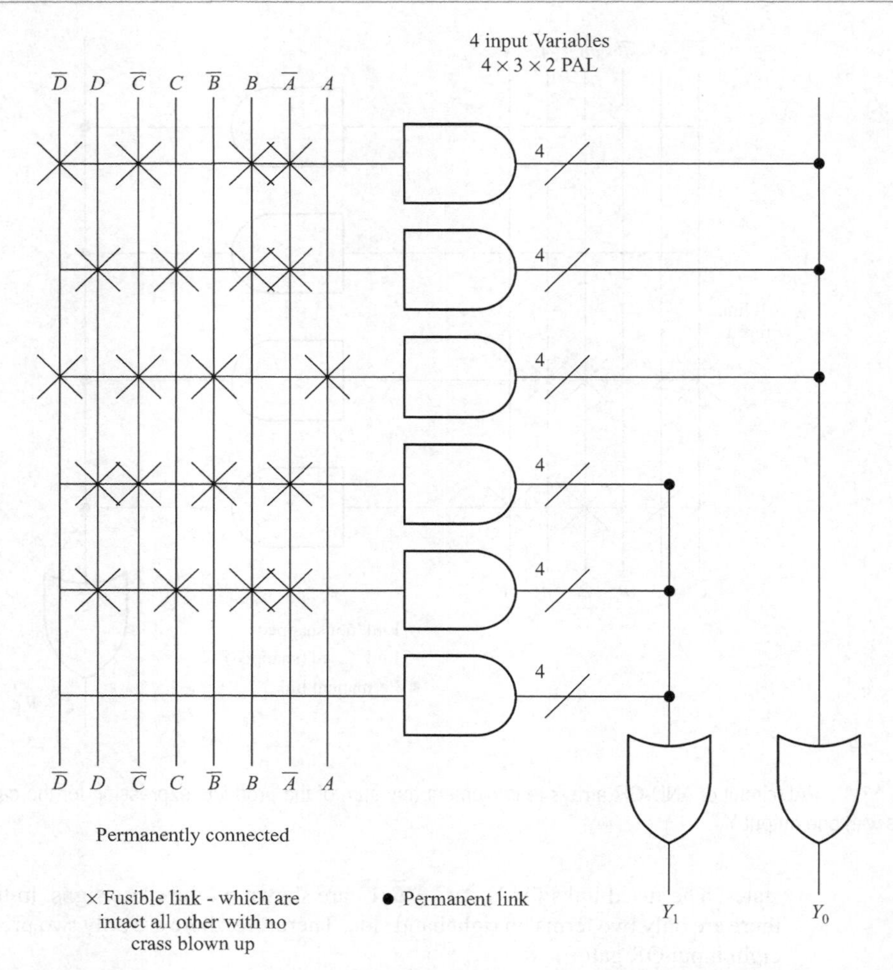

FIGURE 13.4 ($n \times l \times m$) PAL circuit of AND-OR arrays to implement prime implicants at the SOP expressions with only $2nl$ fusible links and generate m outputs.

2. ($n \times m$) PROMs have n input AND gates (2^n in number) and 2^n input m ORs. Each AND output corresponds to a minterm each in the SOP expression. Number of fusible links are $2^n \times m$ only in place of $2n$. $2^n \times m$ maximum possible.
3. An OR gate with no input given at one of its line is assumed to have input 0.
4. An AND gate with no input given at one of its line is assumed to have input 1.
5. Multiple n lines, each with a fusible link can be shown as a single line with the n fusible links.
6. A fusible link, which is not fused (snapped) if shown by a cross or slash sign at an intersection at the input from a previous stage output or from a source of input to the link, then the slash or cross sign is now equivalent of soldered link in the electrical circuit.

7. A fusible link that is fused (snapped) is shown by a missing cross sign at the intersections at the inputs from the previous stage outputs. A missing cross sign is equivalent of un-soldered (detached) link in the electrical circuit.
8. The AND gate or OR gate input line with all links fused with no cross sign any where is placed a cross sign in its center of AND or OR symbol, respectively.
9. A programming is a systemic hardware related procedure implemented by software at a programming system, called programmer or laboratory programmer or PLD programmer and the programming means executing the procedure for fusing (snapping) the needed links in the erased or fresh PLD.
10. Erased PLD means in which fusible links ready for programming.
11. Complete erasing occurs in EPROM by UV. Byte by byte erasing occurs in EEPROM and sector by sector in a flash memory by electrical means alone.
12. Programming of EEPROM or EPROM or flash is writing a byte one by one for fusing the appropriate links at the previously erased byte.

13.2 PAL (PROGRAMMABLE ARRAY LOGIC)

Each AND array corresponds to a SOP minterm or an implicant of Boolean expression. The output of each AND array is given to an m-input OR. An output of OR corresponds to output, Y. In general, there should be a 2^n input OR gate at each output variable, Y. However, in actual practice, these are limited as there are limited numbers of AND gates (only l number) used in the PALs. Let the number of AND gates be l. Each AND has $2n$ inputs. Therefore, the number of fusible links = $(2n.l)$.

PAL is a registered trade name of Monolithic Inc. USA, and Advanced Micro Devices. The OR links are fixed in a PAL, and are not programmable. Each OR gate has (l/m) fixed connections in the PALs. [When l is not an integer multiple of m, then one of the OR gate will have less number of OR gates and others $(l-1)/(m-1)$ or $(l-2)/(m-1)$ fixed connections. For example, if $p = 8$, $m = 3$, then two OR gates will have three each and one OR gate 2 fixed connections only.]

Programmable Array Logic (PAL) is a special programmable logic device (PLD) in which each input of the AND gate in the AND-arrays have a fusible link each, which is fused as per the minimum requirement after a suitable minimization and reduction procedure. The AND-array fusible links are fused as per the requirement of minimized form (prime implicants) after multi output minimization the number of the Boolean expressions (called multi output minimization) to be implemented.

Figure 13.4 showed a $(n \times l \times m)$ PAL A $(n \times l \times m)$ PAL has l number AND gates each with $2n$ inputs and each input has a fusible link and $n = 4$, $l = 6$ and $m = 2$. The fusible links number is just $(2.n.l)$ in place of $(2n.2^n.m)$ maximum possible and $(2^n.m)$ in PROMs.

A PAL needs to implement only the m' number of Boolean expressions by multi output minimization through a procedure the m number of Boolean expressions but generating the required m outputs. Reduction is by considering those expressions, which contain the expressions with a common set or complementary set of prime implicants.

A unit called PAL programmer program a PAL. Using a PAL, any desired logic function to obtain an output bit Y can be implemented. A PAL is programmed like an E^2ROM. A PAL is a general-purpose combinational circuit for the prime implicants of Boolean expressions. User needs to know only the truth table while using a PROM. For using a PAL, user need to know only the sets of the prime implicants of the logic functions needed to obtain the required outputs.

Why does a PAL have less number of ANDs l compared to 2^n in a PROM?

Reason is that usually n = quite high, say 10 and there are computer based minimization procedures available, that makes it feasible to implement a combination logic circuit with prime implements only. Occurrence of common sets and complementary sets of prime implements in the Boolean expressions further reduce the requirements of the ANDs. So the PAL will have less cost (less number of gates) requirements and $l < 2^n$.

For example, a version of PAL called 16L8 contains eight ORs. Therefore, there are maximum eight outputs. Figure 13.5 shows its logic circuit. Each OR has seven inputs. There can be maximum 16 inputs in 16L8. The 16 inputs can be the Boolean variables and also the feedback of the ORs two outputs are dedicated and six outputs pins can be programmed so. There are 32 columns to link eight AND arrays. There are eight AND array as there are eight ORs. Each AND is an 8 inputs AND. Therefore, there are 64 rows total number of fusing links in 16L8 are, therefore, 2048 (= 32 multiplied by 64) which are programmable by an external unit called PAL programmer. Inside this IC for PAL, between each OR output and an output pin, there is a tristate NOT meaning thereby, that the output pins are active 0s. An output pin represents function $Y = 1$ situation when at 0 and in standby (inactive) state, it is at 1.

An unregistered PAL acts like a combinational circuit. A registered PAL acts like a general-purpose sequential circuit (a circuit in which the past sequences (states) also matter). Instead of using discrete gates and FFs, we can use a registered PAL (FFs are described in next chapter). A registered PAL find applications in the state machines (the logic circuits generating the outputs states according to a sequence).

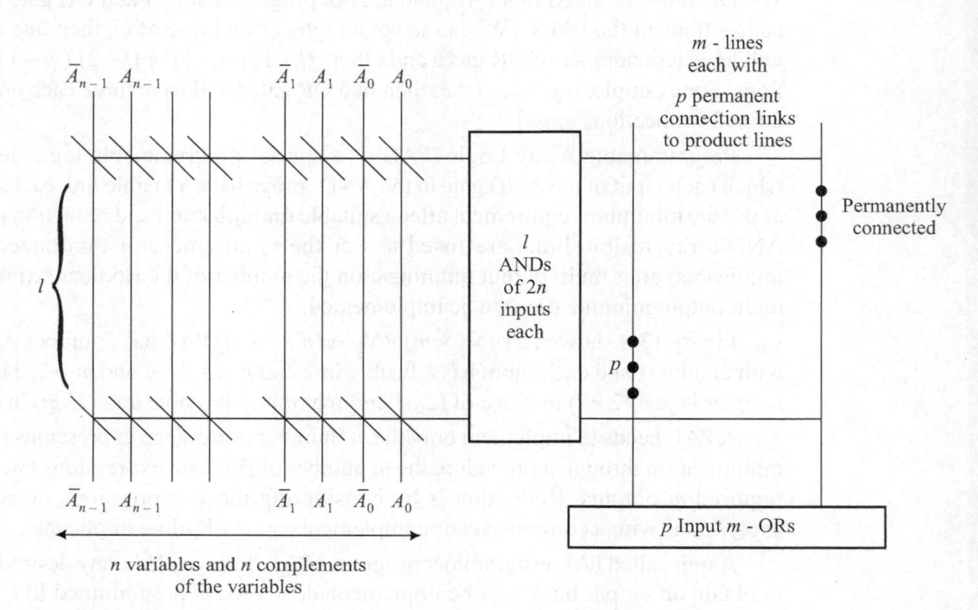

FIGURE 13.5 PAL 16L8 inputs and outputs and logic circuit. $m = 8$, $l = 7$, $n = 8$, $p = 7$.

The PALs are less costly (fewer gates) than PROMs, give an advantage of smaller access times of ~ 30 ns or less compared to ~ 100 ns or more in PROMs. These devices simplify complicated circuits of the gates by at least five times or more. CMOS version of programmable logic devices (PLDs) like PALs have access times 0.015 μs to 0.030 μs more than the bipolar PLDs, but have smaller power consumption per PLD. The HCMOS (High Speed CMOS) versions are also available. An output from a PAL is, however, available within, typically, ~0.030 μs in the HCMOS versions, and about 10 ns in bipolar transistor versions.

[Note: Electrically erasable PALs are also called field programmable PALs and are more costly. Masked PALs can be made like the masked ROMs at the manufacturing stages.]

Remember the Definition

PAL is a special programmable logic device (PLD) in which each input variable and its complement at an AND in the ANDs array have a fusible link, which is fused as per the prime implicants or the common and complementary sets of prime implicants and give the outputs as per the truth table for the combinational circuit.

Points to Remember

1. PAL is a special programmable logic device (PLD) in which each input and its complement at an AND has a fusible link. Total number of AND gates are limited in number to l with $l < 2^n$ maximum possible.
2. The AND-array fusible links are fused as per the requirement of minimized form (prime implicant form) after a multi output minimization of Boolean expressions for the outputs.
3. We implement m' number of Boolean expressions by multi output minimization of the m number of Boolean expressions. Reduction is by considering the expressions containing the common set or complementary set of prime implicants.
4. Inputs at AND array are programmable and inputs at OR array are not programmable in a PAL.
5. $(n \times l \times m)$ PAL has $2n$ input AND gates (l in number) and fixed input (number of inputs $< l$) m number OR gates. Each AND output corresponds to a prime implicant each in the SOP expression. Since number of ANDs are limited in number of fusible links l are $(2n.l)$ only in place of $(2n.2^n.m)$ maximum possible.
6. At the AND inputs, a slash or cross sign shows a fused link, which is not snapped and none just an intersection shows a fused (snapped) link A dot shows the permanently connected link.
7. The AND gate, with all links fused with no sign any where, is placed a cross sign at the center of the AND symbol, respectively.
8. A programming is a systemic hardware related procedure implemented by software at a programming system, called programmer or laboratory programmer or PAL programmer and the programming means executing the procedure for fusing (snapping) the required links in a fresh (unprogrammed) PAL.
9. Electrical erasability in PAL is most often not available.

13.3 PLA (PROGRAMMABLE LOGIC ARRAYS)

Each AND array corresponds to a SOP minterm or implicant. The output of each AND array is given to an m-input OR. An output of OR corresponds to output, Y.

(1) In general, there should be $(2.n)$ input 2^n number AND gates with OR gate for each output variable, Y. However, in actual practice, these are limited as there are limited numbers of AND gates used in the PLAs.

(2) Similarly in general, there should be a 2^n input OR gate at each output variable, Y. However, in actual practice, these are limited as there are limited numbers l of AND gates used in the PLAs and therefore there are (l/m) input m number OR gates in the PLAs. [(l/m) if not an integer, then last OR gate will have $(l-1)/m$ or $(l-2)/m$ inputs and remaining $(l-1)/(m-1)$ or $(l-2)/(m-1)$ input OR gates with each OR gate having fusible links.

Let l is the number of AND gates present in a PLA. Each AND has $2n$ inputs. Therefore, the number of fusible links (like a PAL) is $(2n.l.)$ at the AND plane of the PLAs. The number of fusible links is l $[= (l/m).m]$ at the OR plane of the PLAs. Therefore, the number of fusible links is $(2n.l^2)$.

OR links are also fusible in a PLA (like a PROM) and are programmable. Last OR gate has $(l-1)/m$ or $(l-2)/m$ inputs that have the fusible links and remaining $(l-1)/(m-1)$ or $(l-2)/(m-1)$ input OR gates with each OR gate having fusible links. For example, if $p = 8$, $m = 3$, then two OR gates will have three each and one OR gate two fusible links only.

Programmable Array Logic (PLA) is a special programmable logic device (PLD) in which each input of the AND gate array (like a PAL) as well as in the OR-array (like a PROM) have a fusible link each, which is fused as per the minimum requirement after a suitable minimization and reduction procedure. The AND-OR arrays fusible links are fused as per the requirement of minimized form (prime implicants) after multi output minimization of the number of the Boolean expressions.

Figure 13.6 shows a $(n \times l \times m)$ PLA. The PLA has l number of AND gates each with $2n$ inputs and each input and output of AND has the fusible links. The fusible links number is just $(2.n.l^2)$ in place of $(2n.2^n.m)$ maximum possible and $(2^n.m)$ in PROMs.

PLAs are extremely suitable to implement the random or complex logic functions. If a PLA is made using bipolar transistors, as in a PAL, then the power and therefore current requirements are very high, typically 200 mA to 800 mA in a bipolar version. It is less than 200 mA in the CMOS versions of the PLAs.

A PLA needs to implement only the m" number of Boolean expressions by multi output minimization through a procedure the m number of Boolean expressions but generating the required m outputs. Multi-output minimization is by considering prime implicants at the Boolean expressions and their products. If Y_1 and Y_2 are two Boolean expressions to be implemented then the problem is to find subset of the prime implicants that is common in Y_1, in Y_2 and in $Y_1.Y_2$.

A unit called PLA programmer programs a PLA. Using a PLA, any desired logic functions to obtain the output bit Y_1, Y_2, Y_3, ... can be implemented with minimal number of 'AND' and 'OR' gates. It has two types of inputs: One dedicated, which means directly connected logic inputs and other feedback, which means connected from the outputs of OR.

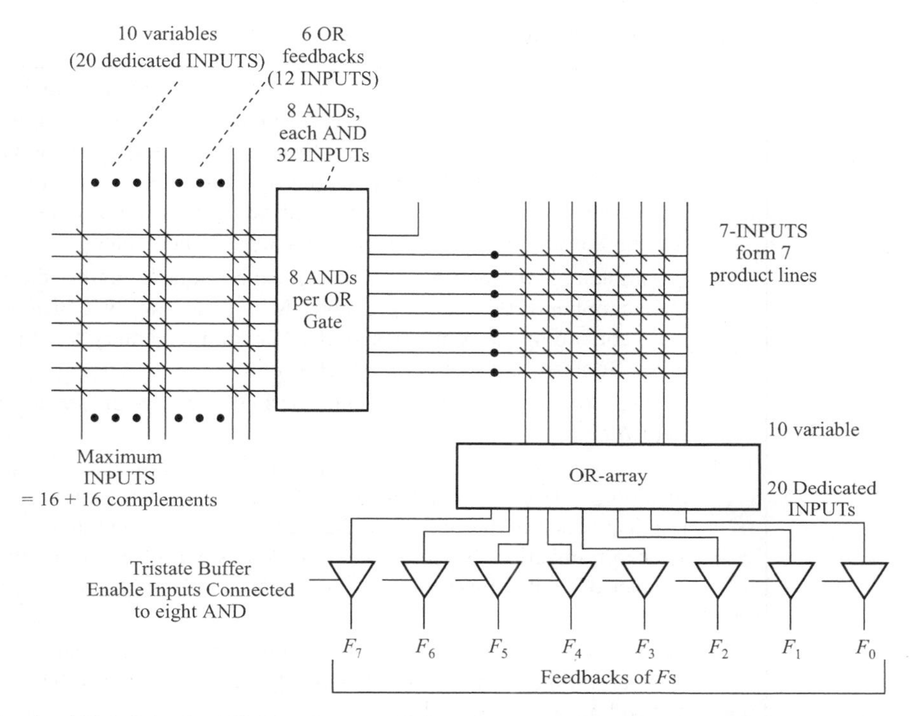

FIGURE 13.6 A ($n \times l \times m$) PLA. A ($n \times l \times m$) PLA has l number AND gates each with $2n$ inputs and each input has a fusible link ($l < 2^n$). ($l = 8$, $n = 10$ dedicated + 6 feed back = 16, $m = 7$).

Why does a PLA have less number of ANDs l compared to 2^n in a PROM?

Reason is that usually n = quite high, say 10 and there are computer based minimization procedures available for the individual Boolean expression minimization and multi output minimization. That makes it feasible to implement a combination logic circuit with fewer prime implement subsets only. So the PLA will have less cost (less number of gates) requirements and $l < 2^n$.

PLAs are extremely suitable to implement the random or complex logic functions. (Note: Electrically erasable PLAs are also called field programmable PALs and are more costly. Masked PLAs can be made like the masked ROMs at the manufacturing stages.)

Remember the Definition

PLA is a special programmable logic device (PLD) in which each input variable and its complement at the AND gate array as well as each input at the OR gate array have a fusible link, which is fused as per the subset of prime implicants needed to obtain the multi outputs realization after individual minimization and multi output minimization and circuit realization is as per the multiple outputs truth table for the combinational circuit.

Points to Remember

1. PLA is a special programmable logic device (PLD) in which each input, its complement at an AND gate and each gate output has a fusible link and total number of AND gates are limited in number to l with $l < 2^n$ maximum possible.
2. The AND-array and OR array fusible links are fused as per the requirement of minimized form (subsets of prime implicants) after multi output minimization of the number of the Boolean expressions to be implemented.
3. We implement *m"* number of Boolean expressions by multi output minimization the *m* number of Boolean expressions. Minimization is by considering the expressions containing the common subset set or complementary set of prime implicants in the Boolean Expressions and their products.
4. AND array inputs are programmable and OR array inputs are also programmable in PLAs.
5. $(n \times l \times m)$ PLAs have $2n$ input AND gates (l in number) and fixed input (number of inputs $< l$) m number OR gates. Each AND output corresponds to a prime implicant each in the subsets of the SOP expressions and their products obtained after minimization. Since numbers of ANDs and ORs are the limited ones, the number of fusible links total is $(2n.l^2)$ only in place of $(2n.2^n.m)$ maximum possible.
6. At the AND inputs, a sign shows a fused link, which is not snapped and no sign just an intersection shows a fused (snapped) link A dot shows the permanently connected link.
7. The AND gate with all links fused with no sign any where is placed a cross sign in its center of AND symbol.
8. A programming is a systemic hardware related procedure implemented by software at a programming system, called programmer or laboratory programmer or PLA programmer and the programming means executing the procedure for fusing (snapping) the needed links in the fresh PLA.
9. Electrical erasing ability in PLA is most often not provided.
10. Programming of PLA means fusing the links after minimization and multi output minimization procedures, most often computer based.

In a general case of PLA, since both the AND and OR arrays are programmable, a PLA is more complex than a PAL or a PROM.

■ EXAMPLES

Example 13.1

Show how to program the fusible links to get a 4-bit Gray code from the binary inputs using a PAL. Also compare the design requirement with a PROM.

Solution

Refer to Example 12.6 for PROM based implementation of the Gray code converter. PAL is used when minimized Boolean expression(s) is available. Therefore, let us recall the Example 11.4. The expressions for code conversion are as follows:

Y_3:

$$Y_3 = A \qquad ...(13.3)$$

Y_2:

$$Y_2 = A.\overline{B} + \overline{A}.B \qquad ...(13.4)$$

Y_1:

$$Y_1 = B.\overline{C} + C.\overline{B} \qquad ...(13.5)$$

Y_0:

$$Y_0 = \overline{D}.C + \overline{C}.D = C\,\text{XOR}\,D \qquad ...(13.6)$$

Step 1: Finding the PAL input bits and data bits and number of fused links to the ANDs that are needed.

Seven number eight-input (A, B, C and D) and ($\overline{A}$, $\overline{B}$, $\overline{C}$ and $\overline{D}$) ANDs are needed to get the four bit Gray code outputs Y_0, Y_1, Y_2 and Y_3. This is because none of the seven prime implicants (one in $Y3$, and two each in $Y2$, $Y1$ and $Y0$) has a common set in these four expressions. When we compare it with PROM, there were 16 AND gates, were needed for all the sixteen minterms (Figure 12.13(b)). We need 7×4 PAL in place of 16×4 PROM. There are $4 \times 2 \times 7 = 56$ fusible links compared to 64 fusible links in PROM. Let OR_0, OR_1, OR_2 and OR_3 give the outputs D_0, D_1, D_2 and D_3, respectively. Let us number the fused links of AND0 as 1, 2, ..., 8 for A, $\overline{A}$, B, $\overline{B}$, C, $\overline{C}$, D and $\overline{D}$. Let us number the fused links of AND1 as 9, 10, ..., 16 for A, $\overline{A}$, B, $\overline{B}$, C, $\overline{C}$, D and $\overline{D}$. Similarly, we number up to 56 for AND6 $\overline{D}$ link. (Figure 13.7 shows a demonstrative $4 \times 7 \times 4$ PAL).

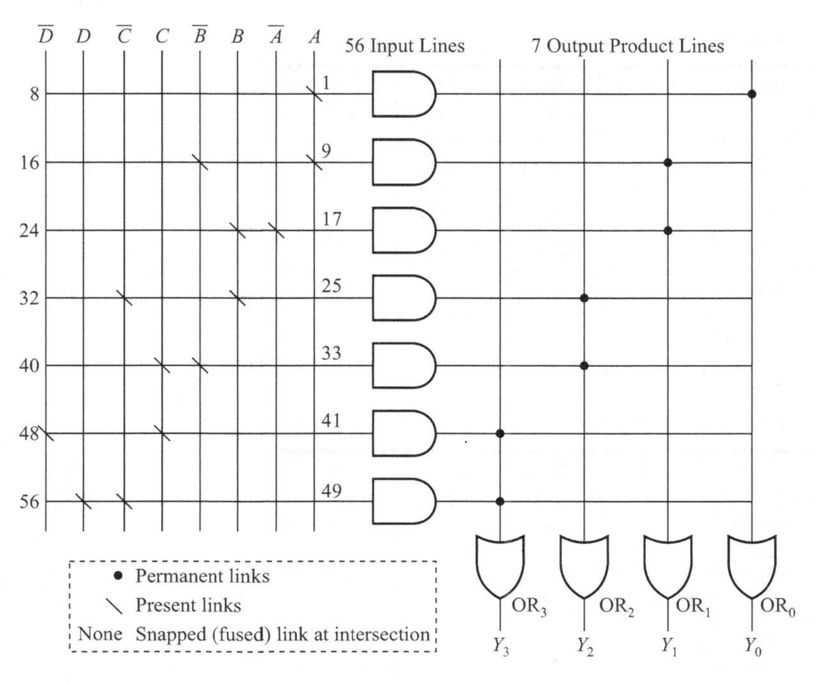

FIGURE 13.7 A demonstrative $4 \times 7 \times 4$ PAL that implements a Gray code converter ($n = 4$, $l = 7$, $m = 4$). Link numbers are also shown.

Step 2: Finding the PLA fused link numbers needed.

Table 13.1 gives the not fused fusible links to implement the Gray code. Figure 13.7 shows the links at the inputs of the ANDs and connections at the ORs that are present.

TABLE 13.1 Not fused links table

Bit	Not Fused Snapped Link Numbers							
	AND0	AAND1	AND2	AAND3	AND4	AAND5	AND6	AND7
Y_0	1							
Y_1			9, 12	18, 19				
Y_2					27, 30	36, 37		
Y_3							45, 48	55, 54

Example 13.2 Show how to program the fusible links to get the outputs as per Boolean expressions correspond to the given sets of combinational circuits implemented using PAL. Compare the number of fusible links and gates needed with respect to the PROM.

$$Y_0 = \Sigma m(2, 4, 5, 6, 7);\ Y_1 = \Sigma m(2, 6, 8, 9);$$

Solution

Recall Chapter 5. There are four variables as there are the terms $m(8)$ and $m(9)$ in Y_2. Four variables Karnaugh Map (Tables 13.2) for maximum 16 minterms in an SOP expression is as follows:

TABLE 13.2 Sixteen minterms map

AB \ CD		$\overline{C}\,\overline{D}$ 00	$\overline{C}D$ 01	CD 11	$C\overline{D}$ 10
$\overline{A}\,\overline{B}$	00	$m(0)$	$m(1)$	$m(3)$	$m(2)$
$\overline{A}B$	01	$m(4)$	$m(5)$	$m(7)$	$m(6)$
AB	11	$m(12)$	$m(13)$	$m(15)$	$m(14)$
$A\overline{B}$	10	$m(8)$	$m(9)$	$m(11)$	$m(10)$

The Karnaugh map for Y_1 is therefore as in Table 13.3.

TABLE 13.3 Map for Y_1 for simplification

A.B \ C.D		$\overline{C}.\overline{D}$ 00	$\overline{C}.D$ 01	$\overline{C}.\overline{D}$ 11	$C.\overline{D}$ 10
$\overline{A}.\overline{B}$	00				1
$\overline{A}.B$	01	1	1	1	1
$A.B$	11				
$A.\overline{B}$	10				

$$Y_0 = \overline{A}.B + \overline{A}.C.\overline{D} \quad ...(13.7)$$

The Karnaugh map for Y_2 is as follows:

A.B \ C.D		$\overline{C}.\overline{D}$ 00	$\overline{C}.D$ 01	C.D 11	$C.\overline{D}$ 10
$\overline{A}.\overline{B}$	00				1
$\overline{A}.B$	01				1
A.B	11				
$A.\overline{B}$	10	1	1		

$$Y_1 = A.\overline{B}.\overline{C} + \overline{A}.C.\overline{D} \quad ...(13.8)$$

Step 1: Finding for the PAL number of fused links, ANDs and ORs needed.

We have a common subset of prime implicant, $\overline{A}.C.\overline{D}$. This can be taken as output Y_2 from an OR gate OR_2. It can be implemented by one AND gate. The equations 13.7 and 13.8 have two other prime implicants. These can be implemented by two AND gates. Therefore, total three ANDs and three OR gates shall suffice. Let OR_0, OR_1 and OR_2 and OR_3 give the outputs F_0, F_1 and F_2 and F_0 give output Y_1, F_1 give output Y_2 and F_2 give output Y_2, respectively.

There are three feedback links also available at inputs E, F and G from OR_0, OR_1 and OR_2, respectively. Let us number the fused links of AND_0, AND_1 up to AND_5.The links are 1, 2, up to 14 for AND_0 A, $\overline{A}$, B, $\overline{B}$, C, $\overline{C}$, D, $\overline{D}$, E, $\overline{E}$, F, $\overline{F}$, G and $\overline{G}$. Let us number the fused links of AND_1 as 15, 16, ..., 28 for A, $\overline{A}$, B, $\overline{B}$, C, $\overline{C}$, D, $\overline{D}$, E, $\overline{E}$, F, $\overline{F}$, G and $\overline{G}$. AND_2 link A up to AND_5 link $\overline{G}$ are numbered from 29 up to 84. This PAL has four variables and three OR feedback input variables. OR0 inputs permanently connect to AND_0 and AND_1, OR_1 inputs to AND_2 and AND_3 and OR_2 inputs to AND_4 and AND_5. (Figure 13.8 shows a demonstrative $7 \times 6 \times 3$ PAL).

Step 2: Finding the fused links needed.

Table 13.4 gives the PAL implementation. (i) Links 58, 61, 64 not fused (snapped) implements $\overline{A}.C.\overline{D}$. (ii) Links 2 and 3 implements $\overline{A}.B$. (iii) Links 29, 32, 34 implements $A.\overline{B}.\overline{C}$. $Y_2 = G$ is feedback link and connects 55 and 27 links of AND_3 and AND_1, respectively.

TABLE 13.4 Not fused links table

Bit	Not fused snapped link numbers					
	AND_0	AND_1	AND_2	AND_3	AND_4	AND_5
$F0\ (Y_1)$	2, 3	27				
$F1(Y_2)$			29, 32, 34	55		
$F2\ (Y_3)$					58, 61, 64	

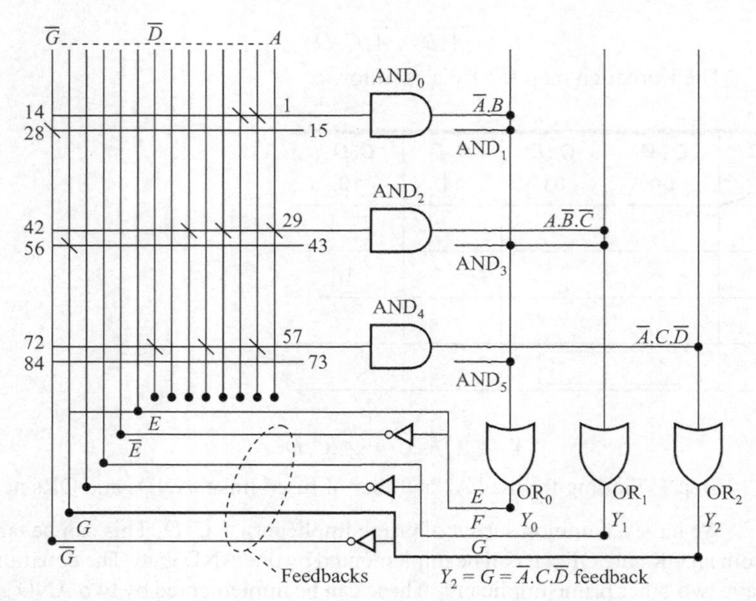

FIGURE 13.8 Demonstrative $7 \times 6 \times 3$ PAL with four input variables and three feedback input variables, six ANDs and three ORs ($n = 7$, $l = 6$, $m = 3$).

Example 13.3 Show how to program the fusible links to get a 4-bit Gray code from the binary inputs using a PLA. Also compare the design requirement with a PROM.

Solution

Refer to Example 12.6 for a PROM and Example 12.1 for PAL based implementation of the Gray code converter. PLA is used when minimized Boolean expression(s) is available, and both ANDs and ORs have fusible links. Therefore, let us recall the Example 11.4.

$\underline{Y_3}$:

$$Y_3 = A \quad \text{...(13.9)}$$

$\underline{Y_2}$:

$$Y_2 = A.\overline{B} + \overline{A}.B \quad \text{...(13.10)}$$

$\underline{Y_1}$:

$$Y_1 = B.\overline{C} + C.\overline{B} \quad \text{...(13.11)}$$

$\underline{Y_0}$:

$$Y_0 = \overline{D}.C + \overline{C}.D = C\,\text{XOR}\,D \quad \text{...(13.12)}$$

Step 1: Finding the PLA input bits and data bits and number of fused links to the ANDs that are needed.

Seven number eight-input (A, B, C and D and $\overline{A}$, $\overline{B}$, $\overline{C}$ and $\overline{D}$) ANDs are needed to get the four bit Gray code outputs Y_0, Y_1, Y_2 and Y_3. This is because none of the seven prime implicants (one in Y_3 and two each in Y_2, Y_1 and Y_0) has a common set in these four expressions we need seven. When we compare it with PROM there were 16 AND gates for all the sixteen minterms. We need $4 \times 8 \times 4$ PLA in place of 16×4 PROM. There are $(4 + 8) \times 8 = 96$ fusible links compared to 64 fusible links in PROM. Let OR_0, OR_1, OR_2 and OR_3 give the outputs D_0, D_1, D_2 and D_3, respectively. Let us number the fused links of AND_0 as 1, 2, ..., 8 for A, $\overline{A}$, B, $\overline{B}$, C, $\overline{C}$, D, and $\overline{D}$, and fused links of AND_0 to OR_0, OR_1, OR_2 and OR_3 as 9, 10, 11 and 12. Let us number the fused links of AND_1 as 13, 14 ..., 24 for A, $\overline{A}$, B, $\overline{B}$, C, $\overline{C}$, D, $\overline{D}$ and 4 OR inputs. Similarly, we number up to 96 from AND_7 OR_3 input link (Figure 13.9 shows a demonstrative $4 \times 8 \times 4$ PLA).

Step 2: Finding the PLA fused link numbers needed.

Table 13.5 gives the not fused fusible links to implement the Gray code. Figure 13.9 shows the links at the inputs of the ANDs and connections at the ORs that are present.

TABLE 13.5 AND-OR not fused links

Bit	Not fused snapped link numbers							
	AND_0	AND_1	AND_2	AND_3	AND_4	AND_5	AND_6	AND_7
Y_3	1, 12							
Y_2		13, 16, 23	26, 27, 35					
Y_1				39, 42, 46	52, 53, 58			
Y_0						65, 68, 69	78, 79, 81	

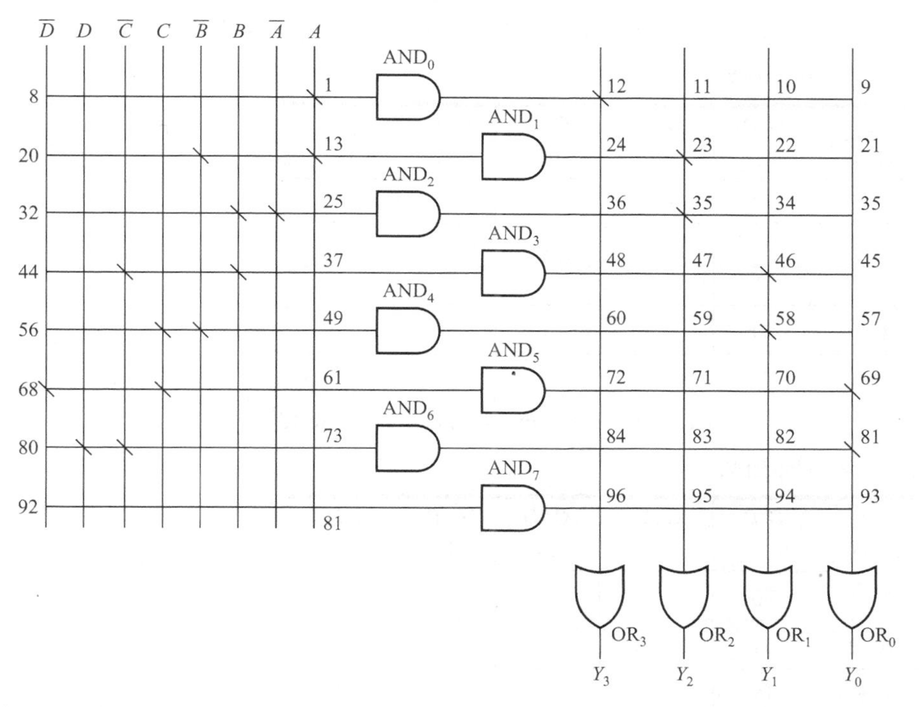

FIGURE 13.9 A demonstrative $4 \times 8 \times 4$ PLA that implements a Gray code converter ($l = 7$, $n = 4$, $m = 4$ only used).

We note that 4 × 7 × 4 PLA is needed to implement Gray code converter and 4 × 7 × 4 PAL was needed for implementing the same using PAL. Total number of AND– OR gates links are 96 for PLA, 56 for PAL and 64 for a PROM.

Example 13.4 Show how to program the fusible links to get the four Boolean expressions correspond to the given sets of 4 combinational circuits implemented using PLA. Compare the number of fusible links and gates needed with respect to the PAL and PROM.

Solution

$$Y_0 = \Sigma m(2, 4, 5, 6, 7);\ Y_2 = \Sigma m(2, 4, 5, 6, 8, 9);$$

Recall Chapter 5. There are four variables as there are the terms $m(8)$ and $m(9)$ in Y_1. Four variables Karnaugh Map (Table 13.6) for maximum 16 minterms in an SOP Expression is as follows:

TABLE 13.6 SOP minterms at map

AB \ CD		$\overline{C}\,\overline{D}$ 00	$\overline{C}D$ 01	CD 11	$C\overline{D}$ 10
$\overline{A}\,\overline{B}$	00	$m(0)$	$m(1)$	$m(3)$	$m(2)$
$\overline{A}B$	01	$m(4)$	$m(5)$	$m(7)$	$m(6)$
AB	11	$m(12)$	$m(13)$	$m(15)$	$m(14)$
AB	10	$m(8)$	$m(9)$	$m(11)$	$m(10)$

The Karnaugh map for Y_1 is therefore as in Table 13.7:

TABLE 13.7 Map for Y_0

A.B \ C.D		$\overline{C}.\overline{D}$ 00	$\overline{C}.D$ 01	$\overline{C}.\overline{D}$ 11	$C.\overline{D}$ 10
$\overline{A}.\overline{B}$	00				1
$\overline{A}.B$	01	1	1	1	1
$A.B$	11				
$A.\overline{B}$	10				

$$Y_0 = \overline{A}.B + \overline{A}.C.\overline{D} \quad \text{...(13.13)}$$

The Karnaugh map for Y_2 is as in Table 13.8.

TABLE 13.8 Map for Y_1

A.B \ C.D		$\overline{C}.\overline{D}$ 00	$\overline{C}.D$ 01	$C.D$ 11	$C.\overline{D}$ 10
$\overline{A}.\overline{B}$	00				1
$\overline{A}.B$	01	1	1		1
$A.B$	11				
$A.\overline{B}$	10	1	1		

$$Y_1 = A.\overline{B}.\overline{C} + \overline{A}.C.\overline{D} + \overline{A}.B.\overline{C} \quad ...(13.14)$$

Karnaugh map of the product $Y_0.Y_1$ is in Table 13.9

TABLE 13.9 Map for $Y_0..Y_1$

A.B \ C.D		$\overline{C}.\overline{D}$ 00	$\overline{C}.D$ 01	$C.D$ 11	$C.\overline{D}$ 10
$\overline{A}.\overline{B}$	00				1
$\overline{A}.B$	01	1	1		1
$A.B$	11				
$A.\overline{B}$	10				

$$Y_2 = Y_1.Y0 = \overline{A}.B.\overline{C} + \overline{A}.C.\overline{D} \quad ...(13.15)$$

We find a subset of two prime implicants into Y_0 and Y_1, both. Hence these can be implemented separately.

$$Y_1 = A.\overline{B}.\overline{C} + Y_2;\ Y_0 = \overline{A}.B.C + \overline{A}.B.\overline{C} + \overline{A}.C.\overline{D} = Y_2 \quad ...(13.16)$$

after rewriting it [because $\overline{A}.B = \overline{A}.B.(\overline{C} + C) = \overline{A}.B.\overline{C} + \overline{A}.B.C$].

Step 1: Finding the PLA number of fused links, ANDs and ORs needed—

We have a common subset of two prime implicant, $\overline{A}.C.\overline{D}$ and $\overline{A}.B.\overline{C}$. Two ANDs are needed for implementing it as well as Y_1. The equations 13.15 and 13.16 have two other prime implicants, $\overline{A}.B.C$ and $A.\overline{B}.\overline{C}$ respectively. These can be implemented by two other AND gates. Therefore, total four ANDs and two OR gates shall suffice in case of PLA.

Let us number the fused links of AND_0, AND_1 up to AND_3.The links are 1, 2, up to 8 for AND_0 A, $\overline{A}$, B, $\overline{B}$, C, $\overline{C}$, D, $\overline{D}$. Two input OR gate links of AND_0 are 13 and 14. AND_1, AND_2 and AND_3 can be numbered up to 56 (fourteen link with each AND product line). This PLA has four variables and two OR feedback input variables. (Figure 13.10 shows a demonstrative $4 \times 4 \times 2$ PLA circuit in a $4 \times 6 \times 2$ PLA).

Step 2: Finding the fused links needed—

Table 13.10 gives the PLA implementation. Links (13, 2, 5, 8) and (16, 17, 20, 27) give the prime implicants of expression 13.15 as well as for y_0 (subset in y_1). Links (29, 32, 34, 42) implements isolated term in equation 13.16. Link 53 and 56 implement y_0 feedback term in expression (13.16).

TABLE 13.10 Not fused links at PLA

Bit	Not fused (snapped) link numbers					
	AND_0	AND_1	AND_2	AND_3	OR_1	OR_0
$F0(Y_0)$				53	42, 56	
$F1(Y_1)$	2, 5, 8	16, 17, 20	29, 32, 34			13, 27

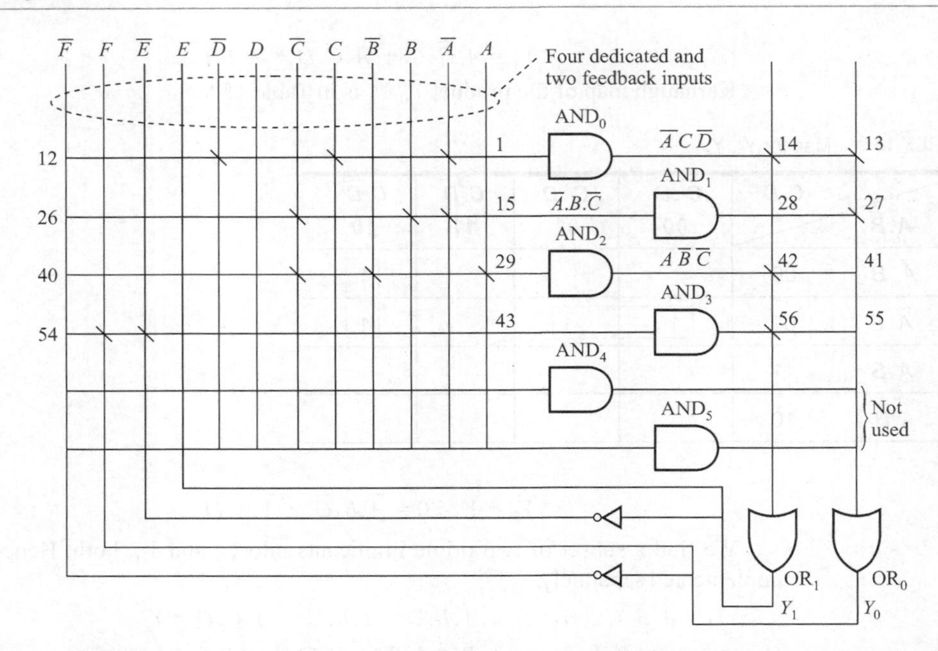

FIGURE 13.10 An implementation by 4 × 4 × 2 PLA with four dedicated and two feedback input variables, 4 ANDs and 2 ORs ($l = 4$, $m = 2$, $n = 4$).

Example 13.5 Show how will we design a hardware lock from an eight input variable PLA having eight number 16 input AND gates to eight OR gates. Lock opens and generates output = 1 when input variables are 10100111 and OR outputs Y_0 to Y_7 are as follows:

$Y_0 = \Sigma m(0, 12, 38, 89)$; $Y_1 = \Sigma m(10, 112, 138, 209)$; $Y_2 = \Sigma m(0, 12, 117, 133, 1769)$; $Y_3 = \Sigma m(37, 49, 74, 189)$; $Y_4 = \Sigma m(53, 92, 108, 128)$; $Y_5 = \Sigma m(8)$; $Y_6 = \Sigma m(83, 1133)$; $Y_7 = \Sigma m(1, 128, 138, 59)$; [Total 25 distinct minterms]

Solution

Maximum minterm is $m(1769)^1$. $2^{10} < 1769 < 2^{11}$. Hence, there are 11 literals maximum. Eight literals are for dedicated inputs. It means there are feedbacks from three ORs. Total number of inputs = 22. Number of ANDs and product lines are 25. Let us number the fuse links, 1 to 22 for AND_0, 23 to 30 for the inputs OR_0 to OR_7 from AND_0. Similarly, number the fuse links, 31 to 60 for AND_1, 61 to 90 for the AND_2 and OR inputs. Now last AND_{24} will have links between 721 and 750. Which link remains not-snapped is left as an exercise to the reader. Figure 13.11 shows the link numbers that can be fused to implement the above.

EXERCISES

1. Show how to program the fusible links to get a 4-bit 7536 code from the binary inputs using a PAL. Also compare the design requirement with a PROM.
2. Show how to program the fusible links to get a 4-bit 8421 code from the binary inputs using a PAL. Also compare the design requirement with a PAL.

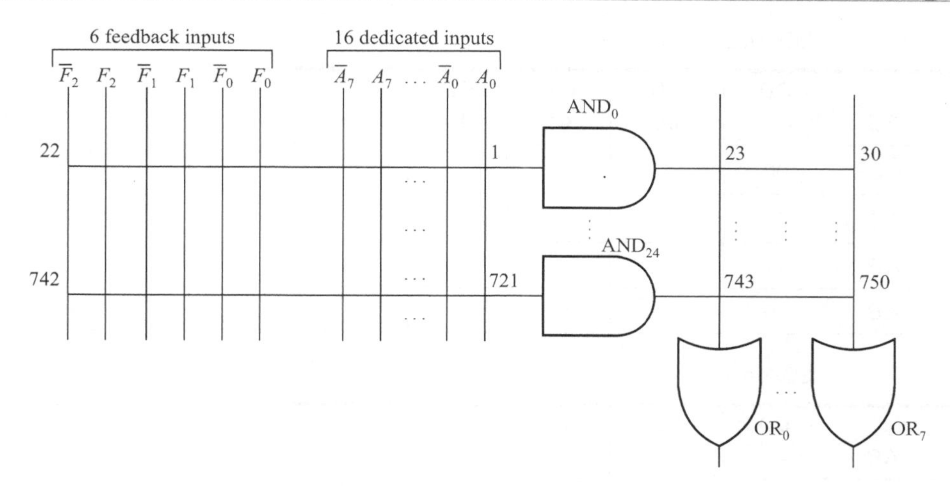

FIGURE 13.11 Hardware lock implementation using an eight input variable PLA having eight number 16 inputs AND gates to eight OR gates.

3. Show how to program the fusible links to get a 4-bit 5421 code from the binary inputs using a PAL. Also compare the design requirement with a PAL.
4. Show how to program the fusible links to get the four Boolean expressions correspond to the given sets of combinational circuits implemented using PLA. Compare the number of fusible links and gates needed with respect to the PROM and PAL. The circuit Karnaugh maps are as in Tables 13.11 and 13.14.

TABLE 13.11 SOPs map in a circuit

AB \ CD		$\overline{C}\,\overline{D}$ 00	$\overline{C}D$ 01	CD 11	$C\overline{D}$ 10
$\overline{A}\,\overline{B}$	00	1			
$\overline{A}B$	01				1
AB	11		1		1
AB	10	1		1	

TABLE 13.12 SOPs map in a circuit

AB \ CD		$\overline{C}\,\overline{D}$ 00	$\overline{C}D$ 01	CD 11	$C\overline{D}$ 10
$\overline{A}\,\overline{B}$	00	1			
$\overline{A}B$	01				1
AB	11				1
AB	10			1	

TABLE 13.13 SOPs map in a circuit

AB \ CD		$\bar{C}\bar{D}$ 00	$\bar{C}D$ 01	CD 11	$C\bar{D}$ 10
$\bar{A}\bar{B}$	00	1	1	1	1
$\bar{A}B$	01				1
AB	11				1
AB	10				

TABLE 13.14 SOPs map in a circuit

AB \ CD		$\bar{C}\bar{D}$ 00	$\bar{C}D$ 01	CD 11	$C\bar{D}$ 10
$\bar{A}\bar{B}$	00	1			
$\bar{A}B$	01				X
AB	11		1	1	X
AB	10	1	1	1	

5. Show how to program the fusible links to get the four Boolean expressions correspond to the given sets of four combinational circuits in Exercise 3 implemented using PROM.
6. Show how will we design a hardware lock from a four input variable PAL having four number eight input AND gates and four OR gates. Lock opens ad generates output = 1 when input variables are 10111101 and OR outputs Y_0 to Y_7 are all 1s.

QUESTIONS

1. Describe the fusible links in AND-OR arrays at a PAL. Describe 16L8 PAL
2. Describe the fusible links in AND-OR arrays at a PLA.
3. Why is it possible to work with l AND gates in case of n input variables when using a PAL? (Hint: Less number of minterms, common or complementary sets of prime implicants).
4. Why is it possible to work with l AND gates in case of n input variables when using a PLA? (Hint: Less number of minterms, subsets of prime implicants in Boolean expressions and their products).
5. Why do we need 2^n AND gates in case of a PROM in case of n input variables? (Hint: 2^n minterms).
6. What is the advantage of a PAL compared to PLA and PROM? (Hint: Simpler minimization, less number of fused links).
7. What is the advantage of a PLA compared to PAL and PROM? (Hint: Complex logic functions implementation with less number of AND-OR arrays).

8. What is advantage of a PROM compared to the PLA and PALs? (Hint: PROM also works as the memory elements for data tables, instructions, data sets, etc.).
9. How do you implement the multi outputs (Boolean expressions) with minimal fusible links in a PAL?
10. How do you implement the multi outputs (Boolean expressions) with minimal fusible links in a PLA?
11. Explain a hardware lock design using a PLA.
12. What is the advantage of tristate outputs provision in a PAL? What is the advantage of a registered output PAL? (Refer Section 21.1)

CHAPTER 14

Sequential Logic, Latches and Flip-Flops

OBJECTIVE

Recall that a combinational circuit is a circuit made up by combining the logic gates such that the required logic at the output(s) depends only on the input logic conditions, both completely specified by either a truth table or by a Boolean expression. Also (i) An output(s) remains constant, as long input conditions do not require the change in the output(s), (ii) An output depends solely on the current input condition(s) and not on any past input condition(s) or past output condition(s), (iii) A combinational circuit has no feedback of the output from a stage to the input of either that stage or any previous stage, and (iv) An output(s) at each stage appears after a delay in of few tens or hundred ns depending upon the type or family of the gate used to implement the circuit. We have learnt the combinational circuits, their logic designs, the problems and ways to implement them by the building blocks, ICs and PLDs in Chapters 9 to 13.

We will learn here another class of important logic circuits, namely sequential circuits in this and succeeding chapters. We will learn following sequential circuits in this chapter: flip-flops- *SR*, *JK*, *T*, *D*, Master Slave *FF*, triggering conditions and the characteristic equations for their analysis.

Often things are done in a sequential manner. For example, in order to prepare tea, firstly water is boiled, and then leaves are added. A sequential job means (i) to remember what steps are to be done next, and (ii) to recall which step has just been finished.

A storage device or a series of storage devices are needed in order to do the things sequentially, so that a step (or steps) that has been previously done can be recalled. The basic unit to store this information in a digital circuit is *Flip-Flop* (FF). In analog electronics, a capacitor can hold (memorize) the earlier applied potential difference (provided the charge in it does not leak due to its leakage resistance). A flip-flop is a similar basic unit in sequential circuits. A flip-flop offers a stable state (logic state not changing with time) at output even if the inputs are withdrawn (unless of course there is a power failure).

A sequential circuit is a circuit made up by combining the logic gates such that the required logic at the output(s) depends not only on the current input logic conditions but also on the past outputs (hence past inputs) and is specified by a table called state table. A state table gives the past, current and future states at the output.

1. An output(s) can remain stable (constant) even after the input conditions change,
2. An output depends on the current input state and past input states (thus past output states),
3. A sequential circuit has a feedback of the output(s) from a stage to the input of either that stage or any previous stage,
4. A sequential circuit may have a clock (gate) input to control the instance or period in which the output gets effected as per the inputs, and
5. An output(s) at each stage appears after a delay in of few ones or tens or hundred *ns* depending upon the type or family of the gate used to implement the circuit in case the inputs or feedbacks change.

Point to Remember

A sequential circuit is a circuit made up by combining the logic gates such that the required logic at the output(s) depends not only on the current input logic conditions but also on the past inputs, outputs and sequences.

Figure 14.1(a) shows two basic digital units in a digital circuit.

14.1 FLIP FLOP AND LATCH

Example 14.1 will give a bi-stable circuit made from the common input NANDs or NORs.

It is possible to have a stable state at an output. Flip-Flop (*FF*) means a digital circuit of two stable states at an output:

1 means, rise on top. It means flip, and

0 means fall to ground. It means flop.

The *FF* is a unit with 1 and 0 as stable states. A *FF* have two definite (discrete) states. It forms a smallest basic memory unit or a one-bit register unit. (Memory means that even if input is withdrawn, the output remains same as before.)

Figures 14.1(b) and (c) show the classifications of the FFs based on configurations and clocking mechanism, respectively.

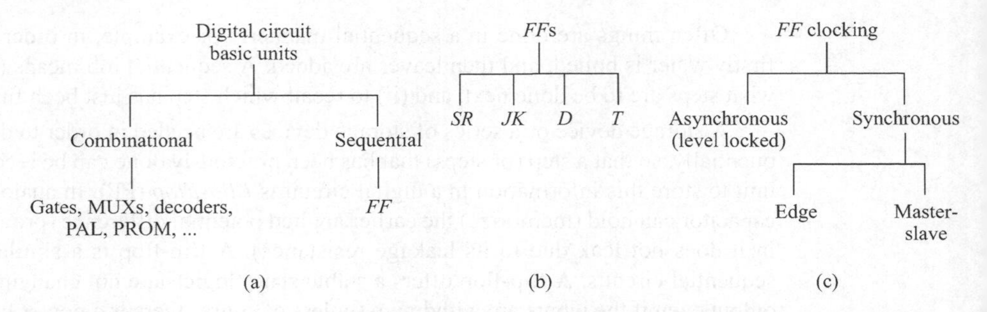

FIGURE 14.1 (a) Division of a digital circuit basic unit. (b) Flip-Flop types (c) Flip-Flop asynchronous and synchronous clocking mechanisms.

A *FF* has two outputs Q and its complement $\overline{Q}$ for a state of flip or flop. A particular combination of Q and $\overline{Q}$ represents one of the two stable states. One of the stable state is $Q = 1$ and $\overline{Q} = 0$, and other stable state is $Q = 0$ and $\overline{Q} = 1$. We may use for interconnections as well feedbacks as the inputs using the outputs, Q and/or $\overline{Q}$. A *FF* has one or two inputs. The logic states at these inputs and the previous Q determine what shall be the current output state.

A sequential circuit has a state table (like truth table in a combinational circuit). A state table of a *FF* describes how for the different input conditions, the output Q (and/or $\overline{Q}$) shall be for a given type of *FF*.

A *FF* is called latch, if the instance at which the output should change has no control input (clock falling or rising edge input). A latch is a *FF* without any edge triggered clocking mechanism for its inputs.

Point to Remember

A *FF* is a circuit, which has two stable (bi-stable) states; 1 (also called *Set*) and 0 (also called *Reset*) and in which output is stable as long as power is not withdrawn or the inputs are not applied such that the output state changes. A timing input (called clock transition or clock edge input) controls the instance at which the output changes.

A latch is a class of *FF* in which the instance at which output changes is uncontrolled. Uncontrolled means not controlled by the timing of the transition at clock input.

14.2 SR LATCH (SET-RESET LATCH) USING CROSS COUPLED NANDS

An *SR* latch is a simplest building block of a *FF*. It is called set-reset latch (*SR*) as it has two input states Set and Reset. It can be made by cross coupling the two NAND gates or by cross coupling the two NOR gates. The coupling of NANDs is done as shown in Figure 14.2(a). Figure 14.2(b) gives the state table for Q and $\overline{Q}$ at the latch. Figure 14.2(c) shows the symbol for the *SR* latch. Q_n means n^{th} sequence before the application of the present inputs.

14.2.1 SR Latch at Various Input Conditions

An *SR* latch have an active state for *S* or *R* as 1. An $\overline{S}\,\overline{R}$ latch have an active state for *S* or *R* as 0. Let us discuss *SR* latch.

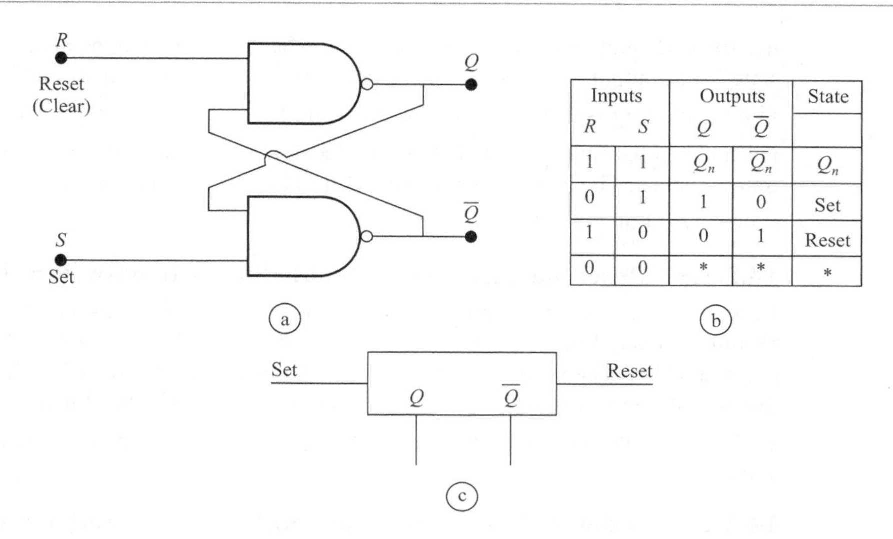

Inputs		Outputs		State
R	S	Q	$\overline{Q}$	
1	1	Q_n	$\overline{Q}_n$	Q_n
0	1	1	0	Set
1	0	0	1	Reset
0	0	*	*	*

FIGURE 14.2 (a) An SR Latch made by cross-coupling two NANDs with *S* input to lower NAND, which has output $\overline{Q}$ (b) State table (c) Logic Symbol of *S-R* Latch.

14.2.1.1 Stable (No change) State when both Inputs, *S* and *R* = 1

Let both inputs, (one is called *S* (means Set) and other *R* (means Reset (clear)) are 1s. [1 means that voltage levels are near to 5 V with respect to ground potential in case when TTL NANDs are used, and near to V_{DD} in case of CMOS NAND).]

If in an nth state *Q* is 1 then $\overline{Q}$ will be 0 due to the basic property of NAND gate, its output = 0 if both inputs (*S* and *Q*) = 1. Now, one input of upper NAND gate in Figure 14.2(a) is at 0 because it is connected to $\overline{Q}$ (output of lower NAND gate). Other input, *R*, is already 1, therefore, *Q* will remain 1. In other words, *Q* will remain the complement of $\overline{Q}$. Here, ($Q = 1$, $\overline{Q} = 0$) is a stable state of the latch. The circuit of Figure 14.2(a) will remain in its stable state at the nth sequence for the given input condition: both *R* and $S = 1$.

Now, if at n^{th} sequence $Q = 0$ and $\overline{Q} = 1$ and let *R* and *S* both are at 1s. In that case, the upper NAND gate output *Q* feeds a 0 input to the lower NAND gate. Therefore, $\overline{Q}$ will remain 1. Here, $Q = 1$ and $\overline{Q} = 0$, remains stable at the latch when $R = S = 1$.

Refer first row of the table in Figure 14.2(b). When $R = 1$ and $S = 1$, then $\overline{Q}$ will be complement of *Q*. It means if $Q = 1$, then $\overline{Q} = 0$, and if $Q = 0$ then $\overline{Q} = 1$. The outputs are as they were before the *R* and *S* inputs changed over to both 1s. [Note: Q_n is usually a symbol to represent a previous *Q* state and means there is no change at $(n + 1)^{\text{th}}$ state from the *n* th state of a *FF* circuit. We will understand meanings of *n* and *n* + 1 stages in our discussions later.]

14.2.1.2 Set State as Stable State of the Latch when *R* = 0 and *S* = 1

If we bring an input terminal to 0 state 0 V to 0.8 V with respect to ground potential for TTL NANDs [V_{SS} to 0.33 ($V_{DD} - V_{SS}$) for CMOS NANDs], then let us assume inactivated. Let us inactivate R input logic state. Let us activate S (Set input) state. It means bring S to the logic state 1. The output Q will become, irrespective of the previous output, 1 and the latch is then said to be set. $\overline{Q} = 0$ in this case. For $R = 0$ and $S = 1$, the outputs are such that Q is set and is 1 and $\overline{Q}$ is complement of Q, and is 0. Second row of state table in Figure 14.2(b) illustrates this fact. The latch Set state means flip state. Set state inputs are $R = 0$ and $S = 1$. Set state is $Q = 1$ and $\overline{Q} = 0$.

14.2.1.3 Reset State as another Stable State when for $R = 1$ and $S = 0$

Let us activate now R state and keep S state in the inactivated logic state. When R reset input terminal is made 1 then output Q will be reset when $S = 0$. It means will go to state 0 even if previous Q is 1. Then Q is the output of the upper NAND gate and the NAND has a property that if both inputs become 1, the output shall become 0. The third row of table in Figure 14.2(b) illustrates this fact. The Reset state is Q is 0 and $\overline{Q} = 1$. Reset state inputs are $R = 1$ and $S = 0$.

14.2.1.4 Indeterminate State when both inputs R and $S = 0$ (Unstable)

If both R reset and S set inputs are inactivated. It means made 0, both Q and $\overline{Q}$ tend to be 1. The logic of the circuit is not satisfied, and the final state at an instant after the inputs R and S changed to 0, is only a matter of chance. The logic state is said to be indeterminate state or racing state. Each state, (Set $Q = 1$ and $\overline{Q} = 0$), and (Reset $Q = 0$, $\overline{Q} = 1$) trying to race through. This causes the unstable state. The fourth row of table in Figure 14.2(b) illustrates this fact.

Points to Remember

(1) SR latch with cross-coupled NANDs or NORs has the following features: (S – input is at that NOR which is giving Q output or S – input is at that NAND which is giving $\overline{Q}$ output).
When $S = 0$ and $R = 0$, the output of SR latch is unstable (meta stable).
When $R = 1$ and $S = 1$, the output of SR latch does not change.
When $R = 1$ and $S = 0$, the output Q resets to 0.
When $R = 0$ and $S = 1$, the output Q sets to 1

(2) $\overline{S}\,\overline{R}$ latch with cross-coupled NANDs or NORs has the following features: ($\overline{S}$ – input is at the NAND giving $\overline{Q}$ output, or $\overline{S}$ – input is at the NOR giving $\overline{Q}$ output).
When $\overline{S} = 1$ and $\overline{R} = 1$, the output of $\overline{S}\,\overline{R}$ latch is unstable (meta stable).
When $\overline{R} = 0$ and $\overline{S} = 0$, the output of $\overline{S}\,\overline{R}$ latch does not change.
When $\overline{R} = 0$ and $\overline{S} = 1$, the output Q resets to 0.
When $\overline{R} = 1$ and $\overline{S} = 0$, the output Q sets to 1.

14.2.2 Difficulties in Using an SR Latch

The difficulties in using an SR latch circuit shown in Figure 14.2(a) or ones described later in the Examples 14.2 to 14.4 are as follows:

1. It has an unstable condition when the input states 0 at *R* and *S* both in Figure 14.2(b).
2. When we are interested in setting (i.e. forcing in logic state 1) *Q* from its reset state (0) or resetting *Q* from the set state, there will be a certain time interval taken in interchanging the *R* and *S* input states. During the intermediate time interval (during the floating of the *R* and/or *S* inputs), what happens we cannot predict. Is it not possible that during this intermediate time interval when the input interchanges at the *S* and *R* inputs, the latch response is in an output change disabling state? Section 14.2.4 will address this question.

14.2.3 Timing Diagrams of an *SR* Latch

State table in Figure 14.2(b) can also be shown in terms of a timing diagram. Different input sets (for examples *S*, *R* and clock inputs) in the different time intervals are chosen and are plotted as a function of time in a diagram. The output *Q* or *Q* and $\overline{Q}$ both are also plotted on the same diagram. In actual practice, the change from 0 to 1 or 1 to 0 is not sharp and showing these in a timing diagram by a vertical line is just an assumption only. It is valid when the intervals instances between the input changes are longer than the transition times from 0 to 1 or 1 to 0.

Figure 14.3(a) shows how to represent an *S* input as a function of time, when *S* = 0 between 0 to *T*1, *T*2 to *T*3, *T*4 to *T*5 and *S* = 1 between *T*1 to *T*2 and *T*3 to *T*4. Figure 14.3(b) shows how to represent an *R* input as a function of time, when *R* = 1 between 0 to *T*1', *T*2' to *T*3', *T*4' to *T*5' and *R* = 0 between *T*1' to *T*2' and *T*3' to *T*4'. Figure 14.3(c) shows the timing diagram, for the outputs *Q* and $\overline{Q}$ for the *SR* latch having state table as per Figure 14.2(b). It also shows the unstable (race condition) region during shaded area, which both *R* and *S* = 0 at the *SR* latch inputs.

1. Propagation delay, *tp* (01) or t_{pLH} is the time interval between *t*' and *t*'', where *t*' is the instance midway between 0 and 1 when an input is changing from 0 to 1 and *t*'' is the instance midway between 0 and 1 when an output *Q* is changing from 0 to 1. (Figure 14.3 right side)
2. Propagation delay, *tp* (10) or t_{pHL} is the time interval between *t*''' and *t*'''', where *t*''' is the instance midway between 1 and 0 when an input is changing from 1 to 0 and *t*'''' is the instance midway between 1 and 0 when an output *Q* is changing from 1 to 0. Average propagation delay, *tp* of a latch or *FF* is the average of *tp* (01) and *tp* (10). (The delays *tp*(01), *tp*(10) and *tp* differs due to different impedances of the output stage transistor. These also depend on the types and family of the gates used in designing an *FF*.)
3. Setup time, t_s is an average of the minimum required time for an input before an enabling input (gate input or clock input) is applied so that the output *Q* is as per the circuit design and its state table. (Refer Section 14.2.4)
4. Hold time, t_h is an average of the minimum required time for an input to hold its logic state unchanged after an enabling input (gate input or clock input) is applied so that the output *Q* is as per the circuit design and its state table. (Refer Section 12.2.4)

If the inputs change during interval t_s before a clock input and t_h after the clock input, the meta stable or unpredictable state may result.

Figure 14.3(d) marks the *FF* propagation delay, *FF* setup and the holding times in the timing plots.

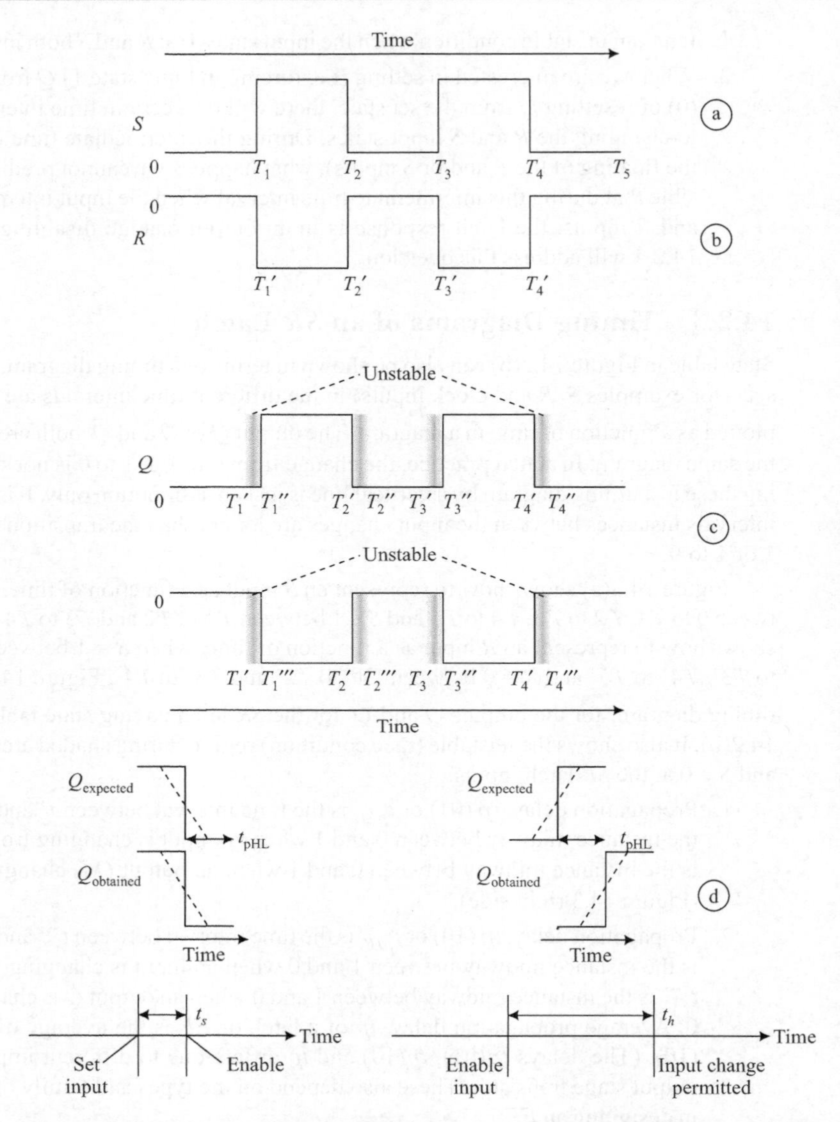

FIGURE 14.3 (a) Timing diagram of S_{input} as a function of time (b) R_{input} as a function of time (c) The timing diagram, for $\overline{Q}$ for the *SR* (d) Understanding the *FF* propagation delay, *FF* setup and the holding times using the timing diagram(s).

14.2.4 Level clocked *SR* Latch

Let us add two NAND gates to the two cross-coupled NANDs (Figure 14.4(a)). We get a clocked *SR* latch. It is explained in detail in example 14.6. It takes care of the difficulty 2 mentioned above. There is an *SR* latch circuit, which has a gating (enabling or clocking). The inputs can thus be first interchanged, and when well defined then only we can make clock input CLK = 1. It means open the gate to make the *S* and *R* inputs transparent in the latch. When CLK = 0, the *S* and *R* are having no effect on the *Qs*.

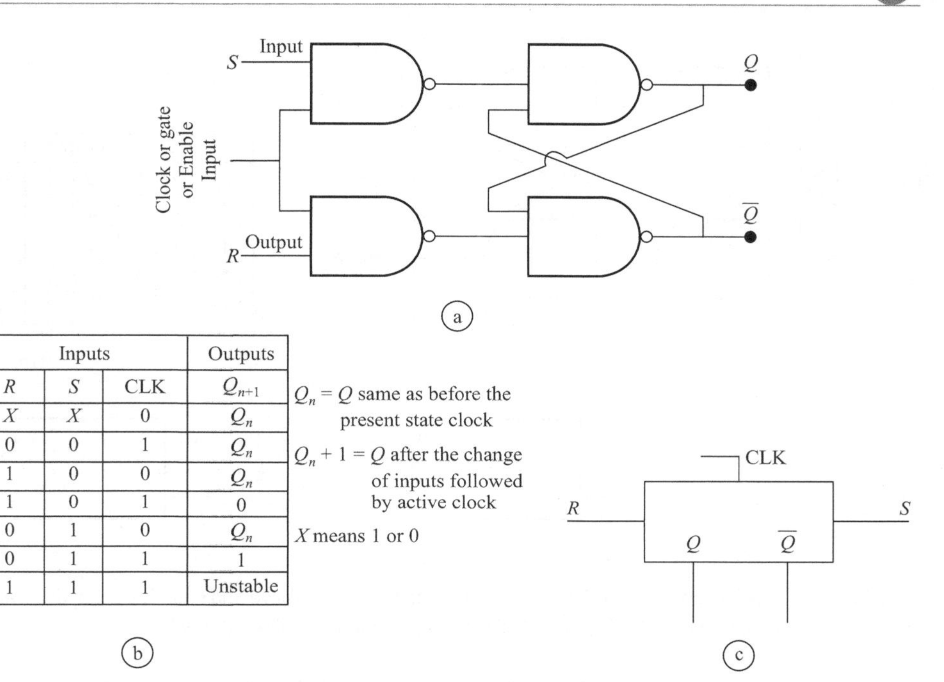

Inputs			Outputs
R	S	CLK	Q_{n+1}
X	X	0	Q_n
0	0	1	Q_n
1	0	0	Q_n
1	0	1	0
0	1	0	Q_n
0	1	1	1
1	1	1	Unstable

FIGURE 14.4 (a) A level-clocked *SR* Latch (b) State tables of clocked *SR* Latch (c) Symbol clocked *SR* Latch.

Points to Remember

SR latch with a clock input has the following features:

When $S = 1$ and $R = 1$, the output of *SR* latch is unstable (meta stable) during the active state of the clock input.

When $R = 0$ and $S = 0$, the output of *SR* latch does not change during the active or inactive state of the clock input.

When $R = 1$ and $S = 0$, the output Q resets to 0 during the active state of th clock input.

When $R = 0$ and $S = 1$, the output Q sets to 1 during the active state of th clock input.

14.3 JK FLIP-FLOP

Taking three input NANDs for the inputs let us change the Figure 14.4(a) Section 14.2.4 circuit as follows:

1. The first stage NANDs S input is now labeled as J input and R input as K input.
2. Second input of both NANDs is common and is the clock input has an additional circuitry to make the J and K inputs transparent at an instance corresponding to an edge at the clock input.
3. Third input of upper NAND connects the $\overline{Q}$ output.
4. Third input of lower NAND connects the Q output.

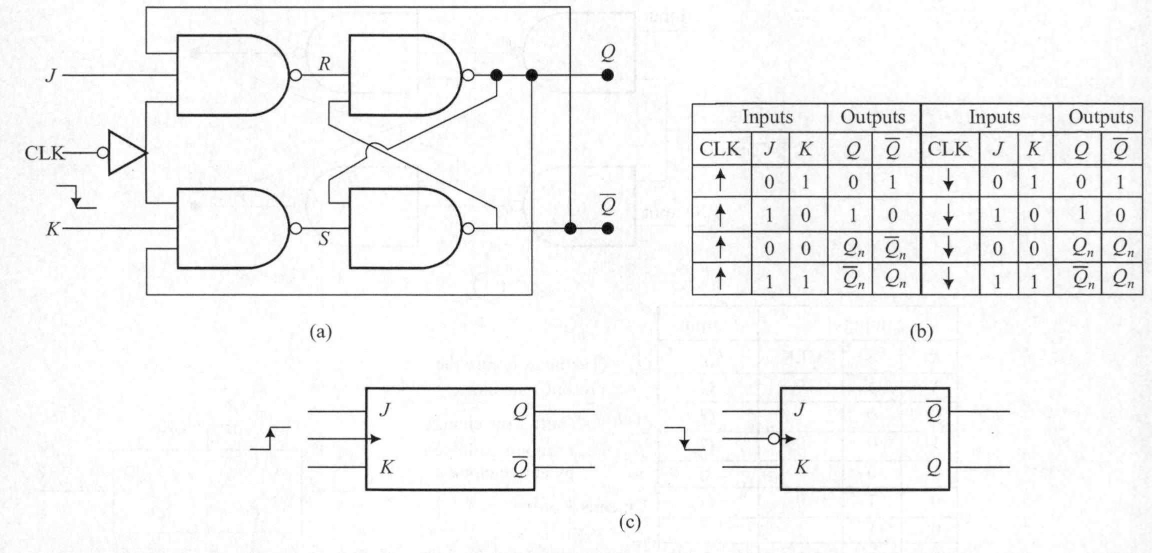

Inputs			Outputs		Inputs			Outputs	
CLK	J	K	Q	$\overline{Q}$	CLK	J	K	Q	$\overline{Q}$
↑	0	1	0	1	↓	0	1	0	1
↑	1	0	1	0	↓	1	0	1	0
↑	0	0	Q_n	$\overline{Q}_n$	↓	0	0	Q_n	Q_n
↑	1	1	$\overline{Q}_n$	Q_n	↓	1	1	$\overline{Q}_n$	Q_n

FIGURE 14.5 (a) An edge triggered *JK* flip flop (b) State tables of positive and negative edge triggered *JK* flip flops (c) Symbols of + ve edge triggered *JK* flip flop (left) and –ve edge triggered *JK* flip flop (right).

Figures 14.5(a) shows the modified circuit. It shows a *JK* flip flop, which has an edge triggered clock input so that output state change only at the instance of the edge. Figure 14.5(b) shows the state tables of positive and negative edge triggered *JK* flip-flops at the left and right sides, respectively. Figure 14.5(c) shows the symbols of +ve edge triggered *JK* (left) and –ve edge triggered (right) *JK* flip flops.

Unlike a transparent latch, the FF circuit given in Figure 14.5(a) responds only at an edge at the clock input. A positive edge means a transition from 0 to 1 at an input. A negative edge means a transition from 1 to 0 at an input.

1. A bubble before a triangle at a clock input represent the fact that the clock input is .negative edge triggered and output will correspond to the *J* and *K* inputs at that –ve edge instance only. At remaining instances of time, Q and $\overline{Q}$ do not get affected (= Q_n and $\overline{Q}_n$).
2. A triangle at a clock input represent the fact that the clock input is positive edge triggered and output will correspond to the *J* and *K* inputs at that +ve edge instance only. At other instances, Q and $\overline{Q}$ are Q_n and $\overline{Q}_n$.

14.3.1 Explanation of the State Table for the Logic Circuit of an Edge-Triggered JK FF

An output $\overline{Q}$ is feedback as one of the inputs at the upper left most NAND. Another output Q is feedback as one of the input at the lower leftmost NAND. These are the important feature of the *J* and *K* inputs flip-flops.

14.3.1.1 $J = 0$ and $K = 0$ Case

Let at the upper NAND gate the input-1 $J = 0$, the output of the first stage NAND will always be 1 whatever may be the other two inputs.

Let at the lower NAND gate the input-1 $K = 0$, the output of first stage NAND will always be 1 whatever may be the other two inputs.

The cross coupled NANDs on the right side of the circuit will have Q and $\overline{Q}$ outputs unchanged as in the circuit of Figure 14.2(a) and first row of state table in Figure 14.2(b). There is no change at the Q output and therefore, $Q_{n+1} = Q_n$. [Q after $(n+1)^{th}$ clock edge is same as one after the nth clock edge.]

TABLE 14.1 ($J = 0$ and $K = 0$) state Q and $\overline{Q}$

Inputs					Outputs		State
J	$\overline{Q}$	CLK	K	Q	Q_{n+1}	$\overline{Q}_{n+1}$	
0	1	X	0	0	Q_n	$\overline{Q}_n$	No Change
0	0	X	0	1	Q_n	Q_n	No Change

*X means any input 0 or 1 or positive edge transition 0 to 1 or negative edge transition.

Result is that the output Q remains same as before even after the clock edge when $J = 0$ and $K = 0$ and unstable condition in S-R latch case no longer exists.

Note that unstable condition for last row of state table in *SR* latch in Figure 14.2(b) does not exist now because all three inputs of upper and lower NANDs are never identical and never all 1s and cross-coupled both S and R NAND inputs are never 0s.

14.3.1.2 $J = 1$ and $K = 0$ Case

When $J = 1$, $K = 0$, the circuit of Figure 14.5(a) behaves as follows: Let lower NAND output $\overline{Q}_n = 1$. Upper three input NAND output = 0, because two other input are also 1s, because J input is 1 and at i^{th} clock edge (+ve edge) CLK_i J is 1. Upper NAND output, which is input to cross coupled NANDs = 0. Now lower NAND output, which is input to cross coupled NANDs = 1 due to $K = 0$. The cross-coupled NAND inputs therefore corresponds to $R = 0$ and $S = 1$ case in the circuit of Figure 14.2(a). Hence output Q_{n+1} will become 1 and Q_{n+1} will become 0. If Q_n is already 0, Q_n is already 1, the clock edge will not cause any change from the original SET state because S is already 1.

Result is that the Output Q sets to 1 after the clock edge when $J = 1$ and $K = 0$.

TABLE 14.2 ($J = 1$ and $K = 0$) state Q and $\overline{Q}$ (+ ve edge case)

Inputs					Outputs		State
J	$\overline{Q}$	CLK_i	K	Q	Q_{n+1}	$\overline{Q}_{n+1}$	$(n+1)^{th}$
1	1	1 at the +ve edge	0	0	1	0	SET
1	1	0	0	0	Q_n	$\overline{Q}_n$	No Change
1	1	1	0	0	Q_n	$\overline{Q}_n$	No Change
1	1	0 at −ve edge	0	0	Q_n	$\overline{Q}_n$	No Change

14.3.1.3 $J = 0$ and $K = 1$ Case

When $J = 0, K = 1$, the circuit of Figure 14.5(a) behaves as follows: Let upper NAND output $Q_n = 1$. Lower three input NAND output = 0, because two other input are also 1s, because K input is 1 and at the clock edge CLK input is 1. Lower NAND output, which is input to cross coupled NANDs = 0. Now upper NAND output, which is input to cross coupled NANDs = 1 due to $J = 0$. The cross-coupled NAND inputs therefore corresponds to $S = 0$ and $R = 1$ case in the circuit of Figure 14.2(a). Hence output Q_{n+1} will become 0 and $\overline{Q}_{n+1}$ will become 1.

TABLE 14.3 ($J = 0$ and $K = 1$) state Q and $\overline{Q}$ (+ve edge case)

Inputs					Outputs		State
J	$\overline{Q}$	CLK_i	K	Q	Q_{n+1}	$\overline{Q}_{n+1}$	$(n+1)^{th}$
0	0	1 at the +ve edge	1	1	0	1	RESET
0	0	0	1	1	Q_n	$\overline{Q}_n$	No Change
0	0	1	1	1	Q_n	$\overline{Q}_n$	No Change
0	0	0 at –ve edge	1	1	Q_n	$\overline{Q}_n$	No Change

Result is that the Output Q resets to 0 after the clock edge when $J = 0$ and $K = 1$.

14.3.1.4 $J = 1$ and $K = 1$ Case when $Q_n = 0$ and $\overline{Q}_n = 1$

The lower NAND gate output, which is input to a cross-coupled lower NAND input = 1 and $Q_n = 0$. The upper first NAND gate output, which is input to a cross-coupled upper NAND = 0 after the clock edge, because $J = 1$, CLK input at the edge = 1 and $\overline{Q}_n = 1$. Therefore, this correspond to the case of $R = 0$ and $S = 1$ in the circuit of Figure 14.2(a) for SR latch. Therefore, Q sets. Hence $Q_{n+1} = 1$ and $\overline{Q}_{n+1} = 0$. The outputs $\overline{Q}$ and Q get reversed.

14.3.1.5 $J = 1$ and $K = 1$ Case when $Q_n = 1$ and $\overline{Q}_n = 0$

The upper NAND gate output, which is input to a cross-coupled upper NAND input = 1 because $\overline{Q}_n = 0$. The lower NAND gate output, which is input to a cross-coupled lower NAND = 0 after the clock edge, because $K = 1$, CLK input at the edge = 1 and $Q_n = 1$. Therefore, this correspond to the case of $R = 1$ and $S = 0$ in the circuit of Figure 14.2(a) for SR latch. Therefore, Q sets. Hence $Q_{n+1} = 0$ and $\overline{Q}_{n+1} = 1$. The outputs $\overline{Q}$ and Q get reversed.

The state table will now be given as follows:

TABLE 14.4 ($J = 1$ and $K = 1$) state Q and $\overline{Q}$

Inputs					Outputs		State
J	$\overline{Q}$	CLK_i	K	Q	Q_{n+1}	$\overline{Q}_{n+1}$	$(n+1)^{th}$
1	0	1 at the +ve edge	1	1	0	1	Toggle $Q_{n+1} = \overline{Q}_n$
1	1	1 at the +ve edge	1	0	1	0	Toggle $Q_{n+1} = \overline{Q}_n$
1	0	–ve edge or 0 or 1	1	1	1	0	Q_n No Change
1	1	–ve edge or 0 or 1	1	1	1	0	Q_n No Change

Cases in Sections 14.3.1.4 and 14.3.1.5 shows that when $J = 1$ and $K = 1$, the output of *JK* flip flops toggles (changes to opposite state) on the clock edge.

A timing diagram depicts a state table of any flip-flop more clearly. Example 14.11 will describe that.

Points to Remember

JK-flip flop has the following features:

When $J = 1$ and $K = 1$, the output of JK flip flops toggles (changes to opposite state) on a clock edge.

When $J = 0$ and $K = 0$, the output of *JK* flip flops does not change on a clock edge.

When $J = 0$ and $K = 1$, the output Q resets to 0 after the clock edge.

When $J = 1$ and $K = 0$, the output Q sets to 1 after the clock edge.

14.4 T FLIP-FLOP

Figure 14.6(a) shows a logic circuit of a *T* flip flop. Figure 14.6(b) shows the state table of positive edge triggered *T*-flip flop. Figure 14.6(c) shows the symbol of a +ve edge triggered *T*-flip flop. Figure 14.6(d) shows the timing diagram for the inputs at the *T*-input and the outputs Q and $\overline{Q}$ in the *TFF*.

T stands for toggling of the state at the output. This flip flop changes on each successive input. It means gives output Q, which is complement of the previous output. It means if previous Q is 1 then it changes state to 0, and if earlier output is 0 then it changes state to 1. This change takes place upon application of a clock transition 0 to 1 at the *T* input. Figure 14.6(d) shows the outputs, Q and $\overline{Q}$, for the given input at the terminal *T* of the *T*-flip-flop. The detailed explanation of this *FF* is as follows.

CASE 1: Consider an instant, when the input *T* becomes 1 from 0. It means instant of first [$(n + 1)$th edge] clock transition from 0 to 1. Assume that $Q = 1$ and $\overline{Q} = 0$ before this instance and after the nth edge. The lower left most NAND gate in Figure 14.6(a) has both inputs 1 at that instant. (One input is a feedback from the Q and other is from *T* input terminal). Therefore, its output is 0. The upper left most NAND gate in this circuit having one of the input 1 but another input is 0 because $\overline{Q}$ is 0. Therefore, this NAND output appears as 1 at *R*. Now, *R* input is, therefore, 1 to the right side cross-coupled NAND gates (a *SR* latch). According to row 4 of the state table Figure 14.4(b), the output Q will become 0 after that instant, and the complemented $\overline{Q}$ will become 1 after the *T* input becomes 1. The *T* is now after very short interval (equal to propagation delay, t_p) forces 0s at leftmost and right most NANDs. This causes $R = 1$, $S = 1$ at the cross-coupled NAND latch, and therefore there is no change after this interval from the edge transition. This is because there should be no change in the Q and $\overline{Q}$ outputs as per row 1 of the state table in Figure 14.4(b). Now even if *T* input becomes 0, the output state will remain as before. This is due to fact that if *T* is '0' then both *S* and *R* inputs remain both 1, and according to row 1 of state table of cross coupled NAND, there should be no change in Q and, therefore, $\overline{Q}$. Table 14.5 gives the sequences followed in the $(n + 1)^{th}$ transition at the *T*-input.

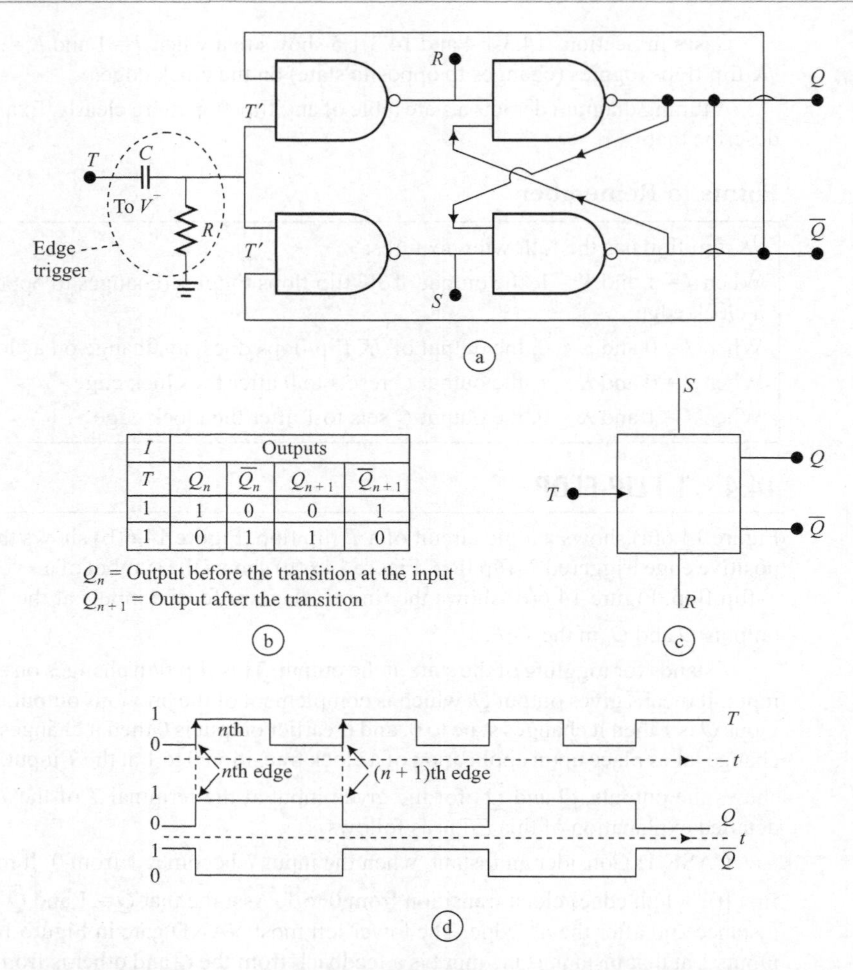

I	Outputs			
T	Q_n	$\overline{Q}_n$	Q_{n+1}	$\overline{Q}_{n+1}$
1	1	0	0	1
1	0	1	1	0

FIGURE 14.6 (a) Logic circuit of a *T*-flip flop (b) State table of positive edge triggered *T*-flip flop (c) Symbol of + ve edge triggered *T*-flip flop (d) Timing diagram for the Inputs at the *T*-input and the outputs Q and Q in the *T-FF*.

TABLE 14.5 *T FF* at $(n + 1)^{th}$ edge

Inputs		Outputs			State
$\overline{Q}_n$	T	Q_n	Q_{n+1}	$\overline{Q}_{n+1}$	
0	1 at the +ve edge	1	0	1	Toggle $Q_{n+1} = \overline{Q}_n$
1	1	0	0	1	No Change
1	–ve edge	0	0	1	No Change
1	0	0	0	1	No Change

CASE 2: Table 14.6 gives the sequences followed in the $(n + 2)^{th}$ transition at the *T*-input. Let us now consider the cycle after $Q = 0$, $\overline{Q} = 1$. It means at a 2nd clock edge $[(n + 2)^{th}$

edge] at the T input. At this second clock edge, when T undergoes transition to logic state 1 from 0, the upper left most NAND gate in Figure 14.6(a) has both inputs 1. So R becomes 0. The lower-most NAND has one input 0 as previous $Q = 0$ and other input is 1 (from the T). So S is 1. Now this state corresponds to 2nd row of the state table in Figure 14.2(b). Therefore, the second clock $(n + 2)^{th}$ edge causes the outputs $Q = 1$ and $\overline{Q} = 0$.

TABLE 14.6 *T FF* at $(n + 2)^{th}$ edge

Inputs			Outputs		State
$\overline{Q}_{n+1}$	T	Q_{n+1}	Q_{n+2}	$\overline{Q}_{n+2}$	
1	1 at the +ve edge	0	1	0	Toggle $Q_{n+2} = \overline{Q}_{n+1}$
0	1	1	1	0	No Change
0	–ve edge	1	1	0	No Change
0	0	1	1	0	No Change

CASE 3: Table 14.7 gives the sequences followed in the $(n + 3)^{th}$ transition at the T-input. Effect of T logic state 1 is removed shortly in a time, called propagation delay, after the 2nd edge also. This happens due to the feedbacks of the new outputs. After now as well as when the T going transition to 1 from 0, the left most NAND has at least one input 0 so R is 1. The lower left most NAND has inputs so as S is 1. Therefore, there is no effect at the right side SR latch. The state at output remains same as previous one according to row 1 of state table in Figure 14.2(b) for the cross-coupled NANDs latch.

TABLE 14.7 *T FF* at $(n + 3)^{th}$ edge

Inputs			Outputs		State
$\overline{Q}_{n+2}$	T	Q_{n+2}	Q_{n+3}	$\overline{Q}_{n+3}$	
0	1 at the +ve edge	1	0	1	Toggle $Q_{n+3} = \overline{Q}_{n+2}$
0	1	0	0	1	No Change
0	–ve edge	0	0	1	No Change
0	0	0	0	1	No Change

We find that if first clock transition to 1 at T input causes transition to $Q = 0$ (and $\overline{Q}$ = 1), then the second clock transition to 1 from 0 at the T input causes transition to $Q = 1$ (and $\overline{Q} = 0$). Each 0 to 1 change at the T input will complement the output Q. The Q is stable to either 0 or 1 between the two successive 0 to 1 transitions.

Figure 14.6(d) gives the effect of successive pulses at the T input. If input frequency is 300 kHz, at Q the output frequency is 150 kHz. The T-flip flop is, therefore, like a counter and a frequency divider. The T-flip-flop is also called, scale-of-two circuit. T-FFs are used in the counters. A binary counter has many T-flip flops in cascade such that each T is connected to the output of the previous T-flip flop (refer Chapter 15). Normally, for the counting applications, the flip flops circuits are converted into the several T-flip flop circuits of Figure 14.6(a).

Point to Remember

T-flip flop has the following feature:
On each successive input, Q and $\overline{Q}$ outputs toggle (complements their previous states).

14.4.1 T Flip-Flop with Clear and Preset

T-flip flop with clear and preset is a circuit in Figure 14.6(a) with *R* and *S* inputs provided to a user of the *T-FF* as the additional inputs. If inputs are given at the points *R* and/or *S* in Figure 14.6(a) circuit, then the circuit shall respond with a priority to the changes at these inputs. For example, if *S* input is activated and made 0, the output $\overline{Q}$ shall stay 1. No matter what are the states occurring at the *T*-input. The *S* input and *R* input are, therefore, having highest priority. Circuit of Figure 14.6(a) also acts as the *SR* latch if its *T* input is not used at all and only the *R* and *S* inputs used.

Point to Remember

T-flip flop with clear and preset is a circuit of *T*-FF with the following additional features:
R input (active 0) is used to clear (reset) the output Q.
S input (active 0) is used to preset the output Q.

14.5 D FLIP-FLOP AND LATCH

14.5.1 D Flip-Flop

Figures 14.7(a) shows a logic circuit of a *D* flip-flop. Figure 14.7(b) shows the state table of positive edge triggered *D*-flip flop. Figure 14.7(c) shows the timing diagram for the inputs at the *D*-input and the outputs Q and $\overline{Q}$ in the *D-FF*. Figure 14.7(d) shows the symbol of a +ve edge triggered *D*-flip flop.

D stands for delay. Let us consider a datum bit is present at an input, called *D* input, just before a clock input exhibits a change to a desired logic level. From the state table in Figure 14.7(b), it can be noted that the bit is transferred to the output after the change (0 to 1 transition) at the clock input with a delay equal to propagation delay. (Remember that (i) *D*-has to be present before a time = setup time from the clock edge. (ii) *D*-has to be present after a time = hold from the clock edge.) Propagation delay is the time interval between occurrence of the clocking event and the appearance of effect at the output, Q.

From Figure 14.7(a), it can be seen that a *D* type *FF* is a modification by using additional gates in the *SR* cross coupled NANDs latch. When *D* is 1 and the clock input, CLK, is 0, the upper left most NAND gate shall give 1 in its output. Therefore, *R* input is 1 to the cross coupled NAND latch formed by the upper and lower NANDs on the right hand side in Figure 14.7(a). The D input after processing through a NOT gate is now 0. Therefore, lower left side NAND gates have both inputs at 0. Therefore, the *S* input, which is the output of this NAND is at 1. The Q output is per row 1 of state table in Figure 14.2(b) will therefore remain as such. It means unchanged.

Now, if the CLK input undergoes to 1 state from 0 state, the *R* input to the right side upper NAND is 0 and as the *S* input is 1, therefore Q output becomes 1. Therefore, Q is now

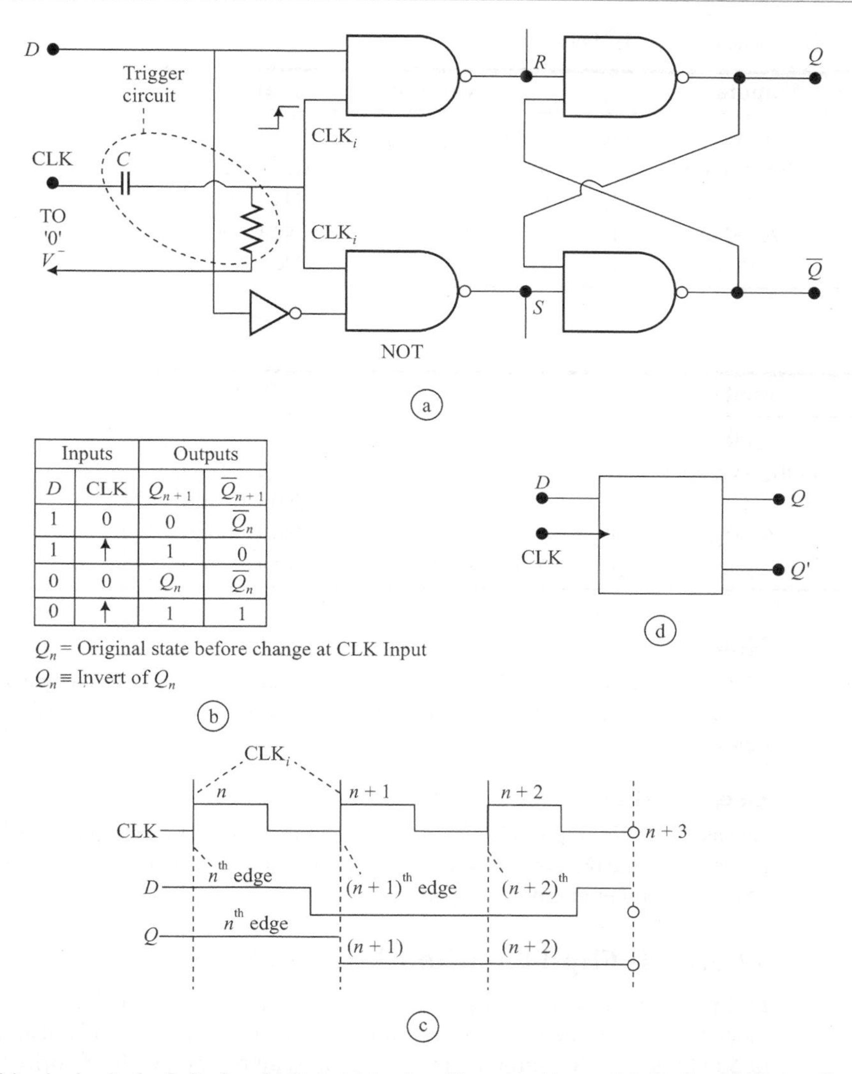

Inputs		Outputs	
D	CLK	Q_{n+1}	$\overline{Q}_{n+1}$
1	0	0	$\overline{Q}_n$
1	↑	1	0
0	0	Q_n	$\overline{Q}_n$
0	↑	1	1

Q_n = Original state before change at CLK Input
$Q_n \equiv$ Invert of Q_n

FIGURE 14.7 (a) Logic circuit of a *D*-flip flop (b) State table of positive edge triggered *D*-flip flop (c) Timing diagram for the Inputs at the *D*-input and the outputs Q and $\overline{Q}$ in the *D-FF* (d) Symbol of + ve edge triggered *D*-flip flop.

identical to the state of D input just before the transition to 1 from 0 at the CLK input. Table 14.8 shows the sequences.

Similar discussion as above can be used to explain the output when $D = 0$ and CLK $= 0$ before a clock edge transition at the CLK. Q will become 0 in such a situation after the transition to 1 from 0 at the CLK input. The presence of the NOT gate shown in the left bottom corner of Figure 14.7(a) ensures that a situation, $R = 0$ and $S = 0$ is never encountered as the later situation makes an *FF* circuit unstable. [Recall the circuit of Figure 14.4(a) and its state table in Figure 14.4(b)]. Table 14.9 shows the sequences.

TABLE 14.8 *D FF* state after $(n + 1)^{th}$ transition

Inputs				Outputs		State
D	CLK	$\overline{Q}_n$	Q_n	Q_{n+1}	$\overline{Q}_{n+1}$	
1	1 at the +ve edge	1	0	1	0	$Q_{n+1} = D = 1$
0 or 1	1	1	1	1	0	No Change
0 or 1	–ve edge	1	1	1	0	No Change
0 or 1	0	1	1	1	0	No Change

TABLE 14.9

Inputs				Outputs		State
D	CLK	$\overline{Q}_n$	Q_n	Q_{n+1}	$\overline{Q}_{n+1}$	
0	1 at the +ve edge	0	1	0	1	$Q_{n+1} = D = 0$
1 or 0	1	1	0	0	1	No Change
1 or 0	–ve edge	1	0	0	1	No Change
1 or 0	0	1	0	0	1	No Change

Note

The clock edge at the CLK input is acting like a shutter of a camera which when clicks (here undergoes transition) causes the photograph at Q of the D input after a delay = propagation delay.

Uses of DFFs

Just as the *T*-flip flop is a basic unit of the counters, the *D*-flip flop is a basic unit of storage registers for the data (bits). Both these *FF* units are widely incorporated into the microprocessor, computer and other circuits.

14.5.2 D Flip-Flop with Clear and Preset

D-flip flop with clear and preset is a circuit of Figure 14.7(a) with *R* and *S* inputs also provided to a user of the *D-FF* as the additional inputs. If inputs are given at the points *R* and/or *S* in Figure 14.7(a) circuit, then the circuit shall respond with a priority to the changes at these inputs. For example, if *S* input is activated and made 0, the output $\overline{Q}$ shall stay 1. No matter what are the states occurring at the *D*-input. The *S* input and *R* input are, therefore, having highest priority. Circuit of Figure 14.7(a) also acts as the SR latch if its *T* input is not used at all.

Points to Remember

1. *D*-flip flop gives the output state = *D* input after the edge transition.
2. *D*-flip flop with clear and preset is a circuit of *D-FF* with the following additional features: *R* input (active 0) is used to clear (reset) the output $Q = 0$, irrespective of previous Q or D states. *S* input (active 0) is used to preset the output $Q = 1$ irrespective of previous Q or D states.

14.5.3 D Latch

A variation of *D-FF* is the *D*-latch shown in Figure 14.8(a). Figure 14.8(a) shows a logic circuit of a *D*-latch. Figure 14.8(b) shows the state table of *D*-latch. Figure 14.6(c) shows the symbol of a *D*-latch. There is no edge triggering circuit. When the clock is at 1, *D*-input is transferred to *Q* output like in a transparent latch, and during this state 1 of clock if *D* input changes then *Q* output also changes. The circuit is transparent to the input D. When the clock is at logic state 0, the *Q* output is frozen (not transparent) to the state of *D*-input prior (how much before, it depends on the setup time) of the 1 to 0 (negative edge) transition. In other words, the *D*-type latch provides unhindered transfer of input to the output during the period when the clock input remains at 1.

The clock input is acting like a valve which when open (here it means clocking logic level) causes the liquid to flow through a pipe.

Table 14.10 shows the sequences.

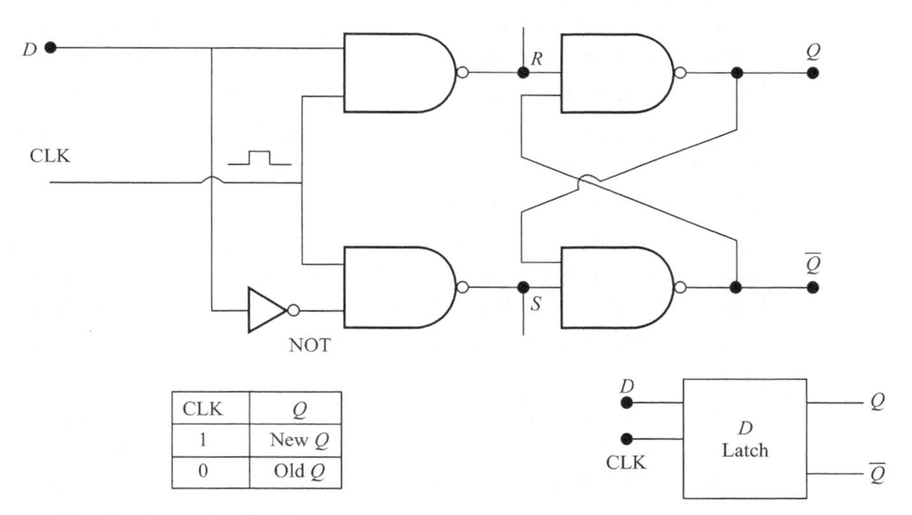

CLK	Q
1	New Q
0	Old Q

New Q = Same after Δt as D.
Old Q = Same as just before last negative edge.

FIGURE 14.8 (a) Logic Circuit of a *D*-latch (b) State table of *D*-latch (c) Symbol of *D* Latch.

TABLE 14.10 *D* latch

Inputs				Outputs		State
D	CLK	$\bar{Q}_n$	Q_n	Q_{n+1}	$\bar{Q}_{n+1}$	
0	1 at the +ve edge	0	1	$Q_{n+1} = 1$	$\bar{Q}_{n+1} = 0$	$\bar{Q}_{n+1} = \bar{Q}_n = 1$
0	1	0	1	0	1	$\bar{Q}_{n+1} = D = 0$
1	Before the −ve edge	1	0	1	0	$Q_{n+1} = D = 1$
1 or 0	At the −ve edge or at 0	1	0	0	1	No Change

14.6 MASTER-SLAVE RS FLIP-FLOP

Figure 14.9(a) shows a logic circuit called master-slave *RS* flip flop. Figure 14.9(b) shows the state table for master-slave *RS* flip flop. Figure 14.9(c) shows logic symbol of *RS MS-FF*. Figure 14.9(d) gives a partial timing diagram. It shows the plots for the pulse clocking input, the periods of master enabling and slave enabling and Q and Q'.

Why is the circuit called master-slave *RS* flip flop will be clear shortly. It is also called pulse triggered *RS* flip flop due to its actions on positive going transition as well as on negative going transition when a clock pulse is applied to it [Refer Figure 14.9(d)].

This flip flop is a combination of the two gated-*SR* latching circuits. This can be easily visualized by comparing the left and right units in Figure 14.9(a) with the circuit in the Figures 14.4(a).

CLK = 0 – Disabling Master Section

The outputs of the attached gated SR NANDs are 1. For the cross-coupled NANDs (Figure 14.2(a), the output from a box marked MT (called master) in Figure 14.9(a), will therefore remain same as before. This follows from first row of state table given at Figure 14.2(b). The master gives intermediate outputs, Q' and $\overline{Q}'$. The outputs of the master are unchanged, therefore, the Q and $\overline{Q}$, the outputs of box marked *SL* (called slave) also remain unchanged as *SL* gets the inputs from the master section.

CLK becomes 1 – Master Section Response

Let us now assume that clock input becomes at logic state 1. Master outputs Q' and $\overline{Q}'$ will now change according to the *S* and *R* inputs. This is analogous to action of a transparent latch in Figure 14.4(a). If $S = 0$ and $R = 1$, then the output according to table in Figure 14.4(b) shall be 0. In other words, Q' and $\overline{Q}'$, the master outputs, shall be 0 and 1, respectively.

The Q and $\overline{Q}$, the outputs of box marked *SL* (called slave) also remain unchanged as during *T'* - *T'* the CLK' input to *SL* section is 0 due to the NOT gate before it.

We find that the first gated NAND flip-flop section and the box marked *ML* is providing the outputs Q' and $\overline{Q}'$ according to *S* and $\overline{R}$, when CLK becomes 1. This is also the reason that the *MT* section is called Master section. Slave is inactive during the transparency of *ML* because *SL* has *T'* CLK input = 0. When master finishes the action, then only slave can act.

CLK becomes 0 – Slave Section Enabled Master disabled

When Q' is 0 and $\overline{Q}'$ is 1, this means CLR is 0 and *PR* is 1. Therefore, according to row 2 of table in Figure 14.10(b), the outputs $Q = 0$ and $\overline{Q} = 1$. Had *S* been 1 and *R* been 0, the outputs, upon clock going to 1 would have been according to row 1 of the table as $Q = 1$ and $\overline{Q} = 0$.

We find that the second gated NAND flip flop section and the box marked *SL* is simply providing the outputs Q and $\overline{Q}$ on CLK becoming 0 according to Q' and $\overline{Q}'$ as CLK' = 1. The Q' and $\overline{Q}'$ are the outputs of the master section when CLK logic state was 1. This is also the reason that the *SL* section is called slave section.

The *RS* flip-flop action arises of the combination of master and slave.

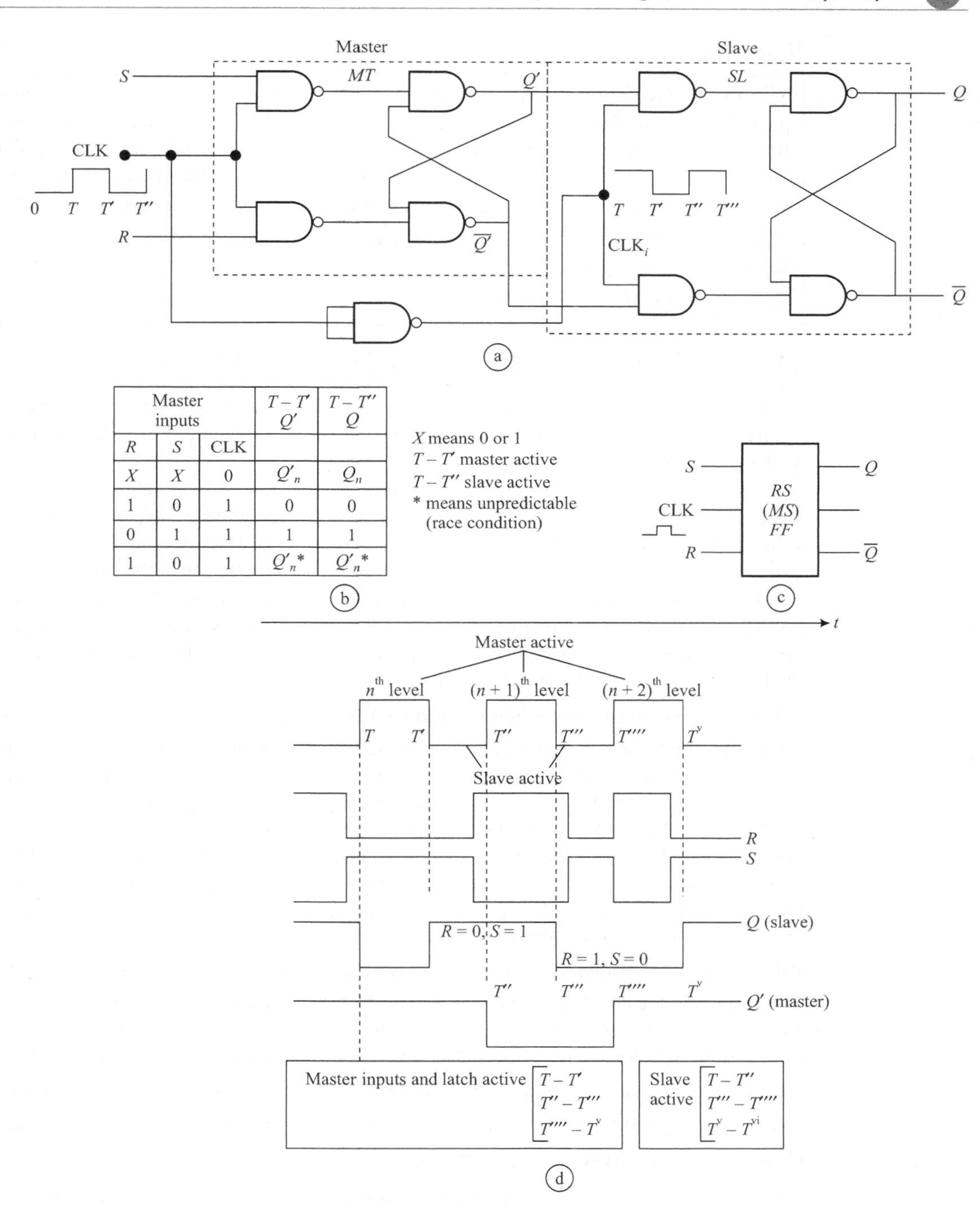

FIGURE 14.9 (a) Master-Slave *SR* Flip Flop logic diagram (b) State table for Master-Slave *RS* Flip Flop (c) Logic Symbol of *SR MS-FF* (d) Timing diagram for the pulse clocking input and periods of master enabling and slave enabling.

Remember

There is no possibility of a false transition at $(n + 1)^{th}$ state on the positive going clock pulse (0 to 1) and followed negative going 1 to 0 in a *SR MS FF*. After positive going from 0 to level 1 transition, the master is triggered and the slave is idle. After the negative going from 1 to 0, the slave responds as per master outputs and the master is idle.
When both S and $R = 1$, the circuit is meta stable and outputs are unpredictable.
When both S and $R = 0$ the previous outputs remain unchanged.
When $S = 1$ and $R = 0$ the $Q = 1$.
When $S = 0$ and $R = 1$ the $Q = 0$.

14.7 MASTER-SLAVE (PULSE TRIGGERED) JK FLIP-FLOP

R-S flip flop has an indeterminate meta-stable state [Refer state tables at the Figures 14.2(b), 14.4(b) and 14.9(b)]. An additional circuitry can avoid this state. For example, Figure 14.10(a) shows one such circuit.

The inputs are now called *J* and *K* instead of *S* and *R*, respectively. The circuit is called master-slave *JK* flip flop (*JK MS FF*) due to reason, which will be clear shortly. It is also called pulse triggered *JK*-flip flop also for the reasons being described shortly. This circuit will give a predictable determined output even when both inputs are at the state 1.

Figure 14.10(b) gives the state table of a *JK MS FF*, and Figure 10.10(c) shows its symbol. This flip flop is a combination of the gated *SR* and *T* inputs latching circuits. This can be easily visualized by comparing the left and right units in Figure 14.10(a) with two circuits in the Figures 14.4(a) and 14.7(a), respectively.

***J-K FF* Circuit Activation when *PR* and CLR = 1** (Inactive and Unconnected)

It means when these two priority inputs are inactivated, then the outputs shall be according to the last four rows of the state table in Figure 14.10(b) and the outputs now depend only on *J*, *K* and CLK inputs. In this case, when only one either *J* or *K* is 0, the action is like the *SR* flip flop after the clock level transition 0 to 1. However, if both *J* and *K* are 0, the outputs after the clock level transition 1 to 0 shall be same as before the clock input change. When both *J* and *K* are 1, the previous output is inverted (complemented). This is unlike the *RS* latch or flip flop in which an output would have been unstable. This fact differentiates, the JK-FF from the *RS-FF* as pointed before.

PR and CLR = 1 (Inactive and Unconnected), and CLK = 0

The outputs of the attached gated *SR* NANDs are 1. For the cross-coupled NANDs (Figure 14.2(a)), the output from a box marked *MT* (called master) in Figure 14.10(a), will therefore remain same as before. This follows from first row of state table given at figure 14.2(b). The master gives intermediate outputs, Q' and $\overline{Q}'$. The outputs of the master are unchanged, therefore, the Q and $\overline{Q}$, the outputs of box marked SL (called slave) also remain unchanged.

CLK becomes 1 – Master Section Response

Let us now assume that clock input becomes at logic state 1. master outputs Q' and $\overline{Q}'$ will now change according to the *J* and *K* inputs. This is analogous to action of a transparent latch in Figure 14.4(a). If $J = 0$ and $K = 1$, then the output according to table in Figure 14.4(b) shall be 0. In other words, Q' and $\overline{Q}'$, the master outputs, shall be 0 and 1, respectively.

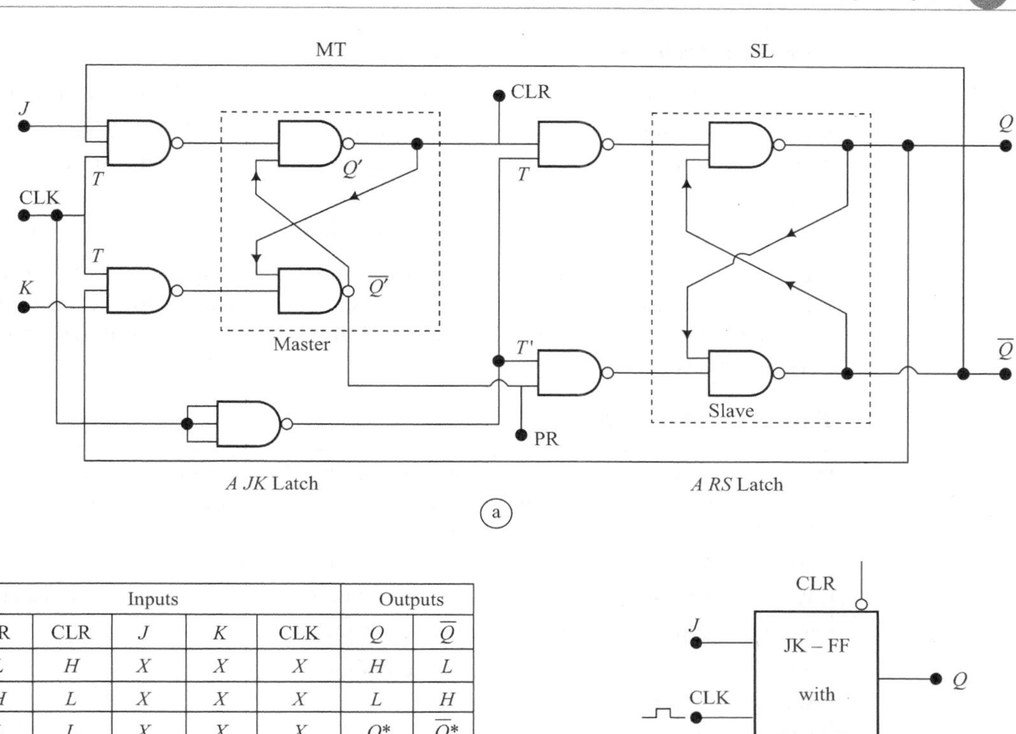

Inputs					Outputs	
PR	CLR	J	K	CLK	Q	$\overline{Q}$
L	H	X	X	X	H	L
H	L	X	X	X	L	H
L	L	X	X	X	Q*	$\overline{Q}$*
H	H	L	L	⎍	Q_n	$\overline{Q}_n$
H	H	H	L	⎍	H	L
H	H	L	H	⎍	L	H
H	H	H	H	⎍	$\overline{Q}_n$	Q_n

* = Unstable

⎍ = Low to high and high to low

Q_n = Pre-state before the clock input change

$\overline{Q}_n$ = Complement of previous state Q_n

H = 1 and L = 0

X = Immaterial H or L

b

FIGURE 14.10 (a) Master-Slave *JK* flip flop logic diagram (b) State table for master-slave *JK* flip flop (c) Logic Symbol of *JK FF* with preset and clear inputs also present.

Here, the J input (in the master section of Figure 14.10(a) is equivalent to the S input in Figure 14.4(a), and K input in this MT section is equivalent to the R input in Figure 14.4(a).

The Q and $\overline{Q}$, the outputs of box marked SL (called slave) also remain unchanged as T' CLK input to SL section is 0.

We find that the first gated NAND flip flop section and the box marked ML is providing the outputs Q' and $\overline{Q}$' according to J and $\overline{K}$, when CLK becomes 1. This is also the reason that the ML section is called Master section. Slave is inactive during the transparency of ML because SL has T' CLK input = 0. When the master finishes the action, then only the slave can act.

CLK becomes 0 – Slave Section Response

When Q' is 0 and $\overline{Q}'$ is 1, this means CLR is 0 and *PR* is 1. Therefore, according to the second row of table in Figure 14.10(b), the outputs $Q = 0$ and $\overline{Q} = 1$. Had *J* been '1' and *K* been 0, the outputs, upon clock going to 1 would have been according to row 1 of the table as $Q = 1$ and $\overline{Q} = 0$.

We find that the second gated NAND flip flop section and the box marked *SL* is simply providing the outputs Q and $\overline{Q}$ on CLK becoming 0 according to Q' and $\overline{Q}'$. The latter are the outputs of master section when CLK logic state was 1. This is also the reason that the *SL* section is called Slave section. The present *JK* flip flop action arises of the combination of master and slave.

Points to Remember

There is no possibility of a false transition at $(n + 1)^{th}$ state on the positive going clock pulse (0 to 1) and followed negative going (1 to 0) in a *JK MS-FF*. After positive going from 0 to level 1 transition, the master is triggered and the slave is idle. After the negative going from 1 to 0, the slave responds as per master outputs and the master is idle.

When both *J* and $K = 1$, the circuit is such that upon a clock pulse, the complements of the previous output states are obtained at Q and $\overline{Q}$.

Both *J* and $K = 0$ the previous outputs remain unchanged.

When $J = 1$ and $K = 0$ the $Q = 1$.

When $J = 0$ and $K = 1$ the $Q = 0$.

Notes

1. Using master-slave we implement equivalent of *JK* edge triggered *FF* by using two latches, one as a master and one as a slave.
2. When the triggering of a *FF* is after the positive and negative edges, it is called pulse triggering or pulse clocking.

14.7.1 MS JK Flip-Flop with Clear and Preset

Following features can be noted from the logic circuit at Figure 14.10(a) and state table in Figure 14.10(b). There are two inputs, called *PR* (preset) and CLR (Clear). These are like set and reset inputs in state table for the Set-Reset latch shown in Figure 14.2(b). When making one of these 0 activates any of these two inputs, the outputs Q and $\overline{Q}$ are there, and then the *J*, *K* and CLK inputs are immaterial. In words, the CLR and *PR* have the higher priorities than the *J*, *K* and CLK.

Points to Remember

JK MS flip flop with clear and preset is a circuit of *JK FF* with the following additional features:

R input (active 0) is used to clear (reset) the output $Q = 0$.

S input (active 0) is used to preset the output $Q = 1$.

14.8 CLOCK INPUTS

14.8.1 Level Clocking of a Clock Input

Recall the circuit of Figure 14.4(a), called a level clocking circuit. Whenever CLK becomes 1 then only the circuit is transparent (responds) to the *R* and *S* inputs.

Level clocking circuit has the following difficulties:

(i) Often the input state or states are not permitted to change at the clock state ↑ or at the clock state ↓ or at the clock state 1. All changes at the inputs *S* and *R* must be over when the clock input is in logic state 0.

(ii) Assume that level 1 activates the clock input. Time interval during which CLK = 1 and CLK undergoes 1 to 0 transition is wasted as in any way the input has already been defined at the instance when CLK became 1.

14.8.2 Edge Triggering at a Clock Input

Incorporating an edge triggering circuit at a clock or gate input or change enable input avoids the difficulties (i) and (ii) pointed in Section 14.8.1. Therefore, a FF, which is a circuit with an additional edge triggering facility at the gate (clock) input [for example, circuits of figures 14.6(a) and 14.7(b)] is free from these difficulties.

Figures 14.11(a) and (b) show the edge triggering clock actions.

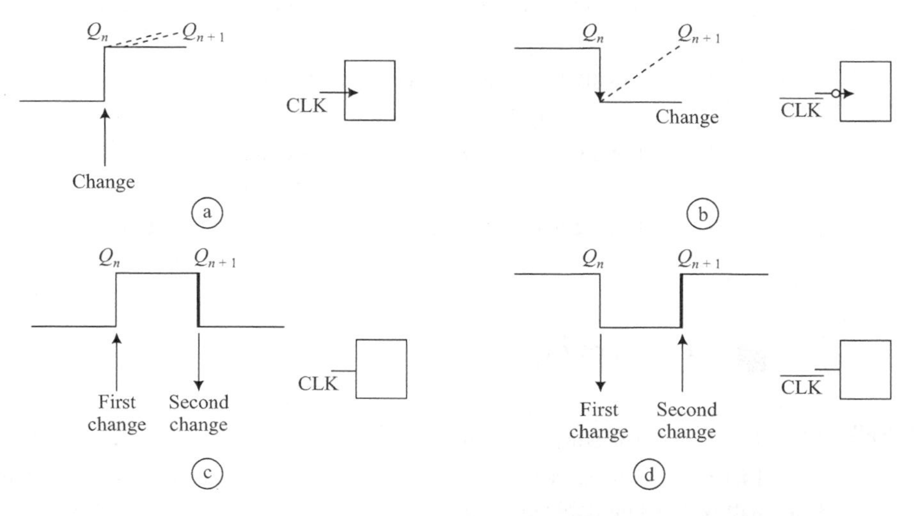

FIGURE 14.11 (a) Positive edge triggering clock action (b) Negative edge triggering clock actions (c) Pulse triggering by positive going followed by negative going transition (d) Pulse triggering by negative going followed by positive going transition.

Figures 14.6 and 14.7 showed a capacitor cum resistor circuit at the clock input to obtain the edge triggering. The refined edge triggering circuits for +ve edged clock and –ve edge clocked circuits are used in an actual integrated circuit chip. A monostable circuit can be used to get positive and negative going edges, and pulses using a push button switch.

14.9 PULSE CLOCKING OF THE LATCHES IN THE FLIP-FLOPS

Let us refer to Figures 10.11(c) and (d). These show the pulse triggering by positive going followed by negative going transition and pulse triggering by negative going followed by positive going transition, respectively.

Master-Slave method gives an alternative and is adopted for a pulse clocking. Pulse clocking means that the outputs are obtained after logic level changes firstly from 0 to 1 and then from 1 to 0. The first change puts the datum or data in the master outputs and the second change transfer the data to the slave outputs which provide the final outputs Q and $\overline{Q}$.

Sign (⎍) at the CLK shows a positive going pulse clocking [Figure 14.11(c)].

It is also possible to design a latch based *FF* such that it responds first when there is a change at clock input from 1 to 0 and the second by the change from 0 to 1 as in Figure 14.11(d). Figure 14.11(d) shows the –ve going pulse clocking by the sign (⎏).

14.10 CHARACTERISTIC EQUATIONS FOR THE ANALYSIS

Combinational circuits are represented by the Boolean expressions and SOPs. Equations can also be defined for the latches and flip flops.

Let Q_{n+1} is next succeeding state after the clock-triggered (level or edge or pulse) transition. Q_n is the state before the transition.

The following is the *SR* latch or *FF* characteristics equation:

$$Q_{n+1} = S + \overline{R}.Q_n \qquad ...(14.1)$$

The following is the *JK FF* characteristics equation:

$$Q_{n+1} = J.\overline{Q}_n + \overline{K}.Q_n \qquad ...(14.2)$$

The following is the *T FF* characteristics equation.

$$Q_{n+1} = \overline{T}.Q_n + T.\overline{Q}_n = T.\text{XOR}.Q_n \qquad ...(14.3)$$

The following is the *D* latch or *FF* characteristics equation:

$$Q_{n+1} = D \qquad ...(14.4)$$

EXAMPLES

Example 14.1

Two NOTs (or common input NANDs or NORs) are cross-coupled as shown in Figure 14.12 by feedback of $\overline{Q}$ to first NOT and of Q to the second NOT. Explain how the circuit will work as bi-stable element.

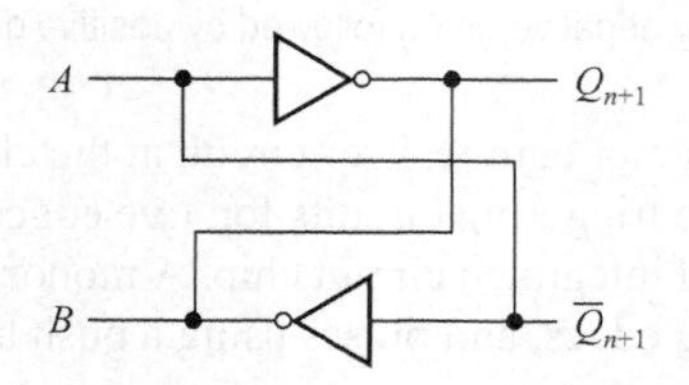

FIGURE 14.12 Cross-Coupled NOTs as a bistable element.

Solution

Let at the upper NOT gate the input A = 1 when the Q = 1 at an nth instance. Q = 0 at the $(n + 1)^{th}$. Then due to cross coupling of Q with B, the B = 0 and hence $\overline{Q}$ remains stable at 1. This is one stable state of the circuit, called Reset state—Q = 0 and $\overline{Q}$ = 1.

Let at the upper NOT gate the input A = 0 when Q = 0 at an mth instance. Q = 1 at $(m + 1)^{th}$. Then due to cross coupling of Q with B, the B = 1 and hence $\overline{Q}$ remains stable at 0. This is another stable state of the circuit called Set state—Q = 1 and $\overline{Q}$ = 0.

Whether A input is 1 or 0 is just a matter of chance or can change due to thermal noise. Circuit can be meta-stable for an unpredictable period and changing from one state to another. The state table will now be given as follows:

TABLE 14.11

Inputs	Outputs		State
A	Q_{n+1}	$\overline{Q}_{n+1}$	
0	1	0	SET (Stable State 1)
1	0	1	RESET (Stable State 2)
Unstable Between 1 and 0	1 or 0	0 or 1	Unpredictable*

*Meta stable state. However a noise, which makes at an instance A = 1 or a noise that makes at an A = 0 will bring back the circuit in stable state, reset or set, respectively.

Example 14.2 If in Figure 14.2(a) circuit S denotes the R input and R denotes S input, show that we get an $\overline{S}$-$\overline{R}$ latch. Why?

Solution

Circuit works as $\overline{S}$-$\overline{R}$ latch. Reset state activates on $\overline{R}$ = 0 and S = 1 and set state activates on S = 0 and R = 1. Figure 14.2(b). State table columns for two inputs will interchange and the table will now be given as follows:

TABLE 14.12 A SR latch with 0 state activating an R or S

Inputs		Outputs		State
S	R	Q_{n+1}	$\overline{Q}_{n+1}$	
0	0	Q_n	$\overline{Q}_n$	No Change
1	0	1	0	RESET
0	1	0	1	SET
1	1	*	*	Unstable

*Unpredictable behaviour, when inputs R and S simultaneously becomes 0 immediately afterwards and meta stable state results.

Example 14.3 Explain the working of an S-R latch with circuit similar to Figure 14.2(a) made from the NORs. (Set occurs on S input = 0 and Reset occurs on R input = 0).

Solution

Circuit works as *SR* latch with reset state activating on $R = 0$ and $S = 1$ and set state activating on $S = 0$ and $R = 1$. The state table will now be given as follows:

TABLE 14.13 State table for cross coupled NORs

Inputs		Outputs		State
S	R	Q_{n+1}	$\overline{Q}_{n+1}$	
0	0	Q_n	$\overline{Q}_n$	No Change
1	0	1	0	RESET
0	1	0	1	SET
1	1	*	*	Unstable

*Unpredictable behaviour, when inputs *R* and *S* simultaneously becomes 0 immediately afterwards and meta stable state results.

Example 14.4

Explain the working of an *SR* latch with circuit similar to Figure 14.2(a) made from the NORs but the *S* input is at the upper NOR, which give *Q* output and *R* input is at the lower NOR, which gives $\overline{Q}$ output.

Solution

The state table will now be given as follows:

TABLE 14.14 State table for cross coupled NORs with *S* input to NAND providing *Q* output

Inputs		Outputs		State
S	R	Q_{n+1}	$\overline{Q}_{n+1}$	
0	0	Q_n	$\overline{Q}_n$	No Change
1	0	1	0	SET
0	1	0	1	RESET
1	1	*	*	Unstable

*Unpredictable behaviour, when inputs *R* and *S* simultaneously becomes 0 immediately afterwards and meta stable state results.

Example 14.5

Why does a meta-stable (unstable for a short period) state exist in the circuit of examples 14.1 to 14.4 for a very brief but for an unknown amount of period?

Solution

Meta stable state occurs either due unpredictable 1 or 0 in example 14.1 and or due to both *R* and *S* input becoming identical and bring the circuit to unstable state and next to that *R* and *S* both becomes identically to 'no change state'. However, a noise always occurs in a circuit, may be for a very brief period. For example in circuit of example 14.1, when a noise makes at an instance $Q = 1$ or a noise that makes at an instance $\overline{Q} = 0$ will bring back the circuit in stable state, Reset or Set, respectively. *Q* becoming 1 or 0 at an instance will also bring the circuit back to stable state.

Example 14.6

Explain a clocked (gated) *SR* latch (an *SR* latch with a clock input). Why is it called a transparent latch?

Solution

Let us add two NAND gates to the two cross-coupled NANDs in circuit of Figure 14.2(a). Figure 14.4(a) showed this new circuit. It is an *R-S* latch circuit, which has a gating. It means provide a clocking facility. The *S* and *R* inputs are at the different NANDs. *S* input is now to upper left most NAND, and *R* to lower left most. It is also called a *RS* transparent latch or a clocked or gated *SR* latch.

The state tables of Figures 14.2(b) and 14.4(b) are for NANDs based Set Reset latch and a gated or clocked *RS* latch, respectively. These differ in the following respects.

Let Q_n denotes the output at terminal Q before the clock input becomes at a logic state 1, and let Q_{n+1} denotes the output at terminal Q after the clock input sets to 1.

1. Q_{n+1} equals to Q_n for $R = 0$, $S = 0$ and CLK = 1. (No change of state).
2. $Q_{n+1} = 0$ for $R = 1$, $S = 0$ after the clock becomes 1. It means a clear or reset signal 1 makes Q clear (whatever may be previous Q) after the clock input equals 1.
3. Q is Set to 1 for $S = 1$, $R = 0$ after the clock becomes 1.
4. While $S = 0$ and $R = 0$ after a Set or Reset causes no difference as per condition 1 above.
5. $R = 1$ and $S = 1$ after the clock becomes 1 causes an unstable output. It means the output becomes indeterminate or races through at an instant.
6. CLK = 0 inactivates the effect of the *R* and *S* inputs, the output remains same as before, $Q_{n+1} = Q_n$.

The two rows of the state tables of Figures 14.2(b) and 14.4.(b) are exactly opposite for defining the 'No change' (Q_{n+1} is equal to Q_n) and unstable states.

Points to Remember

In a clocked (gated) *SR* latch, the output can change only after the clock logic state transition from 0 to 1 or whenever the clock input is at logic state 1 or before the clock state transition from 1 to 0, but there is no change at any outputs if clock input is at 0. Clocking input is acting like a gate for the S and R inputs. It acts like a change-enable input bit. For this very reason, the present circuit of Figure 14.4(a) is also called a transparent *SR* latch. There is a transparency when the clock input = 1. A gate (or clock) pin controls the period during which input changes can affect the Q and $\overline{Q}$ and latch is transparent.

The gating NANDs latch can be designed using a CMOS IC 4011B. Using cross-coupled NANDs, we get the *SR* latch with *R*, *S*, Q and $\overline{Q}$ terminals.

Example 14.7

Explain the difficulties in using a clocked (Transparent) *SR* Latch?

Solution

The unstable state exists for an unpredictable period for a certain condition of $R = S$ input. (Condition whether $R = S = 1$ or $R = S = 0$ depends on the circuit made by NANDs or NORs).

Example 14.8

Show the timing diagram for a clocked *SR* latch with clock input is active level 1.

Solution

Figure 14.13 shows the timing diagram of a clocked SR latch having state table as per Figure 14.4(a). Q changes only during CLK = 1 and are as per S and R.

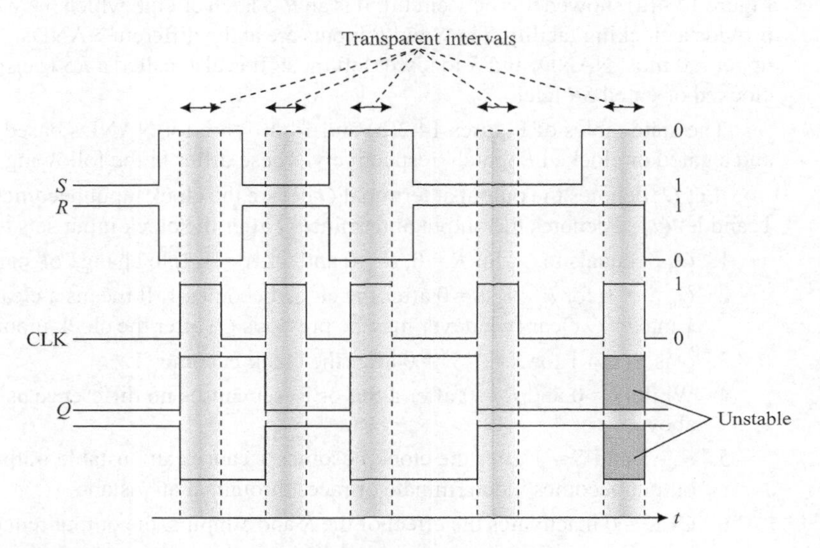

FIGURE 14.13 Timing diagram of a clocked *SR* latch having state table as per.

Example 14.9

Describe edge triggered *SR FF*.

Solution

Figure 14.14 (a) the logic circuit of an edge triggered *SR* latch in which during gating instead of clock at level input 1 in Figure 14.4(a), there is an edge at which the inputs can reflect the change at the output Q and hence at $\bar{Q}$. Figure 14.14 (b) shows the state table for the same. Figure 14.14 (c) shows symbol of the *S-R FF*.

Example 14.10

How does a latch differ from a flip flop in strict sense?

Solution

A latch is a class of flip flop in which the instance at which output changes is uncontrolled. [Uncontrolled means not controlled by the timing input (called clock edge input).] A timing input (called clock edge input) can control the instance at which the output changes. The outputs can reflect change during a transparency interval.

A flip flop on the other hand is expected to have defined by a controlled instance of a clock edge after which the output changes occur. Note: Inputs must be set t_s before the edge and must hold, for t_h the hold period after the edge. Sometimes, a transparent latch is also loosely referred as a flip flop.

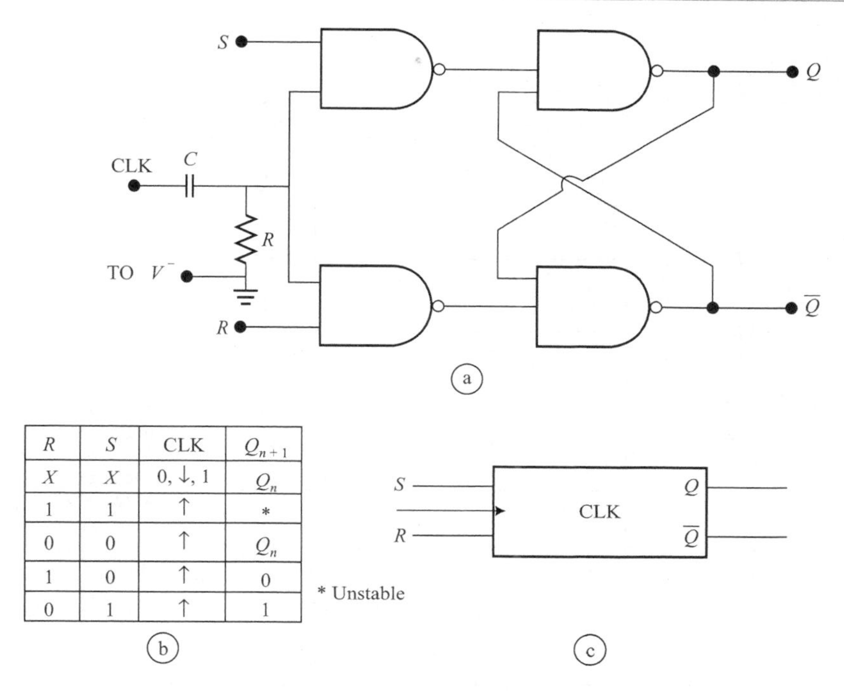

R	S	CLK	Q_{n+1}
X	X	0, ↓, 1	Q_n
1	1	↑	*
0	0	↑	Q_n
1	0	↑	0
0	1	↑	1

FIGURE 14.14 (a) Logic circuit of an edge triggered *SR* flip flop (b) State table of an *SR* flip flop (c) Symbol of a positive edge triggered *S-R* flip flop.

Example 14.11

Show the two timing diagrams concurrently for both the +ve and –ve edge triggered *JK* flip flops.

Solution

A timing diagram depicts a state table of any flip flop more clearly. Figure 14.15 shows two timing diagrams concurrently for both the +ve and –ve edge triggered *JK* flip flops. Two top most plots show the format of the chosen inputs for *J* and *K*. The middle part shows the chosen clock input format. It also shows + edge instance for the case of the positive edge triggered *JK FF*. The two bottom-most plots explain the functions of +ve and –ve edge triggering in the *JK FF*s.

Example 14.12

Explain timing diagram drawn in Figure 14.15 for a logic circuit of an edge triggered *JK FF*.

Solution

Let *J* and *K* inputs vary with time as shown in Figure 14.15. Let CLK input also vary as shown here. This type of CLK input shows positive edges at n, $n + 1$, $n + 2$, $n + 3$ that are denoted in the Figure by up arrows at various instances. It shows negative edges at n, $n' + 1$, $n' + 2$, $n' + 3$, which are denoted in the Figure by down arrows.

Let us first consider *Q* at a positive edge triggered *JK FF*. We find that before the nth edge *J*, *K* = 0, 1, respectively. Therefore, *Q* becomes 0 after this edge and before the $(n + 1)^{th}$ edge, *J*, *K* becomes 1, 0. Therefore, at $(n + 1)^{th}$ edge *Q* becomes 0. Just before the $(n + 2)^{th}$

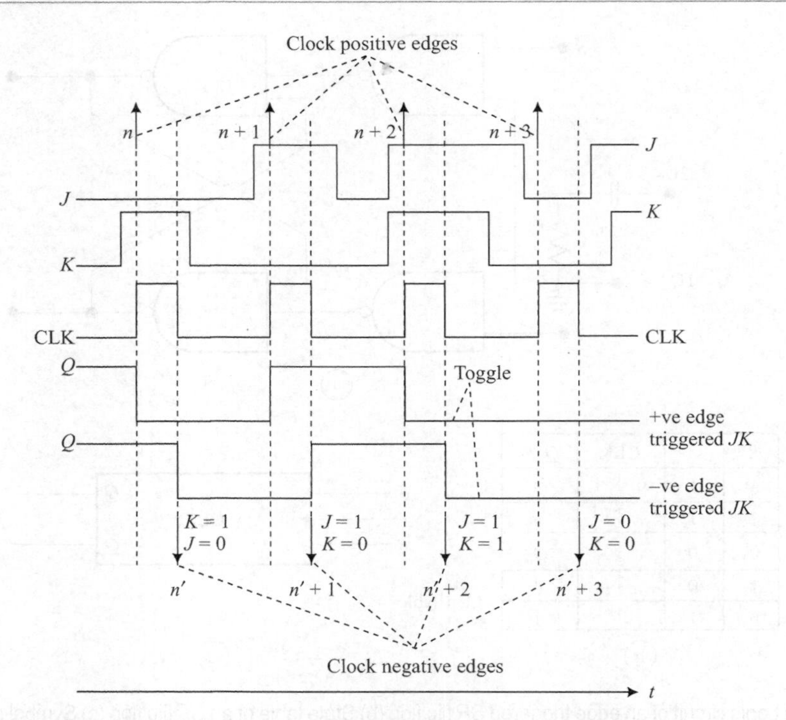

FIGURE 14.15 Timing diagrams for both the + ve and –ve edge triggered *JK* flip flops.

edge and $(n + 3)^{th}$ edges, both *J*, *K* = 1, 1, therefore, *Q* toggles to 0. At $(n + 3)^{th}$ edge, *JK FF* finds both *J*, *K* = 0, 0, therefore, *Q* remains 0.

Now let us consider output expected from a negative edge triggered *JK FF*. *J* and *K* inputs are 0, 1 at the *n'* edge. Therefore, $Q \rightarrow 0$. Just before $(n' + 1)^{th}$ edge, *J* and *K* are 1 and 0 therefore $Q \rightarrow 1$. Just before $(n' + 2)^{th}$ edge, *J*, *K* = 1, 1. Therefore, *Q* toggles to '0'. Before $(n' + 3)^{th}$ edge, both *J* and *K* = 0, therefore, *Q* remains 0.

A negative edge triggered *JK FF* finds extensive application as building block in the counters where the *J* and *K* inputs are pegged to 1, and *JK FF* functioning is then analogous to that of a *T FF* circuit of Figure 14.5(a).

Example 14.13 How will we use the *J-K FF* for designing a *T-FF*?

Solution

A *JK FF* when *J* = 1 and *K* = 1 toggles on the clock input. Therefore, either by leaving the *J* and *K* input floating in case of a TTL gate or connecting these to logic 1 state (necessary for CMOS), we would get a *T FF* and we use the clock edge input as *T* input.

Example 14.14 How will we use the *JK FF* for designing a *D-FF*?

Solution

A *JK FF* when *J* = 1 and $K = \bar{J} = 0$ on the clock input gives the *Q* output = 1. The *JK FF* when *J* = 0 and $K = \bar{J} = 1$ on the clock input gives the *Q* output = 0. For a *D FF*, the *D* input passes to Q output on a clock transition. Therefore, if we connect K input from J through a NOT gate (or a common input NAND), we can use J input as D input. We get a D *FF* when we use the J input as the D-input provided K is made complement of J.

Example 14.15 How will we use the *SR* latch for designing a D latch?

Solution

An *SR* latch when $R = 1$ and $S = \overline{R} = 0$ on the clock input gives the Q output $= 0 = S$. The *S*-*R* latch when $R = 0$ and $S = \overline{R} = 1$ on the clock input gives the Q output $= 1$. For a D *FF*, the D input passes to Q output on a clock transition. Therefore, if we connect S input to R through a NOT gate (or a common input NAND), we can use S input as D input. We would get a D latch when we use the S input as the D input provided R is made complement of S.

Example 14.16 How will we use a T *FF* as a pulse frequency divider and how will we use a chain of T *FF*s as a counter?

Solution

Figure 14.6(d) showed the effect of successive pulses at the T input. If input frequency is 300 kHz, at Q the output frequency is 150 kHz. The T flip flop is, therefore, like a counter and a frequency divider. The T flip flop is also called, scale-of-two circuit. T *FF*s are used in the counters. A binary counter has many T flip flops in cascade such that each T is connected to the output of the previous T flip flop (refer Chapter 15). Normally, for the counting applications, the *JK* flip flops circuits are converted into the cascaded T flip flop circuits of Figure 14.6(a).

Example 14.17 How will we use the MSI ICs for designing a (i) D-FFs and (ii) T FFs based sequential circuit?

Solution

(i) The D *FF*s designed from TTL 74 x 74 family ICs, TTL 74 x 374 family and CMOS 74HC374 family. IC 74 x 374 has eight number D *FF*s with Q_0 to Q_7 as the eight outputs. Each output is in tristate and is enabled (to logic 1 or 0) only when an out enable input is given after the clock transition. This is due to incorporation of a tristate buffer at each of the Q output with a common out enable input for the eight outputs.

(ii) The T *FF*s are designed from IC 74x109 family of gates.

Example 14.18 Describe functioning of a *JK* Master Slave flip flop.

Solution

When the input at CLK is 0, the slave output is fixed according to Q' and $\overline{Q}$'. It means according to the state of outputs from the master circuit. This is so because in that case T' is 1 because when the CLK is 0, the T' is complement of CLK due to a NOT gate between the CLK and T'. The MT section does not respond, and the slave section can be said to be locked when CLK = 0. When the CLK will become again 0 from 1, the T' shall be becoming 1 and the Slave will now respond.

In other words, master section accepts and open locks for the J and K inputs when the clock moves from logic level 0 to 1 and the slave section accepts the master section outputs when the clock moves from 1 to logic level 0.

If both the J and K inputs are 1 the CLK input acts like a toggling input. This can be seen as follows. When CLK is 0, there is no effect and the slave simply accepts the Q' and $\overline{Q}$'. When CLK moves to 1 state, the *MT* box gets both set and reset inputs as 1. According to table in Figure 14.4(b), the outputs Q and $\overline{Q}$ shall remain as before. Upon CLK again moving to level 0, the slave will provide Q and $\overline{Q}$ same as before toggling.

Example 14.19 What are the uses of *JK MS* flip flop clear and preset inputs?

Solution

JK flip flop with clear and preset is a circuit of Figure 14.10(a) with R and S inputs also provided to a user of the *JK FF* as the additional inputs. If inputs are given at the points R and/or S in Figure 14.10(a) circuit, then the circuit shall respond with a priority to the changes at these inputs. For example, if S input is activated and made 0, the output $\overline{Q}$ shall stay 1. No matter what are the states occurring at the T-input. The S input and R input are, therefore, having highest priority. Circuit of Figure 14.10(a) also acts as the *SR* latch if its T input is not used at all.

Example 14.20 How will we use the MSI ICs for designing (i) D latches based (ii) –ve edge triggered *JK* and (iii) Pulse triggered (Master Slave) JK sequential circuits?

Solution

IC TTL 74*x*75 or IC CMOS 74HC75 provides the D latches. D latch is also cheaper than a D flip flop because there are less numbers of gates. The each of the ICs TTL 74373 and CMOS 74HC373 have eight numbers D latches with each output in tristate. All eight outputs are enabled (to logic 1 or 0 as per corresponding D input) only when an out enable input is given after the clock is inactivated. This is due to incorporation of a tristate buffer at each of the Q output with a common out enable input for the eight outputs.

A –ve edge triggered *JK* and a pulse clocked (master slave) *JK FF*s: These are made by using the ICs 74HC108 and 74HC76, respectively.

Example 14.21 Explain master slave (pulse triggered) *D FF*.

Solution

Figure 14.16(a) shows master slave (pulse triggered) D-FF. There is one input called D input in place of S input in *RS- MS FF* of Figure 14.9(a) and J input of Figure 14.10(a). The circuit is called master-slave D flip flop (*D MS FF*) due to reason that a master section drives a slave section when CLK input becomes 0. It is also called pulse triggered D flip flop also due to pulse on going positive and negative transitions both are needed to obtain the output Q and $\overline{Q}$.

Figure 14.16(b) gives the state table of a *D MS FF*, and Figure 10.10(c) shows its symbol. This flip flop consists of a combination of the gated *SR* and T inputs latching circuits. The S input is connected permanently to R input through a NOT gate and is labeled as D input.

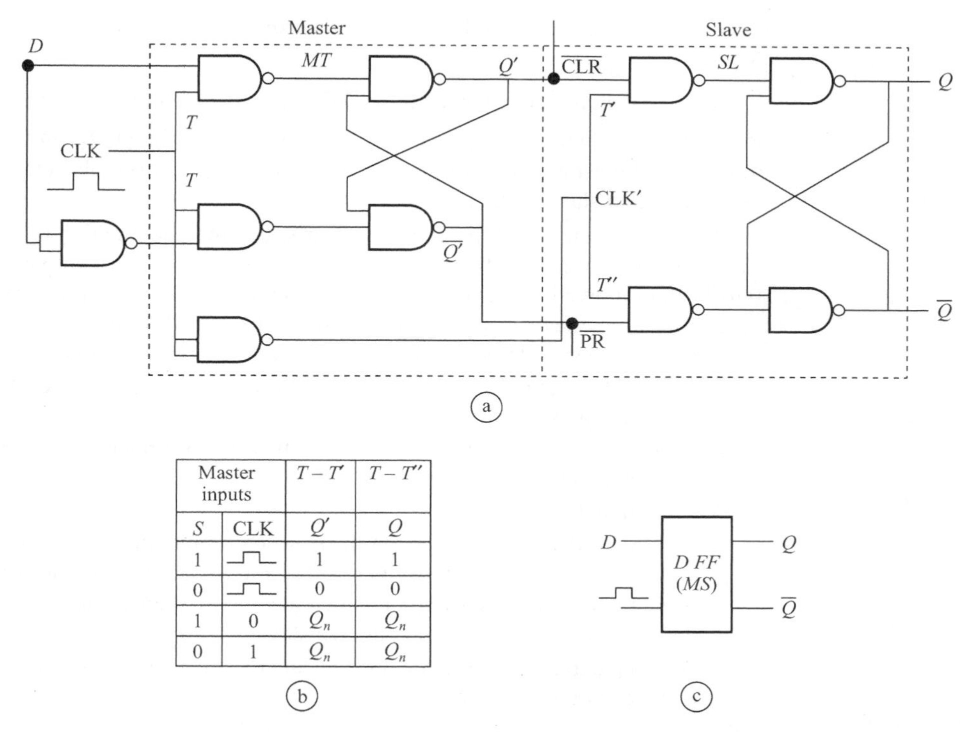

Master inputs		$T-T'$	$T-T''$
S	CLK	Q'	Q
1	⎍	1	1
0	⎍	0	0
1	0	Q_n	Q_n
0	1	Q_n	Q_n

FIGURE 14.16 (a) Master-Slave *D* flip flop logic diagram (b) State table for master slave *D* flip flop (c) Logic symbol of *D-FF*.

Master Section Response

Let us now assume that clock input becomes at logic state 1. Master outputs *Q*' and $\overline{Q}$' will now change according to the *D* input. This is analogous to action of a transparent latch in Figure 14.4(a). If $D = 0$, then the output according to table in Figure 14.4(b) shall be 0. In other words, *Q*' and $\overline{Q}$', the master section outputs, shall be 0 and 1, respectively.

Here, we *D* input [in the master section of Figure 14.16(a) is equivalent to the *S* input in Figure 14.4(a), and inverted *D* input in this *MT* section is equivalent to the *R* input in Figure 14.4(a)].

The *Q* and $\overline{Q}$, the outputs of box marked *SL* (called slave) also remain unchanged as *T*' CLK input to *SL* section is 0 due to NOT before CLK'.

We find that the first gated NAND flip flop section and the box marked *ML* is providing the outputs *Q*' and $\overline{Q}$' according to *D* (and internally created $\overline{D}$), when CLK becomes 1. This is also the reason that the *ML* section is called master section. Slave is inactive during the transparency of *ML* because *SL* has *T*' CLK input = 0. When master finishes the action, then only slave can act.

CLK becomes 0 – Slave Section Response

When *Q*' is 0 and $\overline{Q}$' is 1, this means $\overline{\text{CLR}}$ is 0 and $\overline{\text{PR}}$ is 1. Therefore, according to second row of table in Figure 14.16(b), the outputs $Q = 0$ and $\overline{Q} = 1$. Had *D* been 1, the outputs, upon clock going to 1 would have been according to first row of the table as $Q = 1$ and $\overline{Q} = 0$.

We find that the second gated NAND flip flop section and the box marked *SL* is simply providing the outputs Q and $\overline{Q}$ on CLK becoming 0 according to Q' and $\overline{Q}'$. The latter are the outputs of master section when CLK logic state was 1. This is also the reason that the *SL* section is called slave section.

The present circuit *D* flip flop action arises of the combination of master and slave. There is no possibility of a false transition at $(n+1)^{\text{th}}$ state on the positive going clock pulse (0 to 1) and followed negative going (1 to 0) in a *D MS FF*. After positive going from 0 to level 1 transition, the master is triggered and the slave is idle. After the negative going from 1 to 0, the slave responds as per master outputs and the master is idle.

1. When $D = 1$ the $Q = 1$ after a delay following the application of the positive going the clock pulse.
2. When $D = 0$ the $Q = 0$ 1 after a delay following the application of the positive going the clock pulse.

We also note as follows:

1. *D MS FF* delay = Propagation delay + setup time and hold time and propagation delay = pulse duration between positive and negative going transitions.
2. Using master slave we implement equivalent of *D* edge triggered *FF* by using two latches, one as a master and one as a slave.
3. When the triggering is at the positive and negative edges both, it is called pulse triggering or pulse clocking.

Example 14.22 What is the advantage of edge triggering in the *FF*s the over pulse triggering *FF*s?

Solution

When there is pulse clocking mechanism, the important condition is inputs must be held constant for a time, which is longer than the clock duration between ↑ (0 to 1 transition) and ↓ (1 to 0 transition) or between ↓ and ↑ transitions. This condition puts the pulse-clocking mode of clock input (for changing the outputs as per other inputs) at a disadvantage.

Edge triggering mode obviates this disadvantage. The *FF* accepts the input data and transfers as per state table to the Q and $\overline{Q}$ outputs either upon ↑ as in Figure 14.11(a) at the CLK input or upon ↓ as in Figure 14.11(b). The circuits of Figure 14.6(a) and Figure 14.7(a) are the examples of edge triggering.

The *FF* responds to either rising edge at the CLK input or at the falling edge at the CLK input. The *FF* response is quite fast in the edge-triggering mode. The disadvantage of edge triggering that the necessity of a sharp rise or fall edge generating circuit. [Use of a *RC* pair or a Schmitt trigger circuit at the CLK input. The later is more refined circuit than *C* and *R* based circuit shown in Figures 14.6(a) and 14.7(a). Nowadays the superior circuits have become available for fast edge-triggering clock generation.]

Example 14.23 Show the timing diagram for a *D- MS FF* with Preset and Clear inputs.

Solution

Figure 14.17 shows the timing diagram of a *D MS FF* and having state table as per Figure 14.16(b).

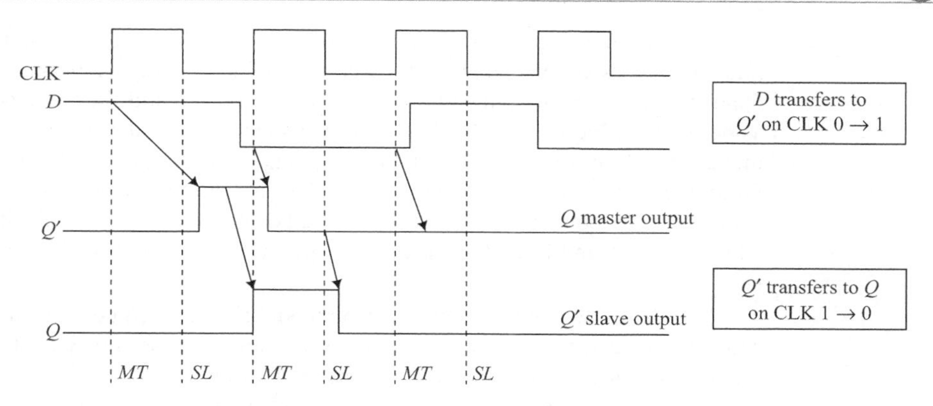

FIGURE 14.17 Timing diagram of a *D-MS FF* having state table as per Figure 14.16(b) (Q' is as per D at $0 \rightarrow 1$ and Q is as per Q' at $1 \rightarrow 0$).

Example 14.24 How will we use an *S-R* latch to Design a Bounce less switch?

Solution

When a switch is pressed or released, there are ups and downs for a short period in the switched current or voltage. Effect of the bounces on the switched voltage appears like one shown in Figure 14.18(a). This effect arises due to the spring actions and reactions inside the switches. An *SR* latch based debouncing circuit is shown in Figure 14.18(b).

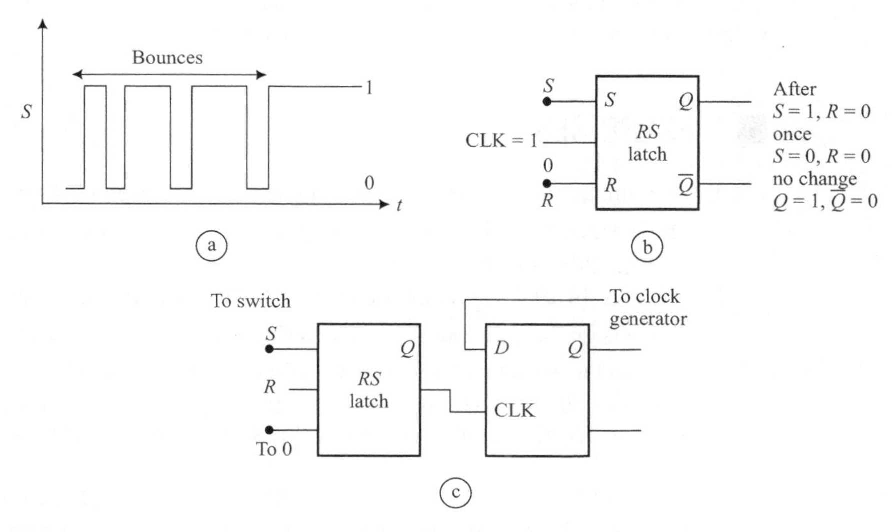

FIGURE 14.18 (a) Effect of the bounces on the switched voltage through after pressing a switch due to the effects of the spring actions and reactions inside the switches. (b) An *SR* latch based debouncing circuit. (c) A clock start and stop circuit using *SR* latch and *D* latch.

Let the state at Q is idle (rest or steady) state of the switch when a switch is in the middle. A cross-coupled gated NAND latch at left side of Figure 14.4(a) is now useful for the debouncing action. In the idle state, the both inputs, R and S are 1, and the output, Q, is as per table in Figure 14.4(b). The Q is unchanged (stable at 1 or 0). When the switch is moved up side at first time contact, the R input is 0 while S input is at 1 (being connected to far of down end of the switch to the supply V_{CC} end. For $R = 0$ and $S = 1$, the Q becomes 1. When switch is bouncing that making repeated contacts for a short while. It means S repeatedly becomes 0 and 1, the Q remains 1. Finally when the switch settles towards up, Q is set to 1.

Now let us consider, situation when the switch is moved downwards to make at first contact $R = 1$ and $S = 0$. Therefore, as per state table of cross-coupled gated NANDs based *SR* latch, Q resets to become 0 (refer fourth row of Figure 14.4(b).

Advantages of the gated cross-coupled NANDs based debouncing:

1. Use of *SR* latch obviates the need of placement of the external components like a capacitor. **2.** Further the latch acts fast. Only first contact will produce desired action in a time just equal to the set up time for $S(t_s)$ plus the propagation delay time (t_p) of the latch.

Example 14.25

How will we design the clock input start and stop circuit using an *SR* latch?

Solution

Refer to the Figure 14.18(a) circuit. A debouncer circuit can also be used to start and stop the clock pulses from a clock generator (sequentially bistable circuit). The generator may be designed using an IC 555 or 7555 or another. The generator pulses its CLK output to a *D FF*. The Q output of *SR* latch is CLK input to a *D*-latch. The output of *SR* latch is the desired output controlled by S. Figure 14.18(c) shows a circuit.

EXERCISES

1. The timing plots of S and R inputs are as shown in Figure 14.19. Copy these on a graph paper. Now show the plots of Q and $\overline{Q}$ from a circuit of figure 14.2(a) on the same graphs and same X-axis.
2. If Figure 14.19 R and S inputs connect a NOR based *SR* and S input is to that NOR, which is giving Q output. Draw the timing diagram for R, S, Q and $\overline{Q}$.
3. Explain the working of an *S-R* latch with a gate (clock) input.
4. Figure 14.20 shows a timing diagram with four plots for A, B, CLK and Q, Find what are A and B? (i) R and S (ii) S and R (iii) S and R. (iv) D and D's complement (v) J and K.
5. Figure 14.21 shows a timing diagram with four plots for A, B, CLK and Q. Name the flip-flop or latch, which you will use to obtain this timing diagram.
6. A *D*-latch with propagation delay = 10 ns has its $\overline{Q}$ output connected to *D*-input. Show that pulses of 20 ns will start appearing as long as clock input remains 1.
7. A *D*-latch with propagation delay = 10 ns has its $\overline{Q}$ output connected to *D*-input. A clock input connects to the pulses 1 for first 75 ns and 0 for next 25 ns, again 1 for

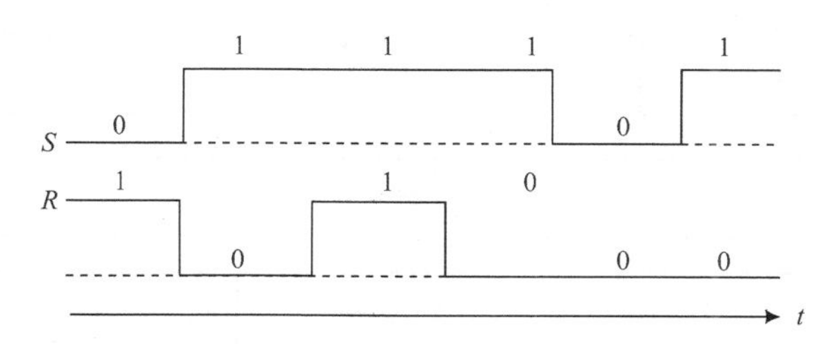

FIGURE 14.19 The timing plots of S and R inputs.

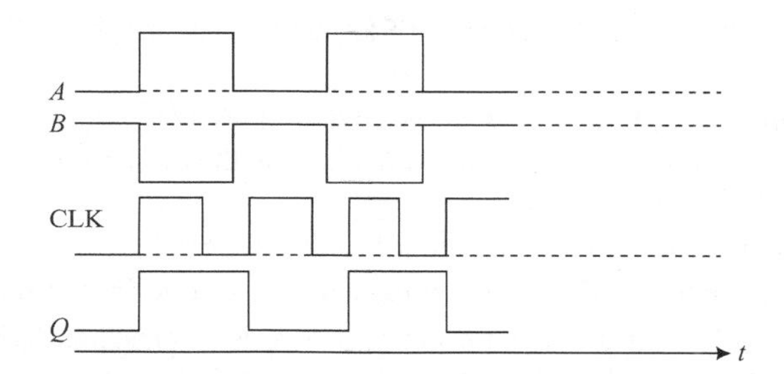

FIGURE 14.20 Timing diagram with four plots for A, B, CLK and Q to find what are the A and B.

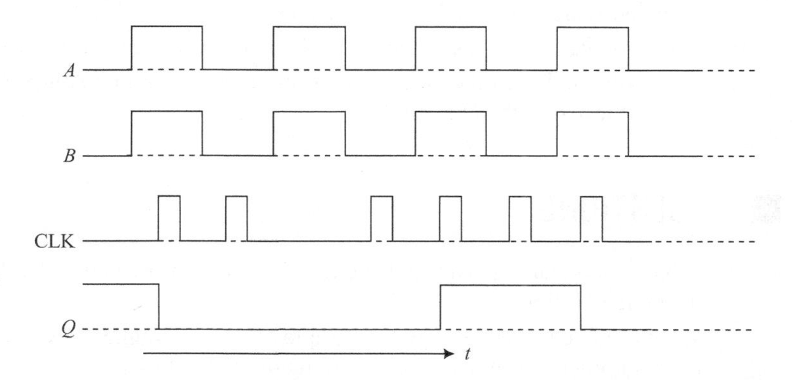

FIGURE 14.21 Timing diagram with four plots for A, B, CLK and Q to find what is the name of the FF or latch.

next 75 ns and 0 for 25 ns. Show the output Q in the total period 0 to 200 ns on a graph paper. (Hint: During first 75 ns, from 10ns onwards, the Q will toggle every 10 ns).

8. Figure 14.22 shows a timing diagram for the preset $\overline{\text{PR}}$, clear $\overline{\text{CLR}}$ and T inputs of a T FF. Draw the time plots for the Q and $\overline{Q}$.
9. Why do you use a D latch if the instance at which input will be arriving is not known exactly, only the interval during which it will arrive is known?
10. Why do you use D FF, when input availability is certain and it is to be registered at the output?

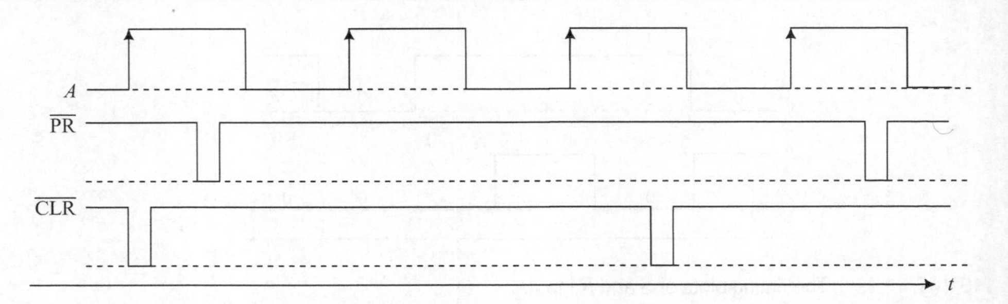

FIGURE 14.22 Timing diagram for the Preset, clear and *T* inputs of a *TFF* for obtaining the time plot for the *Q* and $\overline{Q}$.

11. Show the timing diagram for a negative edge triggered *D FF*.
12. Draw a logic circuit for a master slave *JK* flip flop, the master section of which triggers by a negative going pulse.
13. How will we use the SR latch sub unit of a *MS-JK FF*?
11. How will we use the *JK FF* for designing a negative edge triggered *T FF*?
12. Setup time in a *FF* with propagation delay of 20 ns is 5 ns and hold time is 5 ns. What is the minimum period for which input should be stable?
13. Show the timing diagram for a *D MS FF* with preset and clear inputs with negative going pulse needed for triggering the master.
14. How will we use a NOR based *SR* latch to design a bounce less switch?
15. How will we design the clock input start and stop circuit using a *SR* latch? (Hint: Convert it first to the *D*-latch).

QUESTIONS

1. What is the meaning of a flip flop circuit? What is the meaning of a flip? What is the meaning of a flop?
2. What is a latch? Explain it with example of a cross coupled NANDs? What will be the resulting state table if cross coupling of NORs done?
3. What is difference between the latch, flip flop and master slave flip flop?
4. Why is a flip flop also called a bistable?
5. How does a *SR* latch differ from a gated *RS* latch?
6. How is a *JK FF* freed from unstable condition found in an *SR* latch, whether gated or simple?
7. What is the use of the preset and clear inputs in a *D FF* or a *D* latch or a *JK FF*?
8. What are the differences in a Master Slave *JK FF*, a +ve edge triggered *JK FF* and a –ve edge triggered *JK FF*?
9. Why does a *D* flip flop give a storage register unit and a *T* type flip flop give a basic counting unit?
10. A *JK FF* is with *J* and *K* inputs not provided. When can it work as equivalent of a *T FF*?

11. What is the difference between a *D* Latch and a *D FF*? Explain uses of each.
12. What are the differences between *D*, *T* and *JK FF*?
13. Explain master slave circuit in a *JK FF*?
14. Normally, $\overline{Q}$ is complement of Q. When is these not so? What is that state and in which cases, this condition encountered?
15. What do we mean by a state table in a *FF* or latch? How does it differ for a simple (without output feeding back an input) logic circuits' truth table? It is called state table. Why?
16. What is the meaning of a small circle at the (a) clock input, (b) *PR* input, and (c) CLR input?
17. What is meaning of a triangle at a clock input?
18. When is a circle followed by a triangle at a clock input?
19. What is the meaning of a transparent latch? What is transparent in a transparent latch?
20. Q output of *D FF* or *D* latch is same as *D* input. Then, why is the *D FF* more useful?
21. Explain, when does a *JK FF* act as a divide by 2 circuit?
22. Explain, how does a *T* flip flop act as a divide by 2 circuit?
23. What is the meaning of bounce? What is the type of bouncing signal from a switch noticed in the digital circuits? How is its effect removed from the switched voltage or current?
24. Why is an *SR* latch used for debouncing and it is a better option than an Schmitt trigger circuit based denouncing?
25. How is an *SR* latch after a switch used to start and stop propagation of the digital signals?

CHAPTER 15

Sequential Circuits Analysis, State Minimization, State Assignment and Circuit Implementation

OBJECTIVE

We learnt in Chapter 14 the following concepts:

1. A sequential circuit is a circuit made up by combining the logic gates such that the required logic at the output(s) depends not only on the current input logic conditions but also on the past inputs, outputs and sequences.
2. A sequential circuit has the memory elements like flip flops and a combinational circuit(s) has no memory elements.
3. A clock input (or a set of inputs is used) to cause a transition to next state. The clock signal can be a gate (control) input 1 or control input 0 or a positive edge state (↑) or at the clock state or a negative edge (↓) or a positive going pulse (⎍) or a negative going pulse (⊔⊓).

We have also learnt in Chapter 14 the *SR*, *D*, *JK* and *T* flip flops as the basic memory elements. A flip-flop *next state* is expressed by one of the following expressions:

SR FF: $Q_{n+1} = S + \overline{R} \cdot Q_n$

D FF: $Q_{n+1} = D$

JK FF: $Q_{n+1} = J \cdot \overline{Q}_n + \overline{K} \cdot Q_n$

T FF: $Q_{n+1} = \overline{T} \cdot Q_n + T \cdot Q_n = T.\text{XOR} \cdot Q_n$

Here, Q_{n+1} is the *next state* after the $(n + 1)^{\text{th}}$ transition and Q_n is present state.

There can be in general a sequential circuit consisting of a complex circuit, which combines the combinational circuit(s) and the memory section. We shall learn in this chapter the analysis of the clocked sequential circuits—their design state minimization, state assignment and circuit implementation.

15.1 GENERAL SEQUENTIAL CIRCUIT WITH A MEMORY SECTION AND COMBINATIONAL CIRCUITS AT THE INPUT AND OUTPUT STAGES

A general sequential circuit network has memory section and combination circuits at the memory inputs and outputs. Assume the followings structure of a general sequential circuit.

1. The circuit memory section consists of the m of flip flops. These have a set **Q** with the m present state outputs $Q0$, $Q1$... and $Qm - 1$.
2. There is a set **X** of i inputs $X0$, $X1$, ... Xi. These are applied to a combinational circuit, the j outputs of which are the inputs to the memory section.
3. The present state of **Q** changes on the clock transition(s) to next state **Q′**. The transition is by clock 1 or 0 or ↑ or ↓ or ⎍ or ⊔.
4. Further, a set **Y** with outputs $Y0$, $Y1$, Yj is obtained from the network as per **Q** (or **Q** and **X**) using a combinational circuit at the output stages.

Figures 15.1(a) and (b) show two general sequential circuits with above features, one for Y as per **Q** only and other **Y** as per **Q** and **X**, respectively. The outputs are also called Moore machine and Mealy machine outputs, respectively. *A sequential circuit is like a machine producing the states.*

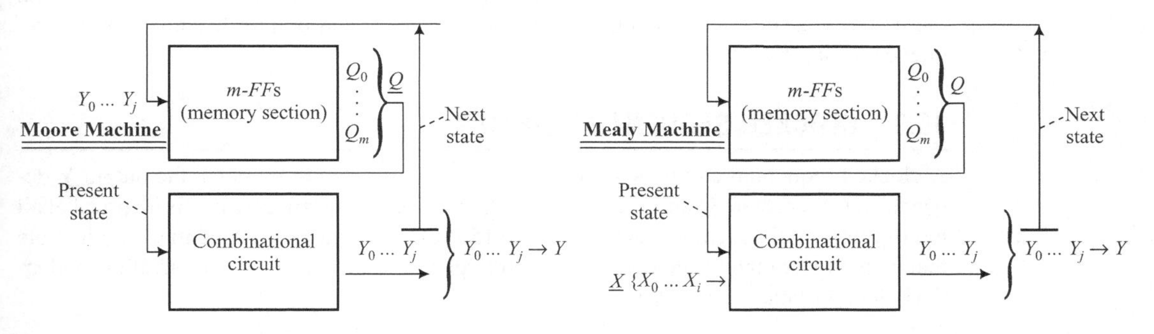

FIGURE 15.1 (a) A general sequential circuit-1 in which the final stage combinational circuit output depends on the **Q′** (Moore Machine Sequential Circuit) (b) A general sequential circuit-2 in which the final stage combinational circuit output depends on the **Q′** *and* **X** (Mealy Machine Sequential Circuit).

15.2 SYNCHRONOUS AND ASYNCHRONOUS SEQUENTIAL CIRCUITS

15.2.1 Synchronous Sequential Circuit

Synchronous sequential circuit is a circuit in which the output **Y** depends on present state **Q** and present inputs **X** at the clocked instances only. These instances can be defined by a clock input ↑ or ↓ or ⎍ or ⊔ and memory section activates to give next state **Q′**. Figure 15.2(a) shows synchronous sequential circuit in there the memory section undergoing change of state at the discrete clock instances c_i, $c_{i'}$, ... at the FFs.

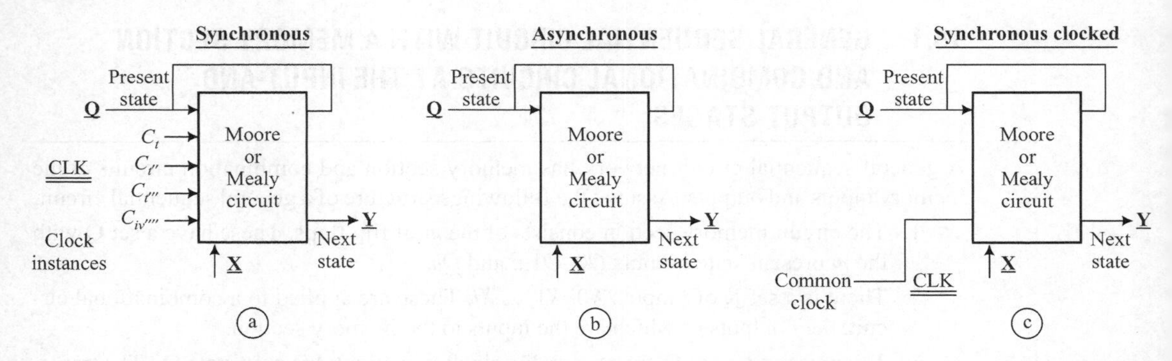

FIGURE 15.2 (a) Synchronous sequential circuit in which the memory section undergoes change of state only at the discrete clock instances (b) Asynchronous sequential circuit in has the memory state changes at undefined instances and output is dependent on sequences of the inputs (c) Synchronous Clocked sequential circuit [Note: Memory section with the identical clock instances at each flip-flop].

15.2.2 Asynchronous Sequential Circuits

Asynchronous sequential circuit is a circuit in which not only the present inputs **X** and present state **Q** but also the sequences of changes affect the output **Y**. The changes can be at the undefined instances of time. Figure 15.2(b) shows asynchronous sequential circuit with the memory section with the undefined clock instances and output dependent on the sequences of inputs.

15.3 CLOCKED SEQUENTIAL CIRCUIT

A clocked sequential circuit is a synchronous sequential circuit in which the output **Y** depends on present state **Q** and present inputs **X** at a clocked instance, which is identical at all the flip flops at the memory section. Figure 15.2(c) shows clocked sequential circuit. Note that in the figure that the memory section undergoes change to next state input after an identical clock instance at each flip flop.

15.4 CLASSIFICATION OF SEQUENTIAL CIRCUIT AS MOORE AND MEALY STATE MACHINE CIRCUITS

15.4.1 Classification of a Sequential Circuit as Moore Model Circuit

Recall the sequential circuit models in Figure 15.1(a). A general clocked sequential circuit in which **Y** final stage combinational circuit output is as per **Q** only is called a circuit implementing a Moore machine and is Moore model of the sequential circuit.

Moore Model

$\mathbf{Q}' = F_Q(\mathbf{Q})$ (Next state outputs **Q′** are the function of *past state* **Q** as *the present inputs* at the clocking instance) and $\mathbf{Y} = F_Y(\mathbf{Q})$ (**Y** is a function of present state outputs **Q** before the clocking instance).

Figure 15.3(a) shows using two D flip flops a clocked sequential circuit-3 in which the final stage combinational circuit output **Y** depends on the **Q** as per Moore model. Following are the inputs, present states, outputs, input functions and output function of circuit-3:

Output Y is taken from AND gate having inputs from the between Q_1 and Q_2 of the *FF*s.

Input D_2 is from an AND gate having inputs X and Q_1. Input D_1 is from the AND-OR array implementing $(X.\overline{Q}_2.\overline{Q}_1 + \overline{X}.\overline{Q}_2.Q_1)$.

15.4.2 Classification of a Sequential Circuit as Mealy Model Circuit

Recall the sequential circuit models in Figure 15.1(b). A general sequential circuit in which the final stage outputs **Y** as per **Q** and **X** is called a circuit implementing a Mealy machine and is Mealy model of the sequential circuit.

Mealy Model

Q′ = F_Q (**X**, **Q**) (Next state outputs **Q**′ are the function of *past state* **Q** and *present inputs* **X** at the clocking instance) and **Y** = F_Y (**Q**, **X**) (**Y** is a function of present state outputs **Q** before the clocking instance).

Figures 15.3(b) shows using one D and one JK flip flops an exemplary clocked sequential circuit 4 in which the final stage combinational circuit output **Y** depends on the **Q** and **X** as per Mealy model. Following are the inputs, present states, outputs, input functions and output function of circuit 3

Output $Y = \overline{X}.Q2 + Q1$.

Input $J = Q_1$ and $K = X$. Input at $D = (\overline{Q}_1 + \overline{Q}_2).X$.

Note

Any sequential circuit for the **Y** can be designed as Mealy machine can be converted to a Moore machine by using appropriate set of (**X**, **Q**, $\mathbf{F_Q}$, $\mathbf{F_Y}$). Similarly any sequential circuit for the **Y** designed as a Moore machine can be converted to a Mealy machine by using appropriate set of (**X**, **Q**, $\mathbf{F_Q}$, $\mathbf{F_Y}$).

15.5 ANALYSIS PROCEDURE

An analysis is important for implementing a sequential circuit. Analysis also provides a tabular description of the circuit. The methods for analysis are as follows:

1. Draw a logic circuit diagram.
2. Perform state variables assignments and excitation (means *FF* inputs) variables assignments.
3. Find the expressions for the excitations from the flip flop characteristics equations as per the excitations and make an excitation table. In other words, find **Q’** = F_Q (**X**, **Q**) and **Y**.

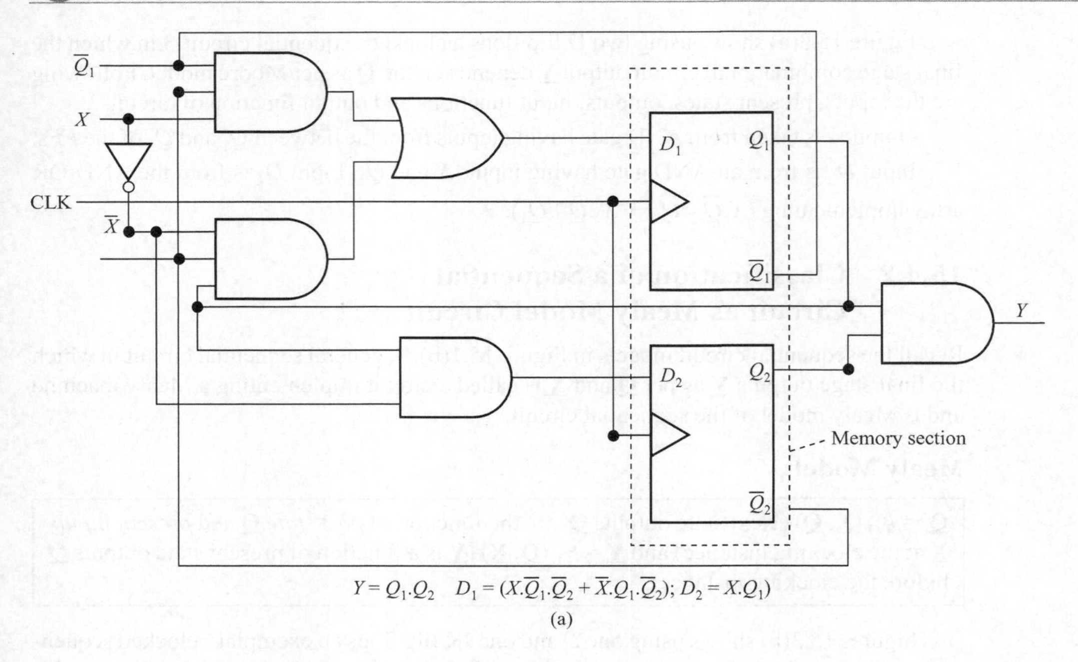

$Y = Q_1.Q_2 \quad D_1 = (X.\overline{Q}_1.\overline{Q}_2 + \overline{X}.Q_1.\overline{Q}_2);\ D_2 = X.Q_1)$

(a)

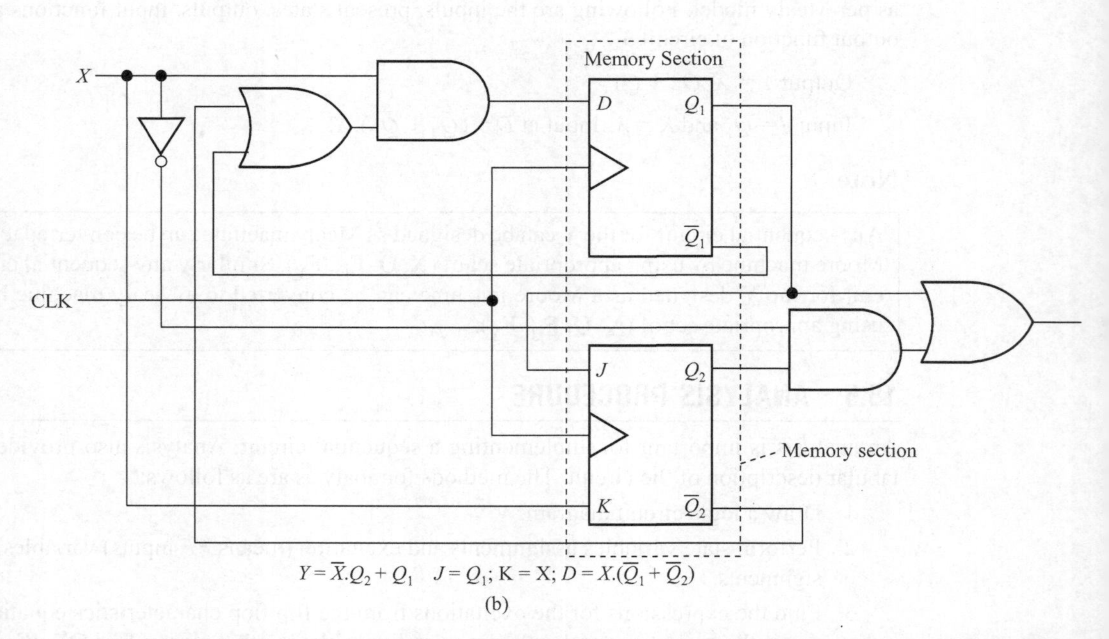

$Y = \overline{X}.Q_2 + Q_1 \quad J = Q_1;\ K = X;\ D = X.(\overline{Q}_1 + \overline{Q}_2)$

(b)

FIGURE 15.3 (a) Using two *D* flip-flops a clocked sequential circuit-3 in which the final stage combinational circuit output depends on the **Q** as per Moore model (b) Using two *JK* flip-flops an exemplary clocked sequential circuit-4 in which the final stage combinational circuit output depends on the **Q** and **X** as per Mealy model.

4. Make a transition table from the expressions for $\underline{\mathbf{Y}} = F_o$ ($\underline{\mathbf{X}}$, $\underline{\mathbf{Q}}$) in case of Mealy model and $\underline{\mathbf{Y}} = F_o$ ($\underline{\mathbf{Q}}$) in case of Moore model.
5. Perform state minimization and make minimal state table.
6. Draw the state diagram.

These steps are explained by taking examples of the circuit-3 and circuit-4 shown in Figures 15.3(a) and 15.3(b), respectively. *Examples 15.3 and 15.4 will explain the procedure.*

Definitions for the different terms used during the analysis of the circuit or during the design and synthesis for implementing the circuit mentioned above are as follows:

15.5.1 Excitation Table

An excitation table is a tabular representation of $\underline{\mathbf{X}}$ and $\underline{\mathbf{Q}}$ at the FFs and of $\underline{\mathbf{Y}}$ as per F_o for the combination circuit at the output stages. It gives present states and the inputs given at the memory section. It also gives the outputs that follow the excitations at the memory section.

Number of rows in each column equals 2^m where m is the number of flip flops because each flip flop has one Q output and m flip-flops can have 2^m different combinations of the states at the Qs. For example, if (Q_1, Q_2) are the Qs of two *FF*s, then $(Q_1, Q_2) = (0, 0), (0, 1), (1, 0)$ and $(1, 1)$ are the four combinations possible for the four different states of the memory-section present outputs.

First column of excitation table gives a present state ($Q1$, $Q2$) in its each row.

The number of columns for the excitation inputs $\underline{\mathbf{Q}}$ equals the number of possible combinations of external inputs in the set $\underline{\mathbf{X}}$. It equals 2^i if there are i distinct literal to represent the inputs when there are i inputs $X_1, X_2, \ldots X_i$. For example, if (X_1, X_2) are the external input to the memory section then $(X_1, X_2) = (0, 0), (0, 1), (1, 0)$ and $(1, 1)$ are the four combinations possible for the four different states of the memory section present outputs. For each set of inputs, there is a set of excitation inputs to the memory section, for example, corresponding to each set of external inputs, there will be four sets of inputs to (D_1, D_2) in case of two D *FF*s at the memory section.

Mealy Model

The number of columns for the output $\underline{\mathbf{Y}}$ also equals the number of possible 2^i combinations of external inputs in the set $\underline{\mathbf{X}}$. Suppose output stage has two outputs, Y_1 and Y_2. Then there will be four columns for the case of four sets of external inputs and each column having entries for pair of outputs of (Y_1, Y_2).

Moore Model

The number of column = 1 for the output $\underline{\mathbf{Y}}$ as in Moore model Y depends on Qs only. The column entries for values of (Y_1, Y_2) as per the combinational circuit between the memory section output Qs and Ys.

15.5.2 Transition Table

A transition table is a tabular representation of F_Q and F_o. It shows how the sequential circuit *FF*s will respond to all the present inputs Xs and Qs and will generate Ys from the Qs.

Number of rows in each column equals 2^m for the m-*FF*s.

First column of transition table gives a present state ($Q1$, $Q2$) in its each row for the case of $m = 2$. This is because there are four combinations possible for the four different states (of Qs) of the memory section present outputs.

The number of columns for the next state **Q'** equals the number of possible combinations of external inputs in the set **X**. It equals 2^i. For each set of inputs, there is a set of memory section outputs after the transition at the memory section, for example, corresponding to each set of external inputs, there will be a set of next state outputs ($Q1'$, $Q2'$) in case of two *FF*s at the memory section.

Mealy Model

The number of columns for the output **Y** also equals the number of possible 2^i combinations of external inputs in the set **X**.

Moore Model

The number of column = 1 for the output **Y** as in Moore model Y depends on Qs only.

15.2.3 State Table

A state table is a tabular representation of the present state in column 1 and next state after the transition in a set of succeeding columns. The outputs are shown in next set of columns. Number of rows in each column equals 2^m for a memory section of m flip-flops.

For example, if ($Q1$, $Q2$) are the Qs of two *FF*s, then let us assume the followings:

$S(Q_1, Q_2) = S(0, 0) = S_0$ for (0, 0) values of (Q_1, Q_2),

$S(Q_1, Q_2) = S(0, 1) = S_1$ for (0, 1) values of (Q_1, Q_2),

$S(Q_1, Q_2) = S(1, 0) = S_2$ for (1, 0) values of (Q_1, Q_2), and

$S(Q_1, Q_2) = S(1, 1) = S_3$ for (1, 1) values of (Q_1, Q_2).

A state in a row on first column is written as either S_0, S_1, S_2 or S_3 for the row 1, row 2, row 3 or row 4, respectively. These are as per four possible values of (Q_1, Q_2) in case of two *FF*s used at the memory section.

The number of columns for the next state **S** equals the number of possible combinations of external inputs in the set **X**. It equals 2^i. For each set of inputs, there is a set of memory section states resulting after the transition at the memory section, for example, corresponding to each set of external inputs, there will be four sets of next states. A state in a row is either S_0 or S_1, S_2 or S_3 as per the post transition values of next state (Q_1', Q_2') in case of two *FF*s used at the memory section.

Moore Model

The number of columns for the output **Y** also equals the number of possible 2^i combinations of external inputs in the set **X**.

Moore Model

The number of column = 1 for the output **Y** as in Moore model Y depends on Qs only.

15.5.4 State Diagram

A state diagram is diagrammatic representation of the state table. A set of present Q_0, Q_1, ... is denoted by a state. There are z (= 2^m) maximum possible states S_0, S_1, ..., S_{z-1} in a sequential circuit with m-*FF*s.

1. The number of nodes = number of rows in the state table.
2. For two flip-flops, there are four states S_0, S_1, S_2 and S_3. So four circles are drawn for the four nodes of a graph.
3. A directed arc from the present state node to the next state node shows a transition.
4. A small diameter circular directed arc marks a transition in which the state remains unchanged.
5. The number of directed arcs equals the number of transitions in which the state changes.
6. The number of directed circular arcs equals the number of transitions in which the state does not change.

Mealy Model

1. States S_0, S_1, S_2 and S_3 are then labeled at the centers of each circle representing a node.
2. Each arc or circular arc is labeled with present input and the output after the transition.
3. Each arc or circular arc can have more then one set of (pre-transition input/post transition output) labeled on it if there are more than one sets of (pre-transition input/post transition output) that are having the same transition from a node to another.

Moore Model

1. States S_0, S_1, S_2 and S_3 as per the present output are then labeled as (state name or representation/present output) at the centers of each circle representing a node.
2. Each arc or circular arc is labeled with present input (from Qs).
3. Each arc or circular arc can have more then one set of (pre-transition input) labeled on it if there are more than one sets of (pre-transition input) that are having the same transition from a node to another.

15.6 CONDITIONS OF STATES EQUIVALENCY

State table may have equivalent pair of states.

Two states S_i and S_j are equivalent in a synchronized clocked sequential circuit, if both the following conditions are fulfilled:

1. The present outputs at S_i and S_j are identical for all possible combinations of the input variables X, and
2. The next states of S_i and S_j after the transitions are also identical for all possible combinations of the input variables X.

15.6.1 State Reduction and Minimization Procedure

State minimization is done by state reduction after finding the pair of equivalent states and then the equivalent classes of the states. The two sequential circuits after state reduction or minimization are said to be equivalent when producing an identical output sequences for the possible input sequences.

15.6.1.1 Equivalency by State Table Inspection

One method of finding equivalency is by finding, what the states, which have the identical present outputs for each possible combination of X and identical next states. For example S_i and S_j has same outputs for all X and have same set of next states for all X, then replace S_j by S_i in the state table. Further continue this process, till no further reduction is feasible.

Consider a state table in Table 15.1.

TABLE 15.1 State table for a Moore model sequential circuit–*i*

Present State		Next State after Transition (Q_1', Q_2', Q_3')				Present Output Y
State	(Q_1, Q_2, Q3)	Input X_1, X_2 = (0, 0)	Input X_1, X_2 = (0, 1)	Input X_1, X_2 = (0, 1)	Input X_1, X_2 = (0, 1)	
S_1	0, 0, 0	S_4	S_3	S_2	S_4	0
S_2	0, 0, 1	S_2	S_1	S_2	S_1	1
S_3	0, 1, 0	S_2	S_1	S_2	S_1	1
S_4	0, 1, 1	S_4	S_3	S_2	S_4	0
S_5	1, 0, 0	S_4	S_3	S_2	S_4	0
S_6	1, 0, 1	S_2	S_1	S_2	S_1	1
S_7	1, 1, 0	S_3	S_4	S_3	S_4	0
S_8	1, 1, 1	S_1	S_2	S_1	S_2	1

We find that output $Y = 0$ is same for $S1$, $S4$, $S5$ and $S7$. This fulfills condition 1 of equivalency. However, S_1 and S_7 have different next states for inputs (0, 0), (0, 1). Hence condition 2 is not fulfilled and these two are not equivalent. However, S_1, S_4 and S_5 have same next states for all the four possible combination of the inputs: (0, 0), (0, 1), (1, 0) and (1, 1). Hence S_1, S_4 and S_5 are equivalent. We replace S_4 by S_1 in state table and reconstruct state table as in Table 15.2.

TABLE 15.2 State table for a Moore model sequential circuit – *i*

Present State		Next State after Transition (Q_1', Q_2', Q_3')				Present Output Y
State	(Q_1, Q_2, Q_3)	Input X_1, X_2 = (0, 0)	Input X_1, X_2 = (0, 1)	Input X_1, X_2 = (0, 1)	Input X_1, X_2 = (0, 1)	
S_1, S_4, S_5	(0, 0, 0) (0, 1, 1) and (1, 0, 0)	S_1	S_3	S_2	S_1	0

Table 15.2 Contd.

S_2	0, 0, 1	S_2	S_1	S_2	S_1	1
S_3	0, 1, 0	S_2	S_1	S_2	S_1	1
S_6	1, 0, 1	S_2	S_1	S_3	S_1	1
S_7	1, 1, 0	S_3	S_1	S_3	S_1	0
S_8	1, 1, 1	S_1	S_2	S_1	S_2	1

We find that output $Y = 1$ is same for S_2, S_3, S_6 and S_8. This fulfills condition 1 of equivalency. We however find that S_2, S_3 and S_6 have identical next states for inputs. Now replace S_3 by S_2 in the state table everywhere we get the Table 15.2 and also delete the rows for S_3 and S_8. We reconstruct Table 15.2 as Table 15.3.

TABLE 15.3 State table for a Moore model sequential circuit – *i*

Present State		Next State after Transition (Q_1', Q_2', Q_3')				Present Output Y
State	(Q_1, Q_2, Q_3)	Input X_1, X_2 = (0, 0)	Input X_1, X_2 = (0, 1)	Input X_1, X_2 = (0, 1)	Input X_1, X_2 = (0, 1)	
S_1, S_4, S_5	(0, 0, 0), (0, 1, 1) and (1, 0, 0)	S_1	S_2	S_2	S_1	0
S_2, S_3, S_6	(0, 0, 1), (0, 1, 0) and (1, 0, 1)	S_2	S_1	S_2	S_1	1
S_7	1, 1, 0	S_2	S_1	S_2	S_1	0
S_8	1, 1, 1	S_1	S_2	S_1	S_2	1

None of the rows have Y and next states for the inputs both identical. Therefore, Table 15.3 is a state minimal table. Next section gives another procedure for a systemic determination of equivalency of pairs and solution of state minimization problem.

15.6.1.2 Equivalency Determination and Minimization of States in State Table by a Procedure of Constructing Implication Table

Following are the steps for building an implication table for synthesis of the circuit after finding the reduced number of states and state minimal table.

Step 1: Construct an implication table as shown in Table 15.14 (Example 15.7) for a five state state-table.

Step 2: Mark the state pair *not to be considered* for equivalency determination in a matrix of cell as follows:

A (S_i, S_i) cell is redundant for equivalency determination and an equivalent state pair (S_i, S_j) is same as pair (S_j, S_i). In a matrix of $n \times n$ cells, $(n^2 - n)/2 = 10$ are the off-diagonal right side cells when $n = 5$ are the cells along the diagonal. Hence total 15 cells are not to be taken into consideration for pairing at the implication table for state table for five states. Cells marked by sign ^ mark are not to be considered.

For example, we put ^ sign in 15 cells in the implication table (S_1, S_1), (S_2, S_2), (S_2, S_1), (S_3, S_3), (S_3, S_2), (S_3, S_1), (S_4, S_4), (S_4, S_3), (S_4, S_2), (S_4, S_1), (S_5, S_5), (S_5, S_4), (S_5, S_3), (S_5, S_2) and (S_5, S_1).

Step 3: Also marke and put a # sign at a cell not having the same set of output for all the input combinations.

Step 4: Fill the unmarked cells by the next state values. For example, if (S_2, S_4) cell is unmarked then put entries of the next state pairs for each combination of *X*s in this cell.

First iteration of cells filling is now over. Now next iteration on marking implication tables is as follows and Table 15.15 shows the result in Example 15.7.

Step 5: Discard the cells not generating equivalent next states (not fulfilling condition 2) by following process of further marking of # signs.

Let an entry be (S_l, S_m) in a cell (S_i, S_j). If cell (S_l, S_m) has a # sign previously placed, then put # sign in cell (S_i, S_j) also. Now ignore any other entry in the cell (S_i, S_j), else look at other entry(ies) also in the cell (S_i, S_j).

Continue this process for all those cells not having a # sign by now.

Step 6: Continue process in Step 5 till no more # sign needs to be placed. All the cells not generating equivalent next states (not fulfilling condition 2) are now free from # sign.

From these cells, which has neither # nor ^ sign, build a new state table and build state minimal table by grouping the states that are equivalent after the above state minimization process.

Example 15.7 and 15.8 explain these steps.

15.6.2 Assignment of Variables to a State

Let there are five states in a state diagram for the three *FF*s. Then there are five rows one each for a in a state table. Assuming that one *FF* can undergo only one change of states, we need eight state assignments for a binary representation of a state at the output of three *FF*s. Even if number of states are eight, we need minimum three bit binary number to represent in order to assign a state at the memory section of the given clocked sequential circuit.

We can assign three binary numbers to the five states. However by increasing the number of binary bits to assign in a state table with 5 rows, it may also be possible to actually reduce the total number of gates. Suppose there are three binary bits for each state. There are $(2^{3)}!/\ (2^3 - 5)! = 8 \times 7 \times 6 \times 5 \times 4$ assignment ways to code the given states.

Few guidelines are as follows: (a) Two or more states at present giving the same output **Y** for a given ***X*** should be made adjacent. (b) Make two or more states adjacent in case they have a same next state for a given input set. (c) Make two next states adjacent for a present sate and two input combinations that are adjacent.

For a Moore model sequential clocked circuit, the state assignment can be made same as number of outputs.

Assuming the five rows only in a state table, Figure 15.4 (a) shows a state assignment map for maximum possible eight and figure 15.4(b) for sixteen states.

15.7 IMPLEMENTATION PROCEDURE

Circuit synthesis or implementation and designing steps are opposite to that of analysis (Section 15.5). The procedure for implementation is as follows:

$Q_1 Q_2$ \ Q_3	$\overline{Q}_1$ 0	$\overline{Q}_3$ 1
$\overline{Q}_1 \overline{Q}_2$ 00	S_1	S_3
$\overline{Q}_1 \overline{Q}_2$ 01	S_2	S_4
$\overline{Q}_1 \overline{Q}_2$ 11	S_5	
$\overline{Q}_1 \overline{Q}_2$ 00		

$S_1 = 000$
$S_2 = 010$
$S_3 = 001$
$S_4 = 011$
$S_5 = 110$

$Q_1 Q_2$ \ $Q_3 Q_4$	$\overline{Q}_3 \overline{Q}_4$ 0 0	$\overline{Q}_3 \overline{Q}_4$ 0 1	$Q_3 Q_4$ 0 1	$Q_3 \overline{Q}_4$ 0 1
00	S_1	S_3	S_2	S_4
01	S_5			
11				
11				

$S_1 = 0000$
$S_2 = 0011$
.
.
.
$S_5 = 0100$

FIGURE 15.4 (a) A state assignment map for maximum 8 rows (states) in a state table (b) A state assignment with maximum 16 rows in a state table.

1. Get specifications of the sequential circuit.
2. Derive the sates and draw the state diagram.
3. Make a state table.
4. Perform state minimization so that there are minimum number of rows in the table.
5. Make a transition table and find the state variables number of rows in the table and excitation variables needed.
6. Find the flip flop characteristics expressions, which implement the transitions as per the table.
7. Implement the circuit as per the expressions.

Example 15.9 will explain the circuit implementation.

EXAMPLES

Example 15.1

Define a Mealy Machine for a general sequential circuit. Why can a Mealy machine produce pulsed outputs? Why is it that a Mealy machine can produce false outputs and glitches?

Solution

(1) Mealy machine can be defined as a machine consisting of set of ($\underline{\mathbf{X}}$, $\underline{\mathbf{Q}}$, $\underline{\mathbf{Y}}$, $\underline{\mathbf{F}}_{\mathbf{Q}}$ and $\underline{\mathbf{F}}_{\mathbf{Y}}$), where $\underline{\mathbf{X}}$ is non-empty finite set of inputs, $\underline{\mathbf{Q}}$ is non-empty finite set of states, $\underline{\mathbf{Y}}$ is non-empty finite set of outputs, the $\underline{\mathbf{F}}_{\mathbf{Q}}$ is state transition function changing $\underline{\mathbf{Q}}$ as a function of $\underline{\mathbf{X}}$ and $\underline{\mathbf{Q}}$, and the $\underline{\mathbf{F}}_{\mathbf{Y}}$ is an output function of states $\underline{\mathbf{Q}}$ and inputs $\underline{\mathbf{X}}$ for producing $\underline{\mathbf{Y}}$.

(2) Since $\underline{\mathbf{F}}_{\mathbf{Y}}$ is an output function of states $\underline{\mathbf{Q}}$ and inputs $\underline{\mathbf{X}}$ for producing $\underline{\mathbf{Y}}$ and changes in $\underline{\mathbf{X}}$ causes the changes in $\underline{\mathbf{Q}}$ later to $\underline{\mathbf{Q}}'$, therefore the transition produces a pulsed $\underline{\mathbf{Y}}$ after the next steady state output $\underline{\mathbf{Q}}'$. This is because the time taken in transition at the memory section is different than the time taken at the combinational section for inputs to the memory section.

(3) The pulses are also called glitches for the outputs between two intervals, the transition instance and the excitation instance.

(4) The outputs can be false during the period before the excitation inputs are applied as these can change and are unstable after the $\underline{\mathbf{X}}$ is applied. False outputs can also occur due to propagation delay at the memory section.

Example 15.2 Define a Moore machine for a general sequential circuit. Why can a Moore machine model sequential circuit produce constant steady state outputs?

Solution

Moore machine can be defined as a machine consisting of set of (**X**, **Q**, **Y**, $\mathbf{F_Q}$ and $\mathbf{F_Y}$), where **X** is non-empty finite set of inputs, **Q** is non-empty finite set of state, **Y** is non-empty finite set of outputs, the $\mathbf{F_Q}$ is state transition function changing **Q** as a function of **X** and **Q** but the $\mathbf{F_Y}$ is an output function of states **Q** only for producing **Y**.

Since $\mathbf{F_Y}$ is an output function of present state **Q** only, **Y** is a steady state output as per the current state only.

Example 15.3 Analyze the sequential circuit-3 shown in Figure 15.3(a).

Solution

Let us analyze the circuit-3 in Figure 15.3(a) by the six-step procedure mentioned in Section 15.5 above.

Step 1: Logic circuit drawing:

Logic circuit drawing is as per Figure 15.3(a).

Step 2: Performing state variables assignments and excitation (FF triggering) variables assignments:

State variables in the circuit are Q_2, $\overline{Q}_2$, Q_1 and $\overline{Q}_1$ at the lower and upper *FF*s, respectively.

Excitation variables are D_1 and D_2.

Step 3: Finding the expressions for the excitations from the flip flop characteristics equations as per the excitations and make an excitation table:

Therefore, the excitation expressions for next state *Q*s are as follows:

$$Q_1' = D_1 \quad ...(15.1)$$

$$Q_2' = D_2 \quad ...(15.2)$$

Now, the steps for finding the transition equation are the followings using the input variables for the excitations of the *FF*s. Two expressions for the excitation inputs for the Figure 15.3(a) combinational circuit are as follows:

$$D_1 = (X.\overline{Q}_1.\overline{Q}_2 + \overline{X}.Q_1.\overline{Q}_2) \text{ and} \quad ...(15.3)$$

$$D_2 = X.Q_1 \quad ...(15.4)$$

Therefore the next state after the state transition equations are as per the excitation expression will be given by

$$Q1' = (X.\overline{Q}_1.\overline{Q}_2 + \overline{X}.Q_1.\overline{Q}_2) \text{ and} \quad ...(15.5)$$

$$Q2' = X.Q_1 \quad ...(15.6)$$

The output equation for the Y is as follows:

$$Y = Q_1.Q_2 \quad ...(15.7)$$

As *Y* depends on *Q*s only, the present circuit is a Moore model sequential circuit. Using the equations (15.3) and (15.4) and (15.7), Table 15.4 gives the excitation table for present states, excitation inputs and the present output *Y* [Post state transition Y = $Q'_1.Q'_2$].

TABLE 15.4 Excitation table for the sequential circuit-3

Present state	Excitation inputs (*D*1, *D*2)		Present output state *Y* before the transition
(Q_1, Q_2)	Input $X = 0$	Input $X = 1$	
0, 0	0, 0	1, 0	0
0, 1	0, 0	0, 0	0
1, 0	1, 0	1, 1	0
1, 1	0, 0	0, 1	1

Step 4: Make transition table from the expressions for **Y** = F_o (**Q**) and **Q'**.

Table 15.5 gives the transition table made by using the Equations (15.5) and (15.6) and (15.7).

TABLE 15.5 Transition table for the sequential circuit-3

Present state	Next state after transition (*Q*1', *Q*2')		Present output state *Y* before the transition
(Q_1, Q_2)	Input $X = 0$	Input $X = 1$	
0, 0	0, 0	1, 0	0
0, 1	0, 0	0, 0	0
1, 0	1, 0	1, 1	0
1, 1	0, 0	0, 1	1

It is same as Table 15.6 because a *D FF*s reflects the *D* at *Q*.

Step 5: Perform state minimization and make minimal state table.

There are four possible states (Q_1, Q_2). These are (0, 0), (0, 1), (1, 0) and (1, 1). Let us designate (assign) these as S_0, S_1, S_2 and S_3. We therefore get the state table as follows. Table 15.6 gives the state table made from the Table 15.5.

TABLE 15.6 State table for the sequential circuit-3

Present state		Next state after transition (Q1', Q2')		Present output state Y before the transition
State	(Q_1, Q_2)	Input $X = 0$	Input X = 1	
S_0	0, 0	S_0 [= (0, 0)]	S_2	0
S_1	0, 1	S_0 [= (0, 0)]	S_0	0
S_2	1, 0	S_2 [= (1, 0)]	S_3	0
S_3	1, 1	S_0 [= (1, 0)]	S_1	1

Reason for next state in column 3 as S_0, S_0, S_2 and S_0 when $X = 0$ is also mentioned to clarify to the reader how to make the state table. Normally the reason is not given, just the states are mentioned in the columns showing the state transitions.

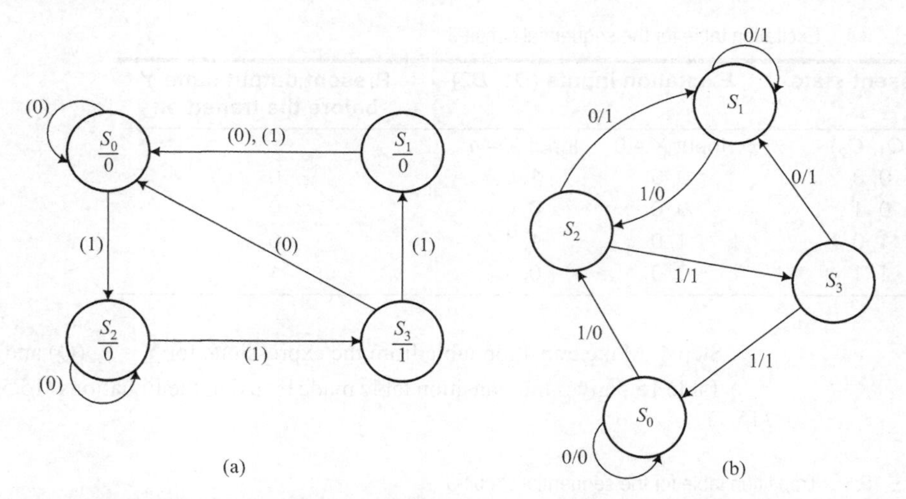

FIGURE 15.5 (a) State diagram for the sequential circuit-3 (b) State diagram for the sequential circuit-4

Step 6: Draw the state diagram

From the state table the state diagram is drawn. Figure 15.5(a) shows the state diagram for the sequential circuit-3. The directed arc represents a transition. It is labeled with present input and the output after the transition. This is clarified in the followings for each state at the rows of the table.

Consider **row 1** of state table in Table 15.6.

1. Transition is from S_0 to S_0 itself when input $X = 0$ and output is $Y = 0$. Node is labeled at the center by $S_0/0$ for the node S_0 in Figure 15.5(a). The arc is labeled 0.
2. Transition is from S_0 to S_2 when input $X = 1$ and output is $Y = 0$. Hence the directed arc in Figure 15.5(a) is from S_0 to S_2 and is also labeled 1.

Consider **row 2** of state table in Table 15.6.

3. Transition is from S_1 to S_0 when input $X = 0$ and output is $Y = 0$. Hence the directed arc in Figure 15.5(a) is from S_1 to S_0 and is labeled 0. Node for state S_1 is labeled at the center by $S_1/0$.
4. Transition is from S_1 to S_0 when input $X = 1$ and output is $Y = 0$. Hence the directed arc in Figure 15.5(a) is from S_1 to S_0 and is also labeled (1).

Since for $X = 0$ and $X = 1$ the directed arcs are identical, we label S_1 to S_0 arc as (0), (1).

Consider **row 3** of state table in Table 15.6.

5. Transition is from S_2 to S_2 itself when input $X = 0$ and output is $Y = 0$. Hence a directed circular arc is shown at the node S_2 in Figure 15.5(a). It is labeled 0. Node for S_2 is labeled at the center by $S_2/0$.
6. Transition is from S_2 to S_3 when input $X = 1$ and output is $Y = 0$. Hence the directed arc in Figure 15.5(a) is from S_2 to S_3 and is also labeled 1.

Consider **row 4** of state table in Table 15.6.

7. Transition is from S_3 to S_0 when input $X = 0$ and output is $Y = 1$. Hence a directed arc is shown from the node S_3 to node S_0 in Figure 15.5(a). It is labeled 0. $S_3/1$ is the label at state S_3 node is placed at the center.

8. Transition is from S_3 to S_1 when input $X = 1$ and output is $Y = 1$. Hence the directed arc in Figure 15.5(a) is from S_3 to S_1 and is also labeled 1. $S_3/1$ is the label at the node center because $Y = 1$ at S_3.

Example 15.4

Analyze the sequential circuit-4 shown in Figure 15.3(b).

Solution

Let us analyze the circuit-4 in Figure 15.3(b) by the six-steps procedure mentioned in Section 15.5 above.

Step 1: Logic circuit drawing:

Logic circuit drawing is as per Figure 15.3(b).

Step 2: Performing state variables assignments and excitation (*FF* triggering) variables assignments:

State variables in the circuit are $Q2$, $\overline{Q}2$, $Q1$ and $\overline{Q}1$ at the lower and upper *FF*s, respectively.

Excitation variables are D, J and K.

Step 3: Finding the expressions for the excitations from the flip flop characteristics equations as per the excitations and make an excitation table:

Therefore the excitation expressions are as follows:

$$Q1' = D \quad ...(15.8)$$

From Equation 14.3 for the *JK FF*, we get the following equation.

$$Q2' = J.\overline{Q}2 + \overline{K}.Q2 \quad ...(15.9)$$

Now, the steps for finding the transition equations are the followings using the input variables for the excitations of the *FF*s. Two expressions for the combinational circuit for the excitation inputs are as follows:

$$D = X.(\overline{Q}1 + \overline{Q}2) \quad ...(15.10)$$

$$J = Q1 \text{ and } K = X \quad ...(15.11)$$

Therefore, the next state after the state transition equations are as per the excitation expression and are given by

$$Q1' = X.(\overline{Q}1 + \overline{Q}2) \text{ and} \quad ...(15.12)$$

$$Q2' = Q1.\overline{Q}2 + \overline{X}.Q2 \quad ...(15.13)$$

The output equation for the Y is as follows:

$$Y = \overline{X}.Q2 + Q1 \quad ...(15.14)$$

Since combinational circuit inputs at the memory section outputs Y is such that the Y depend on X, we have a circuit, which is a Mealy model sequential circuit. Using the equations (15.10) and (15.11) and (15.14), Table 15.7 gives the excitation table for present states, excitation inputs and the present output Y.

TABLE 15.7 Excitation table for the for the sequential circuit-4

Present state	Excitation inputs D, (J, K)		Present output state after the X inputs but before the transition	
(Q_1, Q_2)	Input X = 0	Input X = 1	Input X = 0	Input X = 1
0, 0	0, (0, 0)	1, (0, 1)	0	0
0, 1	0, (0, 0)	1, (0, 1)	1	0
1, 0	0, (1, 0)	1, (1, 1)	1	1
1, 1	0, (1, 0)	0, (1, 1)	1	1

Step 4: Make transition table from the expressions for $Y = F_o(X, Q')$ and Q'.

Table 15.8 gives the transition table made by using the equations (15.12) and (15.13) and (15.14).

TABLE 15.8 Transition table for the sequential circuit-4

Present state	Next state after transition ($Q1'$, $Q2'$)		Output state Y	
(Q_1, Q_2)	Input X = 0	Input X = 1	Input X = 0	Input X = 1
0, 0	0, 0	1, 0	0	0
0, 1	0, 1	1, 0	1	0
1, 0	0, 1	1, 1	1	1
1, 1	0, 1	0, 0	1	1

Step 5: Perform state minimization and make minimal state table.

There are four possible states (Q_1, Q_2): These are (0, 0), (0, 1), (1, 0) and (1, 1). Let us designate (assign) these as S_0, S_1, S_2 and S_3. We therefore make the state table. Table 15.9 gives the state table made from the Table 15.8.

TABLE 15.9 State table for the sequential circuit-4

Present state		Next state after transition ($Q1'$, $Q2'$)		Output state Y	
State	(Q_1, Q_2)	Input X = 0	Input X = 1	Input X = 0	Input X = 1
S_0	0, 0	S_0	S_2	0	0
S_1	0, 1	S_1	S_2	1	0
S_2	1, 0	S_1	S_3	1	1
S_3	1, 1	S_1	S_0	1	1

Step 6: Draw the state diagram.

From the state table the state diagram is drawn. Figure 15.5(b) shows the state diagram for the sequential circuit-4.

Since the circuit is a Mealy model sequential circuit, the nodes S_0, S_1, S_2 and S_3 are labeled at the center of a circle. The directed arc represents a transition. It is labeled with present input and the output after the transition. It is clarified in the following for a state present at each row.

Consider **row 1** of state table in Table 15.9.

1. Transition is from S_0 to S_0 itself when input $X = 0$ and output is $Y = 0$. Hence a directed circular arc is shown for the node $S0$ in Figure 15.5(b). It is labeled 0/0.
2. Transition is from S_0 to S_2 when input $X = 1$ and output is $Y = 0$. Hence the directed arc in Figure 15.5(b) is from S_0 to S_2 and is also labeled 1/0.

Consider **row 2** of state table in Table 15.9.

3. Transition is from $S_1$to S_1 itself when input $X = 0$ and output is $Y = 1$. Hence the directed circular arc in Figure 15.5(b) is shown at the node S_1 and is labeled 0/1.
4. Transition is from S_1 to S_2 when input $X = 1$ and output is $Y = 0$. Hence the directed arc in Figure 15.5(b) is from S_1 to S_2 and is also labeled 1/0.

Consider **row 3** of state table in Table 15.9.

5. Transition is from S_2 to S_1 when input $X = 0$ and output is $Y = 1$. Hence a directed arc i+s shown from the node S_2 to S_1 in Figure 15.5(b). It is labeled 0/0.
6. Transition is from S_2 to S_3 when input $X = 1$ and output is $Y = 1$. Hence the directed arc in Figure 15.5(b) is from S_2 to S_3 and is also labeled 1/1.

Consider **row 4** of state table in Table 15.9.

7. Transition is from S_3 to S_1 itself when input $X = 0$ and output is $Y = 1$. Hence a directed arc is shown from the node S_3 to node S_0 in Figure 15.5(b). It is labeled 0/1.
8. Transition is from S_3 to S_0 when input $X = 1$ and output is $Y = 1$. Hence the directed arc in Figure 15.5(b) is from S_3 to S_0 and is also labeled 1/1.

Example 15.5 When are the two states equivalent in a clocked sequential circuit?

Solution

Two states S_i and S_j are equivalent in a synchronized clocked sequential circuit, if both the following conditions are fulfilled:

1. The present outputs at S_i and S_j are identical for all possible combinations of the input variables X.
2. The next states of S_i and S_j after the transitions are also identical for all possible combinations of the input variables X.

Example 15.6 Perform state minimization for the state table (Table 15.10) for a Mealy model sequential circuit-k by inspection procedure.

TABLE 15.10 State table for a Mealy model sequential circuit-k

Present state		Next state after transition (Q1', Q2', Q3')		Present output Y	
State	(Q_1, Q_2, Q_3)	Input $X = 0$	Input $X = 1$	Input $X = 0$	Input $X = 1$
S_1	0, 0, 0	S_1	S_4	0	1
S_2	0, 0, 1	S_2	S_4	0	0
S_3	0, 1, 0	S_4	S_3	0	1

Table 15.10 Contd.

S_4	0, 1, 1	S_1	S_4	0	1
S_5	1, 0, 0	S_2	S_4	0	0
S_6	1, 0, 1	S_2	S_1	1	1
S_7	1, 1, 0	S_1	S_3	1	1
S_8	1, 1, 1	S_2	S_1	0	0

S_1 and S_4 have all values of Q' and therefore, post transition Y_s same for the inputs $X = 0$ and 1. We can replace S_4 by S_1. S_2 and S_5 states have all values of Q' same for all the inputs and have same next states S_2 and S_4. Reduced six rows at table are now as follows:

TABLE 15.11 Reduced state table for a Mealy model sequential circuit-*k*

Present state		Next state after transition (Q1', Q2', Q3')		Present output Y	
State	(Q_1, Q_2, Q3)	Input $X = 0$	Input $X = 1$	Input $X = 0$	Input $X = 1$
S_1, S_4	(0, 0, 0) and (0, 1, 1)	S_1	S_1	0	1
S_2, S_5	(0, 0, 1) and (1, 0, 0)	S_2	S_1	0	0
S_3	0, 1, 0	S_1	S_3	0	1
S_6	1, 0, 1	S_2	S_1	1	1
S_7	1, 1, 0	S_1	S_3	1	1
S_8	1, 1, 1	S_2	S_1	0	0

S_3 and S_7 now have all values of post transition Y same for the inputs $X = 0$ and 1. We can replace S_4 by S_1. Therefore, reduced table of 4 rows is now as follows:

TABLE 15.12 New reduced state table for a Mealy model sequential circuit-*k*

Present state		Next state after transition (Q1', Q2', Q3')		Present output Y	
State	(Q_1, Q_2, Q3)	Input $X = 0$	Input $X = 1$	Input $X = 0$	Input $X = 1$
S_1, S_4	(0, 0, 0) and (0, 1, 1)	S_1	S_1	0	1
S_2, S_5	(0, 0, 1) and (1, 0, 0)	S_2	S_1	0	0
S_3, S_7	(0, 1, 0) and (1, 1, 0)	S_1	S_3	0	1
S_6	1, 0, 1	S_2	S_1	1	1
S_8	1, 1, 1	S_2	S_1	0	0

Table 15.12 is now a minimized state table for the circuit-*k*. Since the numbers of rows = 5, an implementation will still need three flip flops as two flip flops can implement only four states.

Example 15.7 Perform state reduction or minimization of state table in Table 15.13, if feasible.

Solution

Consider the state table for a Mealy model sequential circuit-l in Table 15.13.

TABLE 15.13 State table for a Mealy model sequential circuit-I

Present state	Next state after transition (Q1', Q2', Q3')		Present output *Y*	
	Input $X = 0$	Input $X = 1$	Input $X = 0$	Input $X = 1$
S_1	S_2	S_1	1	1
S_2	S_1	S_3	0	1
S_3	S_4	S_1	1	1
S_4	S_4	S_5	0	1
S_5	S_3	S_4	0	0

Step 1: Construct an implication table as shown in Table 15.14 for a five states state table.

Step 2: Marked the state pairs not to be considered for equivalency determination in matrix of cells.

A pair (S_i, S_i) is redundant for equivalency determination and an equivalent state transion pair (S_i, S_j) is same as pair (S_j, S_i). In a matrix of $n \times n$ cells, $(n^2 - n)/2 = 10$ are the off diagonal right side cells when $n = 5$ are the cells along the diagonal. We put ^ sign at the 15 cells in implication table (S_1, S_1), (S_2, S_2), (S_2, S_1), (S_3, S_3), (S_3, S_2), (S_3, S_1), (S_4, S_4), (S_4, S_3), (S_4, S_2), (S_4, S_1), (S_5, S_5), (S_5, S_4), (S_5, S_3), (S_5, S_2) and (S_5, S_1).

Step 3: Put # sign for a cell not having the same set of *Y* outputs for all the input còmbinations for *X*. Table 15.14 state table shows that for a set of inputs, which are possible, the *Y* outputs are same *Y*(1, 1) only for (S_2 and S_4) states and same *Y*(0, 1) for (S_1 and S_3) states. Except in cells for (S_1, S_3) pair and for (S_2, S_4) pair, we therefore mark remaining cells by # sign.

Step 4: Fill the entries at the unmarked cells by the next-state values from the state table. It means fill (S_2, S_4) pair for $X = 0$ and (S_1, S_1) pair for $X = 1$ in the cell for (S_1, S_3) and fill (S_1, S_4) for $X = 0$ and (S_3, S_5) for $X = 1$ in cell for (S_2, S_4).

TABLE 15.14 Implication table first iteration for a Mealy model sequential circuit-*I*

	S1	S2	S3	S4	S5
S_5	#	#	#	#	^
S_4	#	(S_1, S_4) (S_3, S_5)	#	^	^
S_3	(S_2, S_4) (S_1, S_1)	#	^	^	^
S_2	#	^	^	^	^
S_1	^	^	^	^	^

First iteration of cells filling is now over. Now next iteration on marking at the implication table is as follows and Table 15.15 shows the result.

Step 5: Remove the cell pairs not generating equivalent next states (not fulfilling condition 2).

Consider cell for ($S1$, $S3$). It has entries ($S2$, $S4$) for $X = 0$ and ($S1$, $S1$) for $X = 1$. None of the paired cells have # sign. Hence leave this entry as such.

Consider cell for ($S2$, $S4$). It has entries ($S1$, $S4$) for $X = 0$ and ($S3$, $S5$) for $X = 1$. The paired cell ($S3$, $S5$) there is # sign. Hence, place the # sign in this cell also as this pair does not fulfill condition 2.

TABLE 15.15 Implication table second iteration for a Mealy model sequential circuit-*I*

	S1	S2	S3	S4	S5
S_5	#	#	#	#	^
S_4	#	$(S_1\ S_4)$# $(S_3\ S_5)$ #	#	^	^
S_3	(S_2, S_4) (S_1, S_1)	#	^	^	^
S_2	#	^	^	^	^
S_1	^	^	^	^	^

Second iteration of cells filling is now over. Only cell pair ($S1$, $S3$) is left without any sign. Now next iteration on marking implication tables is as follows and Table 15.16 shows the result.

Step 6: Again remove the cell pairs not generating equivalent next states (not fulfilling condition 2).

Consider cell for ($S1$, $S3$). It has entries ($S2$, $S4$) for $X = 0$ and ($S1$, $S1$) for $X = 1$. Now in this iteration ($S2$, $S4$) paired cell has # sign. Hence we put the # sign here also.

TABLE 15.16 Implication table third iteration for a Mealy model sequential circuit-*I*

	S1	S2	S3	S4	S5
S_5	#	#	#	#	^
S_4	#	$(S_1\ S_3)$ $(S_4\ S_5))$ #	#	^	^
S_3	(S_2, S_4) (S_1, S_1) #	#	^	^	^
S_2	#	^	^	^	^
S_1	^	^	^	^	^

All cell pairs have a sign. Hence none of the states is found equivalent in circuit-I.

Example 15.8 Perform state reduction or minimization of state table in Table 15.17, if feasible. Use implication table approach.

Solution

Consider the state table for a Moore model sequential circuit-m in Table 15.17.

TABLE 15.17 State table for a Moore model sequential circuit-*m*

Present state	Next state after transition (Q1', Q2', Q3')		Present output *Y*
	Input $X = 0$	Input $X = 1$	
S_1	S_2	S_3	0
S_2	S_4	S_5	1
S_3	S_1	S_6	1
S_4	S_5	S_3	1
S_5	S_7	S_8	0
S_6	S_2	S_8	0
S_7	S_4	S_6	1
S_8	S_6	S_5	0

Step 1: Construct an implication table as shown in Table 15.18 for a five states state-table.

Step 2: Mark the cells for state pairs not to be considered for equivalency determination in matrix of cell-pairs:

A (S_i, S_i) pair is redundant for equivalency determination and an equivalent state pair (S_i, S_j) is same as pair (S_j, S_i). In a matrix of $n \times n$ cells, $(n^2 - n)/2 = 28$ are the off diagonal right side cells and $n = 8$ are the cells along the diagonal. Hence total 36 cells are not to be taken into consideration for pairing at the implication table for eight states table. Cells for the pairs marked by sign ^ marked are not to be considered. We put ^ sign in 36 cells in the implication table.

Step 3: Put # sign for a cell pair, which is not having the same set of *Y* output. Table 15.18 state table shows that for a set of inputs, which are possible, the outputs are same $Y = 1$ for ($S2$ and $S7$) states and same $Y = 0$ for ($S5$ and $S6$) states. We also mark the remaining cells by # sign, which do not show same *Y*.

Step 4: Fill the unmarked cells by the next state values. For example, ($S2$, $S6$) and ($S3$, $S5$) entries in cell ($S1$, $S8$).

TABLE 15.18 Implication table first iteration for a Moore model sequential circuit-*m*

	S_1	S_2	S_3	S_4	S_5	S_6	S_7	S_8
S_8	(S_2, S_6), (S_3, S_5)	#	#	#	(S_6, S_7) (S_5, S_8)	(S_2, S_6) (S_5, S_8)	#	^
S_7	#	(S_4, S_4), (S_5, S_6)	(S_1, S_4), (S_6, S_6)		#	#	^	^

Table 15.18 Contd.

S_6	(S_2, S_2) (S_3, S_8)	#	#	#	(S_2, S_7) (S_8, S_8)	^	^	^
S_5	(S_2, S_7) (S_3, S_8)	#	#	#	^	^	^	^
S_4	#	(S_4, S_5) (S_3, S_5)	(S_1, S_5) (S_3, S_6)	^	^	^	^	^
S_3	#	(S_1, S_4) (S_5, S_6)	^	^	^	^	^	^
S_2	#	^	^	^	^	^	^	^
S_1	^	^	^	^	^	^	^	^

First iteration of filling and marking of the cells is now over. Now next iteration on marking at the implication table is as follows and Table 15.19 shows the new result.

Step 5: Remove the cell pairs not generating equivalent next states (not fulfilling condition 2):

Consider cell for ($S2$, $S7$). It has entries ($S4$, $S4$), ($S5$, $S6$). We mark # for the cells according to following rule: "Let an entry be (S_l, S_m) in a cell (S_i, S_j). If cell (S_l, S_m) has a # sign previously placed, then put # sign in cell (S_i, S_j) also. Now ignore any other entry in the cell (S_i, S_j), else look at other entry(ies) also in the cell (S_i, S_j)."

TABLE 15.19 Implication table second iteration for a Moore model sequential circuit-*m*

	S_1	S_2	S_3	S_4	S_5	S_6	S_7	S_8
S_8	(S_2, S_6), # (S_3, S_5)	#	#	#	(S_6, S_7)# (S_5, S_8)	(S_2, S_6) # (S_5, S_8)	#	^
S_7	#	(S_4, S_4), (S_5, S_6)	(S_1, S_4),# (S_6, S_6)		#	#	^	^
S_6	(S_2, S_2) (S_3, S_8)#	#	#	#	(S_2, S_7) (S_8, S_8)	^	^	^
S_5	(S_2, S_7)# (S_3, S_8)	#	#	#	^	^	^	^
S_4	#	(S_4, S_5) # (S_3, S_5)	(S_1, S_5) (S_3, S_6)#	^	^	^	^	^
S_3	#	(S_1, S_4) # (S_5, S_6)	^	^	^	^	^	^
S_2	#	^	^	^	^	^	^	^
S_1	^	^	^	^	^	^	^	^

Second iteration of cells filling is now over. Only cell pair (S_2, S_7) and (S_5, S_6) are left without any sign. Now next iteration marking of implication table is as follows and Table 15.20 shows the result.

Step 6: Again remove the cell pairs not generating equivalent next states (not fulfilling condition 2).

Table 15.20 Implication table third iteration for a Moore model sequential circuit-*m*

	S_1	S_2	S_3	S_4	S_5	S_6	S_7	S_8
S_8	(S_2, S_6), # (S_3, S_5)	#	#	#	(S_6, S_7)# (S_5, S_8)	(S_2, S_6) # (S_5, S_8)	#	^
S_7	#	(S_4, S_4), (S_5, S_6)	(S_1, S_4), # (S_6, S_6)		#	#	^	^
S_6	(S_2, S_2) (S_3, S_8)#	#	#	#	(S_2, S_7) (S_8, S_8)	^	^	^
S_5	(S_2, S_7)# (S_3, S_8)	#	#	#	^	^	^	^
S_4	#	(S_4, S_5) # (S_3, S_5)	(S_1, S_5) (S_3, S_6)#	^	^	^	^	^
S_3	#	(S_1, S_4) # (S_5, S_6)	^	^	^	^	^	^
S_2	#	^	^	^	^	^	^	^
S_1	^	^	^	^	^	^	^	^

Both (S_5, S_6) and (S_2, S_7) pair cells have no # sign. So this iteration results in no change in Table 15.19. All cell have a sign # or ^ except $(S2, S7)$ cells and $(S5, S6)$ cells. We get the following implication table as Table 15.21 for equivalency determination.

TABLE 15.21 Implication table equivalency cells for a Moore model sequential circuit-*m*

	S_1	S_2	S_3	S_4	S_5	S_6	S_7	S_8
S_7		(S_4, S_4), (S_5, S_6)						
S_6					(S_2, S_7) (S_8, S_8)			

We find $S_2 = S_7$, and $S_5 = S_6$. Remaining states do not appearing here. Table 15.22 gives the state minimal table.

TABLE 15.22 State table for a Moore model sequential circuit-*m*

Present state	Next state after transition (*Q*1', *Q*2', *Q*3')		Present output *Y*
	Input $X = 0$	Input $X = 1$	
S_1	S'	S_3	0
(S_2, S_7) = S'	S_4	S"	1
S_3	S_1	S"	1
S_4	S"	S_3	1
(S_5, S_6) = S"	S'	S_8	0
S_8	S"	S"	0

Example 15.9

Make state table, and transition tables for a clocked sequential circuit-*n* from the state diagram given in Figure 15.6(a).

Solution

For the state diagram in Figure 15.6, we find that there are four states S_0, S_1, S_2 and S_3. Therefore the number of rows in state, transition and excitation tables shall also be four.

The diagram has present/present output on the directed arc. It means we will be having Mealy sequential circuit. It means output *Y*s for different values of *X*.

Now, there is only one external input *X*. Hence there will be two combinations, $X = 0$ and $X = 1$ for the present output *Y* in each of the tables: state, transition and excitation.

Now, there is only one external input *X*. Hence there will be two combinations, $X = 0$ and $X = 1$ for the next state and excitation inputs in tables for the state, transition and excitation, respectively.

Table 15.23 gives a state table for the state diagram in Figure 15.6(a).

TABLE 15.23 State table for the sequential circuit-*n*

Present state		Next state after transition (Q1', Q2')		Present output State *Y* before the transition	
State	(Q_1, Q_2)	Input $X = 0$	Input $X = 1$	$X = 0$	$X = 1$
S_0	0, 0	S_2	S_1	0	0
S_1	0, 1	S_3	S_2	1	0
S_2	1, 0	S_0	S_2	1	1
S_3	1, 1	S_1	S_1	1	0

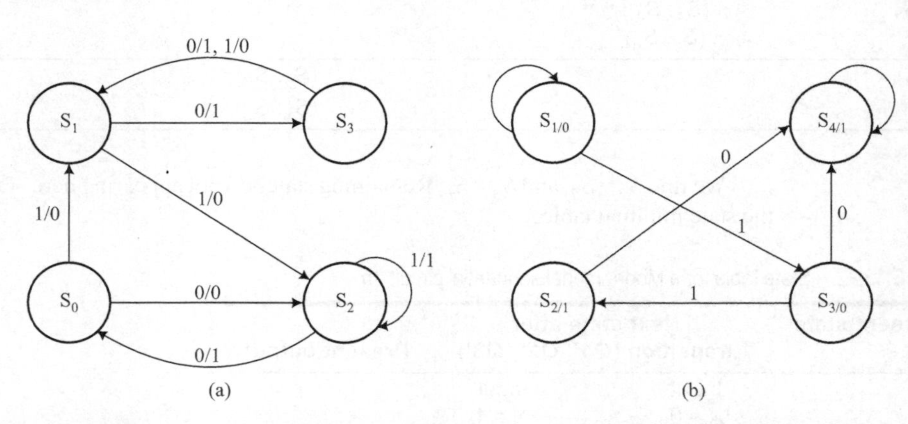

FIGURE 15.6 (a) State diagram for a circuit-*n* implementation (b) State diagram for a circuit-*t* implementation (Exercise 10).

TABLE 15.24 Transition table for the sequential circuit-*n*

Present state		Next state after transition ($Q1'$, $Q2'$)		Present output state Y before the transition	
State	(Q_1, Q_2)	Input $X = 0$	Input $X = 1$	$X = 0$	$X = 1$
$S0$	0, 0	1, 0	0, 1	0	0
$S1$	0, 1	1, 1	1, 0	1	0
$S2$	1, 0	0, 0	1, 0	1	1
$S3$	1, 1	0, 1	0, 1	1	0

Example 15.10 Table 15.24 gives the corresponding transition table.

Let us construct Karnaugh map to simplify the circuit specified state diagram of Figure 15.6(a) and the synthesize the circuit and find the expressions.

Solution

TABLE 15.25 Karnaugh Map for Q1'

Q1Q2 \ X		$\overline{X}$ 0	X 1
$\overline{Q}1\overline{Q}2$	00	1	0
$\overline{Q}1Q2$	01	1	1
$Q1Q2$	11		
$Q1\overline{Q}2$	10		1

Karnaugh Map for Q2'

Q1Q2 \ X		$\overline{X}$ **0**	X **1**
$\overline{Q}1\overline{Q}2$	00		1
$\overline{Q}1Q2$	01	1	
$Q1Q2$	11	1	1
$Q1\overline{Q}2$	10		

Table 15.25 gives the corresponding Karnaugh map.

The Boolean expressions for $Q1'$ and $Q2'$ are as follows:

$$Q1' = \overline{Q}1.\overline{X} + \overline{Q}1.Q2 + Q1.\overline{Q}2.X \quad ...(15.15)$$

$$Q2' = Q2.\overline{X} + Q1.Q2 + \overline{Q}1.\overline{Q}2.X \quad ...(15.16)$$

Example 15.11 Make the excitation table for curcuit-*n*.

Solution

Table 15.26 gives the excitation table when using *D-FF*s. $Q' = D$. Therefore, simply replace $Q1'$ and $Q2'$ in the transition table by $D1$ and $D2$ to get the excitation table.

TABLE 15.26 Excitation table for the sequential circuit-*n*

Present state (Q_1, Q_2)	Excitation inputs before the transition ($D1$, $D2$)		Present output state Y before the transition	
	Input $X = 0$	Input $X = 1$	$X = 0$	$X = 1$
0, 0	1, 0	0, 1	0	0
0, 1	1, 1	1, 0	1	0
1, 0	0, 0	1, 0	1	1
1, 1	0, 1	0, 1	1	0

EXERCISES

1. Draw a state diagram for Mealy Machine for which a general sequential circuit has excitation table (Table 15.27) as follows:

TABLE 15.27 Excitation table for the for the sequential circuit-4

Present State	Excitation Inputs (J, K) (J, K)		Present Output state after the X inputs but before the transition	
(Q_1, Q_2)	Input $X = 0$	Input $X = 1$	Input $X = 0$	Input $X = 1$
0, 0	(1, 0), (1, 1)	(0, 0), (0, 0)	0	1
0, 1	(0, 1), (0, 0)	(1, 1), (1, 0)	1	0
1, 0	(1, 1), (0, 0)	(1, 0), (1, 1)	0	1
1, 1	(0, 0), (1, 0)	(1, 1), (0, 0)	1	1

2. Draw a Moore state machine diagram for which a general sequential circuit has the excitation table (Table 15.28) as follows:

TABLE 15.28 Excitation table for the for the sequential circuit-4

Present state	Excitation inputs (J, K), D		Present output Y before the transition
(Q_1, Q_2)	Input $X = 0$	Input $X = 1$	
0, 0	(1, 0), 0	(0, 0), 0	0
0, 1	(0, 1), 1	(1, 1), 1	1
1, 0	(1, 1), 0	(1, 0), 1	0
1, 1	(0, 0), 1	(1, 1), 0	1

3. Analyze the sequential circuit given by following expressions and draw excitation, transition and state tables.

   $Q1' = D1$

   $Q2' = D2$

$D1 = (X.\overline{Q}1.Q2 + X.Q1.\overline{Q}2)$ and

$D2 = X.\overline{Q}1$

$Y = Q1.XOR.Q2$

4. Analyze the sequential circuit-0 with following characteristics equations. Draw the state diagram.

$Q1' = D1$

$Q2' = D2$

$Y = \overline{Q}1.X + \overline{Q}2.\overline{X}$

$Q2' = J.Q2 + K.Q2$

$J = Q1.X$ and $\overline{K} = \overline{X}$

5. Analyze the sequential circuit-*p* with following transition table. Draw the state diagram. If *JK FF*s are used make the required excitation table (Table 15.29).

TABLE 15.29 Transition table for the sequential circuit-4

Present state	Next state after transition (Q1', Q2')		Output state *Y*	
(Q_1, Q_2)	Input $X = 0$	Input $X = 1$	Input $X = 0$	Input $X = 1$
0, 0	1, 0	1, 1	1	0
0, 1	0, 1	1, 0	1	0
1, 0	1, 1	0, 0	0	1
1, 1	1, 0	0, 1	1	1

6. Draw the state diagram for the following transition table (Table 15.30).

TABLE 15.30 State transition table for the sequential circuit-*q*

Present state		Next state after transition (Q1', Q2')		Output state *Y*	
State	(Q_1, Q_2)	Input $X = 0$	Input $X = 1$	Input $X = 0$	Input $X = 1$
*S*0	0, 0	S2	S0	0	1
*S*1	0, 1	S2	S1	0	0
*S*2	1, 0	S3	S1	1	1
*S*3	1, 1	S0	S1	1	0

7. When are the two states not equivalent in a clocked sequential circuit?
8. Perform state minimization for the State table (Table 15.31) for a Moore model sequential circuit-*r* by inspection procedure.

TABLE 15.31 State table for a Moore model sequential circuit-*r*

Present state		Next state after transition (Q1', Q2', Q3')		Present output *Y*
State	(Q_1, Q_2, Q3)	Input $X = 0$	Input $X = 1$	
*S*1	0, 0, 0	*S*4	*S*1	1
*S*2	0, 0, 1	*S*1	*S*2	0
*S*3	0, 1, 0	*S*3	*S*4	1
*S*4	0, 1, 1	*S*1	*S*1	1
*S*5	1, 0, 0	*S*4	*S*2	0
*S*6	1, 0, 1	*S*1	*S*2	1
*S*7	1, 1, 0	*S*3	*S*1	1
*S*8	1, 1, 1	*S*1	*S*2	0

9. Perform state reduction or minimization of state table in Table 15.32, if feasible. Consider the state table for a Mealy model sequential circuit-*s* in Table 15.32

TABLE 15.32 State table for a Mealy model sequential circuit-*s*

Present state	Next state after transition (Q1', Q2', Q3')		Present output *Y*	
	Input $X = 0$	Input $X = 1$	Input $X = 0$	Input $X = 1$
*S*1	*S*1	*S*2	1	0
*S*2	*S*3	*S*5	1	1
*S*3	*S*1	*S*4	0	0
*S*4	*S*5	*S*4	0	1
*S*5	*S*4	*S*3	0	1

10. Make state table, transition and excitation tables and the synthesize a clocked sequential circuit-*t* from the state diagram given in Figure 15.6.(b)

QUESTIONS

1. Explain meaning of a state in a sequential circuit.
2. What is the difference between the clocked sequential and asynchronous sequential circuits?
3. Describe procedure for analysis of a given logic circuit based on Moore model.
4. Describe procedure for analysis of a given logic circuit based on Mealy model.
5. How will you make a transition table from a given excitation table?
6. An excitation table is given to you for *D FF*s at the memory section. How will you convert it for an excitation table using *JK FF*s?

7. A transition table is given to you. How will you make a state table from it? How will you draw state diagram from it?
8. Explain state minimization procedure.
9. Explain the steps needed for implementing a two-bit counter circuit having four states.
10. Describe rules for the state assignments.

CHAPTER 16

Sequential Circuits for Registers and Counters

OBJECTIVE

We learnt in Chapter 14 the basic memory section elements *SR*, *D*, *JK* and *T* flip flops. A *D* flip flop registers at the *Q* output the bit at *D* input. This fact can be used to design the multi-bit registers. *JK* flip-flop, when both *J* and *K* equal 1, and *T* flip flop toggles the *Q* output on each excitation. This fact can be used to design a divide-by-two element and a multi-bit counter.

We learnt the circuit implementation concepts for the clocked sequential circuits in Chapter 15. Using these, we design a large number of useful circuits, for example, for registers and counters.

We will learn the designs of registers, shift registers, ripple and synchronous counters in this chapter. We will also learn their timing diagrams.

16.1 REGISTERS

Registers can be designed using the *D-FF* or an *SR FF* with *PR* and CLR inputs for presetting all outputs as 1s or 0s, respectively. Note that one type of *FF* converts by adding some suitable circuit to another type of *FF*. For example, a *JK FF* coverts to *D FF*, if a NOT gate is placed before *K* from *J*. An *SR FF* coverts to *D FF*, if a NOT gate is placed before *R* from *S*.

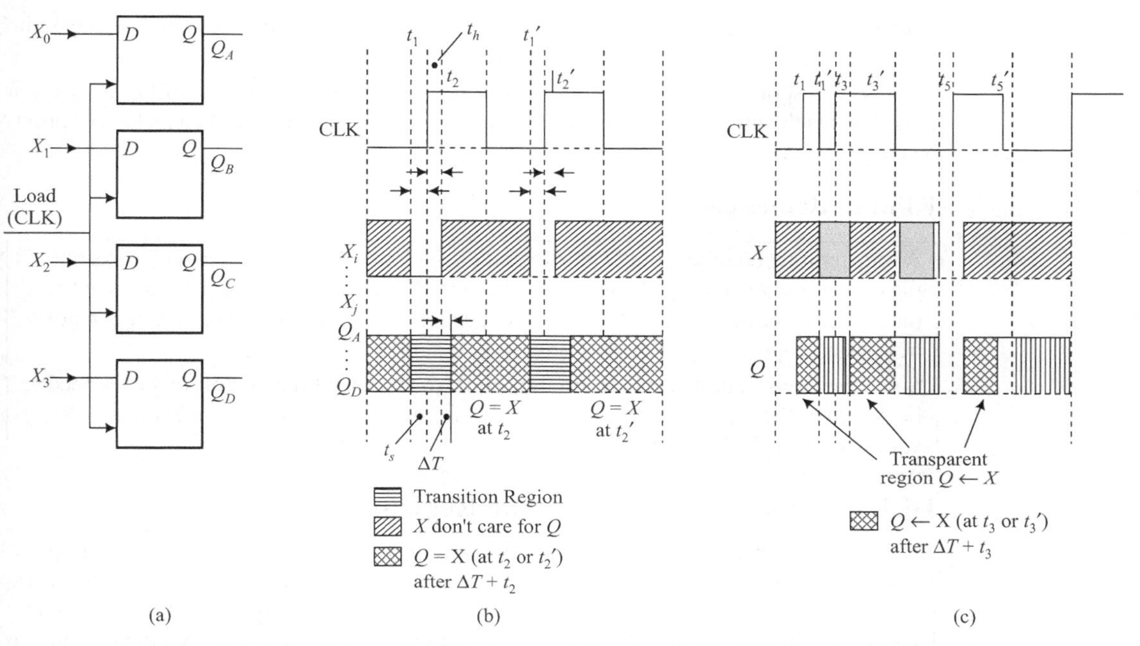

FIGURE 16.1 (a) Parallel—in Parallel—out four-bit register made from four *D FF*s (b) Timing diagram of a register consisting of the flipflops (c) Timing diagram of a register consisting of the latches.

Figure 16.1(a) shows a parallel–in parallel–out four-bit register. A register simplest design is made by a combination of *D FF*s. It is capable for storing logic outputs for a nibble or byte or word (of 4 or 8 or *n* bits). The number of *Q* outputs available as the distinct stable logic states defines the number of bits in a register.

A register is a clocked sequential circuit. The clock inputs of all *D FF*s are interconnected together in the register. At a clock transition at this common input, all *Q*s all become identical with respect to the *D* inputs after ΔT (propagation delay). Let a set of outputs, Q_D Q_C Q_B Q_A define a binary word of four bits. Then the corresponding register shall be called a 4-bit register made from the four *D FF*s. [Figure 16.1(a)]

Let the CLK of all the four *FF*s are simultaneously activated by a load input (also called a strobe input or an enable input). The outputs Q_D Q_C Q_B Q_A get the next state after a certain delay equal to propagation delay, ΔT and these simultaneously get stabilized to the input values D_D D_C D_B D_A just before the clock transition from 0 to 1, if the positive edge triggered *D FF*s are used.

The register shown in Figure 16.1(a) can be extended to the 8-bits or 16-bits or more.

The above register is also called edge triggered load register because the data is only accepted by the register during the very short time between t_1 and t_2 during which clock rises from logic 0 to logic 1 level [refer Figure 16.2(b)].

Figure 16.1(b) shows a timing diagram. D_D D_C D_B D_A must be stable before a time equal to set-up time when the write or load input (clock transition) occurs. D_D D_C D_B D_A must be stable till a time equal to hold time (= t_h) when the write input (clock transition) occurs. Data at the inputs needs to be valid only at time equal to t_s (setup time) before a rising edge at the clock input. The data clocks (registers) at the rising edge into the register after a time equal to propagation delay ΔT. [Note: Certain edge triggered registers may write into it the inputs after a falling (trailing) edge i.e. negative edge.]

Qs correspond to new state after ΔT from the edge. The rising edge of the CLK input thus changes the previous Q outputs in the register and places the new data instead or previous one after ΔT from t_2.

Point to Remember

A register transfers the input D bits to next Qs such that $Q'_i = D_i$. A register "looks upon" the data bits at $D_D\, D_C\, D_B\, D_A$ only at the instant of a rising edge. A register does not care (accept or clock) the data before rising edge and after t_2 and will care only again at the next rising edge.

Since the Ds must be set up before a time equal to the setup time, therefore the Ds need to be set-up at the inputs before t_s (= set up time) from the clock edge. Next-state Qs are valid when the Ds hold up to t_h (= hold time) after the edge.

16.1.1 Bi-stable Latches as the Register

The bi-stable latches, for example, D latches, arranged in place of the flip flops in circuit of Figure 16.1(a), can also store the data. However, a latch differs from the register in the sense that after the activation of the clock (enable or strobe or write) input (CLK), the bi-stable latches are transparent, and allow the data bits at $D_D\, D_C\, D_B\, D_A$ to pass from the D inputs to the $Q_D\, Q_C\, Q_B\, Q_A$ outputs. However, the $Q_D\, Q_C\, Q_B\, Q_A$ stabilizes to the values after the CLK input inactivates. If we send a pulse at an CLK input, then the $D_D D_C D_B D_A$ should be stable during the entire duration of clock activation plus the duration for set-up for the inputs plus the duration needed t_n = hold up time of the inputs. The last Qs become the new registered values. These appear after a time equal to ΔT from the inactivation of the CLK.

Figure 16.1(c) gives the timing diagrams for a transparent latche based register. It shows the timings for the parallel bits $\mathbf{X}$ as inputs and the parallel bits $\mathbf{Q}$ as outputs. Latches are used when the inputs are expected from some another source and a time interval during which those are expected is the transparent region time interval–the time for which transition enable input (CLK) is active.

Point to Remember

Registers consisting of latches accept the data during the full period of active clock.

16.1.2 Parallel-In Parallel-Out Buffer Register

The circuit of Figure 16.2(a) is also called a parallel-in parallel out (PIPO) register when there are the parallel loading mechanism for the excitation inputs to the memory section (of D FFs) and the parallel outputs mechanism for the next state outputs. Parallel inputs mean $D_i = X_i$ and parallel outputs mean $Y_i = Q_i$ where $i = 0, 1, 2$ or 3 for a 4-bit PIPO.

Point to Remember

A PIPO register transfers the input bits $\mathbf{X}$ to next $\mathbf{Q}$s such that $Q'_i = X_i$. A PIPO register loads the external inputs as the excitation inputs and undergoes transition to the next state

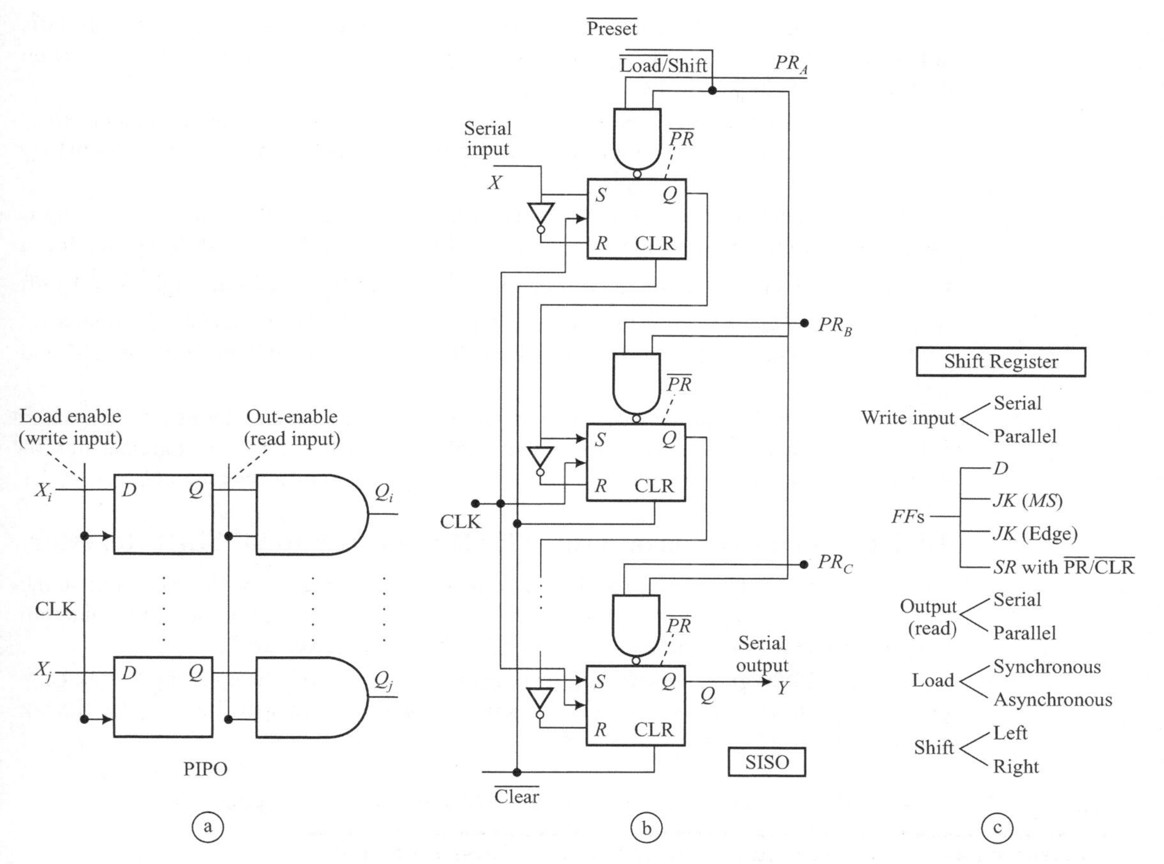

FIGURE 16.2 (a) Parallel—in Parallel—out four bit buffer register made from four *D* flip flops with out-enable and load-enable input (b) 4-bit serial in serial out (SISO) shift register using the *S-R* Flip-flops with preset and clear inputs (c) Various types of shift registers.

on a clock transition. It gives parallel outputs, which are the same as the next state. Parallel inputs mean $D_i = X_i$ and parallel outputs mean $Y_i = Q_i$ where $i = 0, 1, 2$ or 3 for a 4-bit PIPO. Note: $i = 0, 1, 2 \ldots n - 1$ for an n-bit register.

16.1.3 Number of Bits in a Register

Figure 16.2(a) shows a PIPO buffer register. Example 16.3 will explain the circuit of the buffer register.

16.2 SHIFT REGISTERS

Figure 16.2(b) shows the circuit of a *serial in serial out* (SISO) shift register. Figure 16.2(c) gives the different forms of the shift registers. A set of cascaded flip-flops to store the information like the register discussed in Section 16.1 along with an additional circuit gives a *shift register*.

A shift register is a clocked sequential circuit in which stored each binary word bit shifts either towards left or towards right (towards higher place value or lower place value) on each successive clock transition.

In a parallel input shift register with a serial out, called a PISO, at the every clock pulse, one of the bit at a time of the parallel data bits from the left-most (lsb) or right-most (msb) Q is transferred to a serial output pin.

Shift registers need not be made from discrete logic gates or flip-flops due to complicated circuits of the type in Figure 16.2(b). An MSI IC 7495 (TTL) is a shift register. It is a four-bit parallel cum serial input with the parallel output. It is a left cum right ($L/\overline{R}$) shift register. When $L/\overline{R}$ input = 1, then left-shift, else right shift. An IC 74164 TTL is an 8-bit PISO register. The 74HC95 and 74HC164 are the high-speed CMOS versions of 7595 and 74164.

Let us denote in following description a transfer from a Q to another nearby next stage Q by $\rightarrow$ and to a previous stage Q by $\leftarrow$. Let us also assume that the Q_A is of greater significance than Q_B; Q_B of greater than Q_C; and Q_C is of greater significance than Q_D.

16.2.1 Serial-In Serial-Out (SISO) Unidirectional Shift Register

Table 16.1 gives a state table of a SISO unidirectional shift register with serial input at the flip flop with QA output and serial output at the flip flop with QD output. State diagram building from the table is left as an exercise for the reader. (Exercise 2).

Figure 16.2(b) gives a SISO unidirectional shift register, when weight of $Q_A > Q_B$, $Q_B > Q_C$ and $Q_C > Q_D$ and X is a serial input, it is shifting right. Note: In opposite case, it will work as a left shift SISO.

TABLE 16.1 State table for a Moore model sequential circuit for a 4-bit SISO right shift unidirectional register

Present state		Next state after transition (Q_A', Q_B', Q_C', Q_D')		Present output $Y = Q_D$
State	$Q_AQ_BQ_CQ_D$	Input $X = 0$	Input $X = 1$	
S_0	0000	S_0	S_8	0
S_1	0001	S_0	S_8	1
S_2	0010	S_1	S_9	0
S_3	0011	S_1	S_9	1
S_4	0100	S_2	S_{10}	0
S_5	0101	S_2	S_{10}	1
S_6	0110	S_3	S_{11}	0
S_7	0111	S_3	S_{11}	1
S_8	1000	S_4	S_{12}	0
S_9	1001	S_4	S_{12}	1
S_{10}	1010	S_5	S_{13}	0
S_{11}	1011	S_5	S_{13}	1
S_{12}	1100	S_6	S_{14}	0
S_{13}	1101	S_6	S_{14}	1
S_{14}	1110	S_7	S_{15}	0
S_{15}	1111	S_7	S_{15}	1

The Q_A Q_B Q_C Q_D outputs up on a clock edge transition, within a time equal to ΔT_{SR} from the transition, will load the serial input into the Q_A and serial output is from Q_D.

The register changes as a $X \rightarrow Q_A$ shift register as right $Q_A \rightarrow Q_B$, $Q_B \rightarrow Q_C$, $Q_C \rightarrow Q_D$, $Q_D \rightarrow$ serial out and Q_A will become equal to 0 when serial-in = 0 and equal to 1, when serial in = 1. [Assume Q_D is of higher place (weight) value].

Serial input at right-shift register when state is 0010 (S_2) will become 0001 (S_1) when serial bit $X = 0$ and serial-out Y will be = 0. Serial input when state is 0001 (S_1) will become 1000 (S_8) when serial bit = 1 and serial-out Y will also be = 1 because past $Q_A = 1$.

Point to Remember

A right shift SISO loads the inputs bit X into the last highest place value stage D_A and gives the serial output = Q_D on a clock edge. . Its state $Q'_j = D_i$ on the clock edge, Here and $i = 0, 1, 2, ..., n - 1$ in an n-bit register and j a nonnegative value differs from i by 1. The left shift register on the other hand, will load $Y \leftarrow Q_A$. on $Q_D \leftarrow X$.

16.2.2 Serial-In Parallel-Out (SIPO) Right Shift Register

Table 16.2 gives a state table of a right shift SIPO. Figure 16.3(a) shows a right shift SIPO register, when weight of $Q_0 > Q_1$, $Q_1 > Q_2$ and $Q_2 > Q_3$ and X is a serial input. Note: In opposite case, it will work as a left shift SIPO. Figure 16.3(b) shows the state diagram of the right shift SIPO register.

The Q_0 Q_1 Q_2 Q_3 outputs up on a clock edge transition, within a time equal to ΔT_{SR} from the transition, will change in a right shift register as $Q_0 \rightarrow Q_1$, $Q_1 \rightarrow Q_2$, $Q_2 \rightarrow Q_3$, and Q_0

TABLE 16.2 State table for a Moore model sequential circuit for a 4-bit SIPO shift right register

Present state		Next state after transition (Q_0', Q_1', Q_2', Q_3')		Present outputs Y_0, Y_1, Y_2, Y_3
State	$Q_0Q_1Q_2Q_3$	Input $X = 0$	Input $X = 1$	
S_0	0000	S_0	S_8	0000
S_1	0001	S_0	S_8	0001
S_2	0010	S_1	S_9	0010
S_3	0011	S_1	S_9	0011
S_4	0100	S_2	S_{10}	0100
S_5	0101	S_2	S_{10}	0101
S_6	0110	S_3	S_{11}	0110
S_7	0111	S_3	S_{11}	0111
S_8	1000	S_4	S_{12}	1000
S_9	1001	S_4	S_{12}	1001
S_{10}	1010	S_5	S_{13}	1010
S_{11}	1011	S_5	S_{13}	1011
S_{12}	1100	S_6	S_{14}	1100
S_{13}	1101	S_6	S_{14}	1101
S_{14}	1110	S_7	S_{15}	1110
S_{15}	1111	S_7	S_{15}	1111

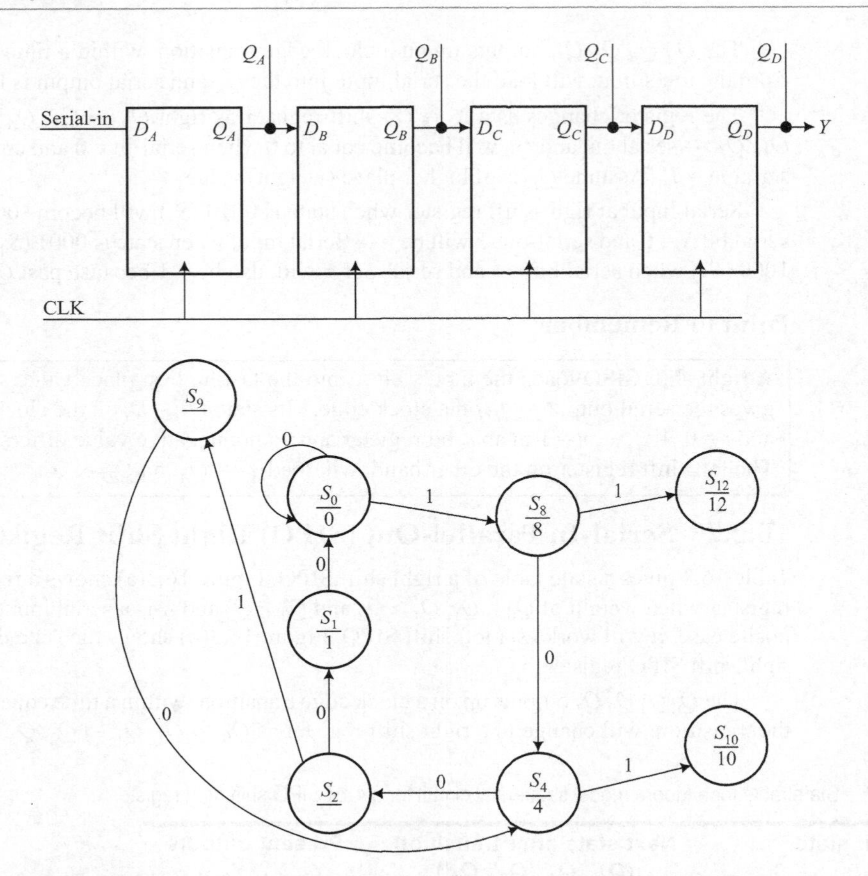

FIGURE 16.3 (a) Right shift SIPO register, when weight of $Q_0 > Q_1$, $Q_1 > Q_2$ and $Q_2 > Q_3$ and X is a serial input and Y is serial output (b) State diagram of the SIPO right shift register.

will become equal to a serial input bit. [Assume Q_0 is of higher place (weight) value]. Figure 16.3(a) shows a 4-bit right shift register in which a serial bit propagate from Q_0 to Q_3 (higher to lower place value) and also appear as output in case of a SIPO. The shifting is done by a clock transition at the CLK input in Figure 16.3(a). Present state 0010 (S_2) will become 0001 (S_1) when serial bit = 0, and 1001, (S_9) when serial bit = 1. [Present Y = 0010.]

Point to Remember

A SIPO shift-right register transfers the input bits X to next **Q**s at each clock edge excitation so that $Q'_j = X_i$, where $j = i - 1; j >= 0$ and $i = 0, 1, 2, ..., n - 1$ in an n-bit register. The msb will become equal to serial bit. It gives parallel outputs, which are the same as the next state, which becomes present state for the next excitation. When serial shift input bit $X = 0$, right shift register divides a binary number by 2. [Then left shift, multiplies by 2.]

16.2.3 Parallel-In Serial-Out (PISO) Right Shift Register

Table 16.3 gives a state table of a right shift PISO; assuming Q_0 is msb (maximum significance bit) and Q_3 as lsb. [Assume that weight of $Q_0 > Q_1$, $Q_1 > Q_2$, $Q_2 > Q_3$.]

TABLE 16.3 State table for a Moore model sequential circuit for a 4-bit PISO right shift register

Present state	Next state after transition (Q0', Q1', Q2', Q3')						Present outputs Y
	$\overline{L}/S$ Serial-in, X_0, X_1, X_2, X_3 = (000000)	$\overline{L}/S$ Serial-in, X_0, X_1, X_2, X_3 = (100000)	$\overline{L}/S$ Serial-in, X_0, X_1, X_2, X_3 = (110000)	...	$\overline{L}/S$ Serial-in, X_0, X_1, X_2, X_3 = (101111)	$\overline{L}/S$ Serial-in, X_0, X_1, X_2, X_3 = (111111)	
$S_i = S_0$ or S_1 or S_2 or ... S_{15}	S_0	S_0	S_8	...	S_7	S_{15}	$Y = Q_3$

Note: There are six inputs are $\overline{\text{Load}}$/Shift ($\overline{L}/S$), Serial-in X_0, X_1, X_2 and X_3. Q_3 is the lsb and Q_0 is msb.

Figure 16.4(a) gives a right shift PISO register, when weight of $Q_0 > Q_1$, $Q_1 > Q_2$ and $Q_2 > Q_3$ and X is a serial input. In opposite case, it will work as a left shift PISO. Figure 16.4(b) shows the state diagram.

The $Q_0\ Q_1\ Q_2\ Q_3$ are outputs after a clock edge transition. Within a time equal to ΔT_{SR} from the transition, the PISO will load the inputs into the Qs when $\overline{\text{Load}}$/Shift = 0. When the $\overline{\text{Load}}$/Shift = 1, the register changes as a right shift register as $D_0 \rightarrow Q_1$, $D_1 \rightarrow Q_2$, $D_2 \rightarrow Q_3$, and Q_0 will become equal to 0 when serial-in = 0 and equal to 1, when serial in = 1. [Assume Q_0 is msb of (higher place (weight) value).]

Figure 16.4(a) shows a 4-bit parallel loading cum right-shift serial-out register in which a serial bit propagate from Q_0 to Q_3 (higher to lower place value).

Parallel input 0010 (X_2) will become 0001 (S_1) when serial bit = 0 and serial-out Y will be = 0. Parallel input 0001 (X_1) will become 1000 (S_8) when serial-in bit = 1 and serial-out Y will be = 1.

Point to Remember

A right shift PISO loads the parallel inputs bits X into the Ds, and the $Q_i = D_i$ on the clock edge, when $\overline{\text{Load}}$/Shift input = 0. Here which $i = 0, 1, 2... n - 1$ in an n-bit register.

When $\overline{\text{Load}}$/Shift input = 1, the PISO transfers the D input bits X to next **Q**s at each clock edge excitation so that $Q'_j = X_i$, where $j = i - 1; j >= 0$. The msb Q_0 will become equal to serial bit X and lsb Q_3 will be the one-bit output Y at the serial-out pin. The parallel Q outputs are the same as the next state. When serial shift input $X = 0$, right shift register divides the input binary number by 2. [The left shift multiplies by 2.]

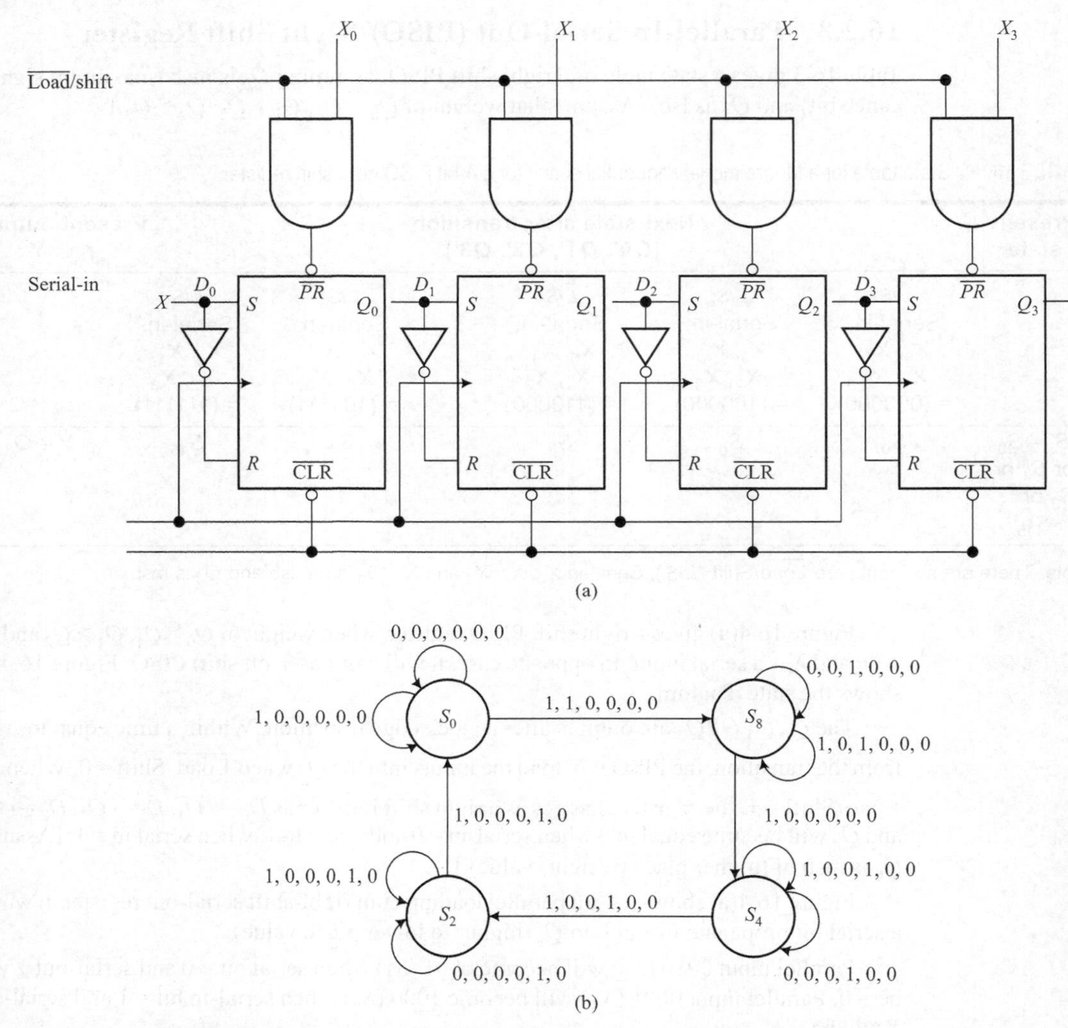

FIGURE 16.4 (a) Right shift PISO register, when weight of $Q_0 > Q_1$, $Q_1 > Q_2$ and $Q_2 > Q_3$ and Y is a serial output and $X0$, $X1$, $X2$ and $X3$ are the serial inputs (b) Sate diagram of PISO right shift register.

16.3 COUNTER

Often there is a need to count the number of pulses or triggering at an input. Counting is an essential circuit in computers. Figure 16.5(a) gives various types of counters and the various features defining a counter. Figure 16.5(b) shows a state diagram of the 16-states in the counter. It changes a state along 16 directed arcs in a cycle.

A circuit with state diagram like in Figure 16.5(b) is a counter. Figure 16.5(c) shows a binary counter. It is the counter when there is additional constraint that a state has the Qs combination corresponding to a binary number, whose output state give a binary number and cyclic changes only can occur either clock wise or anti clock wise to increase or decrease the count number for a up counter or down counter, respectively.

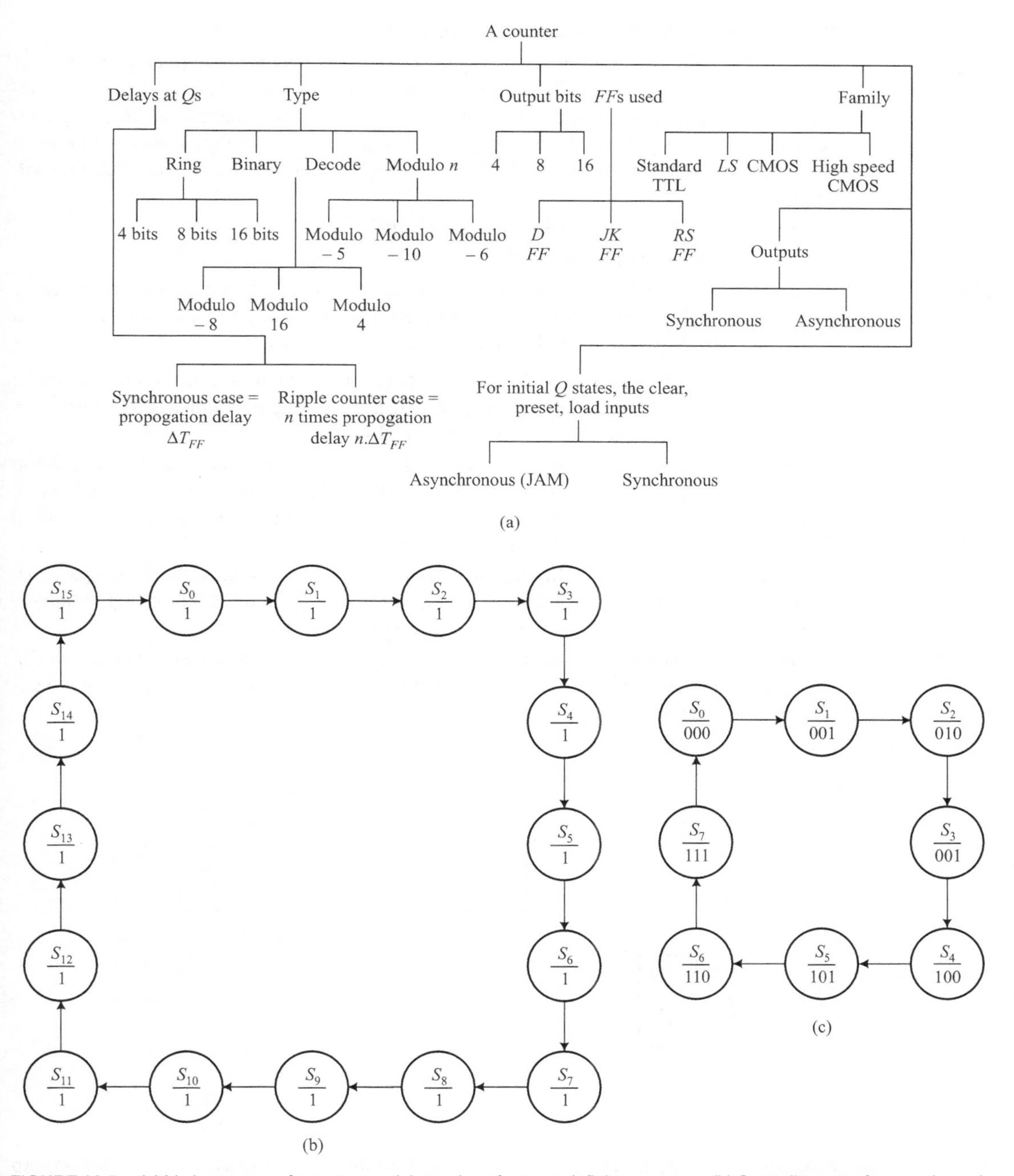

FIGURE 16.5 (a) Various types of counters and the various features defining a counter (b) State diagram of a negative pulse triggered 16-state counter. It changes state along 16 directed arcs in a cycle (c) State diagram of a count input C triggered 8-state binary 3-bit counter. It changes state along the 8 directed arcs in a cycle.

Divide-by-Two Flip-Flop as a Basic Counting Element

A divide-by-2 circuit produces one output pulse for every two pulses applied to its input. A divide-by-two circuit is made from a TFF. We learnt in Chapter 14 that one type of *FF* converts by adding some suitable circuit to another type of *FF*. For example, a *D* flip flop converts to *TFF* if *D* input = (T XOR Q_n). Q_{n+1} will toggle and will be $\overline{Q}_n$ when $T = 1$ before a clock edge. A *JK FF* coverts to *TFF* if both *J* and *K* are made = 1. Q_{n+1} will toggle and will be $\overline{Q}_n$.

Following are the ways to design a divide-by-2 circuit:

1. Use a circuit of *T*-type flip flop
2. Use a single *JK* flip flop with its *J* and *K* inputs made 1. [Refer method *i* in the Figure 16.6(a). The *T* flip flop (*FF*) is designed from *JK* or any other method to act as a divide-by-two circuit and *JK* input is now the *T*-input.]
3. Use a *D* flip flop (not *D* latch) with its $\overline{Q}$ output feedback to the *D* input [refer method *ii* in Figure 16.6(a)]. Alternatively use a *S R* flip flop with a NOT in-between *S* and *R* to get a DFF and then convert a DFF into *T*.

T stands for the toggling. We can say the output *Q* toggles between 0 and 1. Let a *T*-flip flop is positive edge triggered. Consider first two 0 to 1 transitions. At the first clock input, *T* transition, the output is complement of the previous *Q* value, and then at next transition (0 to 1) at the clock input, the output *Q* reverts to its previous value.

Let us now consider a *T*-flip flop, which is –ve edge, triggered. At the *T* input, a 1 to 0 transition is applied twice to obtain the previous value of *Q*. In other words, the output returns to original logic state for every two successive clock input at *T*.

Figure 16.6(b) shows the next state part of the transition table for a divide-by-two flip flop circuit. Two pulses at the *T* input cause the circuit to shift from one stable state to another and then back to first again.

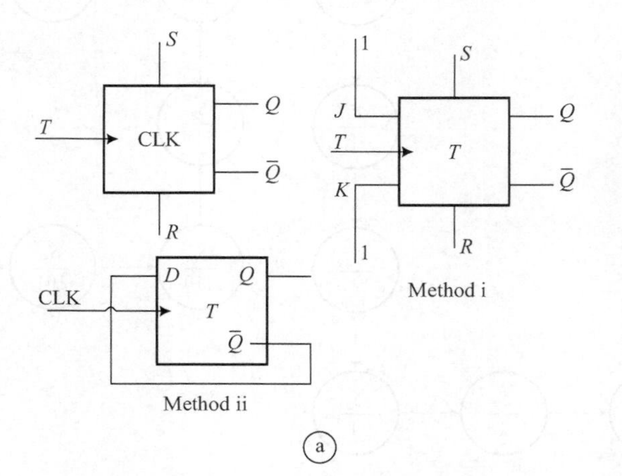

Input	Output	
T	Q'	$\overline{Q}'$
0	Q_o	$\overline{Q}_o$
1	$\overline{Q}_o$	Q_o
0	$\overline{Q}_o$	Q_o
1	Q_o	$\overline{Q}_o$
0	Q_o	$\overline{Q}_o$
1	$\overline{Q}_o$	Q_o

Q_o = Initial state of Q

(b)

FIGURE 16.6 (a) Divide-by-two circuit basic unit using *T*-, *JK*- or *D* flip flops (b) A partial transition table for divide-by-two circuit.

16.4 RIPPLE COUNTER

16.4.1 Cascaded Divide-By-2^n Circuit as a Ripple Counter

We have seen above that a *T FF* acts as a divide-by-2 circuit. If *Q* output of the *FF* connects to the *T* input of the second *FF*, and the *Q* output from the second *FF* connects to T-input of the third flip flop and so on, the *FF*s are said to be in a cascade (Figure 16.7(a)). The output transition partial table of this cascade asynchronous circuit for a 4-bit –ve going input pulse triggered ripple counter is given in Figure 16.7(b).

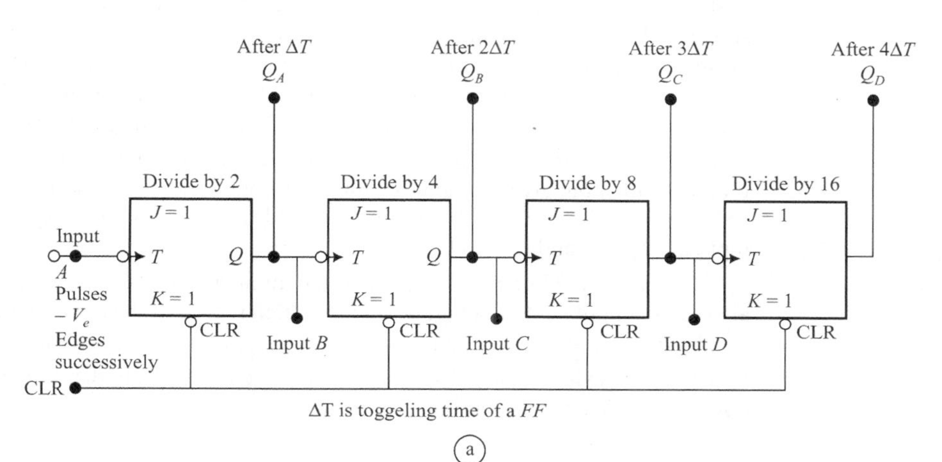

First make CLR = 0 to get $Q_A = Q_B = Q_C = Q_D = 0$					
	CLR = 1, J = 1, K = 1 and T input as under	Outputs after different delays			
		Q_D	Q_C	Q_B	Q_A
1		0	0	0	1
2		0	0	1	0
3		0	0	1	1
4		0	1	0	0
5		0	1	0	1
6		0	1	1	0

(b)

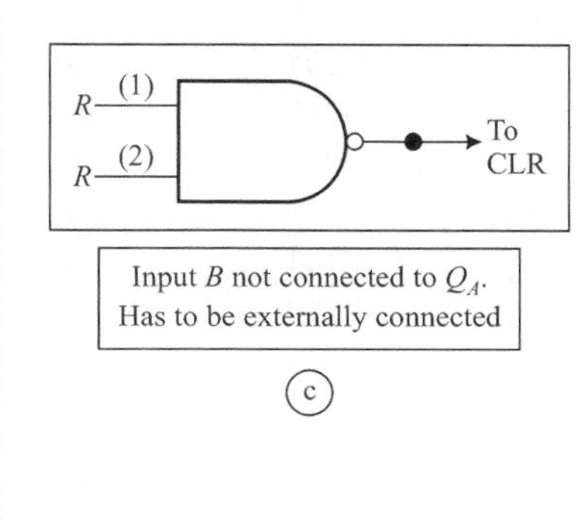

FIGURE 16.7 (a) Cascaded divide-by-two circuits using *T* –ve edge triggering pulse to get a 4-bit binary asynchronous (ripple) counter (b) Partial transition table (c) Additional feature in ripple counter in IC 7493 to enable its conversion to modulo-6 counter.

A cascade of n flip flops gives a divided by 2^n circuit. It is also called a ripple counter because a state is transferred from one stage to next stage after some delay like a ripple in water.

A cascade of four flip flops A, B, C, D is there in Figure 16.7(a) circuit. The outputs are Q_A, Q_B, Q_C, and Q_D after each negative edged successive transition at T. A 0 to 1 followed by an 1 to 0 input at a T input forms one negative edged trigger. By convention, a sequence of writing in a transition table is Q_D, Q_C, Q_B, and QA. This is so because in decimal system, we write a bigger placed value digit on the left. In one hundred, the place value of 1 is biggest. So 1 is written as left most when we write 100 in decimal system. Similarly, in the *FF*s shown in Figure 16.7(a), the Q_D changes after 8 pulses and return to original state after 16 pulses. Therefore, Q_D is written on the left-most side by convention. The output Q_A is obtained after a delay, called propagation delay, ΔT, of a T *FF*. Q_B is obtained after delay of $2\Delta T$, Q_C is obtained after a delay of $3\Delta T$ and Q_D is obtained after a delay of $4\Delta T$.

A combination of the flip flops A and B gives a divide by four circuit, combination of B, C and D with input given at input B with Q_A disconnected gives a divide by eight circuit, and combination of A, B, C and D *FF*s gives a divide by 16 circuit when Q_A not disconnected to input B. Q_B to C and Q_C to D.

The cascade *FF*s act as registers of the number of pulses, act as a ripple counter, and also can manipulate digital logical inputs, which represent the binary number.

Table 16.4 gives state table for a 4-bit –ve going input pulse triggered ripple counter. Figure 16.5(b) showed the state diagram of the counter. Note that it is table for asynchronous circuit. State is also as one of the input.

TABLE 16.4 State Table for a for a 4-bit —ve going input pulse triggered ripple (asynchronous) counter

Present state		Next state after transition (Q_3', Q_2', Q_1', Q_0')		Present output $Y = Q_3Q_2Q_2Q_1$
State	Q_3Q_2 Q_1Q_0	For an Input State with $T = 1$	For an Input State with $T \to 0$	
S_0	0000	S_0	S_1	0000
S_1	0001	S_1	S_2	0001
S_2	0010	S_2	S_3	0010
S_3	0011	S_3	S_4	0011
S_4	0100	S_4	S_5	0100
S_5	0101	S_5	S_6	0101
S_6	0110	S_6	S_7	0110
S_7	0111	S_7	S_8	0111
S_8	1000	S_8	S_9	1000
S_9	1001	S_9	S_{10}	1001
S_{10}	1010	S_{10}	S_{11}	1010
S_{11}	1011	S_{11}	S_{12}	1011
S_{12}	1100	S_{12}	S_{13}	1100
S_{13}	1101	S_{13}	S_{14}	1101
S_{14}	1110	S_{14}	S_{15}	1110
S_{15}	1111	S_{15}	S_0	1111

The counting circuit of Figure 16.7(a) is also called asynchronous up counter. Asynchronous behavior is because of its design as a ripple counter. (Up counter is because its next state binary number increases from all 0s to all 1s). Qs change from 0000 to Qs 1111). Figure 16.7(b) gives its transition partial state table.

Figure 16.7(c) shows the two additional features in a standard TTL IC 7493 or CMOS IC 7493, which has configuration of the type of Figure 16.7(a). This enables modulo 6 or 7 or 10 designs from 16 state counter.

16.4.2 Modulo-6, Modulo-7 and Modulo-10 Counters

If a counter returns to original state after $Q_B = 0$, $Q_A = 1$ and $Q_C = 1$, we say it is modulo-6 counter. Such counter is useful in watches/clocks due to fact that 60s = 1 minute and 60 minute = 1 hour. [Figure 16.8(a)]

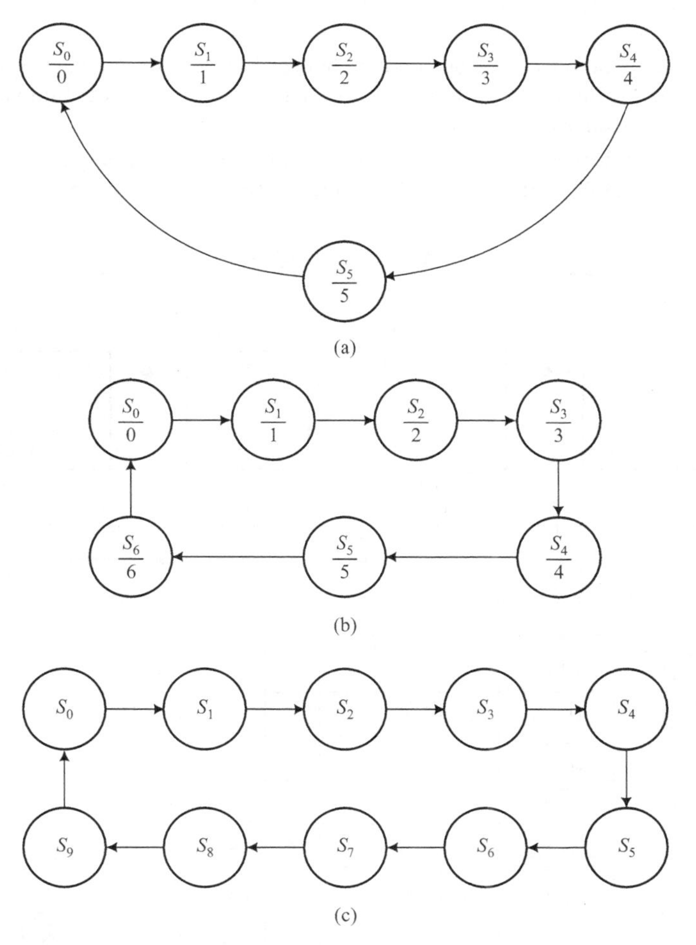

FIGURE 16.8 (a) State diagram of modulo-6 binary counter (b) State diagram of modulo-7 binary counter (c) State diagram of modulo-10 binary counter from a 4-bit binary counter.

If a counter returns to original state after $Q_B = 1$, $Q_A = 0$ and $Q_C = 1$ and has state diagram as shown in Figure 16.8(b), we say it is modulo-7 counter. Such counter is useful in triggering actions on odd number of sequences (action after seven in place of six).

If a counter returns to original state after $Q_D = 1$, $Q_B = 0$, $Q_A = 1$ and $Q_C = 0$, we say it is modulo-10 counter or decade counter. [Figure 16.8(c)]

16.4.3 Ring Counter

A ring counter is shown in Figure 16.9(a). It is a shift register based counter. It is a shift register with serial-input bit permanently from Q_D (msb). It has clear and preset facilities and with no serial output facility.

Its partial transition state stable is given in Figure 16.9(b). The ring counter shifts state 1 from lower significant bit (lsb) end to maximum significant bit (msb) end. The msb is feedback to *D FF-A* to form a ring. It acts like a cyclic rotate right of bit 1 toward left (msb place).

When the outputs of a ring counter connect a buffer of *n*-inputs, the *n*-sequences activate cyclically. Figure 16.10(b) shows how a 4-bit ring counter output to a buffer that activates four sequences cyclically.

The output of the counter is not binary. From the table, output sequences are 0001, 0010, 0100, 1000, 0001, 0010, ... [Refer Figure 16.10(c) state diagram]

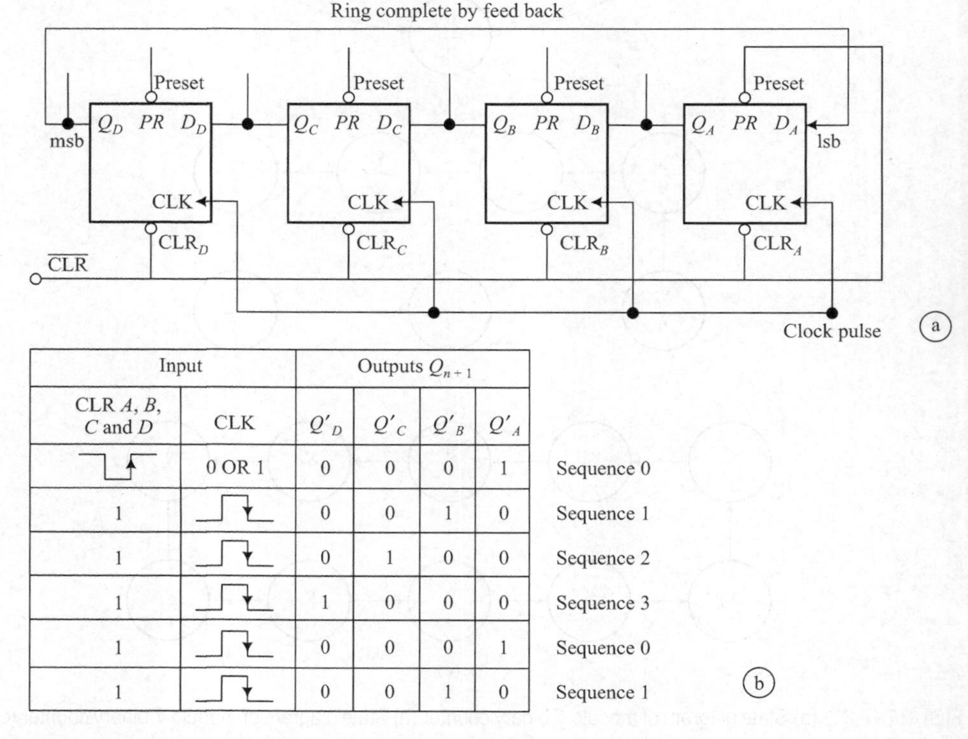

Input		Outputs Q_{n+1}				
CLR *A*, *B*, *C* and *D*	CLK	Q'_D	Q'_C	Q'_B	Q'_A	
↑	0 OR 1	0	0	0	1	Sequence 0
1	↓	0	0	1	0	Sequence 1
1	↓	0	1	0	0	Sequence 2
1	↓	1	0	0	0	Sequence 3
1	↓	0	0	0	1	Sequence 0
1	↓	0	0	1	0	Sequence 1

FIGURE 16.9 (a) Circuit of a ring counter (b) Partial transition table of ring counter.

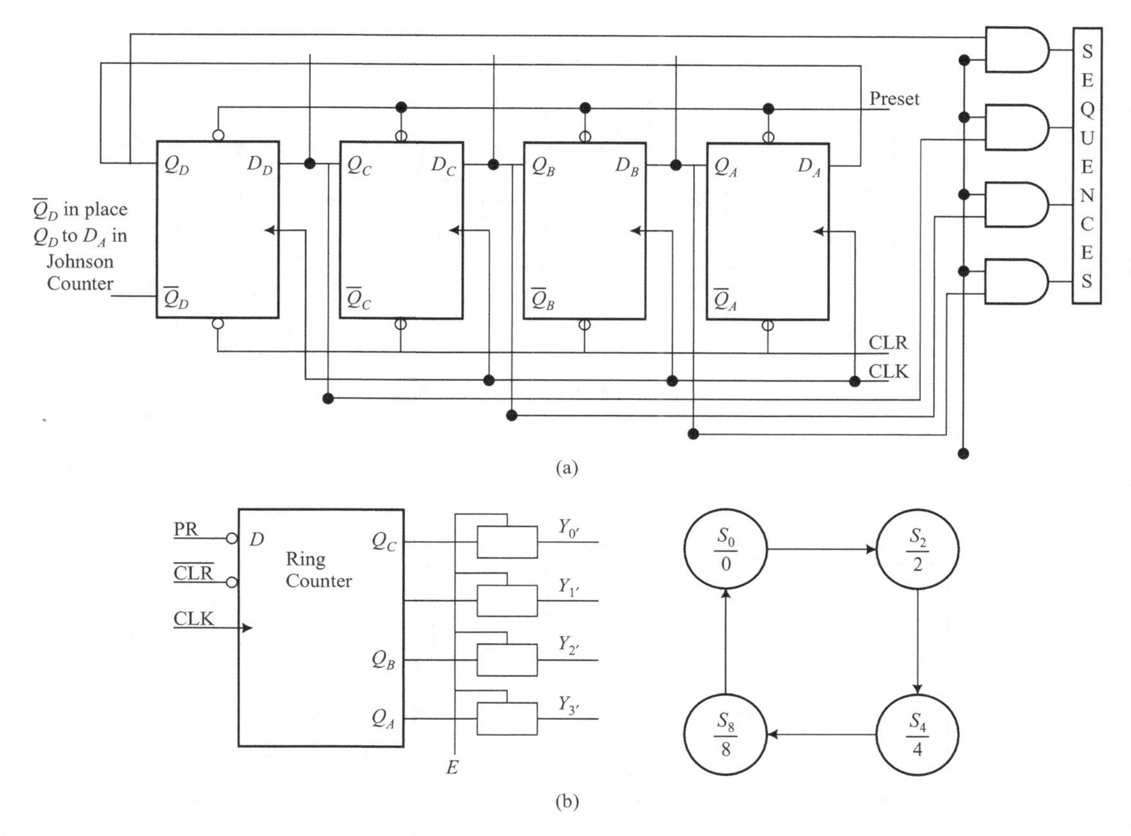

FIGURE 16.10 (a) Activation of four sequences cyclically using a four bit ring counter and method to change it into Johnson counter for activation of eight sequences cyclically using four bits (b) Ring counter block diagram and four sequences.

Such a counter is useful in a computer where several circuits are sequentially enabled (or activated) from 1st stage to the *n*th stage using an *n*-bit ring counter.

16.4.4 Johnson Counter (Even Sequences Switch Tail or Twisted Ring Counter)

A Johnson counter connection changes are shown in Figure 16.10(a). It is a shift register with serial-input bit permanently from last msb state $\overline{Q}$. It has clear and preset facilities and with no serial output facility. Figure 16.11(a) shows the changes in ring counter of Figure 16.10(a) to activate 8 sequences.

Table 16.5 gives the state table. The Johnson counter shifts state 1 from lower significant bit (lsb) end to maximum significant bit (msb) end. The complement of msb $\overline{Q}_D$ is feedback to D_A input of the first stage *D FF* to form a ring. It acts like a shift left of the bit and also rotate the complement of msb.

The output of the counter is not binary. From the table, output sequences are 0001, 0011, 0111, 1111, 1110, 1100, 1000, 0000, 0001, 0011 ...

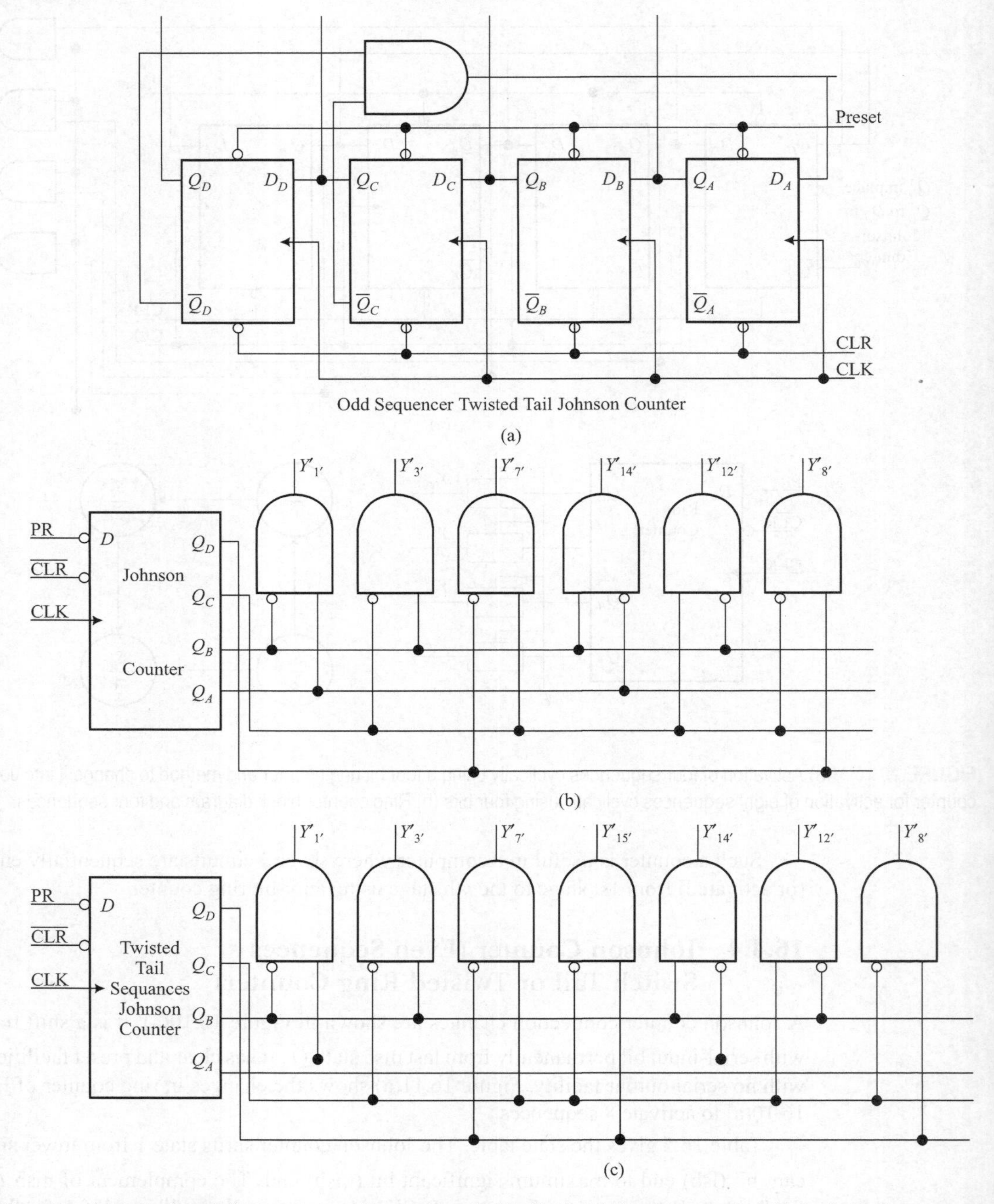

FIGURE 16.11 (a) Activation of eight sequences (state table of Table 16.5 in a twisted ring counter) (b) Cyclically 6 sequence activator using 6-AND array and a four bit Johnson counter (c) A four bit twisted tail odd sequencer Johnson counter—activator of seven sequences cyclically.

TABLE 16.5 State table for a 4-bit Johnson counter to activate eight sequences ((Y'_0, Y'_1, Y'_3, Y'_7, Y'_{15}, Y'_{14}, Y'_{12}, Y'_8)

Present state		Next state after transition (Q_A', Q_B', Q_C', Q_D')	Present output Y
State	$Q_3Q_2Q_2Q_1$		
S_0	0000	$\underline{S_1}$	00$\underline{01}$ (Y'_1)
S_1	0001	$\underline{S_3}$	0$\underline{01}$1 (Y'_3)
S_3	0011	S_7	$\underline{01}$11 (Y'_7)
S_7	0111	S_{15}	1111 (Y'_{15})
S_{15}	1111	$\underline{S_{14}}$	11$\underline{10}$ (Y'_{14})
S_8	1000	$\underline{S_{12}}$	1$\underline{10}$0 (Y'_{12})
S_{12}	1100	$\underline{S_8}$	$\underline{10}$00 (Y'_8)
S_{14}	1110	S_0	0000 (Y'_0)

Note: Q outputs which define a unique sequence are connected to AND gate of the AND array are underlined in the last column to activate outputs in circuit of Figure 16.11(b).

When pair of Q outputs of a Johnson counter connects to a 2– input enable buffer (Figure 16.11(b)), the $2n$-sequences activate cyclically. Figure 16.11(b) shows how a 4-bit Johnson counter pair of FF outputs to the buffers activates 6 sequences Y_1', Y_3', Y_7', Y_8', Y_{14}', Y_{12}' cyclically. [Maximum eight outputs Y_1', Y_3', Y_7', Y_{15}', Y_{14}', Y_{12}', Y_8' and Y_0' can be activated].

Such a counter is useful in a computer where $2n$ circuits are sequentially enabled (or activated) from 1st stage to the $2n^{\text{th}}$ stage using an n-bit Johnson counter.

16.4.5 Odd Sequencer Johnson Counter (Odd Sequencer Switch Tail or Twisted Ring Counter)

An odd sequencer (seven numbers) Johnson counter (switch tail or twisted ring counter) is shown in Figure 16.11(c). It is a shift register with serial-input bit permanently to an AND gate output, which has inputs $\overline{Q}_D$ and $\overline{Q}_C$. It has clear and preset facilities and with no serial output facility.

Table 16.6 gives the state table. The twisted tail odd sequencer counter shifts state 1 from lower significant bit (lsb) end to maximum significant bit (msb) end. The msb Q_D complement and its previous stage msb complement are feedback through a two-input AND gate to D input of the first stage FF_A to form a ring [Figure 16.11(a)]. It acts like a shift left of bit and rotates two stages complement after ANDing.

The output of the counter is not binary. From the table, output sequences are 0001, 0011, 0111, 1111, 1110, 1100, 1000, 0001, 0011...

When pair of Q outputs of a Johnson counter connects to a 2– input enable buffer, the $(2n - 1)$ sequences activate cyclically. Figure 16.11(c) shows how a 4-bit twisted tail modulo-7 counter connect 7 pairs of FF outputs to the seven number two-input AND arrays and activates 7 clock sequences $Y1$', $Y3$', $Y7$', $Y15$', $Y14$', $Y12$' and $Y8$' cyclically.

Such a counter is useful in a computer where 6 or 7 circuits are sequentially enabled (or activated) using an 4-bit Johnson counter.

TABLE 16.6 State table for a for a 4-bit Johnson counte*r to activate seven sequences from circuit of Figure 16.11(c)

Present state		Next state after transition (Q_A', Q_B', Q_C', Q_D')	Present output Y
State	$Q_3Q_2Q_2Q_1$		
S_1	0001	S_1	0 0 0 1 (Y'$_1$)
S_3	0011	S_3	0 0 1 1 (Y'$_3$)
S_7	0111	S_7	0 1 1 1 (Y'$_7$)
S_8	1000	S_{15}	1 1 1 1 (Y'$_{15}$)
S_{12}	1100	S_{14}	1 1 1 0 (Y'$_{14}$)
S_{14}	1110	S_{12}	1 1 0 0 (Y'$_{12}$)
S_{15}	1111	S_8	1 0 0 0 (Y'$_0$)

Note: Q outputs which define a unique sequence are connected to AND gate of the AND array are underlined in the last column which activate outputs in circuit of Figure 16.11(c).

16.5 SYNCHRONOUS COUNTER

16.5.1 Synchronous Counter Using Additional Logic Circuit

Let us consider a typical design shown in Figure 16.12. The *J* and *K* inputs connect together in each flip flop, and are connected to a logic combinational circuit. *J* and *K* inputs are held 0 so that *Q*s don't change up to the final stage gets the counting input. As soon as final stage

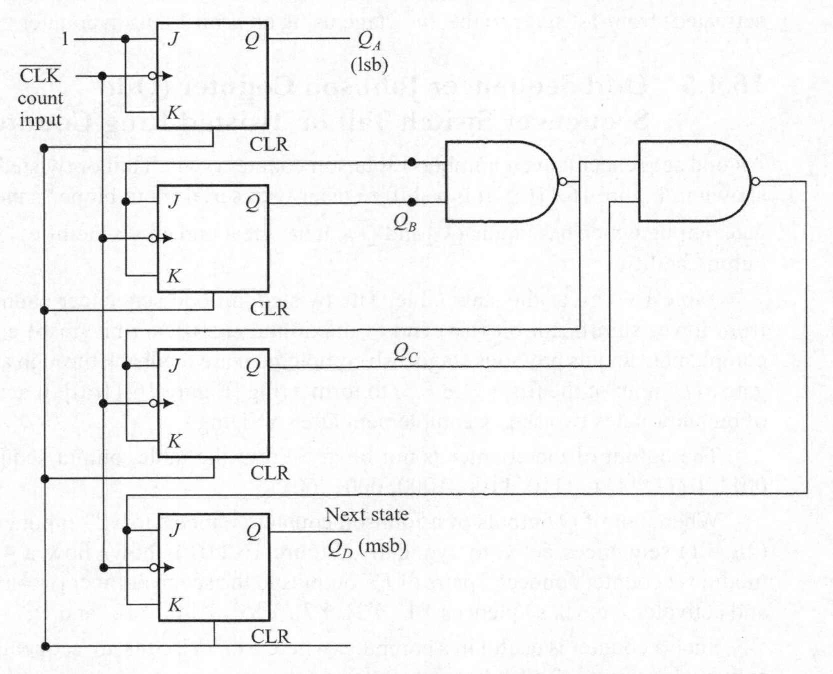

FIGURE 16.12 A synchronous counter using *J-KFF*s and circuit to force state change at all Qs.

gets the input, the J and K of all FFs will simultaneously equal to 1 and toggle as per the inputs. Except first Q_A, all Q outputs are also inputs to this combinational logic circuit. This logic circuit checks the states of Q_A, Q_B, Q_C, and Q_D and gives the different JK FFs a pair of inputs 0 or 1. When both JK inputs to a FF are 0 the output is unaffected but if both JK inputs to a FF are 1, the output is affected, and it's output Q_n is complemented.

The combination logic circuit ensures that all the FFs change the Q outputs precisely at the same time after the clock edge transition.

Whenever the clock input makes a desired transition, that all T inputs are interconnected together. Such synchronous counters are useful for fast counting applications as these provide output valid states fast, and for applications when all Q_D, Q_C, Q_B, QA are to be read simultaneously. False intermediate states are also now not obtained in the output.

16.6 ASYNCHRONOUS CLEAR, PRESET AND LOAD (JAM) IN A COUNTER

1. There may be a need to reset all Qs as 0s at the start of the counter. For example, in a up counter.
2. There may be a need to reset all Qs as 1s at the start of the counter. For example, in a down counter.
3. There may be a need to set certain Qs as 1s and remaining as 0s at the start of the counter. For example, in a timer. Timer is a counter getting counting inputs at regular intervals after presetting a certain value.

How do we do these three operations in a counter? Following description explains these operations.

When we activate CLR input at a counter, without waiting for a clock edge all the output Qs may become 0 after the propagation delay ΔT. This procedure-adopting counter is known as the asynchronous clearing counter. If all output Qs can be preset to 1s without waiting for a clock edge (but of course after a time ΔT). Such a procedure-adopting counter is known as an asynchronous presetting counter.

If without waiting for a clock edge, we can clear certain Qs and preset remaining Qs in a counter simultaneously, then individual FF clearing and presetting inputs are called JAM inputs. JAM inputs asynchronously load with out waiting for a clock edge at the input the Qs of the counter.

Figure 16.13(a) shows resetting = 0s of all Qs by a common CLR input. Figure 16.13(b) shows setting of all Qs = 1s by a common PR input. Figure 16.13(c) shows setting and resetting of all Qs as desired by the JAM inputs.

16.7 SYNCHRONOUS CLEAR, PRESET AND LOAD FACILITIES IN A COUNTER

Figure 16.13(d) shows the circuit for synchronous clear, preset and load.

TTL 74163 or CMOS 74HC163 has synchronous clear and synchronous load facility. In that case, the clear, preset or load inputs must be properly defined at a time, t_s, called setup time, before a clock edge at the input is activated. The effects of the CLR (clear) or PR (preset) or load inputs appear in the outputs only after a time equal to ΔT_{FF} from the clock

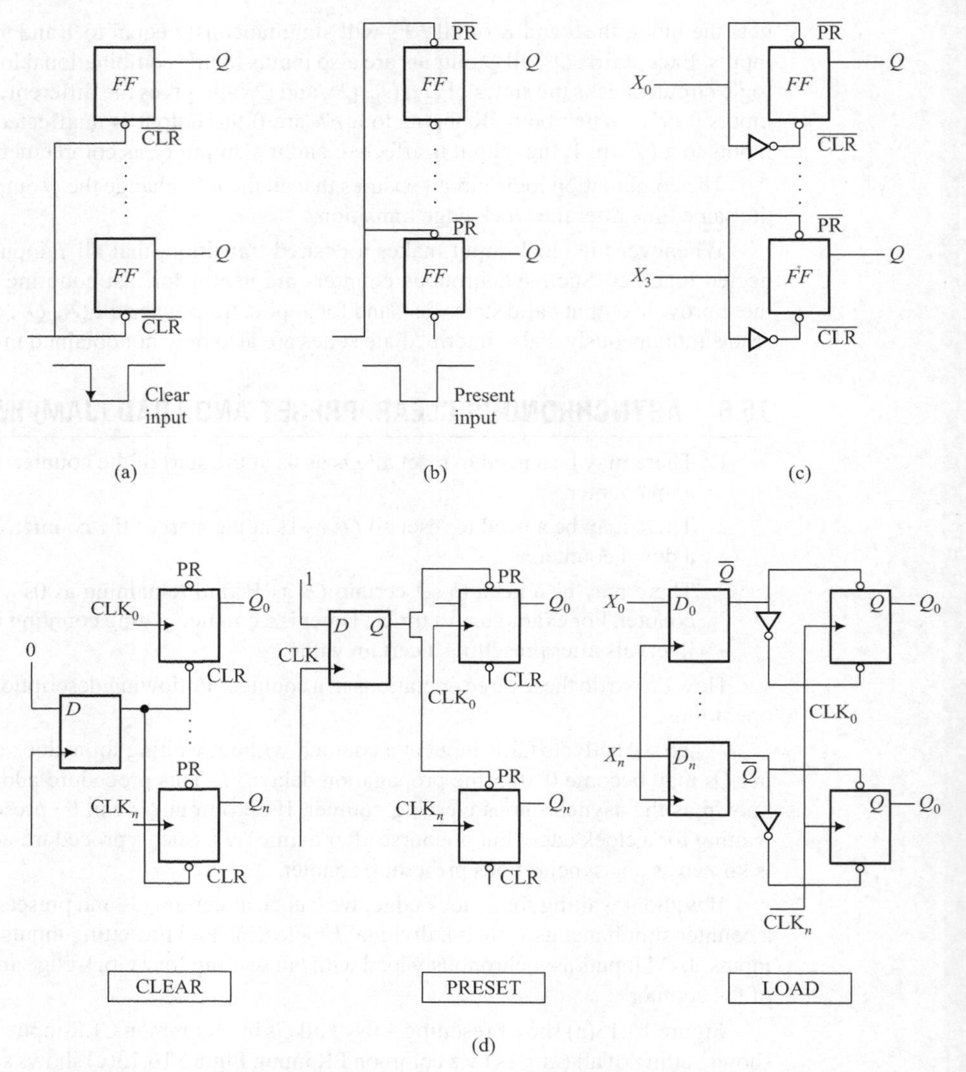

FIGURE 16.13 (a) Clearing all Qs (b) Presetting all Qs (c) Setting and clearing of Qs using the JAM inputs (d) Synchronous clear, preset and load.

input transition i.e. activation (net time taken > $t_s + \Delta T_{FF}$). Here, the ΔT_{FF} is propagation delay within a FF of the counter.

16.8 TIMING DIAGRAMS

Figures 16.14, 16.15, 16.16 and 16.17 show the timing diagrams in a SISO, in a left-shift shift SIPO register, in a ripple counter and in a synchronous counter, respectively.

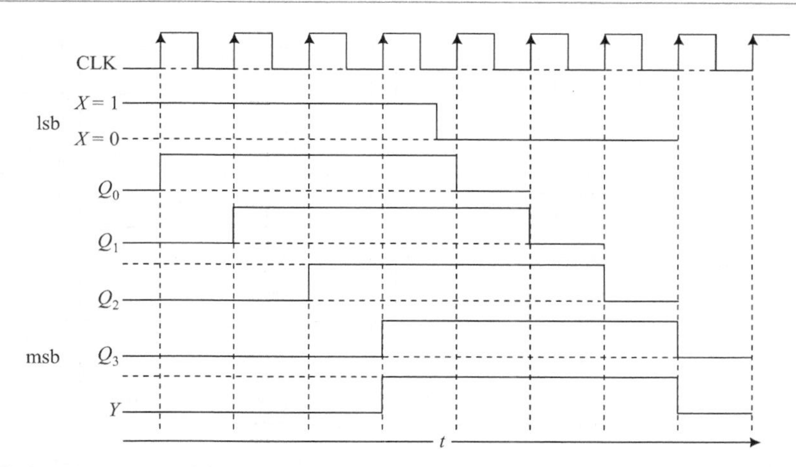

FIGURE 16.14 Timing diagrams in a SISO.

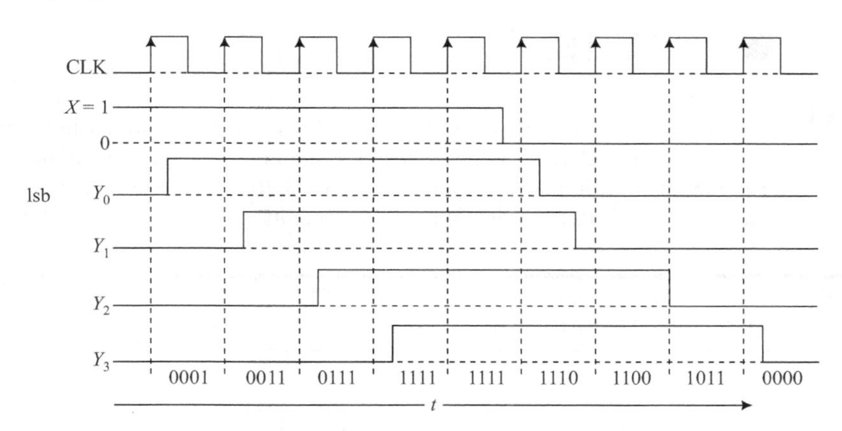

FIGURE 16.15 Timing diagrams in a SIPO left-shift shift register. (ΔT_{FF} delay effects also shown)

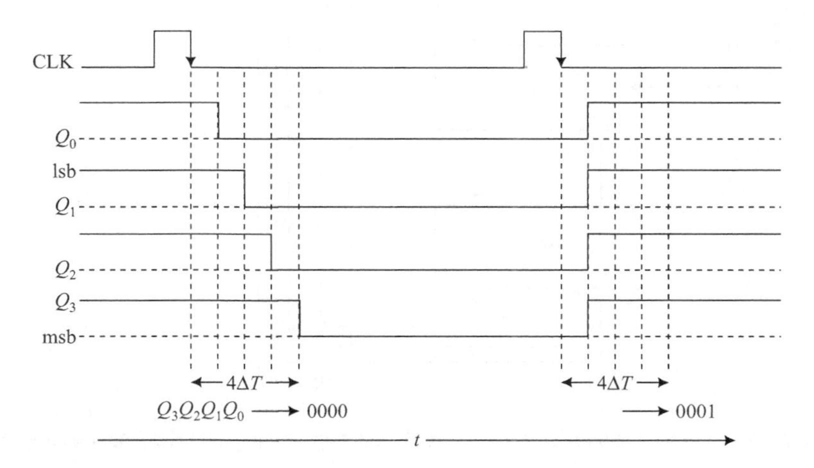

FIGURE 16.16 Timing diagrams in a —ve edge triggered ripple counter. (ΔT_{FF} effects also shown)

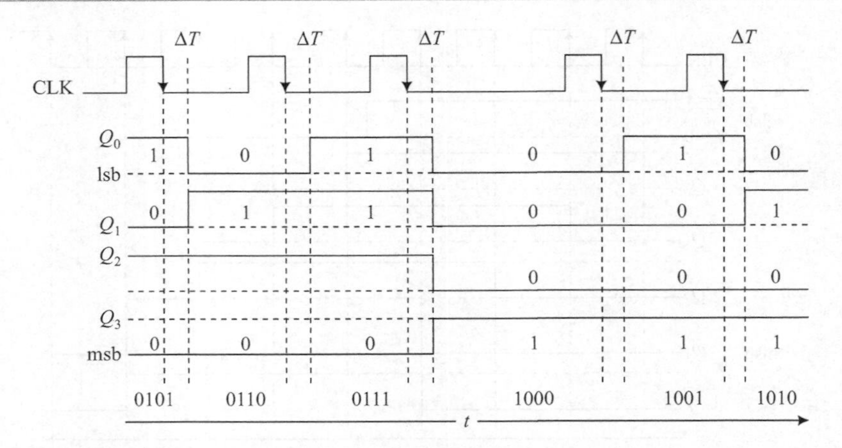

FIGURE 16.17 Timing diagrams in synchronous counter. (ΔT_{FF} effects also shown)

EXAMPLES

Example 16.1

Consider the circuit-*a* shown in Figure 16.18 ***A*** is a 4-bit parallel-in parallel-out register, which loads at each rising edge at the clock input *C*. The input to it is from a four-bit bus, *W*. The output from it acts as an input to a 16 × 4 ROM, whose output is floating when the enable input *E* is 0. A partial table of the contents at the ROM is as follows:

Address	0_d	2_d	4_d	6_d	8_d	10_d	11_d	14_d
data+	0011	1111	0100	1010	1011	1000	0010	1000

+ Four bits

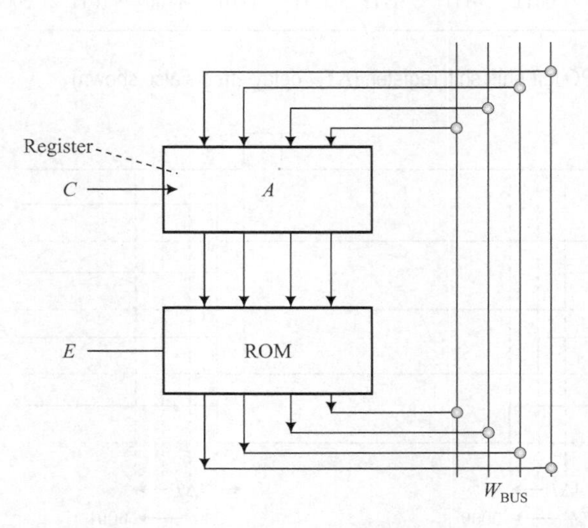

FIGURE 16.18 Circuit-*a* consisting of a register *A* and a ROM with register input and ROM output on a common set of four lines (bus *W*).

The clock to the register is shown, and the data on the *W* bus at first edge time t_1 is 0110. The data on the bus at second edge time t_2 is (A) 16.71 (B) 1011 (C) 1000 (D) 0010.

Solution

A common set of lines, used by a part of a digital circuit as inputs or outputs or both forms a bus. Usually these are four or eight or 16 or 32 lines and carry data or address bits. In present problem, there are four-lines for the 4-bit data.

Consider the time $t1$. $C = 0$. Register input is given to us as 0110 (= 6_d).

After the first +ve edge at the clock, the register output will become 6_d. Therefore the ROM output will be for address = 6_d. From the table given to us, it will be 1010 (= 10_d) at the *W* bus. This now becomes the input to the register.

After the second edge, register output will also become 10_d. Therefore ROM output for address = 10_d from the table given to us will be 1000_b on the *W* bus.

Between t_2 and the second edge, there is no other positive clock edge. Hence the data on *W* bus at t_2 is also 1000_b. Therefore, answer (*C*) is correct.

Example 16.2

Study the problem in Example 16.1 once again. Suppose you use parallel loading serial right shift register, what will be the output on *W* bus at t_2 now. Assume that output for the next adjacent addresses of 0, 2, 4, 6, and 8 is complement of the one given in the table.

Solution

After the first +ve edge at the clock, the shift register output will become 0110 (= 6_d) shifted right, which will now equal 0011 (= 3_d). Therefore, ROM output for address = 3 from the table given to us will be 0000 (complement of the output at address = 2). *W* bus now becomes the input to the register. The register input will now be 0.

After the second edge, register output will also become 0011 shifted right, which will now equal 0001. Therefore, ROM output for address = 1 from the table given to us will be complement of 0011. Therefore, the output after the edge will be 1100.

Between t_2 and the second edge, there is no other positive clock edge. Hence the data on *W* bus at t_2 will be also be 1100.

Example 16.3

Design a 4-Bit buffer Register with parallel output after storing.

Solution

A buffer register has an additional feature. The data is clocked into the register by a CLK or strobe input. However, the data is actually placed on the output lines after the receipt of an input at the output enable (or control) pin. The buffer register is often made by placing an AND gate each between a *Q* output and a datum output (Refer Figure 16.2(a)). The one input terminal of each AND is connected to each *Q* output and the other terminal of all the ANDs coupled together and connected to an output (read) enable (or control or out) pin. If at the control pin, input logic is 0, the outputs from the register are 0s. If 1, then the *Q*s of the internal *FF*s appear at the register outputs. The buffer register is like a small memory unit in which the data bits can be written upon the activation of *store* or *load* or *write* signal at its CLK input, and the same can be read upon the activation of a output enable (or strobe or read or control) signal. Figure 16.2(a) showed a buffer register.

Example 16.4 Give the state table of a 4-bit PIPO register and then draw a state diagram of the PIPO when parallel-loading parallel-output register no shift.

Solution

Table 16.7 shows the state table of a PIPO and Figure 16.19(a) shows the state diagram.

TABLE 16.7 State table for a Moore model sequential circuit for a 4-bit PIPO register

Present state	Next state after transition (Q_1', Q_2', Q_3', Q_4')						Present outputs Y_1, Y_2, Y_3, Y_4
	Input (X_1, X_2, X_3, X_4) = (0, 0, 0, 0)	Input (X_1, X_2, X_3, X_4) = (0, 0, 0, 1)	...	...	Input (X_1, X_2, X_3, X_4) = (1, 1 1, 0)	Input (X_1, X_2, X_3, X_4) = (1, 1 1, 1)	
$S_i = S_0$ or S_1 or S_2 or ... S_{15}	S_0	S_1	...	...	S_{14}	S_{15}	Same as Qs corresponding to present state S_i

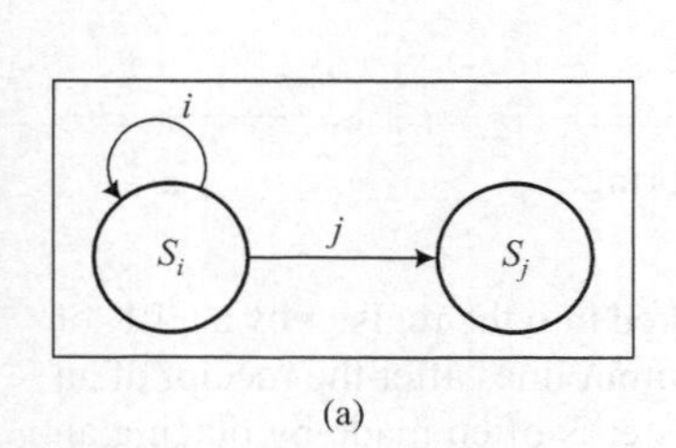

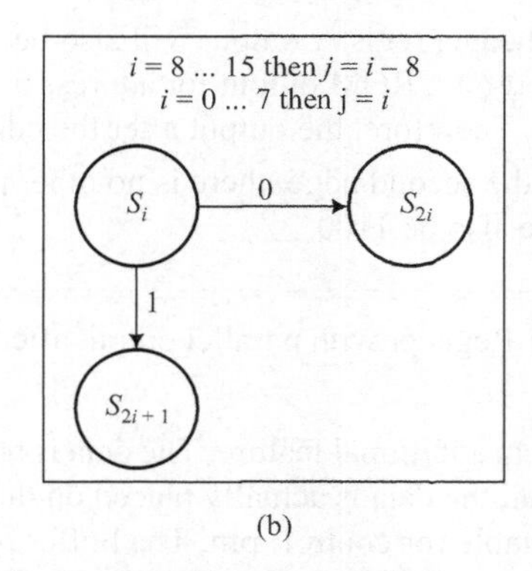

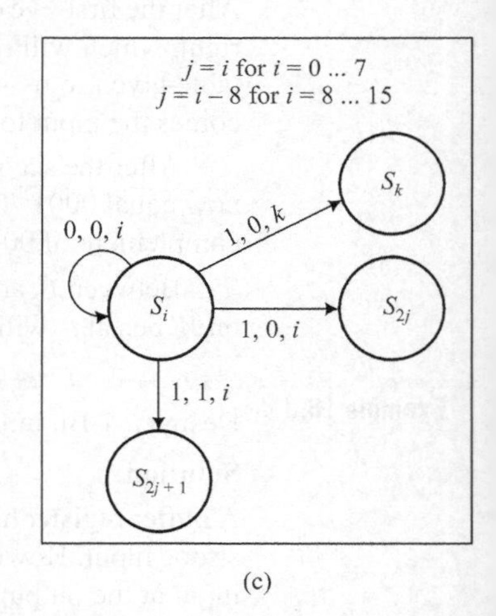

FIGURE 16.19 (a) State diagram of a Parallel-In Parallel-Out (PIPO) no shift register (b) State diagram of a SIPO left shift register (c) State diagra of a PISO left shift register.

Example 16.5 Explain Serial-In Parallel-Out (SIPO) left shift register by a state table.

Solution

Table 16.8 shows the sate table for a left shift SIPO and Figure 16.19(b) shows the state diagram.

TABLE 16.8 State table for a Moore model sequential circuit for a 4-bit SIPO shift left register

Present state		Next state after transition (Q_0', Q_1', Q_2', Q_3')		Present outputs Y_0, Y_1, Y_2, Y_3
State	$Q_0Q_1Q_2Q_3$	Input $X = 0$	Input $X = 1$	
S_0	0000	S_0	S_1	0000
S_1	0001	S_2	S_3	0001
S_2	0010	S_4	S_5	0010
S_3	0011	S_6	S_7	0011
S_4	0100	S_8	S_9	0100
S_5	0101	S_{10}	S_{11}	0101
S_6	0110	S_{12}	S_{13}	0110
S_7	0111	S_{14}	S_{15}	0111
S_8	1000	S_0	S_1	1000
S_9	1001	S_2	S_3	1001
S_{10}	1010	S_4	S_5	1010
S_{11}	1011	S_6	S_7	1011
S_{12}	1100	S_8	S_9	1100
S_{13}	1101	S_{10}	S_{11}	1101
S_{14}	1110	S_{12}	S_{13}	1110
S_{15}	1111	S_{14}	S_{15}	1111

The $Q_0\ Q_1\ Q_2\ Q_3$ outputs will change after ΔT_{SR} as $Q_0 \leftarrow Q_{1,}\ Q_1 \leftarrow Q_2$, and $Q_2 \leftarrow Q_3$ in a left shift register upon a clock transition and serial datum bit is placed at $Q_{3.}$ Present state 0001 (S_1) will become 0010 (S_2) when serial bit = 0 and 0011 when serial bit = 1.

A SIPO shift-left register transfers the input bits X to next **Q**s at each clock edge excitation so that $Q'_j = X_i$, where $j = i + 1, j \Leftarrow n - 1$ and $j > 0$ and $i = 0, 1, 2, ..., n - 1$ in an n-bit register. The lsb Q'_0 will become equal to serial bit. It gives parallel outputs, which are the same as the next state. When serial shift input $X = 0$, left shift register multiplies a binary number by 2. Figure 16.19(b) shows that state $S_i \rightarrow S_{2i}$ when serial in at $Q_3 = 0$.

Example 16.6 Explain a Parallel In Serial Out (PISO) left shift register using a state table.

Solution

Assume $Q0$ is msb and $Q3$ as lsb. Assume that weight of $Q0 > Q1$, $Q1 > Q2$, $Q2 > Q3$.

Table 16.9 shows the sate table for a left shift PISO and Figure 16.19(c) shows the state diagram. When serial-in at $Q_3 = 1$ and $\overline{\text{load}}$/shift = 0 then $S_i \rightarrow S_k$ where $k = i$ if i is even and $k = i + 1$ if is odd. $Y \leftarrow Q_0$.

When serial-in at Q_3 and no shift then S_i does not change. When serial-in = 1 and shift = 0 then $S_i \rightarrow S_{2i}$. When serial-in = 1, shift 1 $S_i \rightarrow S_{2i+1}$. The $Q_0\ Q_1\ Q_2\ Q_3$ outputs up on a clock edge transition, within a time equal to ΔT_{SR} from the transition, will load the inputs into the Qs when $\overline{\text{load}}$/Shift = 0. When the $\overline{\text{load}}$/Shift = 1, the register changes as a left shift register as $Q_0 \leftarrow Q_1$, $Q_1 \leftarrow Q_2$, $Q_2 \leftarrow Q_3$, and Q_3 will become equal to 0 when serial-in = 0 and equal to 1, when serial in = 1. Serial output will be from Q_0 assume Q_0 is of higher place (weight) value.

TABLE 16.9 State table for a Moore model sequential circuit for a 4-bit PISO left shift register

Present state	Next state after transition (Q_0', Q_1', Q_2', Q_3')						Present outputs Y
	$\overline{\text{Load}}$/Shift, Serial-in X_0, X_1, X_2, X_3 = 000000	$\overline{\text{Load}}$/Shift, Serial-in at Q_r X_0, X_1, X_2, X_3 = 100000	$\overline{\text{Load}}$/Shift, Serial-in at Q_0 (X_0, X_1, X_2, X_3) = 110000	...	$\overline{\text{Load}}$/Shift, Serial-in (X_0, X_1, X_2, X_3) = 101111	$\overline{\text{Load}}$/Shift, Serial-in (X_0, X_1, X_2, X_3) = 111111	
$S_i = S_0$ or S_1 or S_2 or ... S_{15}	S_0	S_0	S_1	...	S_{14}	S_{15}	$Y = Q_0$

Note: There are 6 inputs—$\overline{\text{Load}}$/Shift, serial-in from Q_3 and parallel inputs (X_0, X_1, X_2 and X_3). Q_3 is the lsb and Q_0 is msb.

Parallel input 1010 ($X10$) will become 0100 (S_4) when serial bit = 0 and serial-out Y will be = 1. Parallel input 0001 ($X1$) will become 0010 (S_2) when serial-in bit = 0 and serial-out Y will be = 0. Parallel input 0001 ($X1$) will become 0011 (S_3) when serial-in bit = 1 and serial-out Y will be = 0.

Point to Remember

A PISO loads the parallel inputs bits X into the Ds and gives the output $Qj = Di$ on the clock edge, when $\overline{\text{Load}}$/Shift input = 0. Here j and i = 0, 1, 2, ..., $n - 1$ in an n-bit register.

When $\overline{\text{Load}}$/Shift input = 1, a PISO shift-left register transfers the D input bits X to next **Q**s at each clock edge excitation so that $Q'_j = X_i$, where $j = i + 1; j <= n - 1$. The lsb Q_3 will become equal to serial bit and post excitation and transition msb Q_3 will become the one-bit output Y at the serial-out pin. The parallel Q outputs are the same as the next state. When serial shift input = 0, left shift register multiplies the input binary number by 2.

Example 16.7

Explain a serial in serial out (SISO) circular right shift register.

Solution

A circular shift register feeds back the msb (last) stage Q_{n-1} input to the output of (next) stage Q_A. Table 16.10 gives a state table of a SISO circular shift register with serial input at the flip-flop with Q_{n-1} output and serial output at the flip-flop with Q_0 output. Q_0 is also the input S_{n-1}.

Figure 16.20(a) a n-bit SISO circular right shift register and Figure 16.20(b) shows its state diagram of a 4-bit circuit.

The $Q_3\,Q_2\,Q_1\,Q_0$ outputs up on a clock edge transition, within a time equal to ΔT_{SR} from the transition, will load the serial input from Q_{i+1} into the Q_i and give serial output to Q_{i-1} stage.

The register changes as a shift register as $Q_3 \rightarrow Q_2$, $Q_2 \rightarrow Q_1$, $Q_0 \rightarrow Q_0$, $Q_0 \rightarrow$ serial out as well as Q_3.

TABLE 16.10 State table for a Moore model sequential circuit for a 4-bit SISO right shift circular register

Present state		Next state after transition (Q_3', Q_2', Q_1', Q_0')		Present output $Y = Q_0$
State	$Q_3Q_2Q_1Q_0$	Input $\bar{i}/s = 1$	Input $\bar{i}/s = 0$	
S_0	0000	S_0	S_0	0
S_1	0001	S_8	S_1	1
S_2	0010	S_1	S_2	0
S_3	0011	S_9	S_3	1
S_4	0100	S_2	S_4	0
S_5	0101	S_{10}	S_5	1
S_6	0110	S_3	S_6	0
S_7	0111	S_{11}	S_7	1
S_8	1000	S_4	S_8	0
S_9	1001	S_{12}	S_9	1
S_{10}	1010	S_5	S_{10}	0
S_{11}	1011	S_{13}	S_{11}	1
S_{12}	1100	S_6	S_{12}	0
S_{13}	1101	S_{14}	S_{13}	1
S_{14}	1110	S_7	S_{14}	0
S_{15}	1111	S_{15}	S_{15}	1

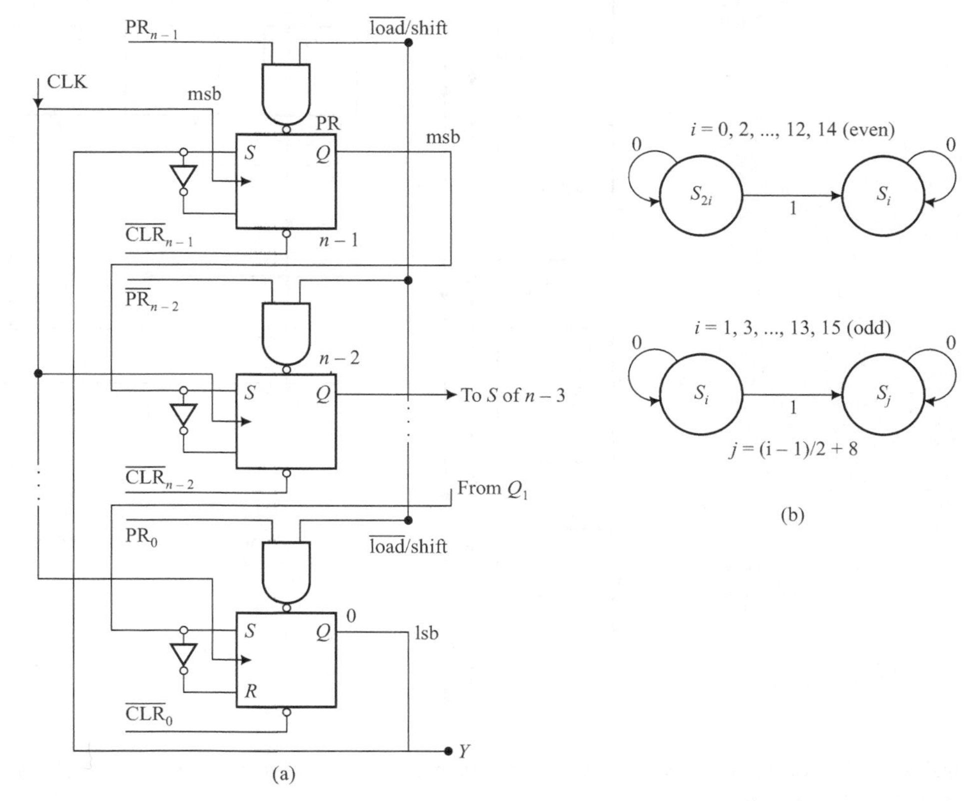

FIGURE 16.20 (a) n-bit SISO circular right shift register (b) 4-bit circuit State diagram.

Serial input when state is 0010 (S_2) will become 0001 (S_1). Serial-out Y will be = 0. Serial input when state is 0001 ($S1$) will become 1000 ($S8$). Serial-out Y became 0 because after the transition $Q_0 = 0$.

When Q_{n-1} has a higher weight than other Qs in the binary number representation of output. A circular right shift SISO loads the inputs bit from Q_1 into the first stage D_0 and gives the serial output at Q_0 on a clock edge. Further, first stage gives the input to the last stage on a clock edge.

Example 16.8

Give the circuit of a left cum right shift cum parallel loading and parallel output universal register using the four multiplexers.

Solution

Figure 16.21 shows the circuit of a universal register using the four multiplexers. Each D flip-flop gets the D-inputs from a multiplexer output. We use four channel multiplexers.

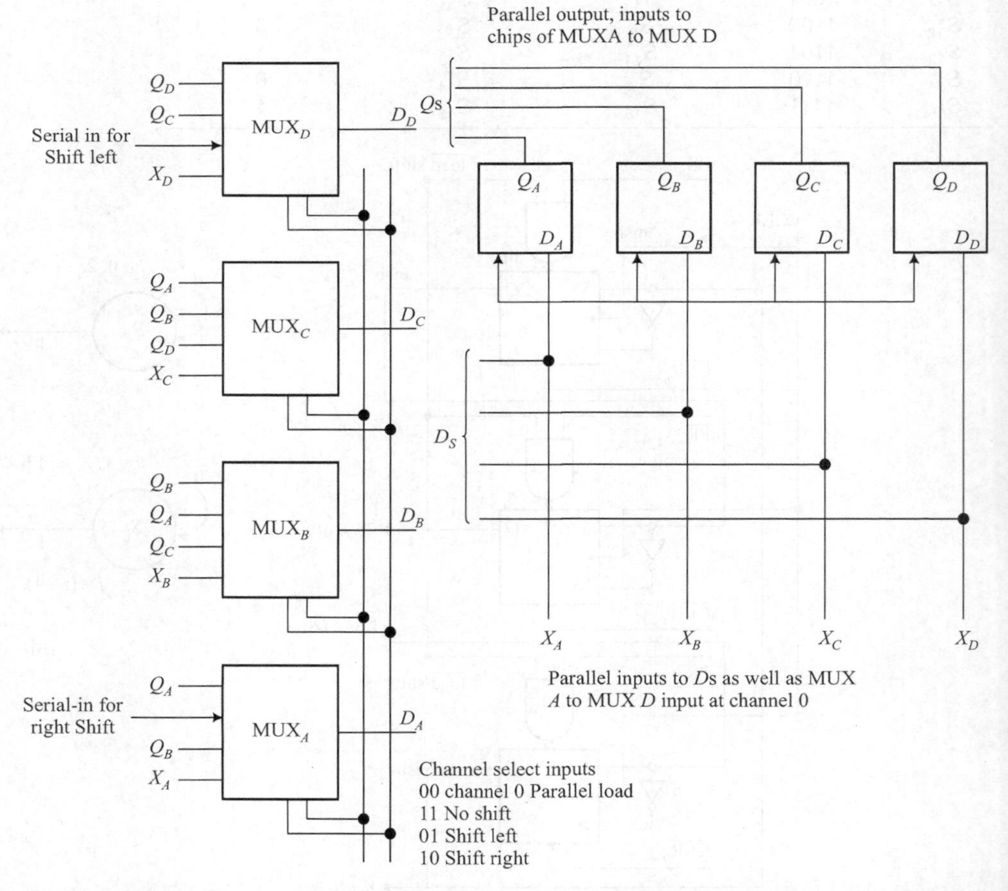

FIGURE 16.21 Circuit of a left cum right shift serial cum parallel loading and serial cum parallel output universal register using four number of multiplexers.

Each multiplexer has a four channel-input lines, one of which is selected at an instant for the output bit connected to *D*-input of a *FF*. Each multiplexer has two channel-address select bits for the four input channels. 00 selects channel *A*, 01 channel *B*, 10 channel *C* and 11 channel *D*.

1. MUX_A, MUX_B, MUX_C and MUX_D are given the parallel inputs X_A, X_B, X_C and X_D, respectively. These inputs are at the channel-A inputs. Therefore, when the channel address select bits are 00, the output on next clock transition will be the Q_A, Q_B, Q_C and Q_D as per D_A, D_B, D_C and D_D and these are equal to X_A, X_B, X_C and X_D, respectively.
2. MUX_A, MUX_B, MUX_C and MUX_D are given the inputs Q_B, Q_C, Q_D and serial-in bit for left-shift needs, respectively. These inputs are at the channel-*B* inputs. Therefore, when the channel address select bits are 01, the output on next clock transition will be as per the Q_B, Q_C, Q_D and serial-in bit at the *D*-inputs, D_A, D_B, D_C and D_D, which register at the Q_A, Q_B, Q_C and Q_D lines, respectively. Therefore, the next states are $Q_D' =$ serial-in left shift bit, $Q_C' = Q_D$, $Q_B' = Q_C$ and $Q_A' = Q_B$, respectively.
3. MUX_A, MUX_B, MUX_C and MUX_D are given the inputs serial-in bit for right shift, Q_A, Q_B and Q_C, respectively. These inputs are at the channel *C* inputs. Therefore, when the channel address select bits are 10, for the output on next clock transition the serial-in bit for right shift, Q_A, Q_B and Q_C, respectively, will be the *D*-inputs, D_A, D_B, D_C and D_D, which register at the Q_A, Q_B, Q_C and Q_D output state lines. Therefore, the next states are $Q_D' = Q_C$, $Q_C' = Q_B$, $Q_B' = Q_A$ and $Q_A' =$ serial-in bit for the right-shift, respectively.
4. MUX_A, MUX_B, MUX_C and MUX_D are given the parallel inputs Q_A, Q_B, Q_C and Q_D, respectively. These inputs are at the channel-*D* inputs. Therefore, when the channel address select bits are 11, the output on net clock transition will remain as the Q_A, Q_B, Q_C and Q_D. This D_A, D_B, D_C and D_D for the next state *Q*s, Q'_A, Q'_B, Q'_C and Q'_D, are same as Q_A, Q_B, Q_C and Q_D, respectively.

Therefore we get a universal shift register, which is a parallel load PIPO when address bits to multiplexers are 00, left-shift PIPO when 01, right shift PIPO when 10 and holds (no-shift) when 11.

Example 16.9 Give a circuit to get a binary synchronous counter.

Solution

Refer the circuit of Figure 16.12. It gives a synchronous divide by 2^n. The *Q* outputs from the each *FF* available at the external output *Q*s, as in us a binary counter. Each output of a *FF* has two states 0 or 1. The binary counter can count from 0 to $(2^n - 1)$ pulses. After every $(2^n - 1)$ pulse, at the next pulse the counter returns to the original state. The original state of all the outputs of the *FF*s can be made 0 by the inputs at the $\overline{CLR}$.

Example 16.10 How will you design a binary counter using the ICs?

Solution

7493 is a standard TTL four stage binary ripple counter. Examples of various binary counters are 4040B (a 12 stage CMOS binary counter) and 4060(a 14 stage CMOS binary counter). 74HC93 or 74HCT93 are high speed CMOS variation of a 7493 counter of four stages.

Example 16.11

How will you make circuit of binary asynchronous up and down ($U/\overline{D}$) counters?

Solution

The up counter has the flip-flops which toggle or change on the negative (1 to 0) transition. When a *JK MS FF* is used for the UP counter, 0 to 1 transition at a clock input simply effects the master section of the *FF* and 1 to 0 transition in the slave section of that *FF*. The flip flop is, therefore, said to trigger at the negative edge of the clock pulse. Using a negative edge triggered *D FF* in method (ii) of Figure 16.6(a) also gives the identical effects.

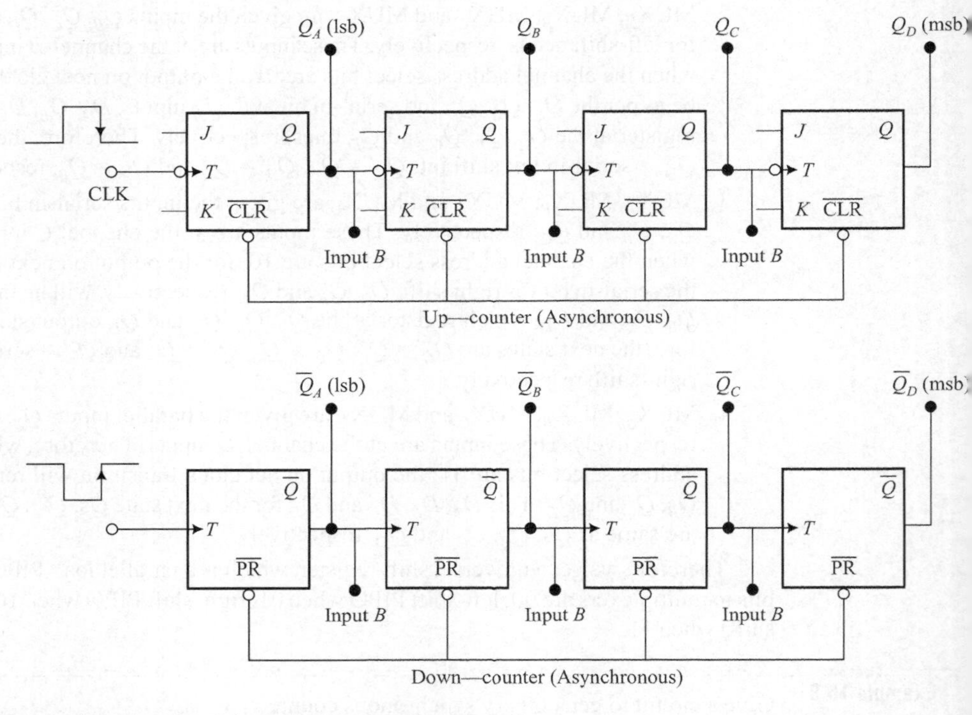

FIGURE 16.22 The circuits of the up and down asynchronous binary counters.

While cascading *FF*s (i) if *T FF*s toggle on positive edge transition (0 and 1), (ii) if '$\overline{Q}$ outputs are connected to the *T* inputs, and (iii) if $\overline{Q}$ outputs be used to indicate the states of the counter instead of *Q*, then we obtain a down counter. When $\overline{PR}$ input is activated, all $\overline{Q}$ outputs become 0s. It is also possible that in an integrated circuit, we may call $\overline{Q}$ as *Q*, CLR as *PR* and may offer it straight away as a down counter. 74HC469 and 744469 are the examples of ICs for the 8 bit (stage) up down counters.

Figure 16.22 shows two counters up and down using *T FF*s.

Example 16.12

Describe an asynchronous counter. What are the asynchronous counter disadvantages and advantages?

Solution

Consider the circuit in Figure 16.22(a). Upon the receipt of 16th input pulse, the flip flops *A* to *D* have to change the state. First Q_A is reset to 0, then after a while Q_B is 0, then after a while Q_C is reset to 0 and then after a while Q_D is 0. Let a flip flop needs time 'ΔT_{FF} to change the *Q* output. When a *T*-input is applied, the total time in changes of *Q*s at the sixteenth pulse at *T*-input is $4\Delta T_{FF}$. The time for the final correct state changes at *Q*s upon the eight pulses is $3\Delta T_{FF}$. The time for the *Q*s final change at the 4th pulse is $2\Delta T_{FF}$.

Disadvantage of using an asynchronous counter is that there is a variable time at which correct *Q* outputs of a counter appear. The Q_A changes at the faster rate than Q_B, the Q_B changes at the faster rate than Q_C, and so on. Readings of the combination Q_D, Q_C, Q_B, Q_A may differ during the period $4\Delta T_{FF}$. For fast changing pulses, the false readings during $4\Delta T_{FF}$ is obtained by such a counter.

An asynchronous counter does not have all clock inputs common in same logic state as in a synchronous counter.

Advantage of an asynchronous counter is that for multi-stage counting (for example, 16-bit) the fewer gates suffice and power dissipation in asynchronous counter is also small.

Example 16.13 How will you make circuit of modulo-10 decade counter?

Solution

IC 7490 or 74HC90 is the decade asynchronous modulo-10 ripple counter, which is very popular. Figure 16.23 shows its circuit diagram. Let Q_A and Q_D are 1, and Q_B and Q_C are 0. This counter has a within it a combinational logic such that when these outputs are present at the Qs at a next clock input pulse, the counter is cleared, and all output *Q*s are reset to 0. This counter has interesting facilities for changing its modulus, as we shall learn while performing an experiment with this counter.

Working of the Modulo – 10 (Decade) Counter

Let us refer to Figures 16.22(a) and (b). Let us join Q_A and input-*B* (CLK of next to lsb *FF*s). Let us also set inputs, R_0 (1) and R_0 (2) at 1. Let us also make at least one of the inputs *Rg* (1) and *Rg* (2) as 0. CLR becomes 0 because of a NAND gate shown at lower end of left hand side and *PR* moves to 1 because of a NAND on the lower end of the left hand side in circuit of Figure 16.22(a). This means that CLR inputs of all *JK* flip flops are activated, and this makes $Q_A = Q_B = Q_C = Q_D = 0$. Now let us make at least one of the R_0 and one of the Rg inputs at 0 logic levels. The CLR and *PR* inputs to all four *FF*s are therefore 1 (inactivated). So the *J*, *K* and CLK inputs are now effective. A –ve edge at the input *A* is now applied which is also the CLK input of FF_A. This will set Q_A to 1. As Q_A is connected to CLK of FF_B, at next input pulse at A, (i) the Q_A toggles to 0 and therefore a –ve edge occurs at CLK of FF_B. At next –ve edge at input *A*, the Q_A will again change its state and does not clocks the FF_B. At next –ve edge at input *A*, the Q_A will again change its state, but now clocks the FF_B, which toggles to 0. This feeds a –ve edge at FF_C which gives $Q_C = 1$. The outputs *Q*s indicate the number of –ve edges at input *A*. This continues up to nine –ve edges at input *A*, which means till $Q_A = 1$, $Q_B = 0$, $Q_C = 0$ and $Q_D = 1$. Q_B and Q_C are connected to a AND gate, the output of which is connected to *K* input at FF_D. As Q_C is connected to *J* input of FF_D, upon receiving 10th edge at input *A*, all outputs will become 0. This means that the configuration shown in Figure 16.22 (a) acts as a decade counter.

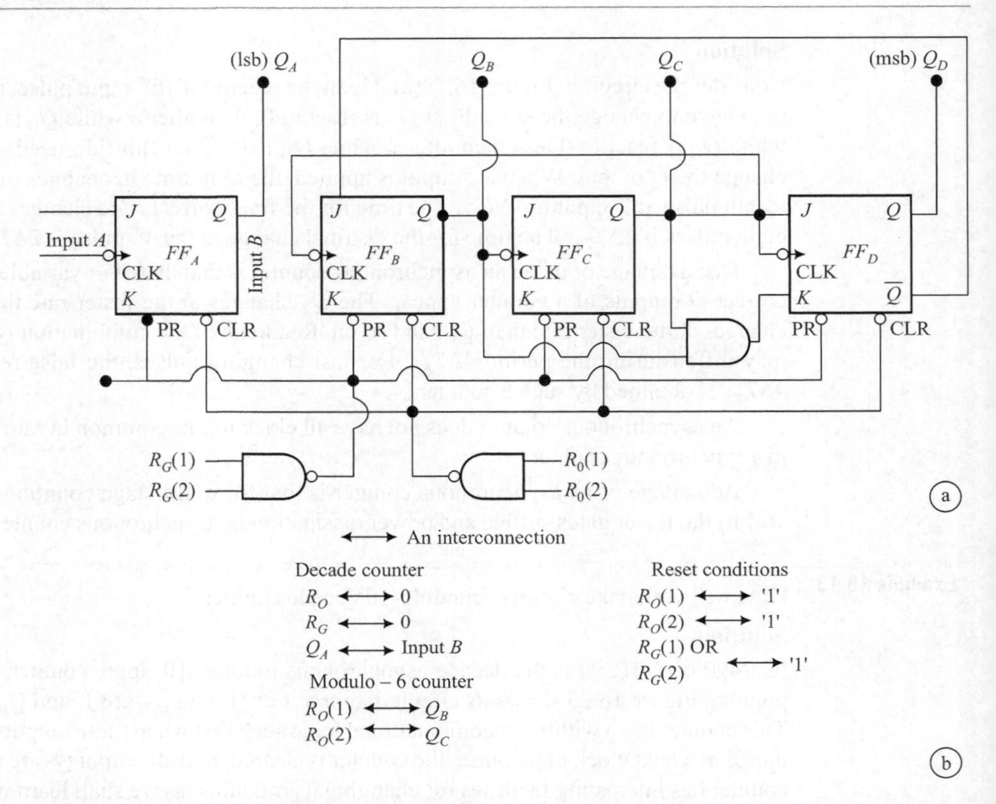

FIGURE 16.23 (a) The circuits of the decade counter using up and down asynchronous binary counters IC 7490 (b) Reset conditions and uses of circuit as a decade counter, modulo-6 counter.

Example 16.14 How will you make circuit of modulo-6 counter?

Solution

A modulo-6 counter can be built from IC-74HC90 as follows due a special logic in it. For modulo-6 counter, we disconnect R_0 (1) and R_0 (2) from level 0s and connect it to Q_B and Q_C [refer Figure 16.22(b)]. As soon as both Q_B and Q_C goes 1 i.e. six clock pulses are completed at the input A from the initial Qs clear condition, the R_0 (1) and R_0 (2) will go 1. So the counter outputs Q_A, Q_B and Q_C are cleared and reset to 0 levels. The Q_D is immaterial in modulo-6-counter.

Example 16.15 How will you make a synchronous counter using MSI ICs?

Solution

TTL 74163 or 74HC163 is synchronous pre-settable binary up counter. 74HC193 or 74LS193 is up-down binary synchronous up-down counter.

Synchronous counter like 40161(b), 40162(b), 40163(b), 74HC161(b), 74HC193(b) are designed such that it has all of its T flip-flops change its Q outputs simultaneously. 74HC590 and 74HC592 are 8-bit synchronous counters (each of eight FF stages).

EXERCISES

1. Design a PIPO, which is a 4-bit buffer register with parallel in (loading) and parallel output (storing). [Hint: refer figure 16.2(a), use the AND buffers at both the ends.]
2. Show the state diagram from a state table corresponding to a SIPO. (Hint: Refer sate table at Table 16.2).
3. Design a multipurpose behavior of a left shift register, which shift left when channel inputs, are 01 and circularly shift left when 10, synchronous clear when 11 and no-shift when 00. (Hint: Refer Example 16.8).
4. Solve the problem in Example 16.2 once again for a left-shift register.
5. Design a SISO circular left shift register.
6. Draw and compare the state diagrams of (i) a SIPO of 4 bits (ii) a 4-stage ripple counter.
7. Give a state table and state diagram of a binary synchronous counter.
8. Give a state table and state diagram of a asynchronous down modulo 6 counter.
9. How much is the time interval for *Q*s to stabilize in a sixteen bit asynchronous counter? Assume flip flop propagation delays = 10 ns.
10. How will you make circuit of modulo-7 counter?
11. Show state diagram of a modulo-4 counter?
12. What should be the count value for a 8-bit counter to time out in 1 ms if input clock pulses at 100 μs intervals?
13. Show the timing diagram of a ring counter.
14. Show the circuit of a 3-bit Johnson counter.
15. Show the timing diagram of a Johnson counter modified for maximum 7-sequence clock generator.

QUESTIONS

1. For a combinational circuit, the all-possible outputs are given in a truth table. For the clocked sequential circuits like counters, and registers. A state table describes the all-possible outputs. When do we use truth table and when do we use a state table? Explain by examples of the multiplexer and shift register.
2. What do we mean by state machines? Can we explain an exemplary state machine in which *D FF*s plus several logic gates put together gives us the various states at the *FF* outputs? (Hint: Shift register at every clock edge generates another state. Similarly a counter does so.)
3. What do we mean by a register?
4. How do we clear (erase) in the register? (Hint: Using $\overline{\text{CLR}}$ input at the *SR* latches or *FF*s).
5. How do we store in the register?
6. What do we get at the outputs from the register when the outputs are in the tristates?
7. A *D FF* is a 1 bit register. Explain, Why can it be said so?

8. Explain circuit of a shift register, say, 74 HC175 or 74HC75.
9. What do we mean by shifting left (i) a binary word 1001 0101 (ii) word 0010 0101 1001 0100? If we shift twice then what will be the result.
10. Can we consider shift left four times as equivalent of multiplication by 16?
11. What do we mean by shift right? If we shift on right twice the following words, what will be resulting number? (i) 0100 0100 (ii) 0000 0011 1100 0011.
12. Can we consider shift right once means division by 2?
13. What is the application of 8-bit shift register? (Hint: In accepting serial data bits over a telephone cable and storing these in a computer or in transmitting a binary word digitally over a cable like a telephone cable).
14. What do we mean by clear input in a register?
15. What do we mean by SIPO, PISO, PIPO and SISO Shift registers? Explain with timing diagram (i) shift left in each (ii) shift right in each.
16. A ripple counter is very valuable as circuit for a frequency divider and when a slow response is tolerable outputs like in a digital clock. Why?
17. What is the main disadvantage in a ripple counter?
18. What are the differences between an asynchronous and a synchronous counter?
19. A synchronous counter provides gives output after, say, ΔT. An asynchronous counter like ripple counter of n FF stages will give a correct output after how much time.
20. What is the difference between an up counter and a down counter?
21. How is the different *T* flip-flops (or *JK FF* with $J = 1$ and $K = 1$) cascaded to obtain a counter?
22. What is a binary counter? Explain its circuit
23. What are the applications of a binary counter?
24. What is a decade counter? Explain its circuit.
25. What are the applications of a decade counter?
26. How do we convert a binary counter into a decade counter and into a divide-by-6 counter? How can such circuits be used in a digital clock?
27. What do we mean by (i) a ripple counter and (ii) a ring counter? Explain their circuits.
28. How will we divide input pulses by four?
29. What do we mean by the JAM inputs to a counter or a register?
30. Suppose we wish to connect a mechanical switch undergoing up down transition several times. Why do we connect a debouncing circuit in between contact switch and a counter? What will be advantage of a monostable circuit in-between? Can an Schmitt Trigger be also used in-between?
31. A counter can also be said to act as a register, which holds the number of input pulses with respect to the contents at the beginning. In a computer, a program counter is such a register or a stack pointer is another such register why?
32. What is the use of carry in and carry out pins in an IC of a synchronous counter?
33. Why do we sometimes incorporate a lockout time (dead time) count inhibit circuit at the count pulses at the counter input by using a monostable?

34. A counter can also be considered as a state machine (a sequential device which provides at an instant a set of outputs followed by another instant another set of outputs). Explain, why can we consider so?

35. Why can't a D latch with $\overline{Q}$ feedback to its D input be used as a TFF? In fact, we get oscillations of several tens of MHz at Q in such a circuit. Why? (Hint: After each time interval equal to propagation delay between D input and Q output, we get a toggling till such time when clock input level is active).

36. Why are asynchronous counters made from -ve edge triggered FFs and synchronous counters for +ve edge triggered FFs? (Hint: Easy interconnectivity between two FFs in an asynchronous between two FFs).

37. A ripple counter can be used to generate long duration pulses instead of using a monostable. How does it happen? Explain it. (Hint: For example, 14 stage binary counter like CMOS 4060 will give 2^{14} T duration pulse at its last stage if T is time interval between two successive pulses at its T input).

38. Describe synchronous counter advantages.

39. Describe ring counter applications.

CHAPTER 17

RAM, Address and Data Buses, Memory Decoding, Semiconductor Memories

OBJECTIVE

We have learnt in Chapter 13 that a memory stores the look-up tables for the bits or bytes at the outputs for each of the inputs, called address inputs. A ROM with 10 address input lines has 2^{10} addresses and each address there are 8 bits in a 1 k × 8 ROM (1 k = 1024 for the purpose of memory space representation). Also a complex combinational circuit can be easily assembled by the programmable logic memories (ROM, EPROM, EEPROM or Flash or OTP). ROM is non-volatile memory (retains the bits even after power-off).

We will learn an important form of semiconductor memory called RAM in this chapter. Concept of address and data buses is important for using the memories. We will learn that also. Decoding of signals for a memory chip placed on the common set of buses is done. This enables the read of the addressed byte. We will learn the decoding procedure.

17.1 DEFINITION OF RAM

RAM stands for random access memory. It is a random accessible read-and-write memory.

A random access means an access can be made from any addressed location in equal interval of time. An access means a read or write. A random access is contrast to a serial access, for example in a tape from which the memory bits (cells) are accessed serially. The cells at beginning of a tape can be accessed faster than the cells at the end. When searching a song on a videotape, the search for first song is faster than the last song. Consider another example. If the letters are arranged in a bundle, then search of a letter for your address is done serially (in sequence from the top). If each address letter is sorted into different mail-boxes (pigeon holes) then the search is random access. For each address, the time taken will be the same. One is permitted to define any random address for data.

Ability to externally write data bits with ease is one feature and loss of data on a power-supply interrupt is another. These two are important features of the RAMs. This feature makes a RAM, a volatile memory. [Ease means there is no need of any high voltage and/or external pulse or any erase cycle before any write operation.]

Point to Remember

> RAM is defined as a random accessible read-and-write memory in a semiconductor chip or system. It is volatile. It consists of cells made up of MOSFETs.

17.1.1 Cell in the Static and Dynamic RAMs

The basic element of a semiconductor RAM is known as a cell.

A cell can be made using a complementary pair of metal oxide semiconductor FETs (CMOS) technology as in an IC. Alternatively a cell can be made from a single MOSFET. MOSFET RAMs have therefore two versions, static and dynamic depending upon the cell design. Figures 17.1(a) and (b) give the circuit for a cell of static and dynamic RAM, respectively.

Static RAM (SRAM) Cell: Bottom left and right sides MOSFETs are two pair of MOSFETs, the *drain* of one feeding charge to *source* of another so that the charge on a MOSFET and thus the voltage drop between drain and source remains static (constant). It works as a memory cell. External charge is given to transistor Q, when it is in non-conducting state. The output from it will be 1. When a discharging path is externally provided, its logic-state becomes 0. MOSFETs at the left and right topsides in Figure 17.1(a) circuit control the current through the cell by action as the active resistor.

Dynamic RAM (DRAM) Cell: Figure 17.1(b) gives the circuit cell of dynamic RAM. It has only one MOSFET. It does not have a cross-coupled connection like static RAM cell. Therefore, once it is charged and acquires state 1, it needs to be charged again due to non-zero leakage current (non-conducting path never has infinite resistance). Reading it and then writing it afresh must be done in a specific period of a few *ms*. A dynamic RAM should be read and written again (a process called refresh) within ≈ 4 ms or less in order to retain a bit in a cell of it. However, a dynamic RAM has an advantage of a high memory density per chip due to less number of MOSFETs needed per cell in it. [Refer Figures 17.1(a) and (b) for a comparison of number of MOSFETs needed per cell in a static and dynamic RAM]. Using a DRAM is no problem. A special circuit can perform read-write job for each cell array sequentially every ms.

A cell can also be made by *n*-channel metal oxide FET gates (called *n*-MOS gates) unlike CMOS where both *n* and *p* channel gates are present. The *n*-MOS operations are speedier.

A cell can also be made using ECL or TTL technologies, which uses bipolar transistors. The speed is very high in these cases and such memories are used as cache (a fast access

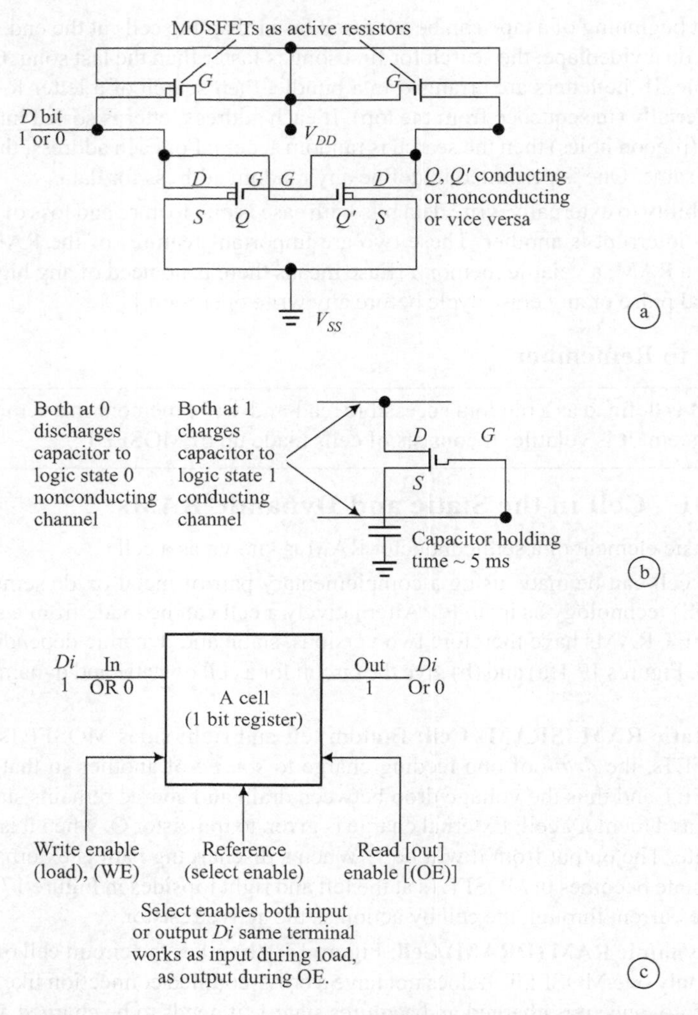

FIGURE 17.1 (a) Cell of static RAM consisting of four MOSFETs (b) Cell of dynamic RAM consisting of single MOSFET (c) Cell *read* and *write* state and select signals.

form of memory) and a high-speed data transfer is possible without wait unlike in conventional MOSFET memories.

However, CMOS RAMs are popular due to very low power dissipation compared to caches.

Figure 17.1(c) shows a cell like a one bit register. The output a datum bit of it can be read as well as can be altered. At an input clock pulse, it registers an input bit. The output *D* bit is available from this cell to a user (say, external circuit) only when read or out enable ($\overline{\text{OE}}$) input is supplied at cell input.

Point to Remember

> Static memories do not need refreshing and can relied upon to retain their memory bits in all the cells until such time as new data is written or the power supply is interrupted or switched OFF.

17.1.2 Cell-Array at the RAM

Figure 17.2(a) shows an array consisting of an octet of cells, which forms a reference address point in a RAM. Eight cells form an array for one byte of storage. Each cell array has a unique address. An array corresponds to an address location. When a word from a memory is of eight bits in that case there are eight cells in an array at an address.

Each cell array can therefore write or read one byte at an instance. When simultaneously clocked for loading or reading, the eight cells array inputs or outputs will be simultaneously written (changed) or simultaneously read, respectively. Instead of saying 8-bits per cell-array, we say that there are as eight bits at an addressed location for that cell array.

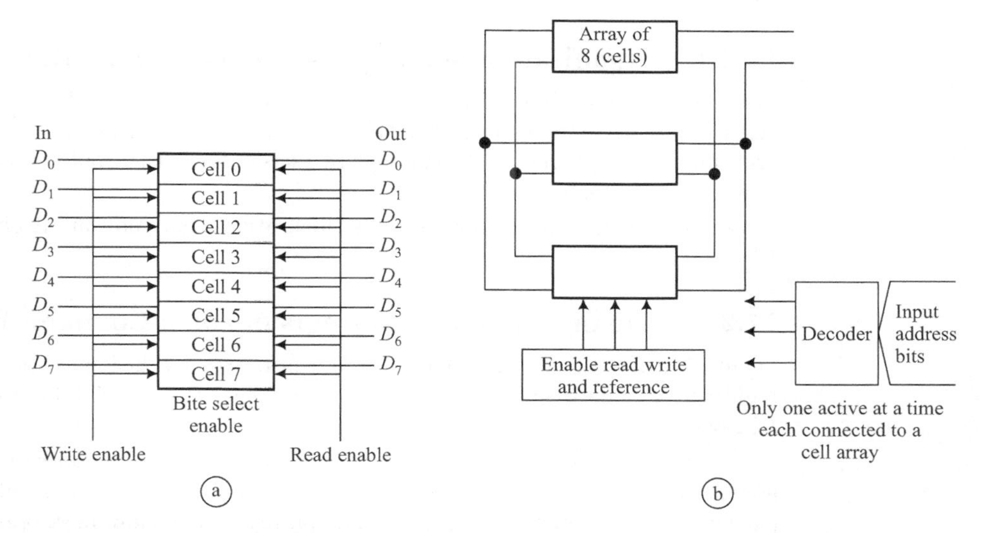

FIGURE 17.2 (a) Cell-array of RAM consisting of eight input-output data lines with select enable, read enable and write enable inputs (b) Cell-arrays on a common internal data bus with the common read enable and write enable inputs.

Point to Remember

> A cell array is of 8-cells when there is one byte 'read' or 'write' at an instance in a RAM. Each cell array has a unique address.

17.1.3 Cell-Arrays at the Addressed Locations in the RAM

An array of an octet of cells has an associated logic circuitry [Figure 17.2(b)] such that only one octet can be referenced (clocked for write or clocked for read at an instant). This forms an addressable byte.

Point to Remember

> There are large number of addressable locations, one location for each cell array. Only one octet can be referenced (addressed for write or clocked for read at an instant) out of these large numbers of locations.

17.2 TRISTATE OUTPUTS-CUM-TRISTATE INPUTS FOR A COMMON BUS ORGANIZATION OF THE BUFFER REGISTERS

Example 17.1 will explain the fabrication of a 4-bit buffer register with tristate outputs. Let us modify the circuit of a buffer register having AND gates [refer Figure 16.2(a)] at the outputs with tristate gates instead. Let us also modify this buffer register with tristate inputs also. Use of the modified circuit of the register is shown in Figure 17.3(a). Here, the circuit incorporates the addition tristate buffers at the input stages also (before the *D*-inputs). Figure 17.4(a) gives a symbol of the buffer register with load-enable input, out-enable input and clock input.

17.2.1 Organization of a Register on a Common Bus

The *Q* outputs and *D* inputs have through a common bus (set of parallel wires), called data bus. Figure 17.3(b) shows a data bus of four bits, D_0, D_1, D_2 and D_3. The data bus interconnects the circuit of Figure 17.3(a) through two sets of four tristate buffers each at the inputs as well as the outputs.

Figure 17.4(a) shows an equivalent block of a register unit in the circuit of Figure 17.3(a).

17.2.2 Organization of the *n*-registers on a Common Bus

Figure 17.4(b) shows how a set of *n* registers can now receive the inputs for the case of $n = 8$. It gives the outputs employing same set of wires, which form the data bus by a sequence of L_i, clock and O_i.

Example 17.3 will explain this organization. It will explain how does a register read and how does it write using the tristate buffers through a common bus. Example 17.4 will explain the connections and signals for the multiple registers and how these organize on a common bi-directional data bus. Example 17.5 will explain how can we read from one register and write to another.

17.3 COMMON BUS ORGANIZATION OF BYTES AT THE CELL ARRAYS IN RAM

The circuit of Figure 17.4(a) also demonstrates an important application of Figure 17.4(b) circuit. Through a common internal data bus, a cell array (like a register) among the multiple arrays in a RAM can get inputs when the write operation is required and can deliver the output when read is required. The contents of one array can be copied (called moved or transferred) to another array by an appropriate sequence of inputs. (Refer Example 17.5)

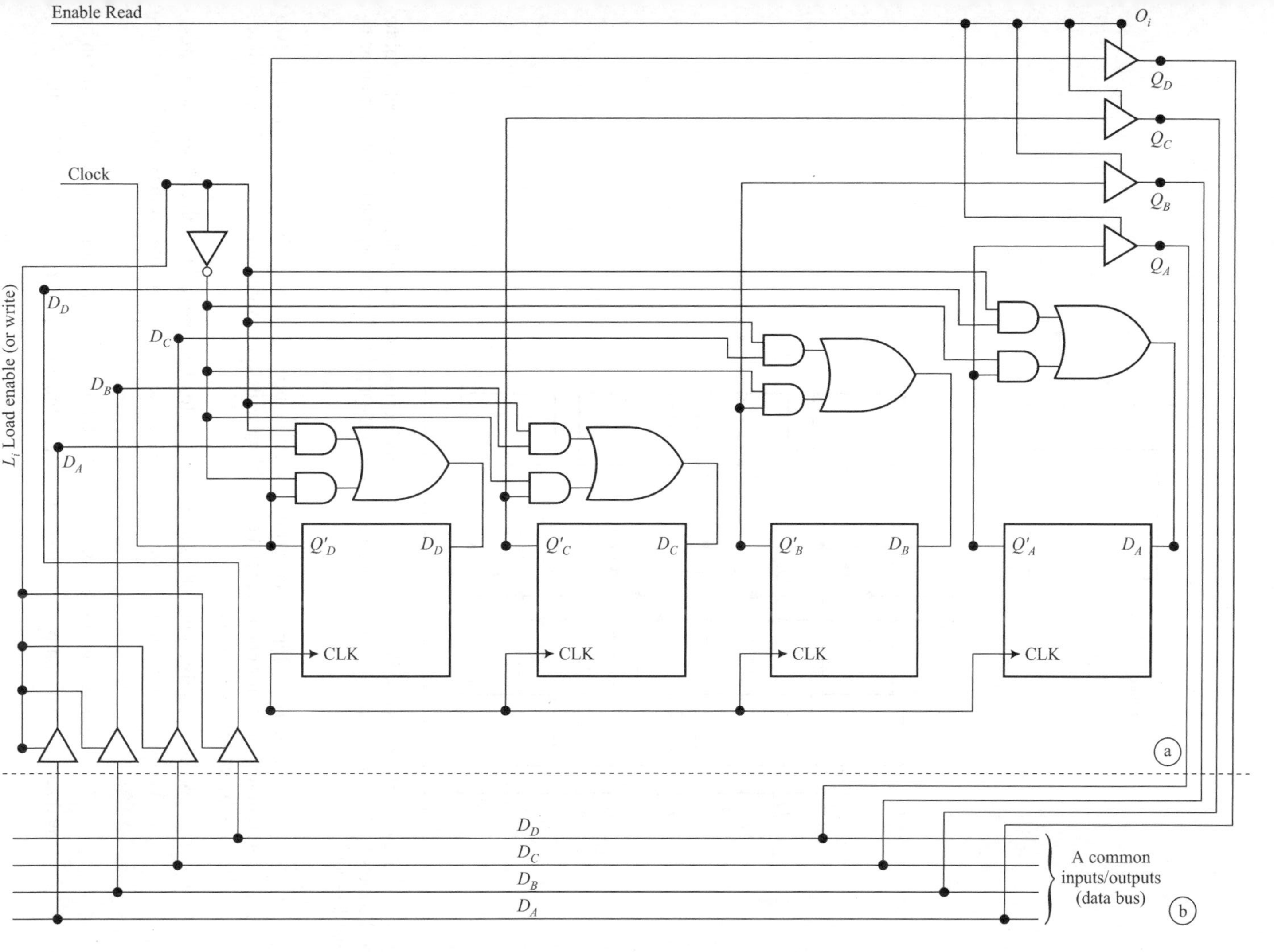

FIGURE 17.3 (a) Parallel—in Parallel—out four bit buffer registers, each consisting of 4 D flip flops with tristate buffers at the inputs as well as outputs using two inputs: one for the store (called load) input and other for out enable (called read) enable (b) Placing (organizing) a single register on a bi-directional data bus.

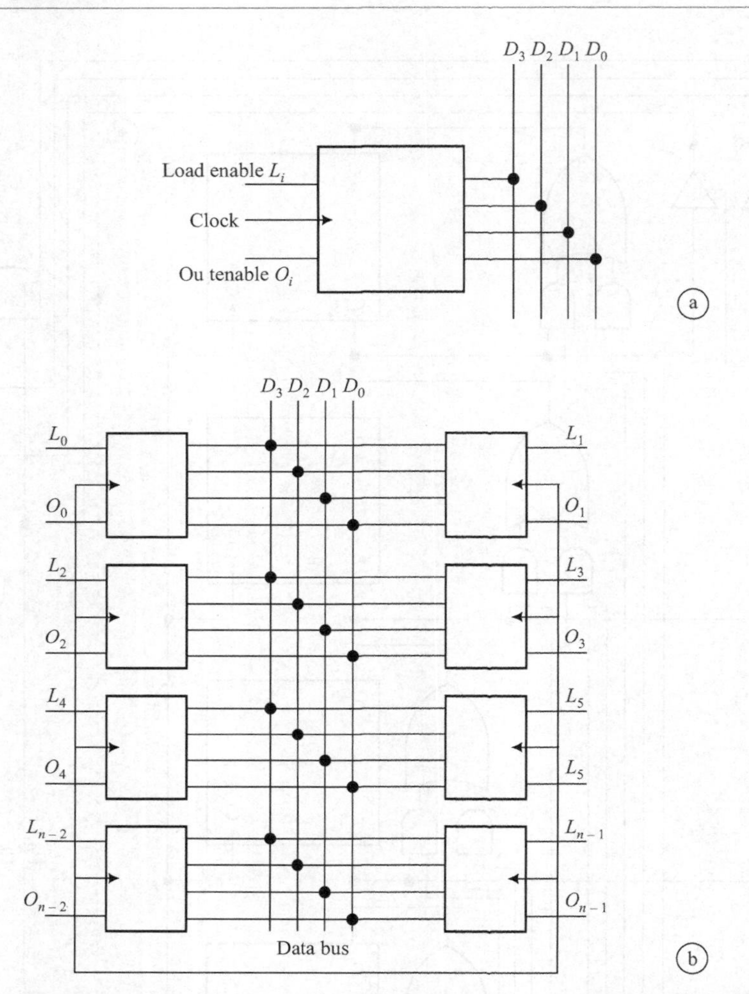

FIGURE 17.4 (a) Symbol of a Parallel—in Parallel—out four bit buffer register made from 4 *D* flip flops with tristate buffers at the inputs and outputs for the store or load input and out enable or read enable input (b) Organizing of the *n*-registers on the bi-directional data bus.

Assume (i) a chip-select (array–select) input $\overline{\text{CS}}$ is active, (ii) $\overline{\text{WR}}$ and $\overline{\text{RD}}$ are two common inputs for enabling read or enabling write for any cell-array, and (iii) for a cell array byte there is an address *j*.

When *j* is active and $\overline{\text{WR}}$ is active, the L_j (load enable) to the *j*-th cell array activates.

When *j* is active and read enable is active, the O_j (load enable) to the *j*-th cell array activates.

Therefore, using four inputs $\overline{\text{CS}}$, $\overline{\text{WR}}$, $\overline{\text{RD}}$ and j, j-th byte (array memory cells) can be read and j-th byte can be written or read depending on whether $\overline{\text{WR}} = 0$ or $\overline{\text{RD}} = 0$.

17.4 INTERNAL ADDRESS DECODER IN A RAM

Suppose $j = 0, 1, \ldots, 2047$ in a $2k \times 8$ memory. It is not possible to have 2048 external inputs to active one byte read or write. There is a internal decoder. The decoder has only m-input address lines actives one output at an instance out of 2^m outputs. For 2048 byte addresses (2k) at the RAM, external inputs $m = 11$. For 4096 byte addresses (4k) $m = 12$, When $m = 20$, 1024×1024 (1M) byte addresses at the RAM can be accessed. When $m = 32$, $4 \times 1024 \times 1024 \times 1024$ (= 4G) byte address inputs to the internal decoder of a RAM are needed. A single chip 64 MB RAM or above is nowadays available.

The m-address inputs to the internal decoder circuit of RAM are from a bus called address bus. Address bus width connecting to the RAM can be 11, 12, 16, 20 or 26 lines to a RAM, depending upon whether the RAM has 2 kB or 4 kB, 64 kB, 1 MB or 64 MB memory.

Let address bus inputs to RAM internal decoder circuit are 0000 1000 0011. The byte from 132th address (counting from 0) can be read using the data bus lines $D0$ to $D7$ when both CS and RD activate. (CS first, then RD).

Let address bus inputs to RAM internal decoder circuit are 1000 0000 1000 0011. The byte from 32900th address can be written using the data bus lines D_0 to D_7 when both $\overline{CS}$ and $\overline{WR}$ activate. ($\overline{CS}$ first, then $\overline{WR}$).

17.5 INTEGRATED CIRCUIT FOR A RAM

A RAM IC has large number of addressable locations (for example, in 64 MB chip). Figure 17.5 gives a model of a RAM IC chip numbered 6116. IC 6116 has 2048 arrays of octet of eight cells each. It is therefore called $2k \times 8$ RAM.

The number of external address bus inputs. The address connects to internal decoder. The 2048 decoder output lines each connect internally to an array. Each one of it is an internal reference select to a byte. Internally in the IC, one of the outputs of this decoder is active at a time and there can be only one but output from the total 2048 cell-arrays at a time on D_0-D_7 bus lines. The IC has hexadecimal addresses 0000H or 07FFH, binary 000 0000 0000 to 0000 0111 1111 1111. Each word equals a byte and is accessible in equal period, whatever may be hexadecimal address because the access to the 6116 is random access.

There is a write input of 1 bit, read input of 1 bit and the decoder outputs are enabled inside the IC by a master command bit, which selects the chip itself. It is called $\overline{CS}$ signal. [$\overline{CS}$ when active logic state is when it is 0.] The internal circuit inside the chip will be immaterial (dead!) to the outside circuits to which it is interconnected in case $\overline{CS}$ bit is 1. (i) It will make internal D_0-D_7 bi-directional lines (input lines cum output lines) in tristates with respect to the external $D0$-$D7$ data bus lines. The external data bus is then free for use by the other chips in a computer system or by the other RAM chip in the system. (ii) The address bus is also in tristate to the inputs of internal address decoder in the RAM chip.

The read bit is also called out enable bit. The read operation is possible only when this bit is activated by state 0 in 6116.

The write operation is possible only when write bit also called load-enable activated by state 0 in 6116.

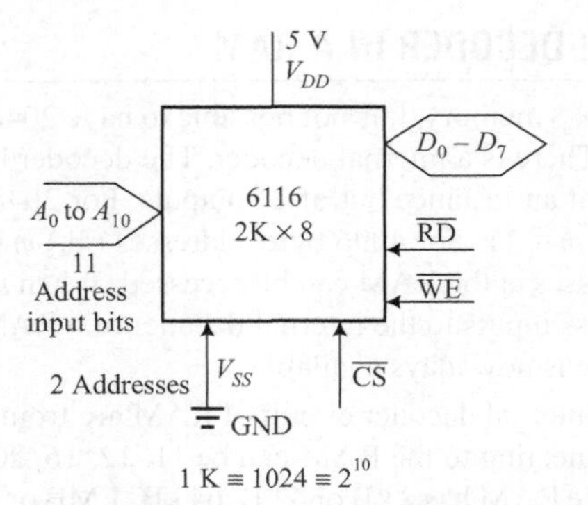

FIGURE 17.5 Model of a RAM IC chip numbered 6116 having 2048 arrays of octet of eight cells each and called 2k × 8 RAM.

The 6116 has 24 pins: (i) 11 address input pins to its internal decoder, (ii) two supply pins V_{DD} and V_{SS} (5 V and GND), (iii) eight data input cum output pins, and (iv) three control pins, $\overline{RD}$, $\overline{WR}$, $\overline{CS}$ Figure 17.5 gives pins used in an IC semiconductor memory 6116 This IC is used as a static RAM of 2 KB.

Points to Remember

> Address bus carries the address in the encoded format. It connects to a internal decoder of the RAM and that decodes an address to make the appropriate byte available.
>
> Data bus connects to internal data bus, which internally carries the bits for reading and writing into a cell array.
>
> Control bus has signals ($\overline{RD}$, $\overline{WR}$) for the read and write operations.
>
> An external mechanism (decoding circuit or logic circuit) selects a RAM chip among may interconnected on a common set of address, data and control buses in a computer.

17.6 INTEGRATED CIRCUITS OF THE RAMS, ROMS AND OTHER DEVICES ON A COMMON SET OF CONTROL, ADDRESS AND DATA BUSES

1. In a system, there are number of ICs for the memories, which can be written with as much ease as can be read. For example, we need 256 MB RAM and use the 64 MB RAM ICs.
2. In a system, there are PROMs (or ROMs). These are also random access. This read/write memory, by a misnomer called random access memory (RAM) though in fact, both type of the memories, RAM and ROM, are randomly accessible. RAM access (read) with equal ease (an access time for a read or for a write is independent of a memory address location), while PROM permits only the read by the system and write is only after erase.
3. In a system, there can be other devices also. Devices may have separate read and write control signals ($\overline{RD}$ and $\overline{WR}$) in certain systems and may have it same ($R/\overline{W}$) in other systems.

All the RAM ICs, ROM ICs and device ICs can be organized and connected to and fro a common set consisting of the unidirectional address and bi-directional data buses and a control bus.

Control bus consists of the signals for read and write, memory and IO, address latch (in case of existence of partially or fully multiplexed address-cum data lines) and reset.

17.7 EXTERNAL DECODING OF THE ADDRESS BUS TO THE INTEGRATED CIRCUITS OF A SYSTEM

Let us learn decoding by the following example. Let us design a decoding circuit for the following memories at the following assigned addresses.

1. Address bus lines from A_0 to A_{15} (total 16 lines).
2. Data bus lines D_0 to D_7 (total eight lines).
3. Control bus consisting of $\overline{\text{RD}}$ and $\overline{\text{WR}}$ for read and write operations, respectively.

Assume the following address-assignments in the example:

(a) PROM chip assigned addresses between 0000H to 1FFFH.

(b) RAM-1 chip between 8000H to BFFFH. (First RAM having addresses, distinct from any other memory addresses).

Figure 17.6 (a) shows the memory addresses and the needed signals for each chip. Right side upper and lower corners of a box (the one depicting a memory or device) can be used to show the lower and upper bounds of the addresses assigned in a RAM or ROM.

Figure 17.6(b) shows A_0 to A_{12} lines to PROM and A_0 to A_{13} lines to RAM-1 because there are 8k (8192) and 16k (32768) internal addresses, respectively.

The fugure shows D_0 to D_0 bus to PROM and RAM-1 memories.

The fugure shows $\overline{\text{RD}}$ control signal to both ROM and RAM. It shows $\overline{\text{WR}}$ as well as $\overline{\text{RD}}$ control signals to RAM-1.

The individual memory chips or devices access internal addresses, each using an internal decoder of its own. For example, ROM internal decoder selects which address to access out of 8 k (= 8 × 1024) addresses between 0000H and 1FFFH. RAM in the present example has 16 k addresses.

Each chip on common set of buses must have a $\overline{\text{CS}}$ select signal each. This isolates the internal decoder and internal data bus from the system buses when the chip is not to be selected. Let $\overline{\text{CS}}$ signals for ROM, RAM-1 are CS_0 and CS_1, respectively.

External decoding is needed to activate one of the CS_0 and CS_1 when the address on the address bus corresponds to the internal addresses in these memories.

Table 17.1 is designed to find using the steps F_1 to F_5 what are the inputs needed and not needed (conditions) for the design of circuit for external decoding.

Step F_1: Find out what are memory chips and devices given for external decoding. Column 1 gives these.

Step F_2: Find out what are the control lines and address-lines available at the system and write these in columns 2 and 3 for each memory and device.

TABLE 17.1 Memory and bus signlas table

System			Memory or device used signals		Needed select conditions from memory or device unused signals and therefore the decoder used signals		
Memory or device	**Control signals**	**Address bus signals**	**Enabling signals all 0s**	**Address bus signals s**	**Disabling signals all 1s**	**Address lines unused**	**Address conditions**
PROM	$\overline{RD}$	A_0 to A_{15}	$\overline{RD}$	A_0 to A_{12}	WR	A_{13} to A_{15}	$A_{13} = 0$, $A_{14} = 0$, $A_{15} = 0$[a]
RAM-1	$\overline{WR}$, $\overline{RD}$	A_0 to A_{15}	RD, WR	A_0 to A_{13}	-	A_{14} to A_{15}	$A_{14} = 0$ and $A_{15} = 1$[b]

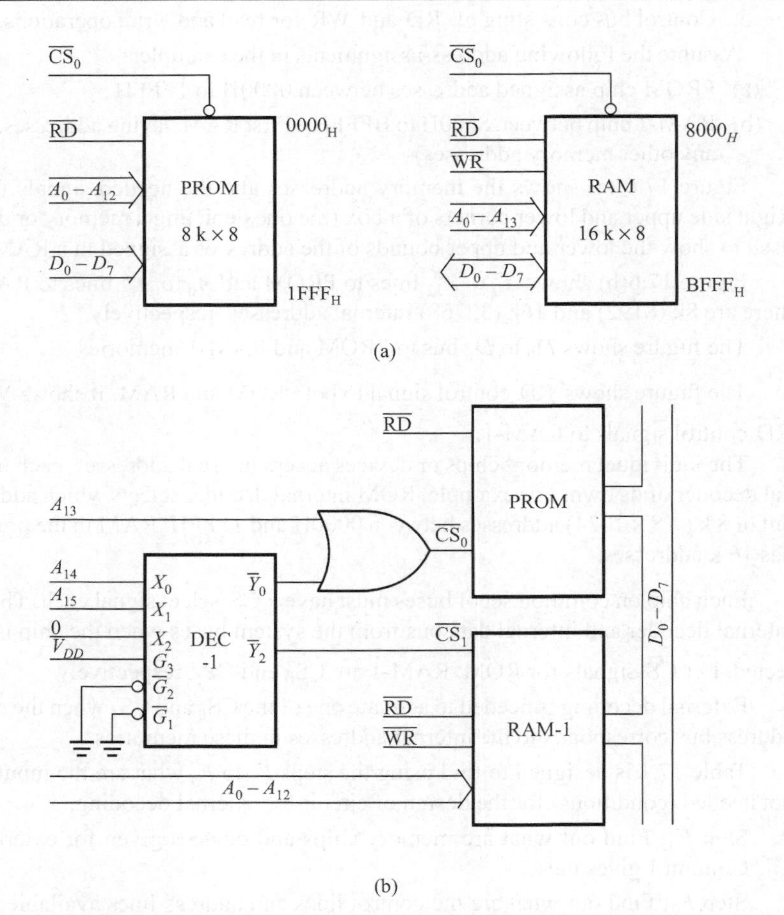

FIGURE 17.6 (a) The exemplary memory addresses and the signals system (b) Decoding circuit (DEC-1) using IC 74138 when PROM and RAM are jointly placed on the buses.

Step F_3: Find out what are the enabling-signals used or not used at each memory or device and write these in columns 4 and 6, respectively. These columns, therefore, define the enabling and disabling conditions of the control bus when selecting a memory or device (used and unused lines at a memory or device are to be used by the external DEC-1 decoder).

Step F_4: Find out what are the address lines not used at each memory and device. Write these in column 7. These must then be used for finding the conditions to select a memory or device.

Step F_5: Find the conditions on these unused address lines using the Figure 17.6(a) address assignments [(0000H-1FFFH) and (8000H-BFFFH)]. These are the conditions needed when selecting a memory or device. Write these also in column 8. The conditions are found as follows:

[a] Refer Figure 17.6(a). PROM addresses are between 0000H to 1FFFH: **000**0 0000 0000 0000 to **000**1 1111 1111 1111. Mark by bold and underline bits that are same for all the memory map addresses shown for PROM in the figure. Therefore, $A_{13} = 0$, $A_{14} = 0, A_{15} = 0$ are the select conditions. $\overline{WR}$ inactive and $\overline{RD}$ active (enabling condition) in column 4 are other conditions, when PROM may select.

[b] Refer Figure 17.6(a). RAM-1 addresses are between 8000H to BFFFH: **10**00 0000 0000 0000 to **10**11 1111 1111 1111. Mark by bold and underline bits that are same for all the memory map addresses shown for RAM-1 in the figure. Therefore, $A_{15} = 1, A_{14} = 0$ are the conditions to distinguish it from PROM.

By a memory or chip, the unused address lines which are also unused by other memory or devices are utilized for decoding the addresses because a chip should be selected under particular condition of these and should not be selected in other conditions of these. For example, PROM should be selected by A_{13} to A_{15} when A_{13} to A_{15} all 0s. RAM-1 should be selected when $A_{14} = 0$ and $A_{15} = 1$.

Table 17.2 is designed to find using the steps G_1 to G_5 the circuit for external decoding.

Step G_1: First decide how many decoders to use. Only one decoder DEC-1 is needed in the present example, becauses, there are only two memory chips.

TABLE 17.2 Decoder enabling inputs and address-select inputs table

Decoders			List of used signals			List of unused signals, if any		
Decoder	memory or devices	Decoder outputs needed for chip select inputs	Enable inputs	Disable inputs	Address select inputs	Enable inputs	Disable inputs	Address bus inputs
DEC-1	PROM, RAM-1	Two CS_0, CS_1	-	-	A_{14} to A_{15}	$\overline{RD} = 0$ or $\overline{WR} = 0$ #	$\overline{RD} = 1$ and $\overline{WR} = 1$ #	A_{13}

(1) Column 6 is for fulfilling the conditions of Table 17.1 column 8 of rows 1 and 2. (2) A_{13} is ORed with $Y2$ output of DEC-1 as shown in Figure 17.6(b) to fully decode the buses. (3) $\overline{RD}$ and $\overline{WR}$ not used at the decoding unit.

Step G_2: Write the memories connected to the decoder in column 2. Write the number of DEC-1 outputs needed and chip select signals needed in column 3 (using the conditions of Table 17.2 columns 4, 6 and 8).

Step G_3: Write the enabling and disabling conditions in columns 4 and 5 for designing the logic circuit for the gate (any decoder output enabling) pin(s) at the decoder.

Step G_4: Find the DEC-1 address and gate select pin inputs. The decoder activates one of the $\overline{Y}$ output as per the combination of inputs at these pins. Write in column 6, the conditions for connecting the address select pins of the decoder. Use here only common unused address lines for decoding. Consider DEC-1 to decode the memories: PROM, and RAM-1. PROM conditions, which are specified for A_{14}, A_{15} as well as A_{13}. But RAM-1 condition for A_{13} is not specified in this column. This is because RAM-1 connects A_{13} directly to address bus. Therefore, in Table 17.2 column 6, first row for DEC-1 gives A_{14} and A_{15}.

Step G_5: Write in columns 7, 8 and 9 the lists of unused enabling signals, disabling signals and address signals not used in the columns 4 to 6. A_{13} is not used by DEC-1 in columns 4 to 6. Therefore, A_{13} is specified in column 9.

Certain signals may be left out in design, as these are not needed presently for decoding. Following are the examples. Let a system have only PROMs or ROMs connected to the buses. $\overline{\text{WR}}$ is now an unused decoder-enabling signal. $\overline{\text{RD}}$ is also an unused decoder-enabling signal in the present example. Reason is that the computer processor itself will have a provision to activate $\overline{\text{RD}}$ or $\overline{\text{WR}}$ and these are also connecting directly to the memories.

Purpose of writing columns 7, 8 and 9 is not immediately apparent. However, it is necessary step to completely decode the system buses for the addresses of memories and devices.

Table 10.6 gave the truth table of 74138. It is a 3 to 8 decoder. Tables 17.1 and 17.2 conditions for DEC-1 show that there is a need for $A14$ and $A15$ two address signals only. Figure 17.6(b) shows the decoding circuit for the memories and IO port devices at the addresses assigned in the present example.

17.8 SEMICONDUCTOR MEMORY

There are different types of semiconductor memories.

1. Static RAM: A RAM in which a stored byte once stored is not refreshed by a read-write operation at the byte address.
2. Dynamic RAM: A RAM in which a stored byte once stored need to be refreshed by a read-write operation at each byte-addresse every few milliseconds.
3. ROM: A masked Read Only Memory prepared by silicon masking as per the look-up table, data or codes to be burned. It is done at the manufacturing site from the files for burning designed by the circuit designer.
4. EPROM: Erasable and Programmable read only memory. It is erasable by UV light exposure of exposed silicon for few minutes. A laboratory programmer tool programs the EPROM.
5. EEPROM: Electrically Erasable and Programmable read only memory. It is erasable byte-by-byte by writing 1s at an addressed location. Writing a required byte at a given address programs an erased address.

6. Flash Memory: Flash is electrically erasable one sector at a time. A sector is erasable by writing 1s at the addressed sector. Writing of the byte required at a given address programs the initialy erased flash address.

EXAMPLES

Example 17.1

Design a 4-Bit buffer register with tristate outputs.

Solution

We replace the AND gates in the circuit of Figure 16.2(a) by the tristate buffers and these are put before the register outputs and the internal Qs. An out-enable input (can be called read input) upon its activation releases Q's at the register outputs from third states. Figure 17.3(a) upper right side showed how a register buffers with the help of four tristate buffers at their outputs.

Example 17.2

Show a representation of a register with load enable and out enable inputs.

Solution

Figure 17.4(a) showed a representation of the buffer register with load-enable (for a write from the bus) and out-enable (for a read from the bus) inputs for the case when there are tristate inputs and tristate outputs of the register.

Example 17.3

Show how do we read a register with tristate buffers through a common bus.

Solution

The buffer register circuit of Figure 17.3, showed as a circuit on a common bus organized with four such buffer registers in Figure 17.4(b). Figure 17.3(a) upper right side showed a register buffered with the tristate buffers at its outputs. The figure lower left side showed how the register is buffered with the tristate buffers at its inputs.

A tristate buffer gives any output L_i state 1 or 0 only when it is enabled by activation (at 0 in IC 74374 or 74HC374) else it disconnects the input(s) from appearing at the output(s).

A Register Write (Load) Operation

When four tristate buffers on the left side of the Figure 17.4(b) are enabled by a load enable state L_i, the data bus logic states become available at the D_A, D_B, D_C and D_D inputs as per the active L_i. If load enable is inactive then no current is sourced or slinked from the data bus wires to the ith the register.

If i^{th} load enable is active, then D_0, D_1, D_2 and D_3 of the data bus are present at the ith register inputs D'_D, D'_C, D'_B and D'_A, respectively. After a common clock input (also called write enable) is activated by a transition (say a +ve edge), the D'_D, D'_C, D'_B and D'_A are transferred, after a delay equal to propagation delay, to the Q'_D, Q'_C, Q'_B and Q'_A at the ith register.

A Register Read (Output) Operation

When four tristate buffers on the upper right side are enabled by an ith out enable logic state, we get the Q output states (1 or 0) onto the data bus wires. If disabled then we have tristate outputs (i.e. in high impedances with no current sourcing and sinking capabilities to or fro another circuit) to which these are connected at the data bus wires.

If out enable (also called read enable) = active, then Q'_D, Q'_C, Q'_B and Q'_A are present at the D_0, D_1, D_2 and D_3 wires of the data bus, respectively. All three-load enable logic input, input clock input, and out enable input, are activated in a sequence during a write read sequence.

Write or read Inactive in a Register

Unless either the load-enable or the out-enable is inactivated:

1. Register *FF*s remain in idle state like in a 4-bit memory unit that will keep all the four previous *Q* values unaltered (till of course power fails), and
2. The data bus can neither input the data to the register nor can get the data outputs from the register.

Example 17.4 How do you connect multiple registers on a common data bus?

Solution

Figure 17.4(b) showed the multiple registers on a common data-bus. L_i is load enable input to an ith register and O_i is out enable input to the register.

Clock input is common to all registers $i = 0 \ldots (n - 1)$, which for the inputs as well as the outputs are connected to a common data bus.

Register *D* inputs are transferred as the new *Q* outputs after the inputs are clocked in the selected register.

An appropriate sequence of the enable and clock inputs must be ensured while using a data bus. Either only one of the L_i is activated at a time or one of the O_i only is active at a time. Therefore, the data bus is used by only one of the register at a time. The bus is used either for accepting the input data bits by a selected register or for providing the output data bits on the bus from the selected register.

Example 17.5 How do you use the bus for reading from one register and writing to another when there multiple registers connected to a common bi-directional data bus?

Solution

The circuit of Figure 17.4(b) showed an important application. Let value of $\boldsymbol{i}$ specify an address of a register. We can say that there is a register each at an address i, and there are total n registers with n distinct addresses. We can copy (i.e. transfer) a binary word in one register R_i to another register R_j as follows. Let us in three sequences, first activate O_i, then activate L_j and then activate clock input. Binary word is stored at Q'_D, Q'_C, Q'_B and Q'_A of register R_j from Q'_D, Q'_C, Q'_B and Q'_A of register R_i through the data bus.

Example 17.6 How does a circuit use the buses for reading from a 4 kB RAM?

Solution

The 4 kB RAM means there are 4 k byte addresses with one byte at each address. For 4 k (4 × 1024), there will be 12 input lines for the addresses. [$2^{12} = 4 \times 1024$.] The circuit connects these 12 lines to the address bus in a system. It also generates $\overline{CS}$ signal by appropriate mechanism (external decoding circuit or logic circuit) for the given RAM.

When $\overline{CS}$ and $\overline{RD}$ both activate, the byte for the address j becomes available. The j is as per the present input combination at the address lines. The byte from j^{th} address is available at D_0-D_7 data bus pins, which connect the internal data bus of RAM. It is available from the RAM as long as $\overline{RD}$ and $\overline{CS}$ are active.

Example 17.7

How does a circuit uses the data and address buses for writing into a 64 kB RAM?

Solution

The 64 kB RAM means there are 64 k byte addresses with one byte at each address. For 64 k (64×1024), there will be 16 input lines for the addresses. ($2^{16} = 64 \times 1024$). The circuit connects these 12 lines to the address bus in a system. It also generates $\overline{CS}$ signal by an appropriate mechanism (external decoding circuit or logic circuit) for the given RAM.

When $\overline{CS}$ and $\overline{WR}$ both activate, the byte for the address k becomes available. The k is as per the present input combination at the address lines. The byte for k^{th} address is less then from the D_0-D_7 data bus pins, which connect the internal data bus of RAM. Byte writes into the jth cell-array during the period $\overline{WR}$ and $\overline{CS}$ are active.

Example 17.8

Design a decoding circuit for the following memories and port devices at the following assigned addresses.

A given circuit has the distinct control inputs for the memories and devices. Assume that the circuit has following bus lines:

1. Address bus inputs from A_0 to A_{15} (total 16 inputs).
2. Data bus inputs-outputs D_0 to D_7 (total 8 lines).
3. Control bus consisting of memory-read input $\overline{MEMRD}$, memory-write input $\overline{MEMRW}$, IO-device read input $\overline{IORD}$ and IO-device write input $\overline{IOWR}$.

Assume there be the following address-assignments by the designer for the memories and devices in a circuit:

(a) PROM between hex memory addresses 0000_H to $1FFF_H$.
(b) RAM-1 between hex memory addresses 8000_H to $BFFF_H$. (First RAM will have addresses, which are distinct from any other memory addresses).
(c) RAM-2 between hex memory addresses $E000_H$ to $FFFF_H$. (Second RAM will have addresses, which are distinct from any other memory addresses).
(d) Device between hex-port addresses 00_H to 03_H.
(e) Device between $2F8_H$ to $2FF_H$.

[Device means a circuit like in a keyboard interface chip. Port address means as address at the device used for the buses to select during a read or write.]

Solution

First draw the Figure 17.7 to show the circuit for memory and IO device addresses and the signals for the system. Right side upper and lower corners of a box (the one depicting a

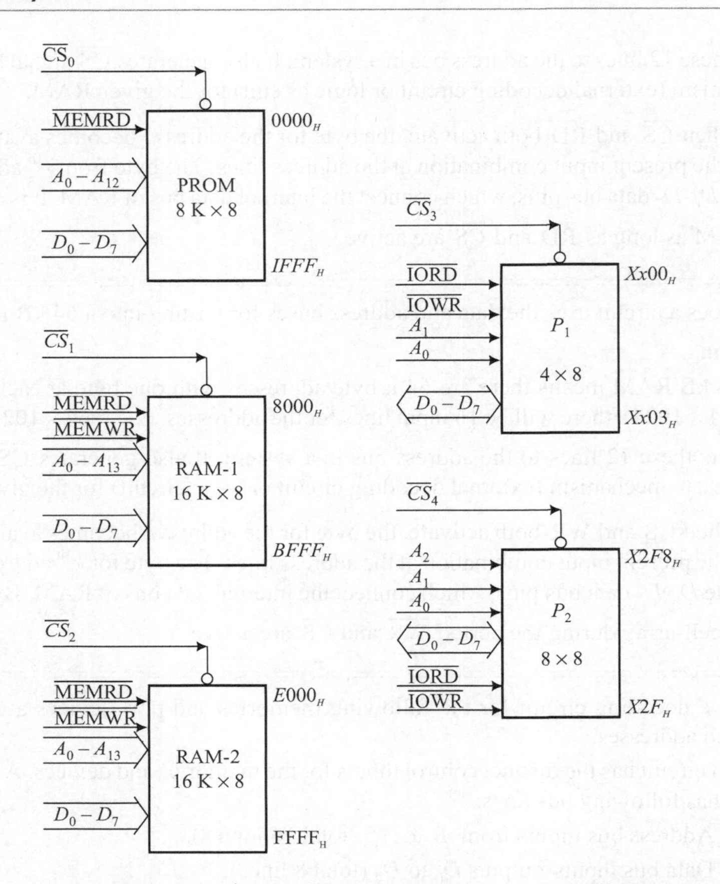

FIGURE 17.7 Memory and IO devices addresses and the inputs and outputs for the memory and device parts.

memory or device) is used to mark the lower and upper bounds of the addresses assigned by the system designer.

The figure shows A_0 to A_{12} lines to PROM, A_0 to A_{13} lines to RAM-1 and A_0 to A_{12} lines to RAM-2 memories because of presence of 8 k, 16 k and 8 k internal addresses, respectively. It shows A_0 to A_1 bus to P_1 and $A0$ to $A2$ bus to P_2.

The figure shows D_0 to D_0 bus to ROM, RAM-1 and RAM-2 memories and same D_0 to D_7 to port devices P_1 and P_2 because of presence of a byte each at each address in these.

The figure shows $\overline{\text{MEMRD}}$ control input to the ROM, RAM-1 and RAM-2 and the $\overline{\text{IORD}}$ control input to P_1 and P_2. It shows $\overline{\text{MEMWR}}$ control signal to RAM-1 and RAM-2 and $\overline{\text{IOWR}}$ to P_1 and P_2. $\overline{\text{MEMWR}}$ is not there in PROM.

The individual memory chips or devices access internal addresses, each using an internal decoder of its own. For example, ROM internal decoder selects which address to access out of 8 k (= 8 × 1024) addresses between 0000_H and $1FFF_H$. The P_2 internal decoder selects which address to access out of 8 addresses there.

TABLE 17.3 Address, data and control bus signal conditions for the design

	System Bus Signals		Memory or Device Used Inputs		Needed Select conditions from and Therefore Decoder Used Signals		
Memory or Device	Control Inputs	Address Bus Inputs	Enabling Inputs All Active 0s	Address Bus Inputs	Disabling Inputs All Inactive 1s	Address Inputs unused	Addressing Conditions
PROM	@	A_0 to A_{15}	$\overline{\text{MEMRD}}$	A_0 to A_{12}	$\overline{\text{MEMWR}}$ $\overline{\text{IORD}}$, $\overline{\text{IOWR}}_a$	A_{13} to A_{15}	$A_{13} = 0$, $A_{14} = 0$, $A_{15} = 0_a$
RAM-1	@	A_0 to A_{15}	$\overline{\text{MEMRD}}$ $\overline{\text{MEMWR}}$	A_0 to A_{13}	$\overline{\text{IORD}}$, $\overline{\text{IOWR}}_b$	A_{14} to A_{15}	$A_{14} = 0$ and $A_{15} = 1_b$
RAM-2	@	A_0 to A_{15}	$\overline{\text{MEMRD}}$ $\overline{\text{MEMWR}}$	A_0 to A_{12}	$\overline{\text{IORD}}$, $\overline{\text{IOWR}}_c$	A_{13} to A_{15}	$A_{13} = 1$, $A_{14} = 1$ and $A_{15} = 1_c$
P_1	@	A_0 to A_{15}	$\overline{\text{IORD}}$ $\overline{\text{IOWR}}$	A_0 to A_1	$\overline{\text{MEMRD}}$, $\overline{\text{MEMWR}}_d$	A_2 to A_{15}	A_3 to A_7 and $A_9 = 0$; A_8 and A_{10} to $A_{15} = 0$ $A_2 = 0_d$
P_2	@	A_0 to A_{15}	$\overline{\text{IORD}}$ $\overline{\text{IOWR}}$	A_0 to A_2	$\overline{\text{MEMRD}}$, $\overline{\text{MEMWR}}_e$	A_3 to A_{15}	A_8 and A_{10} to $A_{15} = 0$, A_3 to A_7 and $A_9 = 1_e$

@ Four control signals $\overline{\text{MEMRD}}$, $\overline{\text{MEMWR}}$, $\overline{\text{IORD}}$ and $\overline{\text{IOWR}}$. Meaning of other signs are explained in the text.

Each chip on common set of buses must have a $\overline{\text{CS}}$ select signal each. This is isolate the internal decoder and internal data bus from the system buses when $\overline{\text{CS}}$ is inactive. Let $\overline{\text{CS}}$ signals for ROM, RAM-1, RAM-2, P_1 and P_2 are $\overline{\text{CS}}_0$, $\overline{\text{CS}}_1$, $\overline{\text{CS}}_2$, $\overline{\text{CS}}_3$ and $\overline{\text{CS}}_4$, respectively.

External decoding is needed to activate one of the $\overline{\text{CS}}_0$, $\overline{\text{CS}}_1$, $\overline{\text{CS}}_2$, $\overline{\text{CS}}_3$ and $\overline{\text{CS}}_4$.

Table 17.3 is designed to find using the steps F_1 to F_5 what are the inputs needed and not needed (conditions) for the design of decoding circuit for external decoding.

Step F_1: Find out what are memory chips and devices given for external decoding. Column 1 gives these.

Step F_2: Find out what are the control lines and address-lines available for the inputs to the device parts and memories. Write these in columns 2 and 3 for each memory and device.

Step F_3: Find out what are the enabling-signals that are used or not used signals at each memory or device and write these in columns 4 and 6, respectively. These columns, therefore, define the enabling and disabling conditions of the control bus when selecting a memory or device (The unused lines for a memory or device are for the use by the decoder).

Step F_4: Find out what are the address lines not used at each memory and device. Write these in column 7. These must then be used for finding the conditions to select a memory or device.

Step F_5: Find the conditions on the unused address lines using the Figure 17.7 address assignments. These are the conditions needed when selecting a memory or device. Write these in column 8.

By a memory or chip, the unused address lines which are also unused by other memory or devices are utilized for decoding the addresses because a particular device or chip should be selected under particular condition of these and should not be selected in other conditions of these. For example, PROM should be selected by A_{13} to A_{15} when A_{13} to A_{15} all 0s $\overline{\text{MEMWR}}$ and $\overline{\text{IORD}}$ and $\overline{\text{IOWR}}$ inactive. RAM-1 should be selected when $A_{14} = 0$ and $A_{15} = 1$ and $\overline{\text{IORD}}$ and $\overline{\text{IOWR}}$ inactive. RAM-2 should be selected when $A13$, A_{14} and $A_{15} = 1$ and $\overline{\text{IORD}}$ and $\overline{\text{IOWR}}$ inactive. Following explains these conditions and the procedure adopted in the step F_5 when writing all the conditions mentioned in columns 8 for the address lines and columns 4 and 6 for the control lines.

a Refer Figure 17.7. PROM addresses are between 0000H to 1FFFH: **<u>000</u>**0 0000 0000 0000 to **<u>000</u>**1 1111 1111 1111. Mark by bold and underline bits that are same for all the memory map addresses shown for PROM in the figure. Therefore, $A13 = 0, A14 = 0, A15 = 0$ are the select conditions. $\overline{\text{MEMWR}}$, $\overline{\text{IOWR}}$ and $\overline{\text{IOWR}}$ not active (disabling conditions) in column 6 and $\overline{\text{MEMRD}}$ active (enabling condition) in column 4 are other conditions, when PROM should select.

b Refer Figure 17.7. RAM-1 addresses are between 8000H to BFFFH: .**<u>10</u>**00 0000 0000 0000 to **<u>10</u>**11 1111 1111 1111. Mark by bold and underline bits that are same for all the memory map addresses shown for RAM-1 in the figure. Therefore, $A15 = 1, A14 = 0$ are the conditions to distinguish it from PROM and RAM-2. IOWR and IOWR not active (disabling conditions) and $\overline{\text{MEMRD}}$ and $\overline{\text{MEMWR}}$ active (enabling conditions) active are other conditions in columns 4 and 6, when RAM-1 should select.

c Refer Figure 17.7. RAM-2 addresses are between E000H to FFFFH: .**<u>111</u>**0 0000 0000 0000 to **<u>111</u>**1 1111 1111 1111. Mark by bold and underline bits that are same and true for all the memory map addresses shown for RAM-2 in the figure. Therefore, $A_{13} = 1, A_{14} = 1$ and $A_{15} = 1$ are the conditions to distinguish it from PROM and RAM-1. $\overline{\text{IOWR}}$ and $\overline{\text{IOWR}}$ not active (disabling conditions) and $\overline{\text{MEMRD}}$ and $\overline{\text{MEMWR}}$ active (enabling conditions) are other conditions in columns 4 and 6, when RAM-2 should select.

d Refer Figure 17.7. *P*-1 addresses are between 00H to 03H: *<u>0000 0000 0000 00</u>*00 to *<u>0000 0000 0000 00</u>*11. Mark by italics and underline bits that are same for all the IO device map addresses shown for P_1 in the figure. Therefore, A_2 to $A_{15} = 0$ are the conditions to distinguish it from P_2. $\overline{\text{IOWR}}$ and $\overline{\text{IOWR}}$ active (enabling conditions) and $\overline{\text{MEMRD}}$ and $\overline{\text{MEMWR}}$ inactive (disabling conditions) are other conditions in columns 4 and 6, when P_1 should select.

[e] Refer Figure 17.7. P_2 addresses are between $2F8_H$ to $2FF_H$: *0000 0010 1111 1000* to *0000 0010 1111 1111*. Mark by italics and underline those bits that are same for all the IO device part addresses shown for P_2 in the figure. Therefore, A_{10} to $A_{15} = 0$, $A_9 = 1$, $A_8 = 0$, A_3 to $A_7 = 1$ are the conditions to distinguish it from P_1. $\overline{IOWR}$ and $\overline{IOWR}$ not active (disabling conditions) and $\overline{MEMRD}$ and $\overline{MEMWR}$ active (enabling conditions) active are other conditions in columns 4 and 6, when P_2 should select.

Table 17.4 is designed to find using the steps G_1 to G_5 the circuit for external decoding.

TABLE 17.4 Enabling signals and input address-select signals at the decoders

Decoders			List of used signals			List of unused Signals, if any		
Decoder	memory or devices	Decoder outputs needed for chip select signals	Enable signals	Disable signals	Address select signals	Enable signals	Disable signals	Address bus signals
DEC-1	PROM, RAM-1, RAM-2	Three [CS_0, CS_1, $CS2$[a]]	Either $\overline{MEMRD} = 0$,or $\overline{MEMWR} = 0$	$\overline{IORD} = 0$ and $\overline{IOWR} = 0$	A_{14} to A_{15}	-	-	A_{13}[e]
DEC-2	P_1, P_2	Two [CS_3, CS_4[b]]	$\overline{IORD}$, $\overline{IOWR}$, and A_8 and A_{10} to $A_{15} = 0$[c]	$\overline{MEMRD} = 0$, $\overline{MEMWR} = 0$	A_3 to A_7 and A_9[d]	-	-	A_2[f]

(1) [a]When fulfilling the conditions of columns 4, 6 and 8 in column 8 of previous table rows 1, 2 and 3.. (2) [b] When fulfilling the conditions of columns 4, 6 and 8 of previous table rows 4 and 5. (3) [d] A_8, A_{10} to $A_{15} = 0$ are common conditions for the P_1 and P_2. Hence these are taken as the enabling conditions like the control signals. (4) [e] A_2 is ORed with Y_2 output of DEC-2 as shown in Figure 17.8 to fully decode the system buses. (5) [f] A_{13} is ORed with Y_0 output of DEC-1 as shown in Figure 17.8 to fully decode the system buses.

Step G_1: First decide how many decoders to use. Since P_1 and P_2 have large number of address lines unconnected, there is a need for separated decoder for P_1 and P_2 with external logic circuitry for the signals that can't connect to given decoder IC. Therefore external decoding needs (i) one memory-chips decoder DEC-1 and one IO devices decoder DEC-2 with an external logic circuitry. The $\overline{MEMRD}$ and $\overline{MEMWR}$ and IORD and IOWR are used as gate input for both memory decoder and IO decoder. Write the decoders needed in column 1.

Step G_2: Write the memories and devices connected to each decoder in column 2. Write the number of decoder outputs needed and chip select signals needed in column 3 (using the conditions of Table 17.3 columns 4, 6 and 8).

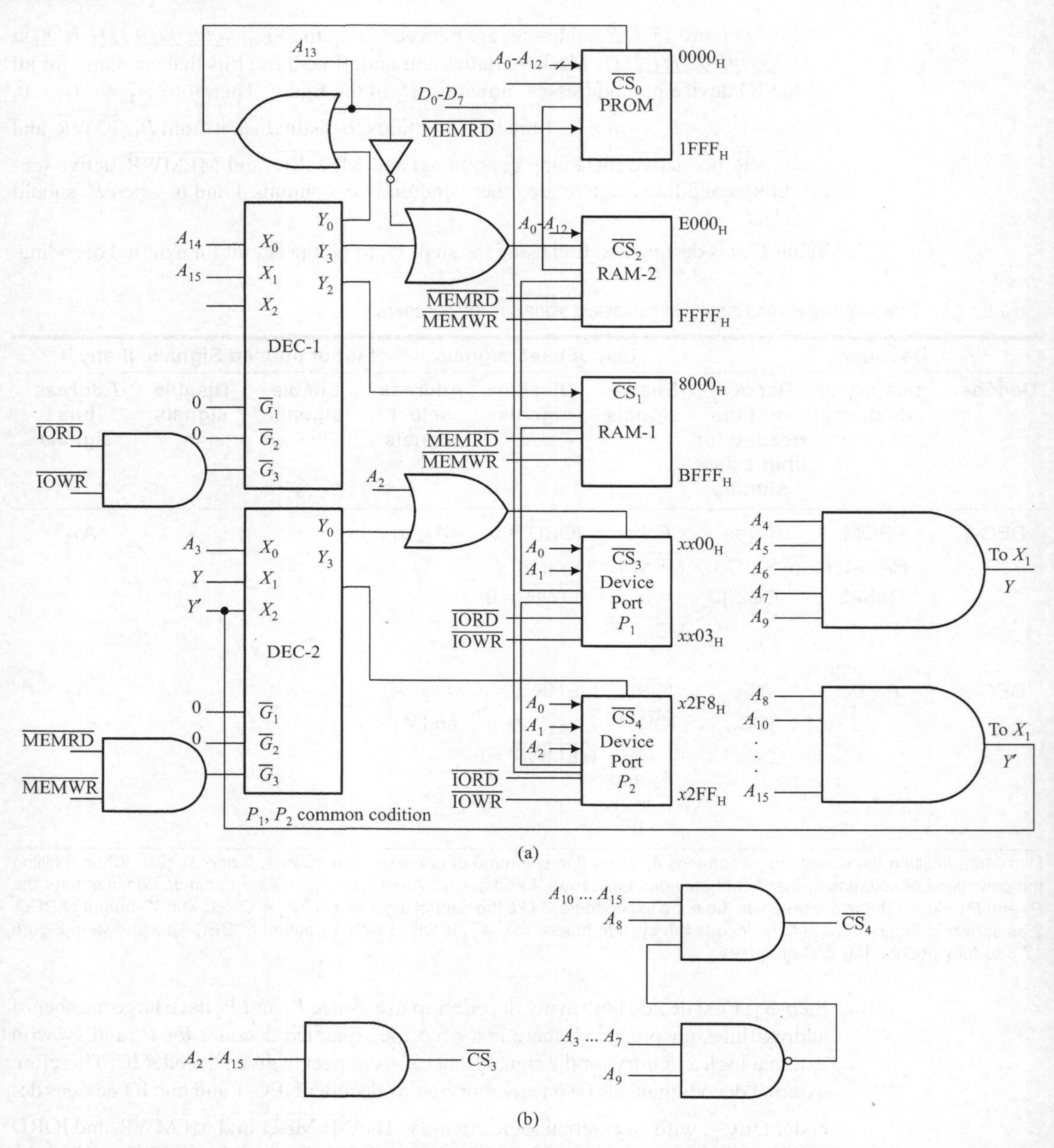

FIGURE 17.8 (a) Decoding circuit for the memories and IO port devices at the addresses assigned in the problem (b) Alternative circuit in place of 74138 for DEC-2 for the IO port devices P_1 and P_2.

Step G_3: Write the enabling and disabling conditions in columns 4 and 5 for designing the logic circuit for the gate (any decoder output enabling) pin(s) at the decoder.

Step G_4: Find the decoder address select pin inputs. The decoder activates one of the Y output as per the combination of inputs at these pins. Write in column 6, the conditions for connecting the address select pins of the decoder. Use here only common unused address lines for decoding. Consider DEC-1. It decodes the memories: PROM, RAM-1 and RAM-2. RAM-2 and PROM conditions are specified for A_{14}, A_{15} as well as A_{13}. But RAM-1 condition for A_{13} is not specified in this column. This is because RAM-1 connects A_{13} directly to address bus. Consider DEC-2. It decodes the P_1 and P_2. The address signal conditions are in rows 4 and 5 of column 8 in Table 17.4. P_1 conditions are specified for A_8 and A_{10} to A_{15}, A_3 to A_7 and A_9 as well as A_2. But P_2 conditions are specified for A_8 and A_{10} to A_{15}, A_3 to A_7 and A_9 only not for A_2. This is because P_1 connects A_2 directly to address bus. Therefore, in Table 17.5 column 6, first row for DEC-1 gives A_{14} and A_{15} and second row for DEC-2 gives A_3 to A_7 and A_9 (A_8 and A_{10} to A_{15}, A_3 to A_7 and A_9 and then excluding the common conditions for both P_1 and P_2).

Step G_5: Write in columns 7, 8 and 9 the lists of unused enabling signals, disabling signals and address signals not used in the columns 4 to 6. For examples, (i) A_{13} is not used by DEC-1 in columns 4 to 6 but RAM-2 and PROM needs it for decoding. (ii) A_2 is not used for decoding in DEC-2 by columns 4 to 6 but $A_2 = 0$ is a condition for decoding P_1.

Certain signals may be left out in design, as these are not needed in present problem for decoding. Following are the examples. Let a system has only PROMs or ROMs connected to the buses. $\overline{\text{MEMWR}}$ is now an unused decoder-enabling signal. Let a system have only the ROMs and RAMs, no IO port devices. $\overline{\text{IOWR}}$ and $\overline{\text{IORD}}$ are now the unused signals. Let ROM-2 addresses are not used in the system. Then $A14$ and $A15$ can be left unused.

Purpose of writing Table 17.3 columns 7, 8 and 9 is not immediately apparent. However, it is necessary step to completely decode the system-buses for the addresses of memories and devices.

Table 10.6 gave the truth table of 74138. It is a 3 to 8 decoder. Table 17.4 conditions for DEC-1 show that there is a need for A_{14} and A_{15} two address signals and four disabling and enabling signals. Therefore, IC74138 is one of the best choices to DEC-1. IC 74138 can also design the DEC-2 circuit, though it is not the best choice. This is because most of pins of 74138 are not being used and logic circuit before DEC-2 has to use many gates. It is better to replace 74138 for the DEC-2 by logic gates only. This reduces the need of the gates. Figure 17.8(b) gives an alternative of DEC-2.

Final solution of the problem is the following:

Figure 17.8(a) shows the decoding circuit for the memories and IO port devices at the addresses assigned. Decoders DEC-1 and DEC-2 are using the ICs 74138. DEC-2 decodes the buses for the IO port devices P_1 and P_2. Best choice for DEC-2 is to implement it by logic gates alone. Figure 17.8 (b) shows this circuit. It is a design, which implements the conditions in rows 4 and 5 of columns 8 and 4 of Table 17.3.

■ EXERCISES

1. Design an 8-bit buffer register having the tristate outputs and inputs.
2. Show how do you read a register A and write to register B with tristate buffers through a common bus.

3. How do you connect four registers *A*, *B* and *C* on a common data bus? Write sequences of signals to be activated to swap the bytes from *A* to *B*. (Hint: Swapping is by A stored in *C*, *B* transferred to *A* and *C* transferred to *B*).
4. How do you connect four registers *A*, *B*, *C* and *D* on a common data bus? Write sequences of signals to be activated to transfer a byte from *B* to *C*.
5. How do you use the buses for writing to a 32 kB RAM?
6. How do you use the buses for a 256 MB system from four 64 MB RAMs?
7. What are bus signals when IO port devices and 1 MB memories *write* and *read* by the same set of bus signals? Why should you assign different addresses for the memory and port addresses?
8. Design a decoding circuit for the following memories at the following assigned addresses:
   (1) Address bus lines from A_0 to A_{15} (total 16 lines).
   (2) Data bus lines D_0 to D_7 (total 8 lines).
   (3) Control bus consisting of IO devices as well as $\overline{RD}$ and $\overline{WR}$ for read and write operations, respectively.

   There are following address-assignments in the system for the memories:
   (a) PROM between hex memory addresses 0000H to 0FFFH.
   (b) RAM between hex memory addresses 1000H to 1FFFH. (First RAM will have addresses, which are distinct from any other memory addresses).
9. Design the memory map for a circuit for the following signals.
   (1) Address bus lines from A_0 to A_{12} (total 13 lines) connect to a RAM.
   (2) $\overline{CE}$ from Y_2 of a 74138 decoder.
   (3) Decoder has *A*, *B*, *C* address select input lines connected to A_{13}, A_{14}, A_{15}, respectively.
10. Write truth table of a 2 to 4 decoder with output polarity control and built with discrete gates and with an 8 × 4 ROM.

QUESTIONS

1. What do you mean by a 1 MB RAM randomly accessibility? Is an EPROM not accessible randomly?
2. What are the uses of RAM?
3. What do you mean by an address bus? What do we mean by a data bus?
4. What do you mean by a bus organised computer system?
5. Why are the data bus bi-directional and address bus unidirectional in a system? Why are the $\overline{RD}$ and $\overline{WR}$ lines unidirectional?
6. Why and how do you use several RAMs together in a computer? (Hint: to get a higher memory RAM capacity).
7. How many ICs of 64 MB will be needed to design a 1 GB RAM?

8. Why do you use both the memories RAMs and PROMs in a system?
9. A dynamic RAM needs refresh every few milliseconds and a static RAM need not. Why?
10. Search on the net or the IC manuals and answer the following. What are the access times for the various RAMs and EPROMs known to us? What is difference in access time of bipolar TTL circuit based memory and DRAM?
11. A data bus is of 64-bits, D_0-D_{63} in a 64-bit computer system. What is the role of this data bus?
12. An address bus is 32 bits, A_0-A_{31}. It is accessing 32 MB maximum memory addresses. How many address lines are idle when the memory addresses begin from 0000 0000_H address. How many when begin from 0000F 000_H?

CHAPTER 18

Fundamental Mode Sequential Circuits

OBJECTIVE

In this chapter, we shall learn the asynchronous sequential circuit operations in a mode called fundamental mode—their stable and unstable states, analysis, output specifications, feedback cycles, races, race-free assignments in their design and circuit implementation.

We learnt in Chapter 15 the following concepts:

1. A sequential circuit is a circuit made up by combining the logic gates such that the required logic at the output(s) depends not only on the current input logic conditions but also on the past inputs, outputs and sequences.
2. A sequential circuit has the memory elements like flip flops and a combinational circuit(s) has no memory elements.
3. A general sequential circuit network has a memory section and the combination circuits at the memory inputs and outputs.

We learnt in Chapter 15 two classifications of the sequential circuits:

(a) **Synchronous sequential circuit** is a circuit in which the output $\underline{\mathbf{Y}}$ depends on present state $\underline{\mathbf{Q}}$ and present inputs $\underline{\mathbf{X}}$ at the clocked $\uparrow$ or $\downarrow$ or $\sqcap$ (MS) or $\sqcup$ (MS) instance(s) only. Memory section activates a transition as per the excitation inputs to the next state $\underline{\mathbf{Q}}'$. A clock input (or a set of inputs) is used, which activates the transition to next state. [There is a special case, called clocked sequential circuit. There is a master clock in the circuit, which activates the memory section.] *Synchronous sequential circuit clock input (s) controls the instance(s) at which an output(s) changes.*

(b) **Asynchronous sequential circuit** is a circuit in which not only the present inputs $\underline{\mathbf{X}}$ and present state $\underline{\mathbf{Q}}$ but also the sequences of changes affect the output $\underline{\mathbf{Y}}$. Asynchronous means that the changes can be at the undefined instances of time. *There are no controls for the instance at which the output(s) change.*

18.1 GENERAL ASYNCHRONOUS SEQUENTIAL CIRCUIT

A general asynchronous sequential-circuit has a section consisting of the (i) latches without clock inputs (or an equivalent delay device) and the (ii) combinational circuits at the input and output stages. It means a memory section of the clocked sequential circuit replaces the latches without the clock inputs (or an equivalent delay device). Assume the followings structure of a general sequential circuit.

1. There is a set $\underline{\mathbf{X_i}}$ of m present input variables X_0, X_1, ..., X_{m-1}. These are applied along with the present $\underline{\mathbf{x_q}}$ to a combinational circuit. The output of $\underline{\mathbf{X_i}}$ along with $\underline{\mathbf{x_q}}$ is a set of present state outputs $\underline{\mathbf{Y}}$ of which $\underline{\mathbf{y_q}}$ is a subset. ($\underline{\mathbf{Y}}$ consists of two subsets; $\underline{\mathbf{Y_i}}$ with outputs Y_0, Y_1 ... Y_{j-1} with no feedback and $\mathbf{y_q}$ with feedback to the input via the memory section).
2. The circuit memory section consists of the m latches without any clock edges or pulse inputs to control the instance or period of their operations. These have a set $\underline{\mathbf{x_q}}$ of variables from the m present state outputs. The $\underline{\mathbf{x_q}} = x_{q0}, x_{q1}...x_{qn-1}$ from the $\mathbf{y_q}$ inputs after a delay.
3. The present state of $\underline{\mathbf{Y}}$ and therefore, its subset $\underline{\mathbf{y_q}}$ changes by the change in $\underline{\mathbf{x_q}}$ after a delay at the memory section. Changed $\underline{\mathbf{Y}}$ can change again after the next feedback-cycle of delay (in the second feedback-cycle at the memory section). The feedback-cycles after first feedback-cycle may continue or may not continue; there may be unstable or stable $\underline{\mathbf{Y}}$ due to unstable or stable $\underline{\mathbf{y_q}}$, respectively.

Figures 18.1(a) and (b) show (i) a representation for the general asynchronous sequential circuits, which has above features and a special case of fundamental mode operation (ii) A sequential asynchronous circuit is also like a machine producing the states without a clock controlling that.

Point to Remember

> Asynchronous sequential circuit is circuit in which there are no control(s) for the instance(s) at which the output(s) change like a control by a clocked instance in synchronous sequential circuit.

18.2 UNSTABLE CIRCUIT OPERATION

Step 1: Let us also assume that the input X change applied only at stable $\underline{\mathbf{x_q}}$. Let the present state $\underline{\mathbf{Y_0}}$ and therefore its subset $\underline{\mathbf{y_q}}$ changes when a bit changes $\underline{\mathbf{X_i}}$ to $\underline{\mathbf{X'_i}}$. Therefore, $\underline{\mathbf{Y_0}}$ changes to $\underline{\mathbf{Y}}$ after a normal gate propagation delays (not in consideration).

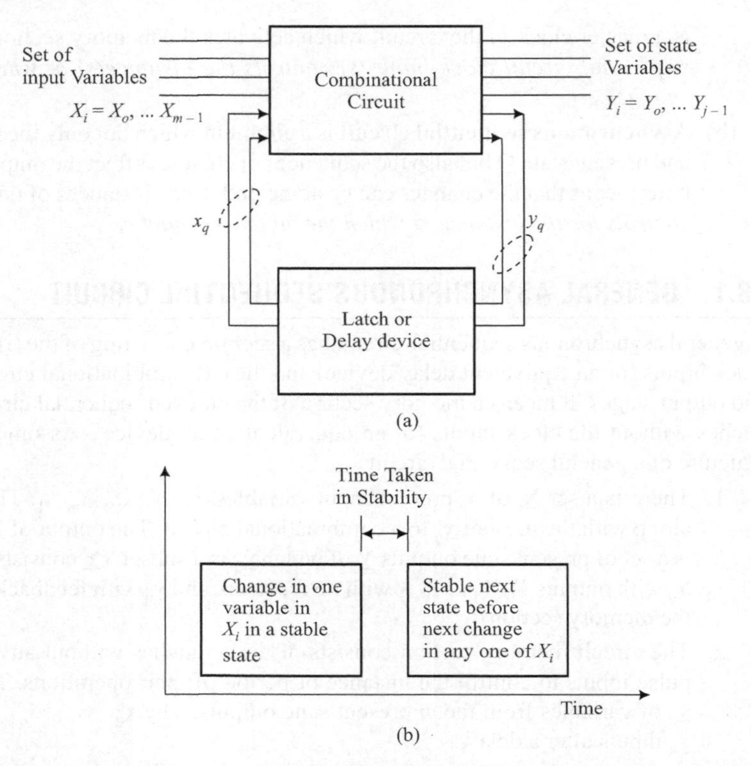

FIGURE 18.1 (a) General asynchronous sequential circuit with a section consisting of the (i) Latches without clock inputs and the (ii) Combinational circuits at the input and output stages (b) Above circuit fundamental mode operation with an assumption that one variable changes at an instance.

Step 2: However, in the first feedback-cycle, a delayed change in $\underline{\mathbf{x_q}}$ occurs due to change in $\underline{\mathbf{y_q}}$ to $\underline{\mathbf{y'_q}}$, being a subset of $\underline{\mathbf{Y}}$. The feedback $\underline{\mathbf{x'_q}}$ after a memory-section delay combined with $\underline{\mathbf{X'_i}}$ changes $\underline{\mathbf{Y}}$ to next state $\underline{\mathbf{Y'}}$.

Step 3: However, in next feedback-cycle a delayed change in $\underline{\mathbf{x'_q}}$ may also occur due to change to $\underline{\mathbf{y'_q}}$, a subset of $\underline{\mathbf{Y'}}$. The feedback $\underline{\mathbf{x''_q}}$ after the feedback-cycle of delay combined with $\underline{\mathbf{X_i}}$ may therefore change $\underline{\mathbf{Y'}}$ to next state $\underline{\mathbf{Y''}}$.

After the step 2 feedback-cycle if there are changes to the next state(s), the $\underline{\mathbf{Y}}$ is said to be unstable till after in any further feedback-cycle there is no change.

Figures 18.2(a) shows a circuit one D lach and two buffer gates that remains unstable in successive feedback-cycles, when X_0 change from 0 to 1.

Let instance when X_0 changes from 0 to 1 is after stable $x_Q = 1$. The buffer gives the output $Y_0 = 1$ and $y_1 = 1$. The y_1 in first feedback-cycle after a delay gives the output $= x'_Q = 0$. Therefore, now y_1 now becomes y'_1 and is 0. Now when $y_1 = 0$, the x''_Q will become = 1. The x_Q continues to toggle after successive delays. The circuit is unstable till infinity.

If after first feedback-cycle or an $(n + 1)^{\text{th}}$ feedback-cycle, if the Y does not change, then the circuit is said to be stable. The n can be 0 to infinity–the circuit can be permanently unstable.

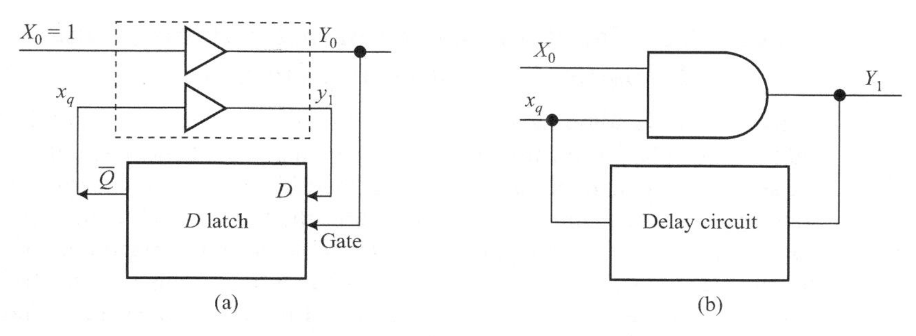

FIGURE 18.2 (a) Unstable asynchronous sequential circuit (b) Stable asynchronous sequential circuit.

Point to Remember

Asynchronous sequential circuit is called unstable when the next state changes repeatedly without settling to a stable state.

18.3 STABLE CIRCUIT ASYNCHRONOUS MODE OPERATION

Figures 18.2 (b) shows a circuit that becomes stable after first feedback-cycle itself. Let when X_0 changes from 0 to 1 when there is stable $x_q = 1$. The AND gives the output $y'_1 = 1$. The y'_1 in first feedback-cycle after a delay gives the output $= x'_q = 1$. Therefore, now y''_1 will be 1 after the first feedback-cycle. In subsequent feedback-cycles also, there is no change unless the input X_0 again changes to 0 again.

Point to Remember

Asynchronous sequential circuit is called a stable circuit when after an input change the next settles to a stable state in first or known number of memory section (latches) feedback-cycles.

18.4 FUNDAMENTAL MODE ASYNCHRONOUS CIRCUIT

Asynchronous sequential circuit operates as a fundamental mode operating circuit when no input variable(s) $\underline{\mathbf{X_i}}$ is changed unless the $\underline{\mathbf{x_q}}$ and $\underline{\mathbf{y_q}}$ are stable. An assumption useful for an analysis of fundamental mode circuit is that only one input variable (one input bit) change at an instance for changing one state of $\underline{\mathbf{Y_0}}$ to the next state of $\underline{\mathbf{Y}}$. Figure 18.1(b) showed a fundamental mode circuit operation with this assumption.

Point to Remember

Fundamental mode of an asynchronous sequential circuit is a mode in which the input(s) changes only when the circuit is in stable state.
An assumption useful for an analysis of fundamental mode circuit is that only one input variable (one input bit) change at an instance for changing one state to the next state.

18.4.1 Tabular Representation of Excitation-cum-Transitions of States and Outputs

Consider the transition table for a clocked sequential circuit in Table 15.8 once again for another exemplary asynchronous circuit operating in fundamental mode. Let us forget about the actual circuit, and the values shown in the table. We just first learn how to show a stable state, transition to an internal state(s) and transition to next stable state. Let us just also learn how to show a stable output undergoing transition to an internal output (s), and transition to next stable output. Table 18.1 shows the excitation-cum-transitions by markings by squares or box and directed arcs to show the changes from one stable state S (and output Y) to another though the intermediate instability feedback-cycles.

TABLE 18.1 Excitation-cum-transition table for an exemplary asynchronous sequential circuit operating in fundamental mode

Present State	Next State after Transition (x'_{q0}, x'_{q1})		Present Output Y_0	
(x_{q0}, x_{q1})	Input $X = 0$	Input $X = 1$	Input $X = 0$	Input $X = 1$
0, 0	[0, 0] →	1, 0 ↓	[0] →	0 ↓
0, 1	0, 1	1, 0 ↓	1	0 ↓
1, 0	0, 1	[1, 1]	1	[1]
1, 1	0, 1	0, 0	1	1

A stable state is marked by a square or circle or box over the state variables or state names. Horizontal directed arc(s) with arrow shows the action at the combinational circuit. Vertical directed arc(s) with arrow shows the action at the memory section by the latches. Number of vertical arcs equal the number of feedback-cycles after which stability is achieved.

A stable state shows by an square or box or circle surrounding the (x'_{q0}, x'_{q1}) in a column. The table shows that (0, 0) is an initial stable state. On changing $X = 0$ to $X = 1$, it changes to state (1, 0) at first feedback-cycle combinational circuit action shown by horizontal arc. Assume that state 1,0 is unstable after the memory section first feedback-cycle transition to next state (x'_{q0}, x'_{q1}). It changes in the feedback-cycle to (1,0) unstable (intermediate or internal) state. Now after the second feedback-cycle transition to another next state (1, 1), the state is stable as shown by the box or circle over it.

A stable output Y_k $(k = 0)$ shows by a box (or circle) surrounding the (Y_k). The table row 1 shows that (0) is an initial stable Y_0. On changing $X = 0$ to $X = 1$, it changes to state (0) at first feedback-cycle combinational circuit action by horizontal arc. Assume that output Y_0 is unstable after the memory section first feedback-cycle transition to next state. The Y_0 changes in the feedback-cycle to (0) unstable (intermediate or internal) case output in row 2. Now after the second feedback-cycle transition to another next output (1)—the output is stable as shown by the box or circle over it in row 3 last column.

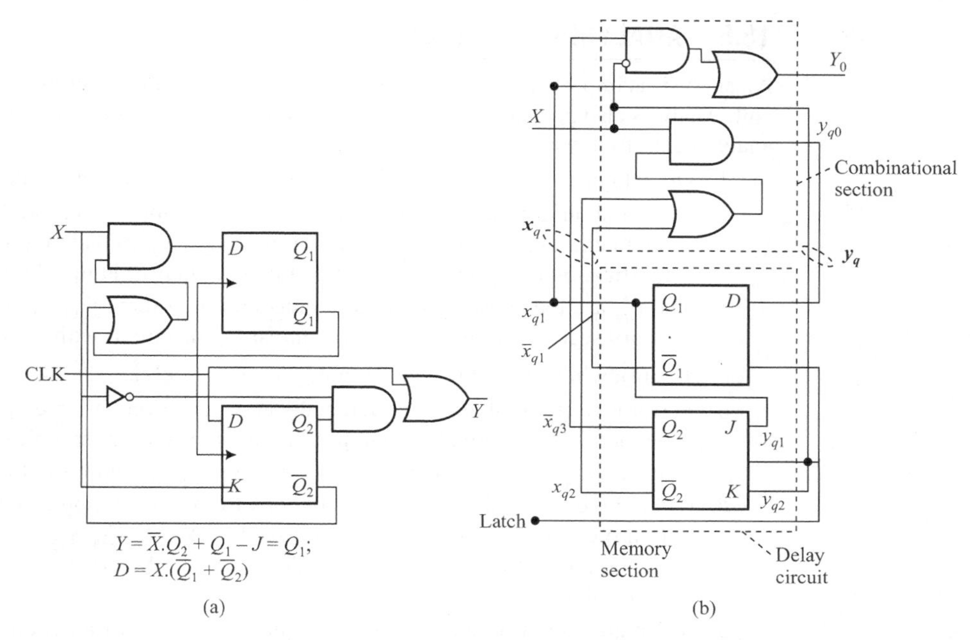

FIGURE 18.3 (a) Circuit-4 of Figure 15.3 (b) Redrawn as asynchronous sequential circuit-5 by withdrawing the clock input and replacing the JK FF by a JK latch.

Table 18.1 showed that in the given circuit whose table it is, the stability in the state or output is shown to reach in two feedback-cycles of actions at the combinational circuit (horizontal arc) followed by memory section in each feedback-cycle (vertical arc). Starting point is always a stable state. Therefore, the operation is the fundamental mode operation because in fundamental mode an input is changed when the circuit is in stable state. It is also seen that the end-state and end-output in the present case one also stable after the excitation cum transitions.

Recall the sequential circuit–4 in [Figure 15.3(b)]. Now, if the control by the clock edge inputs are removed, the circuit will be asynchronous mode circuit. Let *J-K* in Figure 18.3(a) become a *J-K* latch now. Figure 18.3(a) shows circuit-4 for an analysis later in fundamental mode. Figures 18.3(b) shows using one *D* latch and one *J-K* latch an exemplary un-clocked sequential circuit-5.

Following are the new inputs, present states, outputs, input functions and output function for the circuit-5:

Four outputs: $Y = \overline{X}.\overline{x}_{q2} + x_{q1}$; $y_{q0} = (\overline{x}_{q1} + x_{q2}).X$; $y_{q1} = \overline{x}_{q1}$; $y_{q2} = X$ ...(18.1)

Three inputs: X, $x_{q1} = D$ and $\overline{x}_{q2} = J.\overline{x}_{q2} + \overline{K}.\overline{x}_{q2}$, ...(18.2)

Delay section: $D = y_{q0}$; $J = y_{q1}$; $K = y_{q2}$ ...(18.3)

There are four outputs Y, y_{q0}, y_{q1} and y_{q2}. There are three inputs; one external input and two sate-output as feedback inputs $\overline{x}_{q1}$ and $\overline{x}_{q2}$.

18.5 ANALYSIS PROCEDURE

An analysis is important for understanding the behavior of an asynchronous sequential circuit. Analysis also provides a tabular representation of the circuit. A seven step method for analysis is as follows:

1. Draw logic circuit diagram to be analyzed. [For example, circuit in Figure 18.3(a)].
2. Circuit should be drawn such that the feedback-lop is clear and feedback loop should be minimized to have minimum feedback inputs and outputs. This helps in simplification of the analysis. [For example, circuit in Figure 18.3(b) only the x_{q1} and $\bar{x}_{q2}$ are the state representing inputs and only three y_{q0}, y_{q1} and y_{q2} inputs to memory section are there. See the contrast to this circuit with the circuit in Figure 18.3(a), which has more state outputs Q_1, $\bar{Q}_1$, Q_2, $\bar{Q}_2$].
3. Perform state variables $\underline{\mathbf{x_Q}}$ assignments and excitation variables $\underline{\mathbf{y_Q}}$ (means latche and other delay-circuit element inputs) assignments. [For example state variable assignments x_{q1} and $\bar{x}_{q2}$ and excitation variables y_{q0}, y_{q1} and y_{q2} in Figure 18.3(b)].
4. (i) Find the expressions for the excitations from the flip flop characteristics equations as per the excitations. In other words, find $\underline{\mathbf{Y}} = F_o$ ($\underline{\mathbf{X_i}}$, $\underline{\mathbf{x_Q}}$) and thus $\underline{\mathbf{Y_i}}$ and $\underline{\mathbf{y_Q}}$. [For example, find expressions given Equation 18.1 for circuit-5.]. (ii) Make an excitation table.
5. Make an excitation-cum-transition table (or simply a transition table) from the expressions for next state inputs $\underline{\mathbf{x_Q}}' = F_Q$ ($\underline{\mathbf{X_i}}$, $\underline{\mathbf{x_Q}}$, $\underline{\mathbf{y_Q}}$) (For example, Equation 18.2 for circuit-5).
6. Draw a flow diagram by making the state, flow and primitive flow tables.
7. Find reduced flow table for the restricted inputs.

Examples 18.1 to 18.3 will explain the analysis procedure for circuit of Figure 18.3– how to make the excitation table, transition table and state table, respectively. Example 18.4 will explain why is it that the flow table can't be constructed for this circuit.

Definitions for the different terms used during the analysis of the circuit or during the design and synthesis for implementing the circuit mentioned above are as follows:

18.5.1 Excitation Table

An excitation table is a tabular representation of the present state $\underline{\mathbf{x_Q}}$, $\underline{\mathbf{y_q}}$ and $\underline{\mathbf{Y_i}}$ at the delay and combinational sections. The table is a tabular representation of present state functions F_Q and output function F_O. Present state defines by a specific combination of state variables $\underline{\mathbf{x_Q}}$. It gives present state inputs given from the memory section as per column 1.

First column of excitation table gives a present state (x_{q0}, x_{q1}) in its each row. Number of rows in each column equals 2^n where n is the number of delay-section outputs. Delay section may consist of either latches (*FF*s without clock for synchronizing) or simple set of gates to create propagation delays for $\underline{\mathbf{x_Q}}$ longer than the combinational circuit section $\underline{\mathbf{y_Q}}$ and $\underline{\mathbf{Y_i}}$ in Figure 18.1. For example, if (x_{q0}, x_{q1}) are the states at the memory section, then (x_{q0}, x_{q1}) = (0, 0), (0, 1), (1, 0) and (1, 1) are the four combinations possible for the four different states of the memory section.

For each state at a row, the excitation inputs $\underline{\mathbf{y_Q}}$ are at the columns 2 to column $(1 + 2^m)$, where 2^m is the number of possible combinations in set of inputs $\underline{\mathbf{X_i}}$. For each state at a row, the outputs $\underline{\mathbf{Y}_i}$ are at the columns $(2 + 2^m)$ to column $(1 + 2^{m+1})$, It also gives the outputs that precede the excitations at the memory section.

Example 18.1 will give an exemplary excitation table.

18.5.2 Transition Table

A transition table is a tabular representation of next state output as per functions F_Q and F_O. It differs from excitation table in the middle section columns; columns 2 to column $(1 + 2^m)$, where 2^m is the number of possible combinations in set of inputs $\underline{\mathbf{X}_i}$. It shows how the delay circuit or latches will respond to all the present inputs $\underline{\mathbf{y}_q}$ and will generate $\underline{\mathbf{x}'_q}$.

Number of rows in each column equals 2^n for the n-delay section outputs.

First and columns $(2 + 2^m)$ to column $(1 + 2^{m+1})$ of transition table are same as excitation table.

The number of columns for the next state equals the number of possible combinations of external inputs in the set $\underline{\mathbf{X}_i}$. It equals 2^m. For each set of inputs, there is a set of memory section next state variables after the transition at the memory section. For example, if (x_{q0}, x_{q1} and $xq2$) are the states at the memory section, then (x_{q0}, x_{q1} and $xq2$) = (000), (001), (010), (011) (100), (101), (110) and (111) are the eight combinations possible for the eight different states of the memory section. [If there are two external inputs, there will be 32 cells in columns 2 to 5.]

Table 18.2 shows transition table of the Table 18.1, which was the excitation-cum-transition table for an exemplary asynchronous sequential circuit operating in fundamental mode.

TABLE 18.2 Transition table for an exemplary asynchronous sequential circuit operating in fundamental mode

Present state	Next state after transition $(x'_{q0}, x'_{q1})^+$		Present output Y_0	
$(x_{q0}, x_{q1})^*$	Input $X = 0$	Input $X = 1$	Input $X = 0$	Input $X = 1$
0, 0	[0, 0]	1, 0	0	0
0, 1	0, 1	1, 0	1	0
1, 0	0, 1	[1, 1]	1	1
1, 1	0, 1	0, 0	1	1

$^+$Stable state is marked by box over the state variables. $^*x_{q1}$ and $\bar{x}_{q2}$ in Figure 18.3(b) (equation 18.2)

Example 18.2 will show construction of transition table using state variables (x_{q1}, $\bar{x}_{q2}$) and excitation table for circuit-5 (Figure 18.3(b)).

18.5.3 State Table

All columns of state table are same as transition table except that instead of state variables, the states are given.

Each set of (x_{q0}, x_{q1} and x_{q2}) is assigned a state by a circuit designer. For example, S_0, S_1, S_2, ..., and S_7 in a three-state variable case. A state table is a tabular representation of the present state as per the assignments of state variables to a state.

Present states are at the column 1. Next states after the transition are in a set of succeeding 2^m columns where m is the number of possible sets of the external inputs. The outputs are shown in next set of 2^m columns.

Number of rows in each column equals 2^n for a memory section of n state-variables.

For example, if (x_{q0}, x_{q1}) are the state variables at the memory section, then let us assign the followings:

$$S(x_{q0}, x_{q1}) = S(0, 0) = S_0 \text{ for } (0, 0) \text{ values of } (x_{q0}, x_{q1}).$$
$$S(x_{q0}, x_{q1}) = S(0, 1) = S_1 \text{ for } (0, 1) \text{ values of } (x_{q0}, x_{q1})$$
$$S(x_{q0}, x_{q1}) = S(1, 0) = S_2 \text{ for } (1, 0) \text{ values of } (x_{q0}, x_{q1}), \text{ and}$$
$$S(x_{q0}, x_{q1}) = S(1, 1) = S_3 \text{ for } (1, 1) \text{ values of } (x_{q0}, x_{q1}).$$

A state in a row on first column is now either S_0, S_1, S_2 or S_3 for the row 1, row 2, row 3 or row 4, respectively. [These are as per four possible values of (x_{q0}, x_{q1}) in case of two state variables used at the memory section.]

The number of columns for the next state **S** equals the number of possible combinations of external inputs in the set **X**. It equals 2^m. For each set of inputs, there is a set of memory section states resulting after the transition at the memory section, for example, corresponding to each set of external inputs, there will be four sets of next states. A state in a row is either S_0 or S_1, S_2 or S_3 as per the post transition values of next state $(x'_{q0}, x'q_1)$.

Table 18.3 shows state table corresponding to the Table 18.2 transition table.

TABLE 18.3 State table for an exemplary asynchronous sequential circuit operating in fundamental mode

Present state		Next state after transition (x'_{q0}, x'_{q1}) S*		Present output Y_0	
(x_{q0}, x_{q1})		Input $X = 0$	Input $X = 1$	Input $X = 0$	Input $X = 1$
0, 0	S_0	S_0 (boxed)	S_2	0	0
0, 1	S_1	S_1	S_2	1	0
1, 0	S_2	S_1	S_3 (boxed)	1	1 (boxed)
1, 1	S_3	S_1	S_0	1	1

* Stable state is marked by circle or box over the state name.

Example 18.3 will show the construction of state table from the state table for circuit-5.

18.5.4 State Diagram

A state diagram is diagrammatic representation of the state table. Each element of a set of present $(x_{q0}, x_{q1}, ..., x_{qn-1})$ is denoted by a state. There are z $(= 2^n)$ maximum possible states $S_0, S_1, S_2, ..., S_{z-1}$ in asynchronous circuit with z-memory or delay section circuits. (The 2^n states can be reduced by state minimization method of finding the equivalent states. In that case state table will have less number of rows and state diagram less number of nodes).

1. States S_0, S_1, S_2 and S_3 are labeled at the centers of each circle representing a node.

2. Each arc or circular arc is labeled with present input and the present output at the transition.
3. Each arc or circular arc can have more then one set of (pre-transition input/post transition output) labeled on it if there are more than one sets of (pre-transition input/ post transition output) that are having the same transition from a node to another.
4. The number of nodes = number of rows in the state table.
5. For two flip flops, there are four states S_0, S_1, S_2 and S_3. So four circles are drawn for the four nodes of a graph.
6. A directed arc from the present state node to the next state node shows a transition.
7. A small diameter circular directed arc marks a transition in which the state remains unchanged.
8. The number of directed arcs equals the number of transitions in which the state changes.
9. The number of directed circular arcs equals the number of transitions in which the state does not change.

Figure 18.4(a) shows the state diagram corresponding to an exemplary state table in Table 18.3. Figure 18.4(b) shows its flow diagram and 18.4 (c) shows the races due to feedacks. [Sections 18.5.8 and 18.6.]

18.5.5 Flow Table

A state table gives the following:

1. A state table gives the stable next state and output before this state. This state is that which finally exists after one or more feedback-cycles of next states from the memory section after change of an input X_k. That is fine when the finally what state(s) is achieved that matters.
2. However, a state table gives the intermediate state(s) also, which are unstable (refer vertical directed arcs in column 3 of Table 18.1) and leads to other unstable state(s) to which the asynchronous circuit is passing through in case of more than one feedback-cycles at the memory section.
3. A state table does give the output at the intermediate unstable state(s) also (refer vertical directed arcs in column 5 of Table 18.1).
4. A state table gives all the next state(s) whether a certain set of inputs is specific as nonexistent after a certain sequence of set on inputs or existent. For example, when two simultaneous input bit changes not permitted the 01 will not follow 11 for X_0 and X_1. However, state table will mention that state also.

A state table does not give the following:

1. A state table does not give the information about state that will not exist due to certain specified input constraints as it gives all cases entries.

   State table therefore gives too much. It does not provide the information about the flow (next sequence) of the stable states and outputs for the final stable existence after one or more feedback-cycles of next states from the memory section after change of an input X_k. For a condensed view of the asynchronous circuit behavior, the flow part is required.

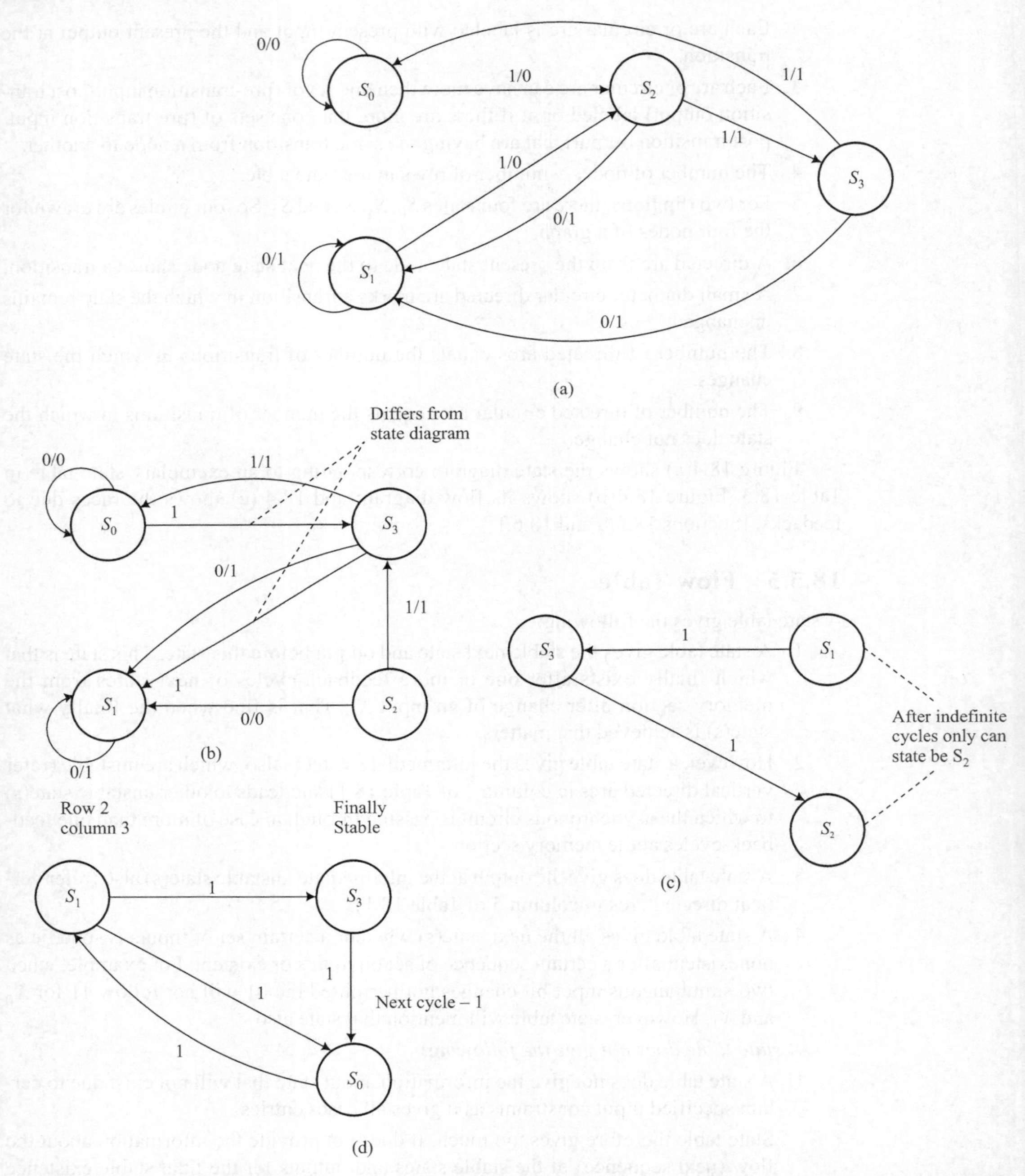

FIGURE 18.4 (a) State diagram for asynchronous circuit having state table for as per Table 18.3. (b) Flow diagram using flow table 18.4 constructed for circuit-5 of excitation-cum-transition table in Table 18.1 (c) Races arising due to the feedback next-state variables, which don't occur simultaneously when two bits change at the delay section during change when a state changes. (d) An alternate flow diagram.

A flow table gives the following:

A flow table gives the present state, for the different input combinations next states (stable) and present outputs, which corresponds to ones leading to stable state. These states exist after one or more feedback-cycles to next stable states at the memory section after change of an input X_k.

However the unstable states, which lead to other unstable states, are not shown in flow table and at a table next state entry only the eventually occurring states are shown. Flow table gives the condensed view of the circuit.

Number of columns and rows of the flow table are the same as state table except that a flow table row(s) does no longer exist if at least one of the next states given at the row is the finally stable one (encircled or boxed one) for any set of inputs at the state able. In other words, each row will show at least one encircled or boxed state.

Entries in columns and rows of the flow table are the same as corresponding rows (now these can be less) at the state table except the following:

1. A flow table row(s) does not show an entry of the unstable state (recognized by no encircling over it in state table) that is going to lead to another unstable state and a stable state entry name (without encircling) replaces it and that is the state that will eventually occur in the next feedback-cycle(s). In other words, each row will give in the columns the next stable states, that are going to occur and will change each state-table unstable state entry if that leads to another state which is stable (Table 18.4).
2. A flow table row(s) show a next state entry by dash if there is another stable state in the row and the transition to this entry from that will need more number of simultaneous input changes than permitted to occur. In other words, it does not show two identical entries for the stable state if both have the inputs differing at two or more bit places when only one input bit change is the only change permitted in a circuit as per constraint specified. For example, for there are encircled C and C' at two input conditions (0, 0) and (0, 1). Then none of them is replaced by the dash and encircled C and C' entries are at two input conditions (0, 0) and (1, 1), then one of them is replaced by the dash. (Refer explantion given later in a foot note of Table 18.6.)
3. A flow table row(s) output entry is shown only for the outputs, which lead to stable next states in the state table, else an output is shown by the dash.
4. Number of rows in each column equals 2^n for a memory section of n state-variables *minus* the number of rows having no input condition with an encircled state *minus* the state that are equivalent if the reduction is performed for minimization.

Table 18.4 gives the flow table constructed from excitation-cum-transition table in Table 18.1. Comparison of the table shows that the output that preceded the unstable state is shown by the dashes. Also only the finally achieved state is shown and intermediate states are not shown. [Note that row for S_1 has been removed as Y for $X = 1$ is –.]

18.5.6 Example of an Excitation-cum-Transition Table

Consider another excitation-cum-transition table, Table 18.5. Assume that it is for a fundamental mode operating circuit-6. Figure 18.5(a) shows the state diagram.

TABLE 18.4 Flow table for an exemplary asynchronous sequential circuit operating in fundamental mode

Present state	Next state after transition (x'_{q0}, x'_{q1})		Present output Y_0	
(x_{q0}, x_{q1})	Input $X = 0$	Input $X = 1$	Input $X = 0$	Input $X = 1$
0, 0 S_0	S_0	S_3	0	–
1, 0 S_2	S_1	S_3	0	1
1, 1 S_3	S_1	S_0	1	1

TABLE 18.5 Excitation-cum-transition table for an exemplary circuit-6 asynchronous sequential circuit operating in fundamental mode

Present state	Next state after transition (x'_{q0}, x'_{q1})*				Present output Y_0			
(x_{q0}, x_{q1})	Input $X_1X_0 = 00$	Input $X_1X_0 = 01$	Input $X_1X_0 = 10$	Input $X_1X_0 = 11$	Input $X_1X_0 = 00$	Input $X_1X_0 = 01$	Input $X_1X_0 = 10$	Input $X_1X_0 = 11$
0, 0	0, 1	0, 0	1, 0	1, 0	0	0	1	1
0, 1	0, 1	0, 1	0, 0	1, 0	1	0	1	0
1, 0	1, 1	1, 1	1,0	1, 0	0	1	0	0
1, 1	1, 1	0, 1	1, 1	1, 0	1	1	0	1

* Stable state is marked by circle or box over the state variables. Horizontal directed arc(s) with arrow shows the action at the combinational circuit. Vertical directed arc(s) with arrow shows the action at the memory section by latches. Number of vertical arcs equal the number of feedback-cycles after which stability is achieved.

18.5.7 Flow Table from Excitation-Transition Table

Table 18.6 gives the flow table costructed from the Table 18.5 transition table, that which showed the directed arcs to give the transitions. Flow table therefore represents the stable behavior of a fundamental mode circuit in contrast to state table.

Flow table in Table 18.6 represents the condensed view of the flow from one set of states to another set of stable states in different input X conditions.

18.5.8 Flow Diagram

Figure 18.4(b) shows a flow diagram constructed from circuit-5 of excitation-cum-transition table in Table 18.1. A flow diagram is diagrammatic representation of the flow table. States S_0, S_1, S_2 and S_3 are then labeled at the centers of each circle representing a node.

1. Each arc or circular arc is labeled with present input and the present output at the transition. If there is dash, then slash sign and outputs for that are not shown.
2. Each arc or circular arc can have more then one set of (pre-transition input/post transition output) labeled on it if there are more than one sets of (pre-transition input/post transition output) that are having the same transition from a node to another.
3. The number of nodes = number of rows in the flow table.

TABLE 18.6 Flow table for an exemplary circuit-6, asynchronous-sequential circuit operating in fundamental mode

Present state	Next state after transition (x'_{q0}, x'_{q1})				Present output Y_0			
(x_{q0}, x_{q1})	Input $X_1X_0 = 00$	Input $X_1X_0 = 01$	Input $X_1X_0 = 10$	Input $X_1X_0 = 11$	Input $X_1X_0 = 00$	Input $X_1X_0 = 01$	Input $X_1X_0 = 10$	Input $X_1X_0 = 11$
S_0	S_1	[S_0]	–	S_2	0	0	–	1
S_1	[S_1]	S_1	S_0	S_2	1	–	1	0
S_2	S_3	S_1^	[S_2]	[S_1^]	0	–	0	1
S_3	[S_3]	S_1	[S_3]	S_2	1	1	0	1

Note the followings:

1. * Stable state is marked by a circle or box over the state.
2. ^ S_1 replaces the S_2 in Table 18.5 at third row column 5, because S_1 is the final stable state after two arcs to S_1. Similarly S_1 replaces by S_3 in row 3 column 3 because S_1 is the final stable state for the given state of inputs and outputs after one arc to S_1.
3. – Dash signs in columns between columns 6 and 9 rows 2 and 4 shows that that output with the given X_1X_0 lead to unstable intermediate states.
4. – A dash sign in row 1 column 4 shows that S_2 will be constrained to non-existance because two input changes are not allowed simultaneously. In row 1 S_0 is the present state. When input changes to 01, then also the state is S_0. Now at the Table 18.6, the states are S_2 for both column 4 and column 5 row 1. Both are S_2. Only 01 to 11 input changes are permitted due to input constraints. Hence a dash sign replaces S_2 in column-4 row-1.
5. Also the column 8 row output also shows a dash because the corresponding state is shown as non-existent.

4. A directed arc from the present state node to the next state node shows a transition to the stable state after one or more feedback-cycles through the memory or delay section.
5. A small diameter circular directed arc marks a transition in which the flow does not appear to occur because the state remains unchanged after one or more feedback-cycles.
6. The number of directed arcs equal the number of transitions in which the state changes and the flow occurs to next stable state.
7. The number of directed circular smaller diameter arcs equals the number of transitions in which the flow does not occur.

Figure 18.5(b) shows an example of flow diagram constructed from circuit-6 flow table in Table 18.6 using the above guidelines (Figure 18.5(a) shows state diagram).

Note that a state diagram in Figure 18.5(a) for circuit-6 shows two directed arc for S_0 to S_2, one is for 10/1 and other is for 11/1, present inputs/output. Flow diagram drawn from it removes one arc 10/1 because two simultaneously bit changes are assumed to be non-existent (Figure 18.5(b)).

A comparison of Figures 18.5(b) and 18.5(a) (flow diagram and state diagrams) shows the followings:

1. The directed arc for the case of two simultaneous input changes at figure left most side at the middle from S_0 to S_2 has been removed.
2. A directed arc from S_2 to S_1 for the cases of the inputs 01 and 11 is shown in place of unstable state transition through S_3 for 01/1 in the state diagram.

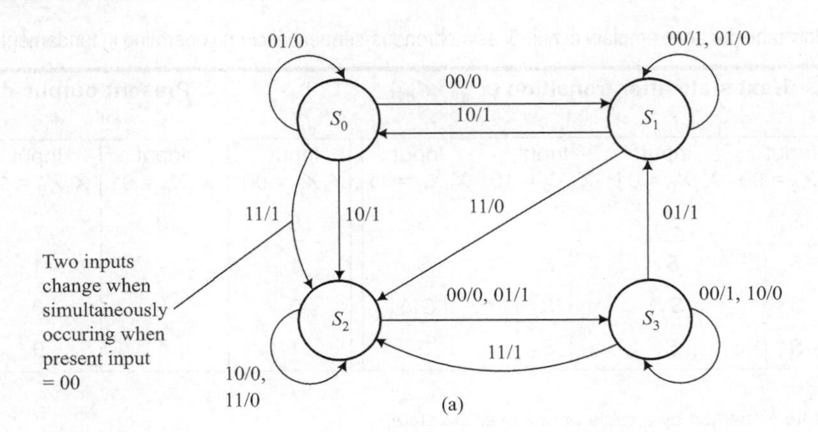

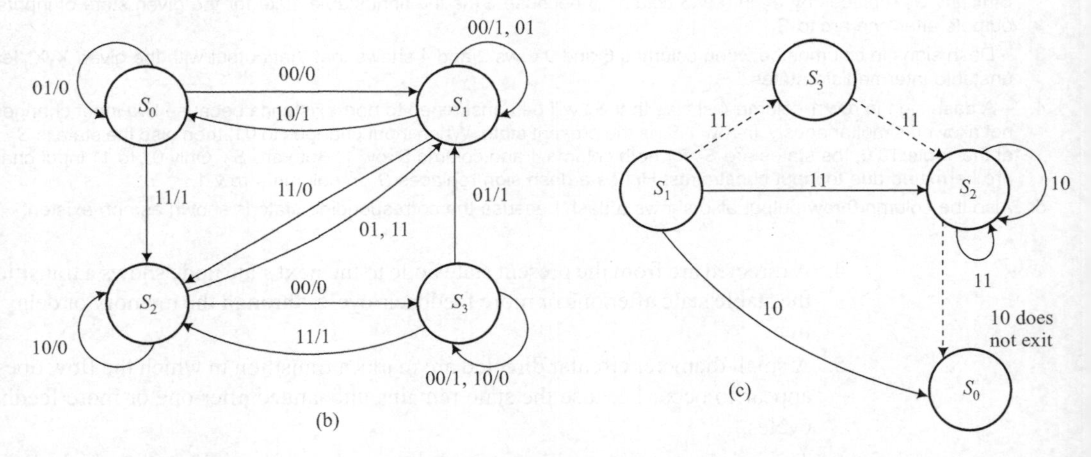

FIGURE 18.5 (a) State diagram of circuit-6 having state table as per Table 18.5. (b) Flow diagram of circuit-6 having state table as per circuit-6 Table 18.5. (c) The race arising due to the feedback of next state variables, which don't occur simultaneously when two bits change at the delay section during change when a state changes.

18.6 RACES

Example Circuit 5

Closely examine the Table 18.1 and look for the entries of (x'_{q0}, x'_{q1}) which differ for both the bits with respect top present state (x_{q0}, x_{q1}) places. Examination reveals the followings:

1. (x_{q0}, x_{q1}) and (x'_{q0}, x'_{q1}) [the present and past states] are (0, 1) and (1, 0), respectively. (Refer column 3 row 2).
2. (x_{q0}, x_{q1}) and (x'_{q0}, x'_{q1}) [the present and past states] are (1, 1) and (0, 0), respectively (Refer column 3 row 4).

There are two races at two entries of next state outputs for the present states in the transitions given by Table 18.1. Figure 18.4(c) showed two races due to the feedback next state variables, which don't change simultaneously when two bits change at the delay section during change when a state changes. The following explains this:

There are two-bit changes at the memory or delay sections. There is no guarantee, which will change x_{q0} first or x_{q1} first to x'_{q0}, or x'_{q1}, respectively as two circuits propagation-delays are never exactly identical. Therefore, (0, 1) present state can change to (1,1) for some period (S_1 can temporarily be S_3 in place of S_2). Alternatively, (0, 1) can temporarily change to (0, 0). Therefore, (0, 1) present state can change to (0, 0) for some period (S_1 can temporarily be S_0 in place of S_2). This is called race condition.

Also, (1, 1) present state can change to (1, 0) for some period (S_3 can temporarily be S_2 in place of S_0). Alternatively, (1, 1) can temporarily change to (0, 1). Therefore, (1, 1) present state can change to (0, 1) for some period (S_3 change temporarily be S_1 in place of S_0). This is called race condition.

Example Circuit 6

Closely examine the circuit-6 Table 18.5 and look for the entries of (x'_{q0}, x'_{q1}) in columns 2 to 4 which differ at both the places of the bits with respect the present state (x_{q0}, x_{q1}) in column 1. Examination reveals the followings:

1. (x_{q0}, x_{q1}) and (x'_{q0}, x'_{q1}) the present and next states are (0, 1) and (1, 0), respectively (Refer columns 1 and 5 row 2).

There are two-bit changes at the memory or delay sections. There is no guarantee, which will change x_{q0} first or x_{q1} first to x'_{q0}, or x'_{q1}, respectively as two circuits propagation-delays are never exactly identical. Therefore, (0, 1) present state can change to (1,1) for some period (S_1 can temporarily be S_3 in place of S_2). Alternatively, (0, 1) can temporarily change to (0, 0). Therefore, (0, 1) present state remains unchanged (0, 1) for some period.

This is race condition in first case only. There is only one race for the present state transitions given by Table 18.5. S_1 raced to S_3 in place of S_2 due to (0, 1) $\rightarrow$ (1, 1). Figure 18.5(c) shows the race due to the feedback of two next state variables, which don't occur simultaneously when two bits change at the delay section during change when a state changes.

Point to Remember

A race is said to occur if between next set of states $\mathbf{\underline{x_q}}$ and present states $\mathbf{\underline{x'_q}}$, there are two or more bits change at a memory or delay section.

18.6.1 Cycles of the Races

There are two possibilities when a temporary change to an intermediate state occurs due to differing propagation delays for two bits in a feedback cycle.

Non-Critical Race (Deterministic Number of Cycles of race)

The state momentarily changed can lead to a stable state with one or more additional feedback-cycles and the state eventually turns out to be the same as would have been expected had there been no race. Race condition is not permanent and the race returns to normal after calculatedly number of races. Asynchronous circuit is deterministic when a non-critical race condition exists.

For example, examine Table 18.1 and note the followings:

1. The present state (0, 1) if momentarily changes to (0, 0); the monetarily state will be S_0. The S_0 for $X = 1$ gives the output (1, 0) again and gives the state S_2 in next feedback cycle.

2. The present state (0, 1) if momentarily changes to (1, 1), the monetarily state will be S_3. The S_3 for $X = 1$ gives the output (0, 0) and state S_0 again in next feedback-cycle. S_0 gives the state S_2 in next feedback cycle.

Therefore after two state transitions, the same stable state is obtained in case (2) and after one more state transitions in case (1) as would have been had there been no race to case (1) or (2) above. The race is therefore not critical in its effect on the behavior, except additional delay. Flow table shall remain unaltered.

Critical Race (Indeterminate Number of Cycles of Races)

> The state momentarily changed can lead to a different stable state with any known number one or more additional feedback-cycles and the state will most often turn out never to be the same as would have been expected had there been no race. Such a race is called critical race. Race condition is indefinite and the race returns to normal after indefinite number of races because when two propagation delays exactly match is never known. Asynchronous circuit is probabilistic, not deterministic when a critical race condition exists.

For example, examine Table 18.1 row-4 column-3 and note the followings:

3. The present state (1, 1) if momentarily changes to (0, 1); the monetarily state will be S_1. The S_1 for $X = 1$ gives the stable state S_3 back after two feedback cycles.
4. The present state (1, 1) if momentarily changes to (1, 0), the monetarily state will be S_2. The S_2 for $X = 1$ gives the state (1, 1) S_3 again in next feedback cycle.

Therefore, after two or one state transition, the starting stable state S_3 is obtained in cases (1) and case (2), respectively. Therefore, even after any number of transitions S_0 [(0, 0) state] will never occur unless both delay circuits are exactly identical, which may need an infinite wait. Flow table state (S_0) will rarely be obtained from S_3 when input changes from 0 to 1. Such a race is called *critical race.* Consider Table 18.5 circuit 6 case Figure 18.5(c) shows that (1, 0) → (1, 1) is critical race as there is no path to return to S_0 for inputs 10.

18.7 RACE-FREE ASSIGNMENTS

Memory (or delay) section next-state variables when have the bits differing by more than one during a transition, the critical (indeterminate time for stable state) and non-critical (deterministic) races occur. Critical race must not occur in an asynchronous sequential circuit.

Race free assignments is an assignment of state variables such that next-state variables when don't have the bits differing by one during a transition. Recall the bit changes in adjacent cells in a Karnaugh map. Only a single bit changes between two adjacent cells.

Step 1: Finding Non-Adjacencies in Transitions for Different Input Conditions

Let us build a Karnaugh map type table for each set of X in a flow table in Table 18.4. Tables 18.7 part (a) and (b) are adjacency map for the assignments of present and next state variables for $X = 0$ and $X = 1$, respectively. It shows in a cell the present state by 0 and next state by 1. Dotted lines in table show the non adjacency.

TABLE 18.7 Adjacency map for the assignments of present and next state variables

	← X = 0 →				← X = 1 →			
(x'_{q0}, x'_{q1})	00	01	11	10	00	01	11	10
(x_{q0}, x_{q1})	S_0	S_1	S_3	S_2	S_0	S_1	S_3	S_2
00 S_0	0	-			0		1	
01 S_1	0					-	-	
11 S_3		1	0		1 —	— — —	— 0	
10 S_2		1 —	— — —	— 0			1	0
	← (a) →				← (b) →			

1. It is observed that for the present state S_3, the transition to S_0 is not to the adjacent cell. S_0 and S_3 have to be the neighbors. [Table 18.7(b)]
2. It is observed that for the present state S_2, the transition to S_1 is not to the adjacent cell. S_2 and S_1 have to be the neighbors. [Table 18.7(b)]

Using the map number of races (non-critical plus critical) is thus found to be two only for asynchronous sequential circuit when the flow table is as per Table 18.4.

Let us build a Karnaugh map type table for each set of X_0X_1 in flow table in Table 18.6. Tables 18.8 parts (a) to (d) show the adjacencies (including wrapping adjacencies) and non-adjacencies (by dotted lines) for $X=$ 00, 01, 11 and 10, respectively. It shows in a cell the present state by 0 and next state by 1. The adjacencies of 0 and 1 are looked for. (Cells at left and right borders are adjacent; these have wrapping adjacency).

1. It is observed that for the present state S_2, the transition to S_1 is not to the adjacent cell. S_2 and S_1 have to be the neighbors. [Row 4 Table 18.8(b)]
2. It is observed that for the present state S_1, the transition to S_2 is not to the adjacent cell. S_1 and S_2 have to be the neighbors. [Row 2 Table 18.8(c)]
3. It is observed that for the present state S_2, the transition to S_1 is not to the adjacent cell. S_2 and S_1 have to be the neighbors. [Row 4 Table 18.7(c)]

All three observations above point out that the S_1 and S_2 state have to be neighbors.

Number of races (non-critical plus critical) is thus found to be three for asynchronous sequential circuit with flow table as per Table 18.6.

Step 2: Finding Race free State Assignments for Different Input Conditions

Method 1: Replacement of an unstable non-adjacent state by another adjacent state in flow table so that in the next cycle the stable state is obtained.

Consider circuit example (Table 18.4). The S_3 in row 1 of Table 18.4 can be replaced by S_2, the S_2 will lead to S_3 in next cycle. [Modification of flow table is permitted if it does not change the result finally achieved.]

S_0 present state is unresponsive to any input condition. [$S_0 \rightarrow S_0$ for $X = 0$ and for $X = 1$, the present Y is –] S_0 S_0 can be assigned three state variables, $(x_{q0}, x_{q1}, x_{q2}) = 101$ then the race free condition can be achieved

Consider circuit-6 example in Table 18.6.

TABLE 18.8 Adjacency map for the assignments of present and next state variables

	$X_0X_1 = 00$				$X_0X_1 = 01$			
(x'_{q0}, x'_{q1}) (x_{q0}, x_{q1})	00 S_0	01 S_1	11 S_3	10 S_2	00 S_0	01 S_1	11 S_3	10 S_2
00 S_0	0	1			01			
01 S_1		01				01		
11 S_3			01			1	0	
10 S_2			1	0		1 —	— —	— 0
	(a)				(b)			

	$X_0X_1 = 11$				$X_0X_1 = 10$			
(x'_{q0}, x'_{q1}) (x_{q0}, x_{q1})	00 S_0	01 S_1	11 S_3	10 S_2	00 S_0	01 S_1	11 S_3	10 S_2
00 S_0	0^+	–		1^+	0			–
01 S_1		0 —	— —	— 1	1	0		
11 S_3			01				0	1
10 S_2		1 —	— —	— 0				1
	(c)				(d)			

$^+$ Wrapping adjacency

S_1 and S_2 can be made neighbors by using three variable assignments. S_1 can be assigned three state variables, $(x_{q0}, x_{q1}, x_{q2}) = 101$. S_0, S_2 and S_3 are assigned 000, 100 and 110, respectively. [Now $S_1 \rightarrow S_2$ cause $101 \rightarrow 100$, which means only one variable change.]

Method 2: There is a systematic method, called one-hot method of race free state assignments.

1. *Action 1*: Let number of rows in flow table = m. Use m state variables. Assign each row $(x_{q0}, x_{q1}, ..., x_{qn-1})$ in sequence as 00...01, 00...10, ..., 10... 00. In k-th row, the k-th state variable is 1. For example, if there are four rows in the flow table, S_0, S_1, S_2 and S_3, assign 1000, 0100, 0010 and 0001, respectively to $(x_{q0}, x_{q1}, ..., x_{qn-1})$ with $m = 4$.
2. *Action 2*: Present-state column 1 contains the states as per the state variable assignments given above.
3. *Action 3*: Assign the stable state variables to the state variable as per corresponding state variables used in present-state column and leave presently unstable state as such. For example for the Table 18.6, assign present state and state variables as shown in Table 18.9.
4. *Action 4*: For unstable state, now write the assignment after ORing with the assignment for its next cycle transition. For example, (*i*) now consider row 2 column 4 assignment for input 10. Stable case $S0$ assignment would have been 1000. The state vertical next cycle transition would be to 0010 row-3 column-4. ORing these two gives 1010. Enter this value in the for row 2 column 4. (*ii*) Consider row 1 column 2 unstable S_1 next cycle transition is S_1 for input 01. For S_1 stable state assignment is

TABLE 18.9 Assignments for the present state and stable states after actions 1, 2 and 3 on the flow table in Table 18.6

Present state	Next state after transition (x'_{q0}, x'_{q1})				Present output Y_0			
(x_{q0}, x_{q1})	Input $X_1X_0 = 00$	Input $X_1X_0 = 01$	Input $X_1X_0 = 10$	Input $X_1X_0 = 11$	Input $X_1X_0 = 00$	Input $X_1X_0 = 01$	Input $X_1X_0 = 10$	Input $X_1X_0 = 11$
S_0 1000	S_1	1000	–	S_2	0	0	–	1
S_1 0100	0100	S_1	S_0	S_2	1	–	1	0
S_2 0010	S_3	S_1	0010	0100	0	–	0	1
S_3 0001	0001	S_1	0001	S_2	1	1	0	1

0100 and S_1 stable state it would have been 0100. Assignment after ORing the two will again be 0100. Therefore, enter 0100 assignment in column 2 of row1. (*iii*) Now consider row 3 column 2 assignment for input 00. Stable case S_3 assignment would have been 0001. The state vertical next cycle transition would be 0001 row-4 column-2. Its assignment is also 0001. ORing these two gives 0001 again. Enter this value in the for row 3 column 2. Complete the assignments for all unstable states by this ORing procedure. Now Table 18.9 will get two type of next-state assignments: assignments with only one variable = 1 and assignment with two variables =1.

5. *Action 5*: Now add an extra row for each of those next states, which have two variables as 1s. The assignment for the same vertical column of the state with two variable 1s can now be done so that next state transition occurs to the same when the next state occurs, which was ORed with the unstable state assignment before. Rewriting Table 18.9 as Table 18.10 explains this.

TABLE 18.10 Assignments in one-hot method (method-2) of race free assignment after actions 1 to 5 (Table 18.9 rewritten after incorporating actions 4 and 5 of the method)

Present state	Next state after transition (x'_{q0}, x'_{q1})				Present output Y_0			
(x_{q0}, x_{q1})	Input $X_1X_0 = 00$	Input $X_1X_0 = 01$	Input $X_1X_0 = 10$	Input $X_1X_0 = 11$	Input $X_1X_0 = 00$	Input $X_1X_0 = 01$	Input $X_1X_0 = 10$	Input $X_1X_0 = 11$
S_0 1000	0100	1000	-	0110[b]	0	0	-	1
S_1 0100	0100	0100	1010[a]	0110[c]	1	-	1	0
S_2 0010	0001	0100	0010	0100	0	-	0	1
S_3 0001	0001	0100	0001	0110[b]	1	1	0	1
S_4 1010[b]	-	-	0010	-				
S_5 0110	-	-	-	0100				

[a] New assignment to unstable state after ORing with its next cycle state assignment in column 1 for present states. [b] New row addition for an unstable state assignment in row-2 column-4. Arrow shows the next cycle state 0010. It is same as expected from Table 18.6. [c] New assignment to the unstable state after ORing with its next cycle state assignment in column 5 for present states.

6. *Action 6*: Change the output column entry for the state assigned two variables as 1s. Put the next cycle stable state. This is to prevent two times changes in an output when adding another extra row in the flow table. Rewriting the Table 18.10, Table 18.11 gives final race free assignment table after the actions 1 to 6 of one-hot method. From an unstable state, there is a path now to hot (stable) state always feasible.

It can be noted from Table 18.11 that indeterminate race condition will not be nonexistent.

TABLE 18.11 Assignments in one-hot method (method-2) of race free assignment after actions 1 to 6 (Table 18.10 rewritten after incorporating action 6 for output entries change for the unstable states)

Present state	Next state after transition (x'_{q0}, x'_{q1})				Present output Y_0			
(x_{q0}, x_{q1})	Input $X_1X_0 = 00$	Input $X_1X_0 = 01$	Input $X_1X_0 = 10$	Input $X_1X_0 = 11$	Input $X_1X_0 = 00$	Input $X_1X_0 = 01$	Input $X_1X_0 = 10$	Input $X_1X_0 = 11$
S_0 1000	0100	1000	-	0110	0	0	-	1
S_1 0100	0100	0100	1010[a]	0110	1	-	0[b]	1[b]
S_2 0010	0001	0100	0010	0101	0	-	0	1
S_3 0001	0001	0100	0001	0110	1	1	0	1
S_4 1010	-	-	0010	-				
S_5 0110	-	-	-	0100				

[a] Now there is the next cycle after which there becomes stable state output entry [0010] row 3, column 4. [b] New output entry (entered from the next row of same column).

EXAMPLES

Example 18.1 Make the excitation table for the circuit-5 of Figure 18.3.

Solution

There is one external input and two next state inputs after a feedback cycle. Therefore, we take two variables (x_{q1}, x_{q2}) and four rows (x_{q1}, x_{q2}) = 00, 01, 10 and 11 for the excitation table. Since there is one external input, there will be two columns in the entries for the excitation variables and two columns in the output.

There are three excitation variables in Equation (18.1) for the circuit of Figure 18.3(a). Using Equation (18.1), the entries of the table are now filled up. Table 18.12 gives the excitation table. Let starting values of $Q1$ and $Q2$ are same as X and $\overline{X}$, respectively.

TABLE 18.12 Excitation table for asynchronous sequential circuit-5 operating in fundamental mode circuit

Present state ($\overline{Q}_1$, $\overline{Q}_2$)	Excitation variables before the transition (y_{q0}, y_{q1}, y_{q2})		Present output Y_0	
(x_{q1}, x_{q2})	Input X = 0	Input X = 1	Input X = 0	Input X = 1
0, 0	0, 0, 0	1, 0, 1	0	0
0, 1	0, 0, 0	1, 0, 1	1	0
1, 0	0, 1, 0	0, 1, 1	1	1
1, 1	0, 1, 0	0, 1, 1	1	1

Example 18.2 Make the transition table from the circuit of Figure 18.3(b) using the excitation table made in Example 18.1.

Solution

Table 18.13 gives the transition table obtained after using Equation (18.2). The equations gives the next states (x'_{q1}, x'_{q2}) after a single feedback cycle. Excitation table first variable will be complemented as there is a D latch between y_{q0} and $\overline{x}'_{q1}$. Excitation table y_{q1} and Y_{q2} will give the next state as expected from a JK latch $\overline{Q}$. When these are 10, $xq1 = 1$; when these are 01, $xq1 = 0$; When these are 00, the result is same as present state. Assume that for 11 also, there is complementation like a JK. Mark by circle or ellipse the transitions, which will be changing, in the next cycle. We find that none of the state is stable. This is because use of JK latch, which toggles.

TABLE 18.13 Transition table for asynchronous sequential circuit-5 operating in fundamental mode

Present state	State variables before the transition (x'_{q1}, x'_{q2})		Present output Y_0	
(x_{q1}, x_{q2})	Input X = 0	Input X = 1	Input X = 0	Input X = 1
0, 0	1, 0	0, 0	0	0
0, 1	1, 0	0, 0	1	0
1, 0	1, 1	0, 1	1	1
1, 1	1,1	0, 0	1	1

Note: All states are unstable. This is expected when using a JK latch.

Example 18.3 Make the state table for the circuit-5 (Figure 18.3(b)) using the transition table made in Example 18.2.

Solution

Make Table 18.14 after assigning the states S_0, S_1, S_2 and S_3 to (x_{q0}, x_{q1}) = 00, 01, 10 and 11, respectively, in Table 18.13.

TABLE 18.14 State Table for Figure 18.3(b) asynchronous sequential circuit operating in fundamental mode

Present state	Excitation variables before the transition (y_{q0}, y_{q1}, y_{q2})		Present output Y_0	
(x_{q1}, x_{q2})	Input $X = 0$	Input $X = 1$	Input $X = 0$	Input $X = 1$
0, 0 S_0	S_2	S_0	0	0
0, 1 S_1	S_2	S_0	1	0
1, 0 S_2	S_3	S_1	1	1
1, 1 S_3	S_3	S_0	1	1

Example 18.4 Why is the flow table is not possible for a circuit with state table as Table 18.14?

Solution

This is because all entries are unstable due to *JK FF*. In flow table, at least one stable state must be there in a row. State table shows that Y is changing in both the cases. When X input is 1, $S_2 \to S_1$, $S_1 \to S_0$, S_2 is unstable. Similarly, when X = 0, $S_0 \to S_2$, S_3. The S_0 is unstable when $X = 0$.

Example 18.5 Draw state diagram from Table 18.14.

Solution

Figure 18.6 shows a state diagram from Table 18.14. State diagram shows that when X undergoes transition from 0 to 1, the S_0 may show intermediate state S_2 and S_3 also.

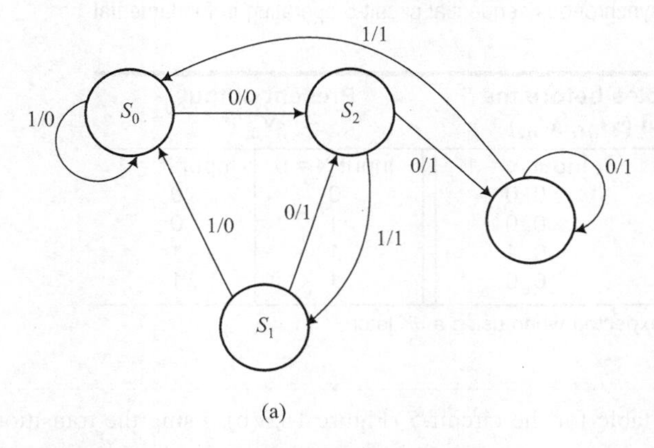

FIGURE 18.6 State diagram for the circuit-5 using table 18.14.

EXERCISES

1. Make the excitation table for the circuit-7 of Figure 18.7.
2. Make the transition table from the circuit-7 of Figure 18.7 using the excitation table made in Exercise 1.

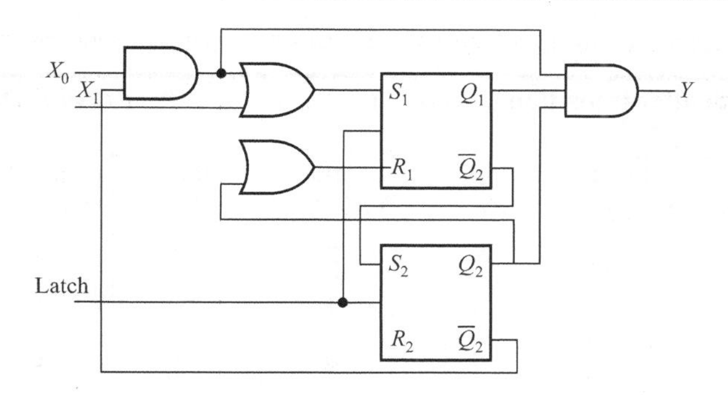

FIGURE 18.7 An asynchronous circuit-7 for Exercise 10.1

3. Make the state table from the circuit-7 of Figure 18.7 using the transition table made in Exercise 2.
4. Make the flow table from the circuit-7 of Figure 18.7 using the transition table made in Exercise 3.
5. Make a flow diagram from flow table in Exercise 4.
6. Make the transition table from the excitation table made in Table 18.15 of a circuit-8. Delay section has just the delay elements.

TABLE 18.15 Excitation table for an exemplary asynchronous sequential circuit-7 operating in fundamental mode

Present state	Excitation variables before the transition (y_{q0}, y_{q1})		Present output Y_0	
(x_{q0}, x_{q1})	Input X = 0	Input X = 1	Input X = 0	Input X = 1
0, 0	1, 0	0, 0	0	0
0, 1	1, 0	0, 0	1	0
1, 0	1, 1	0, 1	0	1
1, 1	1,1	0, 0	1	1

7. Make the transition table from the circuit-8 using the excitation table made in Exercise 6.
8. Make the state table from the circuit-8 using the transition table made in Exercise 7.
9. Make the flow table and flow diagram from the circuit-8 using the state table made in Exercise 8.
10. Find the races in a circuit whose flow table is as per Table 18.16. Examine and tell which of the race is critical.
11. Perform race free assignments for a flow table in Table 18.16 using the systematic method.

TABLE 18.16 Flow table of an exemplary circuit-8; asynchronous sequential circuit operating in fundamental mode

Present state	Next state after transition (x'_{q0}, x'_{q1})				Present output Y_0			
(x_{q0}, x_{q1})	Input $X_1X_0 = 00$	Input $X_1X_0 = 01$	Input $X_1X_0 = 10$	Input $X_1X_0 = 11$	Input $X_1X_0 = 00$	Input $X_1X_0 = 01$	Input $X_1X_0 = 10$	Input $X_1X_0 = 11$
S_0	S_1	S_0	S_3	S_2	0	0	-	1
S_1	S_1	S_1	S_0	S_2	1	-	1	0
S_2	S_1	S_3	S_2	S_3	0	-	0	1
S_3	S_0	S_1	S_3	S_3	1	1	0	1

QUESTIONS

1. When will you call a circuit a asynchronous sequential circuit?
2. Describe what do you mean by operating in fundamental mode in a asynchronous sequential circuit.
3. Why is the assumption that one input bit changes at an instance is a practical one most of the times, explain?
4. Why do you get unstable states in a asynchronous circuit?
5. When do you call a state as stable?
6. List the steps used for analyzing a given asynchronous sequential circuit.
7. Describe procedure to get state table from excitation table in an asynchronous sequential circuit. How does it differ from synchronous sequential circuit? (Hint: By encircling the stable states and by identifying the unstable states).
8. How do you get output specifications from a flow table in asynchronous sequential circuit operating in fundamental mode?
9. When do you get the non-critical races?
10. When do you get the critical races?
11. How do you perform race free assignments by adding extra rows in the flow table?
12. Describe a procedure to identify the races in asynchronous sequential circuit.

CHAPTER 19

Hazards and Pulse Mode Sequential Circuits

OBJECTIVE

We will learn in this chapter the hazards that are present in a logic circuit when two inputs change due to a single logic variable. We will also learn about pulse mode sequential circuits. Hazard means a source of trouble or risk.

We learnt in Chapter 18 the following important points:

1. A general asynchronous sequential circuit is a circuit in which there are no control(s) for the instance(s) at which the output(s) change like a control by a clocked instance in synchronous sequential circuit.
2. The circuit has a section consisting of the (i) combinational circuits at the input and output stages and (ii) latches (without clock inputs) or an equivalent delay devices, which have the circuit outputs as the excitation inputs and after the transitions and delays give the new input states.
3. A mode called fundamental mode is a mode of operation in which external input changes only when the circuit is in stable state. Stable state means further transitions don't create any new state.
4. Asynchronous sequential circuit is analyzed in fundamental mode by first making the tables for the excitations, transitions, states and then finally the flow tables and flow diagrams.

5. Asynchronous sequential circuit becomes unstable when the next state(s) changes repeatedly without settling to a stable state in a finite time.
6. There can be a race condition. It occurs when two or more states, which are the inputs after the delay, changes the states in a cycle. For example, x_q, and x_q' inputs, which are the outputs of the latches or delay elements after a transition cycle, change to 10 from 01 or change from 10 to 11 and then to 01 due to the different amount of delays for obtaining x_q, and x_q' in any two transitions. Note: It is natural that two circuits can rarely be having exactly identical propagation delays. Even temperature in vicinity of identically made gate can cause the difference.
7. Race condition can sometimes be critical. That means the transition cycle leads to an unstable state from which the finally settable state (s) is not recovered in the succeeding cycles. Finally settable state(s) is one that would have been had the two delays are identical when there are two transitions like of the change to 01 from 10 or change is from 10 or to 01 or 11 from 00 or 00 from 11.
8. Race free assignments are possible and that results in the asynchronous circuit free from critical race condition.

19.1 HAZARDS

Consider a NAND gate (of propagation average delay = Δt) in which one of the inputs gets input from a logic variable *X*. Other input of NAND gets input from *X* due to a NOT (of propagation average delay = Δt') between *X* and that input. The NAND characteristics suggest that NAND with *X* and *X* inputs will always give output = 1. In actual practice, the output of NAND will show a transient from 1 to 0, then again to 1 during the period = Δt'. This is because in the Boolean operation $Y = X.\overline{X}$, when *X* changes, the $\overline{X}$ changes only after a period = Δt'. Figures 19.1(a) and (b) show the circuit with two inputs dependent on one variable and the timing diagrams for *X*, $\overline{X}$, and *Y* in this operation. The NOT is a delaying element in this circuit. Figures 19.1(c) and (d) show the use of a delay element, which is some Boolean expression implementing circuit to give the output as a source of a hazard and show the timing diagram, respectively.

Suppose that NOT implementing NAND in circuit of Figure 19.1(a) has much shorter delay compared to output stage NAND. Now, there is no hazard when the response time of the sequential or combinational next stage happens to be longer than that.

From the timing diagrams in Figure 19.1(b) and (c) it appears that the output effects only momentarily for Δt' or Δt''' only. Suppose this output *Y* is input to a counter reset input. Then counter will loose all previous counts. In that case, the hazard creates a malfunction, which is not momentarily. Suppose *Y* is input to a clocked sequential circuit. Then the hazard creates no malfunction because input at the clocking instance only is important.

Point to Remember

> A hazard arises and becomes a potential source of momentary or permanent malfunctioning in a combinational circuit and asynchronous sequential circuit when a variable(s) and its complement(s) are used with complement(s) having a delay element(s) before it, so that for a certain period the complement is not available, leading to a different output either momentarily for the delay period or permanently.

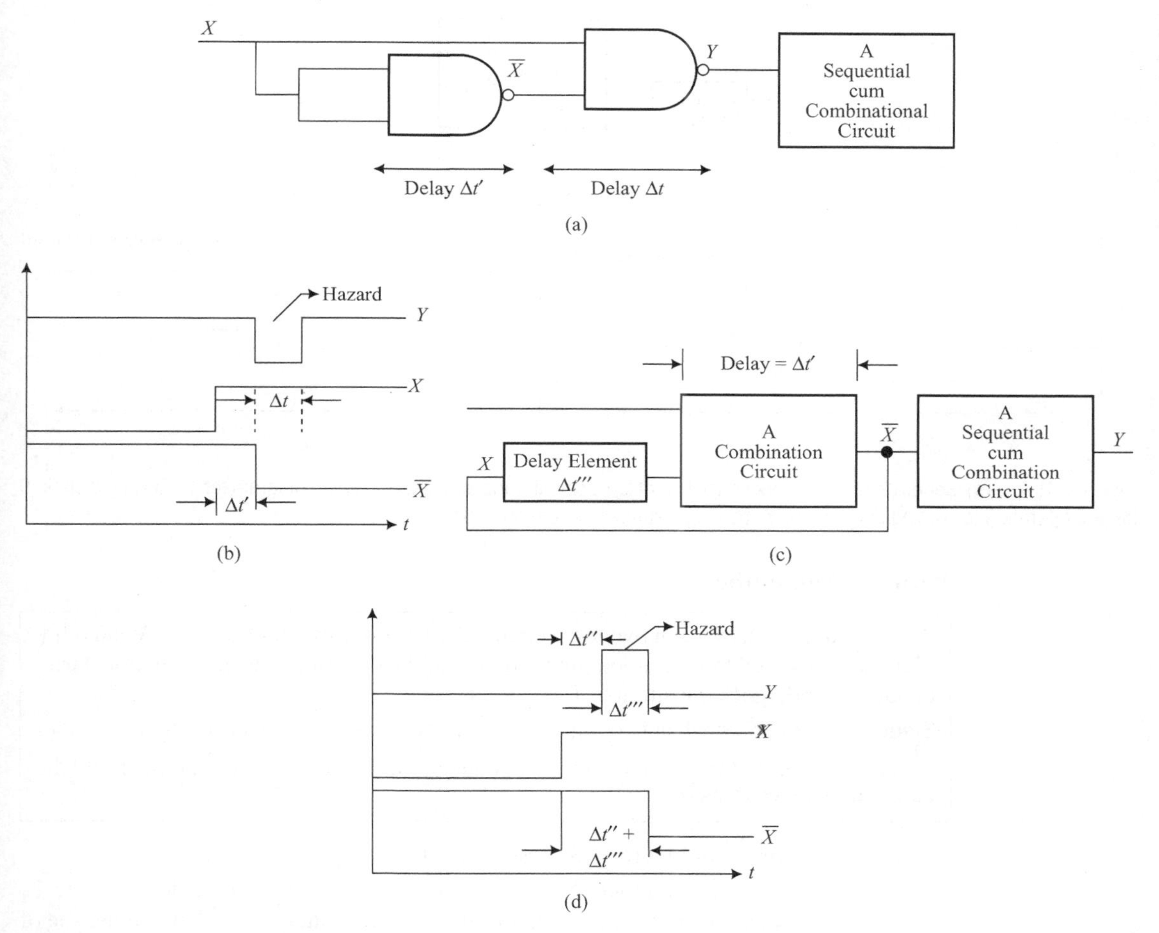

FIGURE 19.1 (a) Two NANDs circuit with two inputs of NAND dependent on one variable (b) Timing diagram for *X*, *X*, and *Y* in this operation. (c) Output, which creates a static-1 hazard.

19.1.1 Static-0 Hazard

Figure 19.2(a) shows a general circuit network consisting of logic circuit-1 and circuit-2. Static-0 hazard is a hazard in which a Boolean algebraic output or next state output expected is 0, but momentarily, it undergoes transition to 1. For example, the circuit of Figure 19.1(c) expected $Y = 0$, but it becomes 1 for a period of propagation delay $\Delta t'''$ of the complementing circuit. Figure 19.2(b) shows the output, which creates a static-0 hazard.

19.1.2 Static-1 Hazard

Static-1 hazard is a hazard in which a Boolean algebraic output or next state output expected is 1, but momentarily it undergoes transition to 0. For example, the circuit of Figure 19.1(a) expected $Y = 1$, but it becomes 0 for a period of propagation delay Δt' of the complementing circuit. Figure 19.2(c) shows the output, which creates a static-1 hazard.

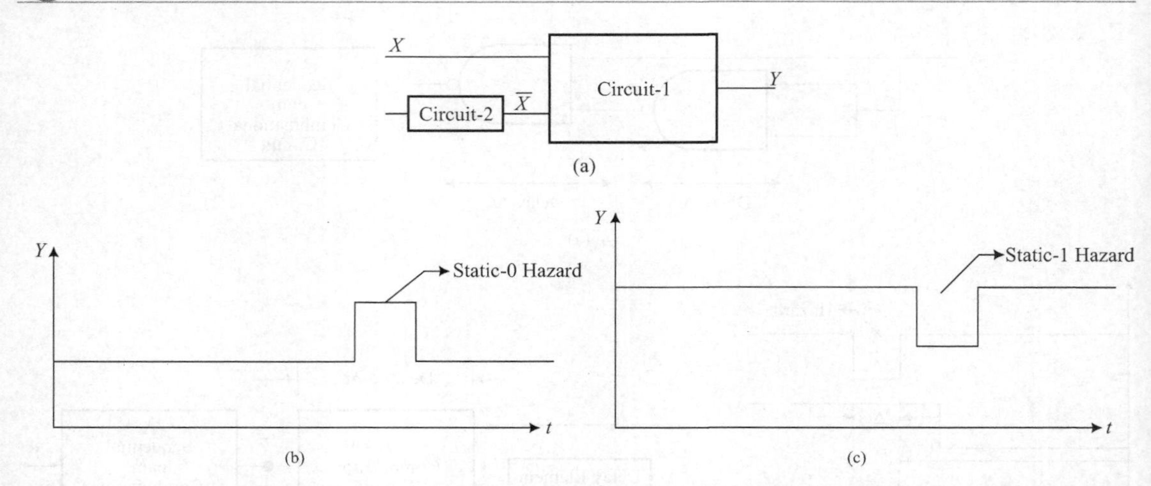

FIGURE 19.2 (a) General circuit network consisting of logic circuit-1and circuit-2 (b) Timing diagram for *Y* showing a static-0 hazard operation (c) Timing diagram for *Y* showing a static-1 hazard operation.

Point to Remember

Static-0 hazard arises when two logic circuit elements one of which is using *X* and other *X* and are expected to give an output 0 but due to the differing delays, the momentarily the output undergoes transition to 1.

Static-1 hazard arises when two logic circuit elements one of which is using *X* and other $\overline{X}$ and are expected to give an output 1 but due to the differing delays, momentarily the output undergoes transition to 0.

Consider a circuit of Figure 19.3. Assume the followings for the circuit:

1. It gives two external inputs X_1 and X_0 to an asynchronous circuit element. The X_1 has combination circuit such that it is at 1 and has static-1 hazard due to the use of two paths one for X_1' and other for $\overline{X}_1'$.

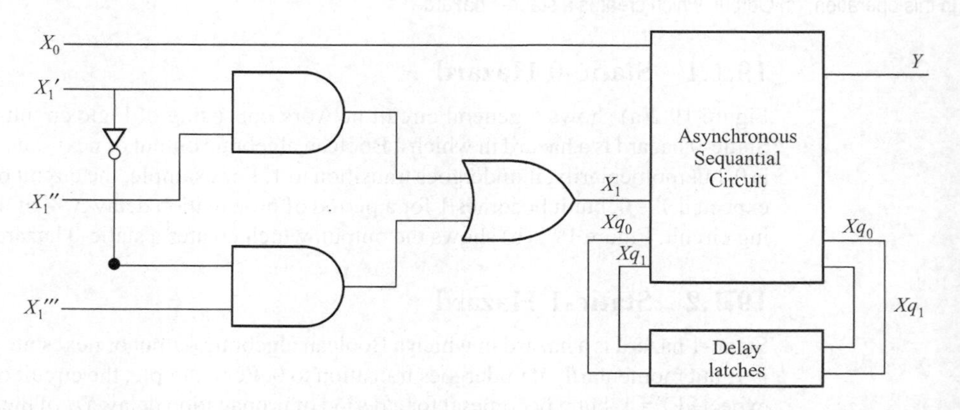

FIGURE 19.3 A circuit with two outputs, which are two external inputs *X*1 and *X*0 to an asynchronous circuit element with *X*1 having a preceding combination circuit with static-1 hazard due to the use of two paths one for *X*1' and other for *X*1'.

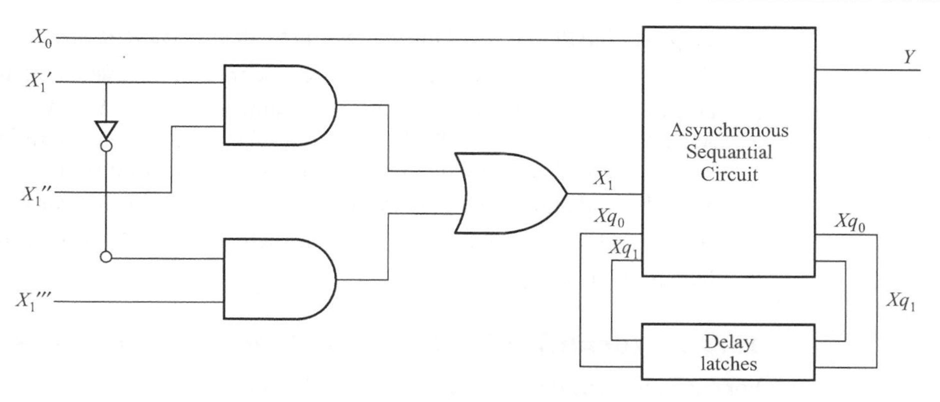

FIGURE 19.3 A circuit with two outputs, which are two external inputs $X1$ and $X0$ to an asynchronous circuit element with $X1$ having a preceding combination circuit with static-1 hazard due to the use of two paths one for $X1'$ and other for $X1'$.

TABLE 19.1 Excitation-cum- Transition table* for an exemplary circuit consisting of two circuit elements one with a very small propagation and other with a significant delay

Present State	Next State after Transition (x'_{q0}, x'_{q1})				Present Output Y_0			
(x_{q0}, x_{q1})	Input $X_1X_0 = 00$	Input $X_1X_0 = 01$	Input $X_1X_0 = 10$	Input $X_1X_0 = 11$	Input $X_1X_0 = 00$	Input $X_1X_0 = 01$	Input $X_1X_0 = 10$	Input $X_1X_0 = 11$
0, 0	0, 1	0, 0	1, 0	1, 0	0	0	1	1
0, 1	0, 1	0, 1	0, 0	1, 0	1	0	0	0
1, 0	1, 1	1, 1	1,0	1, 0	0	1	0	0
1, 1	1, 1	0, 1	1, 1	1, 0	1	1	0	1

2. The asynchronous circuit element is such that it has the transition table as per Table 19.1.

Consider row 1. When X_1 and X_0 are 10 and 11, present state (x_{q0}, x_{q1}) is (0, 0), the present outputs are 1 in both cases of the inputs. Assume that X_1 has a static-1 hazard due to presence of an X_1 output implementing circuit, which has the delaying element. X_1 undergoes a static hazard, such that momentarily it undergoes a transition to 0 from 1. During the period when X_1X_0 are 00 in place of 10, the next state becomes (0, 1) and output becomes 0. It will lead to either (i) another next state unstable (0, 0) with output = 0 in case the X_1X_0 again settle to 10 or (ii) to a stable (0, 1) with output = 1.

19.2 IDENTIFYING STATIC HAZARDS

The question is how can one find that what are the hazards. First, follow the following steps:

Step 1: Write Boolean expression for the logic circuit and assume variable X and its complement as two separate variables.

Step 2a: Find POS form of the expression to identify static-0 hazard without using $X_0 + \overline{X}_0 = 1$ and $X_0.\overline{X}_0 = 0$ or expressions dependent on these, where X_0 is a Boolean variable that is used both as X_0 and $\overline{X}_0$. For example, use $X_1 + X_2 = X_1 + X_1.X_2$ or $X_1.X_2 + X_1.X_3 = X_1.X_2 + X_2.X_3 + X_1.X_3$ or $X_1.(X_1 + X_2) = X_1.X_2$ is not permitted when finding static hazards. Prefer use of DeMorgan theorem or distributive laws.

Step 2b: Find SOP form of the expression to identify static-1 hazard without using $X_0 + \overline{X}_0 = 1$ and $X_0.\overline{X}_0 = 0$ based expressions, where $X0$ is a Boolean variable that is used both as X_0 and $\overline{X}_0$.

19.2.1 Identification from the Boolean Expressions

For example, consider the circuit of Figure 19.1(a).

$Y = \overline{X.\overline{X}}$ ($X.\overline{X} = 0$, not allowed to be used for analyzing transients).

$= \overline{X} + X$; (Using DeMorgan theorem, get an OR expression. Modify further as OR expression not allowed in this form for analysis).

$= \overline{X} + X.X$; (Use rule $X_1.X_1 = X_1$).

$= X_1 + X_2.X_2$; (Assume a variable and its complement as a separate set of two variables).

When $X_1, X_2 = 0, 0$ the output $= 0$.
When $X_1, X_2 = 1, 0$ the output $= 1$.
When $X_1, X_2 = 1, 1$ the output $= 1$
When $X_1, X_2 = 0, 1$ the output $= 1$

We find that there is one condition of X_1 and X_2 in which output will show transient to 0. Hence circuit has static-1 hazard.

19.2.2 Identification from the Karnaugh Map (Only One-variable Input Case)

Use steps 1, 2a and 2b described in Section 19.2.

Step 3: Draw two variable Karnaugh map by assuming the $\overline{X}_0$ and $\overline{X}_0$ as separate variables, X_1 and X_2.

Step 4: Find two pair of diads at right angle (90°) to each other. [Diad means pair of adjacent cells with the 1s or 0s in SOP or POS, respectively.] Show the pairs by dotted lines, as circuit for implementing does not exist. It is needed to implement that path when there is a transition of the variable. When we change from adjacent column to another, there is a single variable change. Diads at the right angle means, when the complement variable changes by following another path to final output, there is a reversal of the state.

Static 1 Hazard

Steps 1 and 2b have already been given in Section 19.2.1. Karnaugh map form SOP form of the Figure 19.1(a) circuit is shown in Table 19.2. It shows two diads at the right angles with dotted envelopes. Hence the circuit has a static-1 hazard for transition 10 to 11.

TABLE 19.2 Two Variable SOP Karnaugh map for the circuit of Figure 19.1(a)

X2 / X1	0	1
0	0	1
1	1	1

Dotted envelope means circuit does not exist to implement it. There is only one variable present. Arrows show the transitions.

Static 0 Hazard

Now let us find static-0 hazard, if any, in circuit of the Figure 19.1(c) assuming NOT gate as the delaying element for the period Δt'''. Combination circuit of the figure be assumed to be AND gate of shorter delay Δt''.

Following are the steps to find static 0 hazard:

Step 1: Boolean expression for the circuit is $Y = \overline{X}_1.X_1$. (Not permitted to be used for transient analysis).

Step 2: Find POS form as follows.

$$\overline{Y} = \overline{\overline{\overline{X}_1.X_1}}$$

$$= \overline{X_1 + \overline{X}_1} \qquad \text{(Use DeMorgan Theorem).}$$

$$= \overline{X_1 + X_2} \qquad \text{(Assume } X_1 \text{ and } \overline{X}_1 \text{ as a set of two separate variables).}$$

When $\overline{X}_1, \overline{X}_2 = 0, 0$ the $\overline{Y} = 0$

When $\overline{X}_1, \overline{X}_2 = 1, 0$ the $\overline{Y} = 0$

When $\overline{X}_1, \overline{X}_2 = 1, 1$ the $\overline{Y} = 1$

When $\overline{X}_1, \overline{X}_2 = 0, 1$ the $\overline{Y} = 0$

Draw the Karnaugh map as per Table 19.3 for POS form (representation of $\overline{Y}$ and on the map) of the expression.

Map shows two diads at right angle to each other. Hence the output has a static-0 hazard.

TABLE 19.3 Two Variable Karnaugh map POS form for the circuit of Figure 19.1(c)

$\overline{X}_2$ / $\overline{X}1$	1	0
1		0
0	0	0

Arrows show the transition

[Example 19.1 will show the static 1 and static 0 hazard detection for the circuit for X_1 in Figure 19.3.]

19.2.3 Identification from the Karnaugh Map (Three-Variable Input)

First use steps 1, 2a and 2b as described in Section 19.2

Step 3: Draw Karnaugh map for the three-variable case.

Step 4: Find a pairs of diads with no overlapping between the two diads.

Consider when we change from adjacent column to another, there is a single variable change. If in the given logic circuit, that path does not exist, then a static hazard arises.

Static 0 Hazard

Find from the three-variable in POS-form map, a no-overlap diad (s) between the pair of diads. No overlap means there is no logic circuit existing to implement that path on a transition of the variable. A pair of complementary variables X_1 and X_2 means X_1 change between 0 and 1 in first diad and X_2 in second diad. A pair of 0s in these adjacent cell diads separate by the two complementary pairs. In case of overlapping (available circuit to implement), the 0s in each differ only by one of the variable change not by two variable changes. Static 0 hazard exists when there exists a pair of 0s in these adjacent cell diads separate by the two complementary pairs. Let the circuit POS form is as follows.

$$\overline{Y} = (X_1 + X_2).(\overline{X}_3 + \overline{X}_2);\ \text{(Without using above mention OR and AND equations)}. \quad ...(19.1)$$

Table 19.4 shows a Karnaugh map with two full lined pair diads *a* and *b* with non-overlapping 0s between them and thereby having a static 0 hazard during transition from to 111 – 100. Consider the shaded envelope for a diad pair *c* over a non-overlapping diad pair. Pair forms *b* and *c* by two adjacent cells in the table by wrapping adjacency. It means the logic circuit $(\overline{X}_1 + \overline{X}_3)$ and therefore the term *C* is not present to implement transition along that path. Full-line envelope over a diad pair means the expression terms and logic circuit are available to implement it.

TABLE 19.4 Three variable Karnaugh map for $\overline{Y} = (X_1 + X_2).(\overline{X}_3 + \overline{X}_2)$

$\overline{X}1\overline{X}2$ \ $\overline{X}3$		1	0
11	$\overline{X}_1 + \overline{X}_2$	a 0	d 0
01	$X_1 + \overline{X}_2$		
00	$X_1 + X_2$		b 0
10	$\overline{X}_1 + X_2$		c 0

When the logic circuit does not have the term $(\overline{X}_1 + \overline{X}_3)$, 111 – 100 the transition has a Static-0 Hazard

Note: Full line box over a diad means the terms exist in equation (19.1) and the shaded box over a diad means term (circuit) does not exist in Boolean expression and that creates a hazard 0 in the present case.

Static 1 Hazard

Three-variable SOP-form map with the no overlap between the pair of diads means, when there is a pair of complementary variables, which change between a 0 in first diad of 1s and a 1 in second diad of 0s. A pair of 1s in these adjacent cell diads separate by the two complementary pairs. In case of overlapping diads, the 1s in each differ only by one of the variable change not by two variable changes. Static 1 hazard exists when there exists a pair of 1s in these adjacent cell diads separate by the two complementary pairs. Static 1 hazard occurs when there is a transition from one diad cell to another diad adjacent cell.

Assume that the Boolean expression for the logic circuit in SOP form is as follows:

$$Y = \overline{X}_2.X_1 + \overline{X}_3.X_2; \text{ (Without using OR and AND equations in Boolean laws).} \quad ...(19.2)$$

Table 19.5 shows the Karnaugh map with two pair diads (a, b) with non-overlapping 1s between them and thereby having a static 1 hazard during transition 110 – 101. Refer the dotted envelope over a non-overlapping diad pair. Pair forms by the two adjacent cells in the table by adjacency or wrapping adjacency. It means the logic circuit and therefore the term c not present to implement transition along that path. Full-line envelope over a diad pair means the expression terms and the logic circuit is available to implement it.

TABLE 19.5 Three variable SOP form Karnaugh map for circuit $Y = \overline{X}_2.X_1 + \overline{X}_3.X_1$

X_1X_2 \ X_3	0	1
00		
10	a 1	1
11	c 1	
01	b 1	

When the logic circuit does not have the term $X_1.\overline{X}_3$ there is a static-1 Hazard

Note: The full line box is because of diad pair $\overline{X}_2.X_1.(\overline{X}_3 + X_3)$ and shaded box is for $X_2.(X_1 + \overline{X}_1).\overline{X}_3$ pair. Equation does not have a term $X_1.\overline{X}_3$ and therefore the shad diad and therefore a hazard is present.

19.2.4 Detecting Absence of Static 1 Hazard from the POS Form of Boolean Expression

If a Boolean expression has a term present in POS or SOP with either a variable or its complement only, then a static 0 or 1 hazard is ruled out, respectively. This however does not mean that if a variable as well as complement occur in the POS or SOP expression, then static 1 or 0 hazard may or may not be present, respectively. Presence can then be detected by Karnaugh map for appropriate form.

For example, the circuit implementation of the following first two expressions do not have static 1 hazard:

$\overline{Y} = (X_1 + \overline{X}_2).(\overline{X}_2 + X_3);$ (X_1, X_3 and $\overline{X}_2$ complements are not existing. So static 1 hazard ruled out).

$\overline{Y} = (X_1 + \overline{X}_3).(X_1 + \overline{X}_2).(\overline{X}_3 + \overline{X}_2);$ (X_1, $\overline{X}_2$ and $\overline{X}_3$ complements don't exist. So static 1 hazard ruled out. Third term is always 1 and hence not taken into account).

$\overline{Y} = (X_1 + X_3).(\overline{X}_1 + X_2 + X_3).(\overline{X}_3 + X_2)$; ($X_1$ and X_3 complements also exist. So static 1 hazard is not ruled out, may be present or may not be present).

19.2.5 Detecting Absence of Static 0 Hazard from the SOP Form of Boolean Expression

If a Boolean expression has a term present in SOP with either a variable or its complement only, then static 0 hazard is ruled out. This is however does not mean that if a variable as well as complement occur in the expression, then static 1 hazard may or may not be present. Presence can then be detected by Karnaugh map for SOP form.

For example, the circuit implementation of the following expressions do not have static 0 hazard:

$Y = \overline{X}_1 + X_2$; ($\overline{X}_1$ and X_2 complements are not existing. So static 0 hazard ruled out).

$Y = X_1.X_3 + X_1.X_2.X_3 + \overline{X}_3.X_3$; ($X_1$, X_2 and X_3 complements don't exist. So static-0 hazard ruled out. Third term is always 0 and hence not taken into account).

$Y = X_1.X_3 + \overline{X}_1.X_2.X_3 + \overline{X}_3.X_3$; ($X_1$ complement also exists. So static-0 hazard is not ruled out, may be present or may not be present).

Points to Remember

1. Step 1: Boolean expression
2. Step 2a and 2b: POS and SOP forms to detect static 0 and static 1 a hazards
3. Step 3: Karnaugh maps for POS and SOP forms
4. Step 4: Analysis by finding the non existant adjacencies and wrapping adjacencies, which are in the map but not in the given Boolean expression.

19.3 ELIMINATING STATIC HAZARDS

A static hazard occurs because a variable and its complement passed through different number of logic operations. For example, in circuit of Figure 19.1(a), *X* input passed through one NAND and *X* passed through two NANDs. If an additional holding circuit adds into the logic circuit, the static hazard eliminates. For example adding an AND gate with common inputs after *X* and before the output stage NAND, will eliminate the static hazard.

What is the systematic way of eliminating a static hazard? We follow steps 1, 2a, 2b, 3 and 4 describe in Section 19.2 and obtain a Karnaugh map as given in Table 19.4 or Table 19.5. Term not available between a pair of diads is shown by a shaded pair. Addition of that term eliminates the hazard (Section 19.2.3).

Eliminating Static 0 Hazard

For example, consider the circuit of Figure 19.4(a), which implements the Boolean expression (19.1) and corresponds to the POS form map shown in Table 19.4.

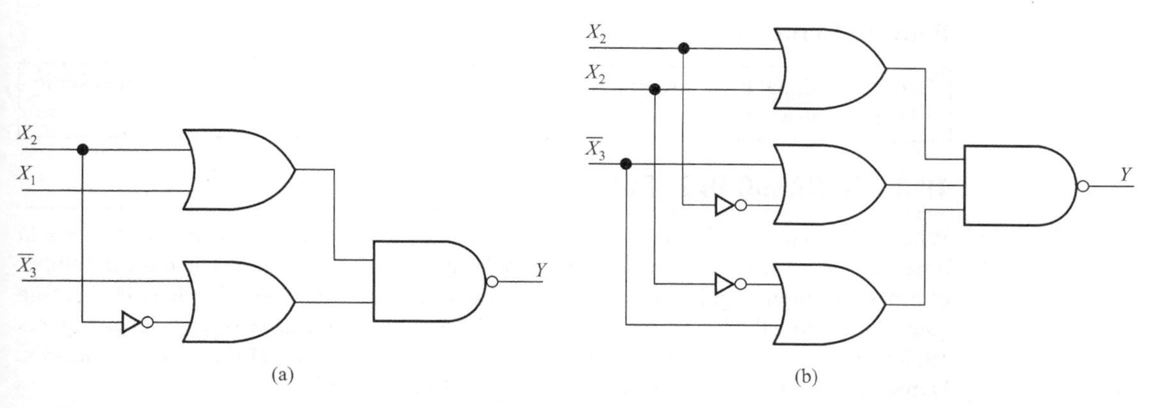

FIGURE 19.4 (a) A circuit, which implements the Boolean expression (19.1) and corresponds to the POS form map (shown in Table 19.4) (b) The circuit after removing static-0 hazard by using expression (19.3).

The expression (19.1) if added with the term $(\overline{X}_1 + \overline{X}_3)$ as follows, then the static-0 hazard eliminates. This term corresponds to dashed-pair implementation by the circuit.

$$Y = (X_1 + X_2).(\overline{X}_3 + \overline{X}_2).(\overline{X}_1 + \overline{X}_3) \quad ...(19.3)$$

Figure 19.4(b) shows the new circuit for static-0 hazard free implementation.

Eliminating Static 1 Hazard

For example, consider the circuit of Figure 19.5(a), which implements the Boolean expression (19.2) and corresponds to the SOP form map shown in Table 19.5.

The expression (19.2) if added with the term $(X_1.\,\overline{X}_3)$ as follows, then the static-a hazard eliminates. This term corresponds to dotted-pair implementation by the circuit.

$$Y = (\overline{X}_2.X_1) + (\overline{X}_3.X_2) + (X_1.\,\overline{X}_3) \quad ...(19.4)$$

Figure 19.5(b) shows the new circuit for static 1 hazard free implementation using the expression (19.4).

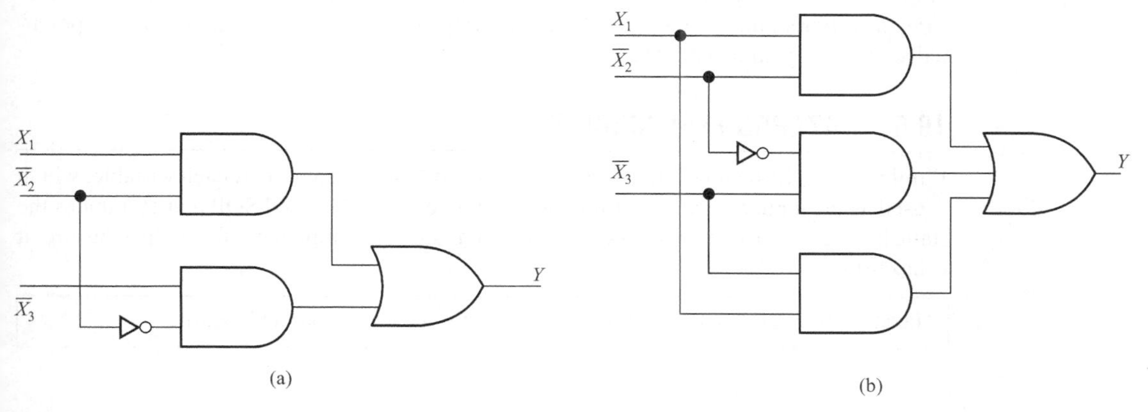

FIGURE 19.5 (a) A circuit, which implements the Boolean expression (19.3) and corresponds to the SOP form map shown in Table 19.5 (b) The circuit after removing satic 1 hazard by using expression (19.4).

Point to Remember

Use of additional logic circuit terms in POS and SOP terms eliminates static 0 and static 1 hazards, respectively.

19.4 DYNAMIC HAZARDS

When an output, though suppose to undergo transition only once, if changes total $(2n + 1)$ times, where $n = 1$ or 2 or 3 and so on, then there is a dynamic hazard. The n is the number of times the output toggle back to the initial value. Dynamic hazards arise due to later stage gates undergoing the transitions slower than the previous stage gates or vice versa. Figures 19.6(a) and 19.6(b) show the output transitions due to presence of the dynamic hazards. Dynamic hazards can also cause logic circuit malfunction.

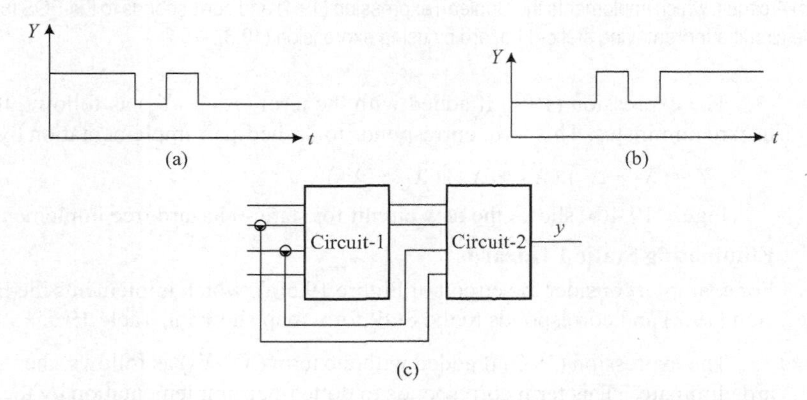

FIGURE 19.6 (a) Output 1 to 0 transition in presence of a dynamic hazard (b) Output 0 to 1 transition in presence of a dynamic hazard (c) Circuit, which is potential source of the dynamic hazards.

Consider a circuit combination of Figure 19.6(c). It consists of the circuit-1 and circuit 2. Let circuit-2 gets part of the inputs from circuit-1 and part of the outputs directly. Therefore, the different inputs of circuit-2 are appearing at the different instances. This is a potential source of dynamic hazards.

19.5 HAZARDS FREE CIRCUITS

Elimination static 0 and static 1 hazards have been described above. If each variable, which is used as the input at any stage (after adding the terms in POS and SOP that eliminates the static hazards), is used in only one form–either as such or complement form, then the circuit is hazard free.

Hazard free circuit is a circuit free from static as well as dynamic hazards.

19.6 ESSENTIAL HAZARDS

Consider asynchronous sequential circuit. Let us assume that (i) Static hazards have been eliminated (ii) Dynamic hazard has been eliminated (iii) Races that are critical have been eliminated (iv) The circuit operates in fundamental mode so that the input is changed only after a stable state is reached.

Asynchronous sequential circuit has following structure:

1. There is a set $\underline{\mathbf{X_i}}$ of m present input variables $X_0, X_1, ..., X_{m-1}$. These are applied along with present $\underline{\mathbf{x_q}}$ to a combinational circuit. The output of $\underline{\mathbf{Y_i}}$ along with $\underline{\mathbf{y_q}}$ is a set consisting of present state outputs $\underline{\mathbf{Y}}$ of which $\underline{\mathbf{y_q}}$ is a subset. ($\underline{\mathbf{Y}}$ consists of two subsets; $\underline{\mathbf{Y_i}}$ with outputs $Y_0, Y_1, ..., Y_{j-1}$ with no feedback and $\underline{\mathbf{y_q}}$ with feedback to the input via the memory section).
2. The circuit memory section consists of the m latches without clock edges or pulse inputs to control the instance or period of their operations. These have a set $\underline{\mathbf{x_q}}$ variables from the m present state outputs $\underline{\mathbf{x_q}} = x_{q0}, x_{q1}, ..., x_{qn-1}$ from the y_q inputs after a delay.
3. The present state of $\underline{\mathbf{Y}}$ and therefore its subset $\underline{\mathbf{y_q}}$ changes by the change in $\underline{\mathbf{x_q}}$ after a delay at the memory section.

Another hazard will still be present. It is essential hazard. It is due to different instances at which change in the excitation variable $\underline{\mathbf{y_q}}$ occurs when there is a change in an input variable X_i. Essential hazard arises out that an input variable affects the different feedback cycle variables at different instances. Before the expected set of all $\underline{\mathbf{y_q}}$ excitation input changes finish, the state variable(s) $\underline{\mathbf{x_q}}$ can change, which may lead to circuit not functioning as expected.

Consider a circuit shown in Figure 19.7(a), for which Table 19.6 gives the excitations-cum-transitions.

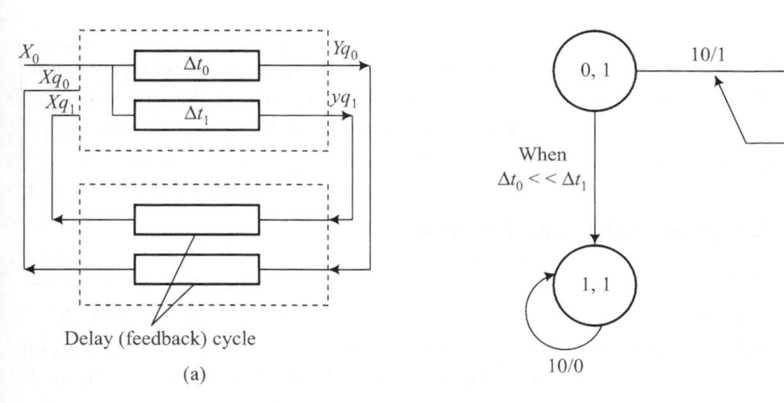

FIGURE 19.7 (a) Structure of an asynchronous sequential circuit, which has excitations and transitions as per Table 19.6 (b) Resulting effects due to essential hazard from $\Delta t_0 << \Delta t_1$ assuming that feed cycle delays are same for both transitions to x_{q0} and x_{q1}.

TABLE 19.6 Excitation-cum- Transition table for an asynchronous sequential circuit exemplary circuit-6, which has (Table 18.5)

Present state	Next state after transition (x'_{q0}, x'_{q1})				Present output Y_0			
(x_{q0}, x_{q1})	Input X_1X_0 = 00	Input X_1X_0 = 01	Input X_1X_0 = 10	Input X_1X_0 = 11	Input X_1X_0 = 00	Input X_1X_0 = 01	Input X_1X_0 = 10	Input X_1X_0 = 11
0, 0	0, 1	0, 0	1, 0	1, 0	0	0	1	1
0, 1	0, 1	0, 1	1, 1	1, 0	1	0	1	0
1, 0	0, 1	1, 1	0, 0	0, 1	1	1	1	1
1, 1	0, 1	0, 0	1, 1	1, 0	0	1	0	1

Suppose inputs X_1X_0 are at 11. Therefore, the stable state as per column 5 row 3 is (0, 1) with output Y= 1. Let X_0 change from 1 to 0. Expected stable state is (1, 0) with output Y= 1 as per column 4 row 1. There will be one additional feedback cycle, which is shown by a vertical transition in the table and then only the stable state reaches.

Assume that the 1-to-0 transition of X_0 affects y_{q0} and therefore x_{q0} earlier in time Δt_0 than y_{q1}, x_{q1} affects in time Δt_1, which is later. ($\Delta t_0 << \Delta t_1$. and assume that feed cycle delays are same for both transitions to x_{q0} and x_{q1}).

Therefore, temporarily, the state becomes (1, 1) with output Y = 0. The stable state achieved is (1, 1) with Y= 0.

Due to essential hazard arising out of quicker excitation of one feedback than another when there is an input change, the different stable state has resulted. Figure 19.7 (b) show the resulting transitions.

Point to Remember

Essential hazard in an asynchronous sequential circuit operation arises out when an input variable affects the different feedback cycle excitation variables at the different instances. Introduction of the additional delays greater than the delays in the combination circuits providing the excitations in the feedback cycles eliminates essential hazard.

19.7 PULSE MODE SEQUENTIAL CIRCUIT

Synchronous sequential circuits undergo the excitations on clock edges. Synchronous circuit operated on the periodic pulses of the clock is called pulse mode sequential circuit.

When the period ΔT of the clock pulse, which excites a synchronous sequential circuit is greater than the time periods during which an input variable(s) affects the different feedback cycle excitation variables, the circuit will be free from essential hazards.

Figure 19.8 shows a pulse mode sequential circuit.

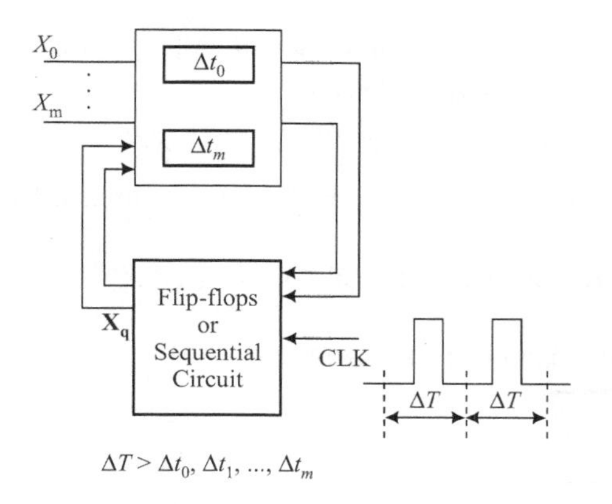

FIGURE 19.8 Pulse mode sequential circuit.

EXAMPLES

Example 19.1 Detect the static 1 and static 0 hazards for X_1 output in the circuit of Figure 19.3 and find the missing circuit to provide transition paths.

Solution

Step 1: First we write the Boolean expression for the logic circuit of Figure 19.3.

$$X_1 = (X_1'.X_1'') + (\overline{X}_1'.X_1''')$$

Step 2a: POS form for static-0 hazards detection

Use DeMorgan theorem as follows:

$$\overline{X}_1 = \overline{(X_1'.X_1') + (\overline{X}_1'.X_1')}$$

$$= \overline{(X_1'.X_1'')}.\overline{(X_1'.X_1''')}$$

$$= (\overline{X_1'} + \overline{X_1''}).(X_1' + \overline{X_1'''}) = (\overline{X_1'} + \overline{X_1''} + \overline{X_1'''}).(\overline{X_1'} + \overline{X_1''} + X_1''').(X_1' +$$

$\overline{X}_1'' + \overline{X}_1''').(X_1' + X_1'' + \overline{X}_1''')$ [On expanding the minimized 2-variable to 3-variable expression] Two terms become four

Step 3: Three Variable Karnaugh map for $(\overline{X}_1' + \overline{X}_1'') . (\overline{X}_1' + \overline{X}_1''')$

TABLE 19.7 Two terms expression expanding to four terms for $(X_1'.X_1''') + (\overline{X}_1.X_1'')$ POS

$\overline{X}_1'\overline{X}_1''$ \ $\overline{X}_1'''$		1 $\overline{X}_3$	0 Z_1
11	$\overline{X}' + \overline{X}''$	0	0 a
01	$X' + \overline{X}''$	0	
00	$X' + X''$	b 0	
10	$\overline{X}' + X''$		

When the logic circuit does not have the term $(X_1' + \overline{X}_1''')$, 110 – 000 transition is a static-0 hazard

There is a static 0 hazard in 110-to-000 transition of $X_1', X_1'', \overline{X}_1'''$.

Step 2b: SOP form for static-1 hazards detection

$$Y_1 = (X_1'.X_1'') + (\overline{X}_1'.X_1''') = X_1'.X_1''.\overline{X}_1''' + X_1'.X_1''.X_1''' + \overline{X}_1'.\overline{X}_1''X_1''' + \overline{X}_1'X_1''X_1'''$$ [On expanding the minimized 2 variable to 3 variable expression.]

Step 3: Three variable Karnaugh map for $(X_1'.X_1'') + (X_1'.X_1''')$

TABLE 19.8 Two terms expression expanded to $(X_1'.X_1''') + (\overline{X}_1.X_1'')$ SOP map

$X_1'X_1''$ \ X_1'''	0	1
00		b 1
10		
11	1	1
01		c 1

When the logic circuit does not exist for $X_1'.X_1''$, 111 to 001 $\overline{X}'X_1''X_1'''$ is a static-1 hazard

Static-1 hazard is present in 011-to-111 transition.

Example 19.2

From the given Karnaugh three variables maps (Table 19.9) find the static 1 and static 0 hazards and show that logic path exists such that no hazards exists in first map.

Solution

TABLE 19.9 Three Variable Karnaugh map for $Y = (\overline{X}_1 + \overline{X}_2).(\overline{X}_3 + \overline{X}_1).(X_2 + \overline{X}_3)$ map (when all variables are 1s in the expression)

$\overline{X}_1\overline{X}_2$ \ $\overline{X}_3$		1	0
11	$\overline{X}_1 + \overline{X}_2$	0	0
01	$X_1 + \overline{X}_2$		
00	$X_1 + X_2$		0
10	$\overline{X}_1 + X_2$		0

Due to $(\overline{X}_3 + \overline{X}_1)$ presence, no static 0 hazard present, as all variables are in the form of 0 only. No static 0 hazard present

Also, no static 0 hazard present as all the three paths (diads) involving single variable change are available in the circuit of the Boolean expression.

Static-1 hazard exists for 001 to 010 transition a path does not exist in which there is single variable change to implement. [There is X_2 as well as $\overline{X}_2$ in the Boolean expression.]

Example 19.3

From the given Karnaugh three variables maps (Tables 19.10 and 19.11) find the static 1 and static 0 hazards and show that no hazard exists. Show the logic existence using Karnaugh map.

TABLE 19.10 Three Variable SOP form Karnaugh map for circuit $Y = \overline{X}_2.\overline{X}_1 + X_3.X_2$

X_1X_2 \ X_3		0	1
00	$\overline{X}_1.\overline{X}_2$	1	1
10	$X_1.\overline{X}_2$		
11	$X_1.X_2$		1
01	$\overline{X}_1.X_2$		1

When the logic circuit does not have the term $\overline{X}_1.X_3$ static-1 hazard

TABLE 19.11 Three variable SOP form Karnaugh map for circuit $Y = \overline{X}_2.\overline{X}_1 + \overline{X}_3.\overline{X}_2$

X_1X_2 \ X_3	0	1
00	1	1
10	1	
11		
01		

Due to an overlapping diad through 000 ($\overline{X}_2.X_1$) term, no static-1 hazard exists in 000 to 001 or 100 transition.

Solution

Table 19.10 map shows static-1 hazard. In SOP form Table 19.11, the variables in the Boolean expression are only the complement forms of X_2, X_1 and X_3 and there is no dual use of variable and its complement. Hence static 0 hazard is also ruled out.

Example 19.4

From the given Karnaugh two variables map (Table 19.12) for $\overline{X}_2$, X_1 + $\overline{X}_1.X_2$, find the static 1 and static 0 hazards and show that a static 1 hazard exists. Find the missing terms, which would have made hazards non existant.

TABLE 19.12 Two variable SOP form Karnaugh map for circuit $Y = \overline{X}_2.X_1 + \overline{X}_1.X_2$

X_1 \ X_2	0	1
0		1
1	1	

Static-1 hazard exists in 10 to 01 as it involves two single variable path changes as the term $\overline{X}_1\overline{X}_2$ does not exist

Solution

In SOP form, the variables used are un-complemented as well as un-complemented complements of X_2, X_1. Hence static-0 hazard can't be ruled out.

Let us find POS form (Table 19.13) of the logic circuit to find the hazard source.

$$\overline{Y} = \overline{\overline{X_2}.X_1 + X_2.\overline{X_1}} = \overline{\overline{X_2}.X_1}.\overline{X_2.\overline{X_1}} = (X_2 + \overline{X}_1).(\overline{X}_2 + X_1)$$

TABLE 19.13 Three variable POS form Karnaugh map for circuit $\overline{Y} = X_2.\overline{X}_1 + X_1.\overline{X}_2$

$\overline{X}_1$ \ $\overline{X}_2$	1	0
1		0
0	0	

Static-0 hazard exists in 10 to 01 as it involves twosingle variable path changes. The missing term $\overline{X}_1.\overline{X}_2$ is the hazard source.

Example 19.5 Find whether static 0 hazard does not exist in implementing the Boolean expression: $Y = X_1.X_3 + X_1.\overline{X}_2.X_3 + X_3.\overline{X}_2$.

Solution

Since complements of X_1, $\overline{X}_2$ and X_3 do not occur in any of the terms, the static 0 hazard does not exist.

Example 19.6 Find whether static-0 hazard may or may not exist in implementing the Boolean expression: $Y = X_1.X_3 + \overline{X}_1.\overline{X}_2 + \overline{X}_2.X_3$. Can you rule out the existence of static-1 hazard when implementing the following POS form circuit?

Solution

$$\overline{Y} = (\overline{X}_1 + \overline{X}_3).(X_1 + X_2).(\overline{X}_3 + X_2);$$

Since X_1 and its complement exist in the terms, the existence of static 1 can't be ruled out, may be present or may not be present.

Example 19.7 $Y = (X_1 + X_3).(X_1 + X_2 + X_3).(X_3 + X_2)$ has no static-1 hazard. Find out.

Solution

Since no variable is present in both forms (variable and its complement) in this POS expression, hence the static 1 hazard is ruled out.

Example 19.8 From the given transition table (Table 19.14) find whether essential hazard exists.

TABLE 19.14 Excitation-cum- Transition table for an asynchronous sequential circuit

Present State	Next State after Transition (x'_{q0}, x'_{q1})		Present Output Y_0	
(x_{q0}, x_{q1})	Input $X_0 = 0$	Input $X_0 = 1$	Input $X_0 = 0$	Input $X_0 = 1$
0, 0	[0, 0]	1, 0	0	1
0, 1	[0, 1] ↔	[0, 1]	0	1
1, 0	0, 0	1, 0	1	0
1, 1	0, 1	[0, 0]	0	1

No essential hazard exists when the state as (0,1) as the X transition does not affect the final stable condition. No essential hazard exists in state (0, 0). Because when X changes from 0 to 1, state will eventually settle to (0, 1) though after a delay in case x_{q0} changes early and x_{q1} later or vice-versa after a delay.

EXERCISES

1. Detect the static 1 and static 0 hazards for X_1 output in the circuit of Figure 19.9.

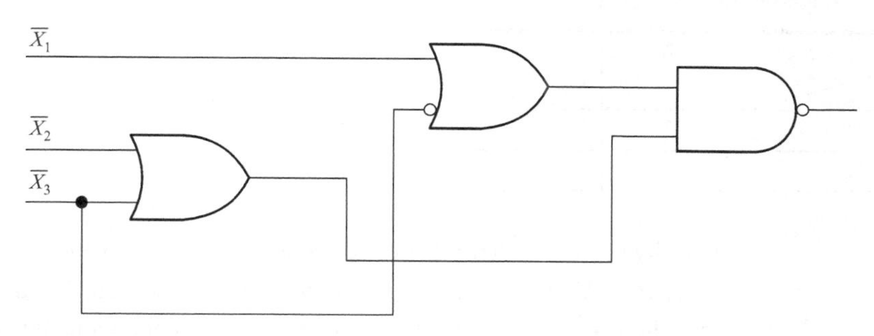

FIGURE 19.9 Circuit for Exercise 1.

2. From the given Karnaugh three variable map, (Table 19.15) write the Boolean expression, draw the circuit and find the static 1 and static 0 hazards

TABLE 19.15 Three variable Karnaugh map

$\overline{X}_3$ / $\overline{X}_1\overline{X}_2$	1	0
11	0	0
01	0	0
00		
10		

3. From the given Karnaugh three variable map (Table 19.16) find the Boolean expression, logic circuit, and static-1 and hazards.

TABLE 19.16 Three variable SOP form Karnaugh map for circuit

X_3 / X_1X_2	0	1
00	1	1
10	1	1
11		1
01		1

4. From the given Karnaugh two variables map (Table 19.17), find the static 1 and static 0 hazards and find that a static 1 hazard exists or not.

TABLE 19.17 A Karnaugh map for two variables

X_1 \ X_2	0	1
0	1	1
1	1	1

5. Find whether static 0 hazard does not exist in implementing the Boolean expression: $Y = X_1.X_3 + X_1.X_2.X_3 + \overline{X}_2.X_3$. If exists, then find the static hazards present.
6. Find whether static 0 hazard may or may not exist in implementing the Boolean expression: $Y = \overline{X}_1.X_3 + \overline{X}_1.X_2.X_3 + X_2.X_3$. If exists, then find the static hazards present.
7. Can you rule out the existence of static 1 hazard when implementing the following POS-form circuit?

$$\overline{Y} = (\overline{X}_1 + X_3).(\overline{X}_1 + X_2).(\overline{X}_3 + X_2)$$

8. $\overline{Y} = (X_1 + X_3).(\overline{X}_1 + X_2 + \overline{X}_3).(X_3 + X_2)$ has static hazards. Trace them.
9. From the given transition table (Table 19.18) find whether essential hazard exists. If existing, list these.

TABLE 19.18 Transition table for an asynchronous sequential circuit

Present state	Next state after transition (x'_{q0}, x'_{q1})		Present output Y_0	
(x_{q0}, x_{q1})	Input X_0 = 0	Input X_0 = 1	Input X_0 = 0	Input X_0 = 1
0, 0	1, 0	0, 1	0	1
0, 1	1, 1	0, 1	0	1
1, 0	0, 0	1, 0	1	0
1, 1	0, 1	0, 1	0	1

10. From the given circuit find in Figure 19.10 whether dynamic hazard exist.
11. What are the additional circuits needed to remove static a hazard in exercise 2 and in 3.
12. Detect all the hazards possible in circuit of Figure 19.11 and find ways to eliminate these.

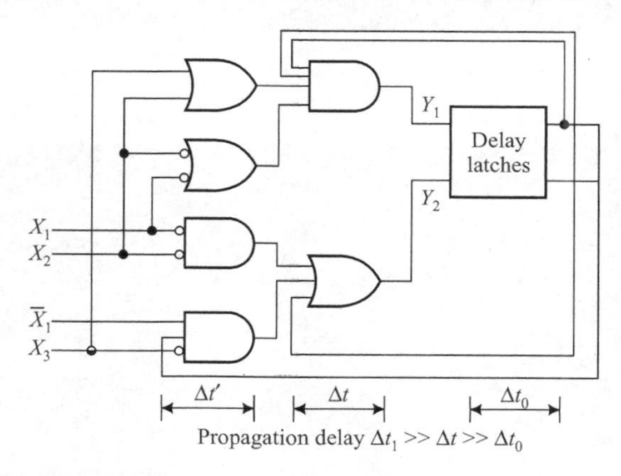

FIGURE 19.10 Circuit for Exercise 10.

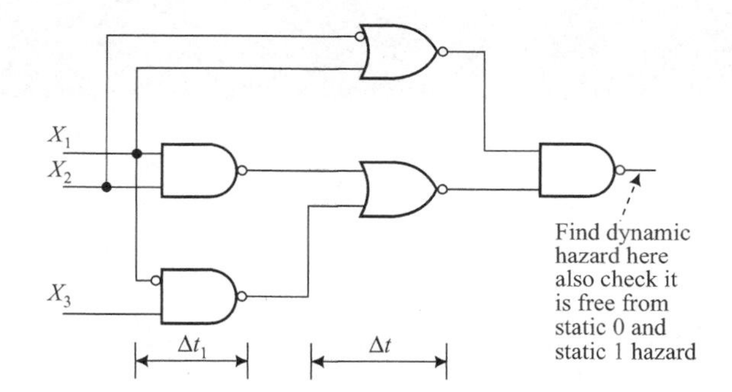

FIGURE 19.11 Circuit for Exercise 12 to find the static 0, static 1, dynamic and essential hazards.

QUESTIONS

1. How do the races differ from the static hazards?
2. What is a static 0 hazard? What are the conditions for arising of a static 0 hazard?
3. How do the races differ from the dynamic hazards?
4. What is a static 1 hazard? What are the conditions for arising of a static 1 hazard?
5. What is a dynamic hazard? What are the conditions for arising of a dynamic hazard?
6. Using a Karnaugh map, how do we detect static 0 hazard in three variable cases and in four variable cases and also find the missing circuit? (Hint in four variable case, use quads in place of diads).
7. What are the steps followed in detecting static hazards?
8. What is the method of detecting essential hazards?
9. How do you eliminate static and dynamic hazards?
10. How do you eliminate essential hazards? (Hint: Additional delays in feedback elements or use of pulse mode sequential circuit).

CHAPTER 20

ADC, DAC and Analog–Digital Mix Interfaces

OBJECTIVE

In this chapter, we shall learn the following topics:

1. Digital circuits are widely used in the control applications. ADC and DAC are used to convert analog to digital and vice versa. We shall be learning these.
2. Interfacing of a gate with the analog device is another important circuit. We shall be learning the analog-digital mix interfaces in this chapter.

20.1 ANALOG TO DIGITAL CONVERTER (ADC OR A/D)

Consider a doctor measuring the body temperature during a sickness. Analog thermometer gives the reading in Fahrenheit as 98°, 99.5°, 100.0°, 100.5°, 101°, 101.5°, ..., at most severe condition 105.0°. *Pray such a condition never comes*!. Thermometer used has markings between 94 °F to 108 °F. There are 32 markings in all. Suppose the result is required digitally. Advantage of the digital output is that it can be processed and stored in a computer memory also. Digital display for the digital outputs shall also have easy readability. Figure 20.1(a) shows three cases: A, B and C for digitally presenting the body temperature.

Case A is that use an 8-bit representation with 94°F decimal shown in binary 0101 1110 and 110 °F shown as in 8-bit binary 0110 1110. Assume that transducer and amplifier give analog input to ADC = 94 mV for 94°F and = 108 mV for 108°F. The resolution is 1 mV/bit = 1°F/bit.

Case B is that the temperature in 8-bit but shows 94.0 °F as binary equivalent of 188 = 1011 1100 (after multiplying by two each input), 94.5 °F as binary equivalent of 189 = 1011 1101 (after multiplying by two each input), and so on. 110 °F is shown as binary equivalent of 220 = 11011100. Assume that transducer and amplifier give analog input to ADC = 94 mV for 94 °F and = 108 mV for 108 °F. Same 8-bit ADC is operating at higher resolution in case B compared to case A. Now, the resolution is 0.5 mV per bit in the digital result. The result is in steps of 0.5 mV = 0.5°F.

Case C is that the temperature in 4-bit number with 94 °F as binary 0000, 95 °F as 0001, and so on. 108 °F is shown as 1111. Assume that the transducer and amplifier give analog input to ADC = 0 mV for 94 °F and = 1500 mV for 108 °F. In case C, a different ADC, which has 4-bit output and is operated at (100) mV/bit ≈ 1.07°F/bit the lowest resolution compared to ADC of the cases A and B–1°F and 0.5°F. Case B gives highest resolution but needs 8 bit ADC.

Figures 20.1(a) and (b) show ADC circuit and results for three cases, A, B and C. A transducer first senses the temperature, generates a signal and then a signal-conditioning amplifier converts the sensor output into an analog voltage. Signal conditioning amplifier is an amplifier with an appropriate offset voltage. Offset is needed when analog 0 V does not correspond to digital all 0s for example in case C.

ADC circuit is a circuit, which coverts the analog input into a digital output consisting of n-bits. Figures 20.1(a) shows the circuit using the analog to digital converter (ADC), which gives the 8-bit, 8-bit and 4-bit outputs for the representation cases A, B and C, respectively. The outputs are shown by $D_{n-1}, D_{n-2}, \ldots, D_1, D_0$ in an n-bit ADC.

Table 20.1 shows the ADC circuit exemplary characteristics for cases A, B and C. Table first row gives number of ADC bits at the output. Table second and third rows give the

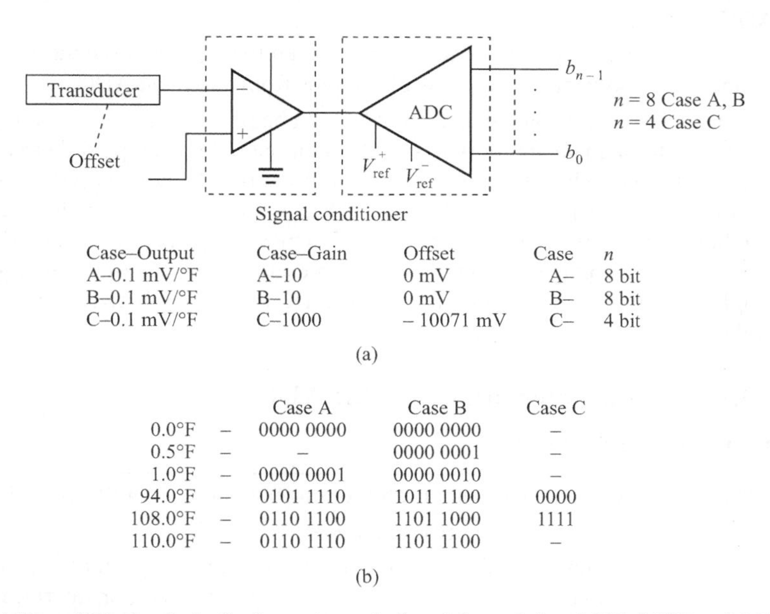

Case–Output	Case–Gain	Offset	Case	n
A–0.1 mV/°F	A–10	0 mV	A–	8 bit
B–0.1 mV/°F	B–10	0 mV	B–	8 bit
C–0.1 mV/°F	C–1000	– 10071 mV	C–	4 bit

(a)

		Case A	Case B	Case C
0.0°F	–	0000 0000	0000 0000	–
0.5°F	–	–	0000 0001	–
1.0°F	–	0000 0001	0000 0010	–
94.0°F	–	0101 1110	1011 1100	0000
108.0°F	–	0110 1100	1101 1000	1111
110.0°F	–	0110 1110	1101 1100	–

(b)

FIGURE 20.1 (a) Three ADC circuits for the three cases, A, B and C resolution 1.0°F, 0.5°F and 1.07°F respectively. (b) Three cases of digital representation $b_{n-1} \ldots b_0$ for the analog readings of body-temperature.

TABLE 20.1 Analog to digital conversion and other characteristics for conversion in circuits of figure 20.1(b) to (d)

Parameter	Case A	Case B	Case C
Number of ADC bits at the output	8	8	4
Exemplary Analog input at ADC for 94°F	94 mV	94 mV	0 mV
Exemplary ADC output 94°F	0101 1110	1011 1100	0000
Lowest Limit V^-_{ref}	0 mV	0 mV	0 mV
Highest Limit V^+_{ref}	255 mV	127.5 mV	1500 mV
Resolution *	1 mV/bit	0.5 mV/bit	100 mV/bit

* Resolution = (Highest measurable analog input limit – Lowest measurable analog input limit) / (2^n – 1), where n is the number of ADC bits at the output. Resolution in °F is 1.0°F, 0.5°F and 1.07°F.

exemplary analog input and ADC outputs. Table fourth and fifth rows give the lowest and highest limits of analog inputs, which can be measured by the given ADC. Table sixth row gives the resolution, which is defined as follows:

Resolution = (Highest measurable analog input limit – Lowest measurable analog input limit) / (2^n – 1), where n is the number of ADC bits at the output.

Let V^+_{ref} is the analog input, which gives output bits = all 1s. Let V^-_{ref} is the analog input, which gives output bits = all 0s. Each ADC has the lowest and highest limits of analog inputs, which is defined by the reference voltage inputs, V^+_{ref} and V^-_{ref} and most ADCs are set the inputs at V^-_{ref} = 0.00 V and V^+_{ref} between 1.2 V to 2.5 V. Some ADCs have $V^+_{ref}/2$ and analog ground pin and some V^+_{ref} and V^-_{ref} pins. A few ADCs have V^-_{ref} settable to –ve values and few to –ve as well as positive value which is always set at voltage less than V^+_{ref}. V^+_{ref} and V^-_{ref} sets the values of highest and lowest measurable analog inputs and defines ADC resolution.

Figure 20.2 shows the relationship between the input and output in a four -bit ADC. Sixteen dots show the sixteen possible outputs from a four-bit ADC.

The external analog inputs to an ADC can be from any source of voltage or current or frequency through an appropriate signal conditioner. For examples, thermocouple in a furnace, ECG electrode with an interfacing amplifier or a dc motor speed-transducer or a sensor of pressure, sensor of blood pressure. We use a frequency to voltage converter signal conditioner whenever the input frequency is a measuring parameter. The ADC converts the analog input in voltage form into the bits representing that analog input. Signal conditioner amplifier setting is such that it gives to the ADC the voltage between the preset lowest and highest limits.

20.1.1 Characteristics of *n*-bit ADC

Points to Remember

The following relationships are useful for understanding an ADC operation.

1. The ADC analog input resolution $R = (V^+_{ref} - V^-_{ref}) / (2^n - 1)$.
2. Range of input signal that is measurable digitally = V^+_{ref} and V^-_{ref}. (That is why we need a suitable amplifier. This amplifier at the input is as an interface circuit. It also has a suitable filter. It has a scaling factor (gain) and/or offset. This provides the

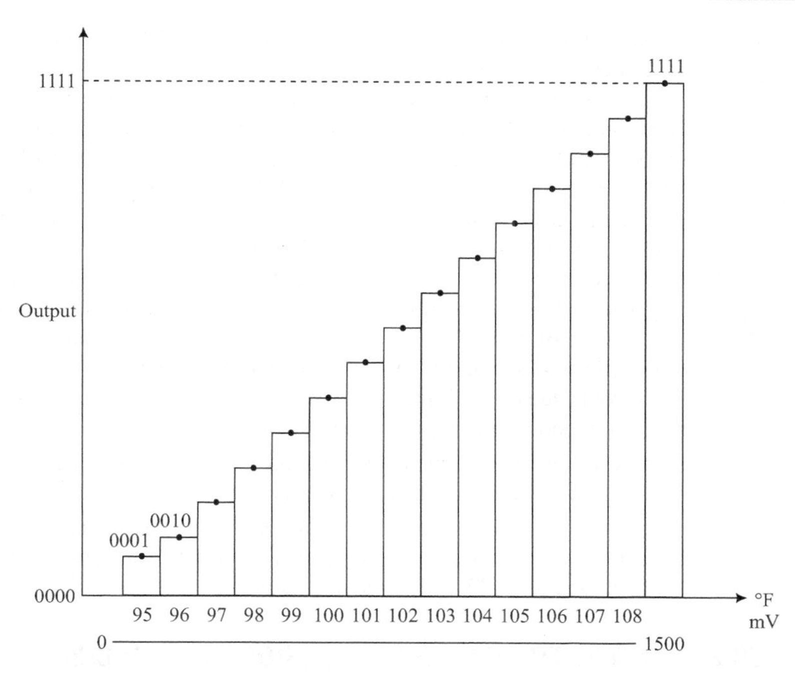

FIGURE 20.2 Relationship between the input and out in a four bit ADC. Sixteen dots show the sixteen possible outputs from a four-bit ADC.

minimum input = V^-_{ref}. This also provides the maximum input = V^-_{ref} for a given analog signal).

3. If digital output of an n bit ADC is decimal m, the input is m times the ADC resolution, when lowest limit is 0.00 V. Digital binary output = decimal m = Binary value of the closes decimal integer for $V_{in}R = V_{in} / (V^+_{ref}$ and $V^-_{ref})$.
4. Output bits $d_{n-1}, d_{n-2}, ..., d_1$ and d_0 are given by the equation:

$d_{n-1} \times w_0^{n-1} + d_{n-2} \times w_0^{n-2} + d_{n-3} \times w_0^{n-3} + ... + d_2 \times w_0^2 + d_1 \times w_0^1 + d_0 \times w_0^0 =$ $m = (V_{in}.R,)$ integer part, where $w_0 = 2$ and $d_{n-1}, d_{n-2}.$, d_1 and d_0 are the output bits from the ADC when the input analog input = V_{in}. ...(20.1)

Besides the resolution, bits n at the ADC output, V^+_{ref} and V^-_{ref}, the following are other important points to remember:

1. An ADC conversion time is a time in which the expected digital output results.
2. An ADC glitch is defined as when a high speed ADC fails to give an output correctly within its conversion time and gives the missing 1s in between. (Appearance of false output bit is called glitch).
3. A linearity of ADC is defined as the maximum deviation from a straight-line relationship. The considered relationship is between the ADC output in decimal vs. the input voltage in Volt.

4. An accuracy of ADC output is defined by the ADC quantization error. A quantized level means a discrete level. Let us assume that we have a 10-bit ADC, which is receiving the analog inputs. The peak measurable voltage is say, 2.0 V. The resolution is then equal to division of 2.0 V by 2^{10} = 1.95 mV/bit. We can then say that there are the 2^{10} = 1024 quantized levels. The levels are from 0 to 2000 mV. These are 0 mV, 1.95 mV, 3.90 mV, 5.85 mV, ..., 2000 mV. If analog input is at a quantized level then only it will be accurately measured. If analog input is 1.98 mV. It will be converted to bits 1.95 mV, the nearest quantized level. If it is 1989 mV, it will be converted as 1980 mV, it's nearest quantized level. That is within 1/2 lsb that the ADC output could be accurate. A quantization error is therefore within 0.5 times of resolution (± 0.5 ADC resolution). It will be ± 0.097 mV in the above example. The quantization error is the analog input range within which the converted bits are the accurate ones. The error is due to the limited number of quantized levels between the minimum and maximum permitted analog measurable input.
5. An analog signal multiplexer can give the analog inputs of the various channels and a digital multiplexer can give the one input at a time to an ADC.
6. Five circuits to design ADCs are—successive approximation, dual; slope, tracking, flesh and two stage flesh. Example 20.12 will explain a successive approximation ADC. Example 20.13 will explain a dual slope ADC.

20.2 DIGITAL-TO-ANALOG CONVERSION (DAC OR D/A)

Figure 20.3 gives three DAC circuits for the cases, D, E and F, respectively. Consider Case D that a 9 V DC motor is to be controlled by giving the analog voltages as 0.0, 0.05, ..., 8.85, 8.90, 8.95, 9.0 V and currents 0 to 250 mA. There are 255 voltage and i_{out} (current) levels in all. Advantage of the digital input is that it can be from a computer or processing circuit. Instead of a motor, we can consider a speaker (Case E). Figure 20.3(a) shows two cases—D and E for a digitally controlled DC motor and s digitally controlled music system speaker.

Case D or E has an 8-bit representation with 0000 0000 binary for the minimum analog threshold current or voltage as the output and 1111 1111 binary for the maximum analog threshold current or voltage as the output. Assume that output current amplifier gives the DAC input is for 1111 1111. V^+_{ref} = 9 V and V^-_{ref} = 0 V.

Case F (Figure 20.2(c)) is that use a 4-bit representation with 0000 binary for the minimum analog threshold current as the output and 1111 binary for the maximum analog threshold current as the output. Assume that the output amplifier gives analog current output from the DAC = 750 mA for 1111 input and –750 nA for 0000 input. The DAC can be preset at V^+_{ref} = 5 V and V^-_{ref} = – 5 V.

DAC circuit is a circuit, which coverts the digital input consisting of *n*-bits into analog output. The inputs are shown by d_{n-1}, d_{n-2}, ..., d_1, d_0 in an *n*-bit DAC.

Table 20.2 shows the DAC circuit exemplary characteristics for cases D, E and F. Table first row gives number of DAC bits at the input. Table second and third rows give the exemplary DAC inputs and analog outputs. Table fourth and fifth rows give the lowest and highest limits of analog outputs, which are obtained from the given DAC. Table sixth row gives the DAC resolution.

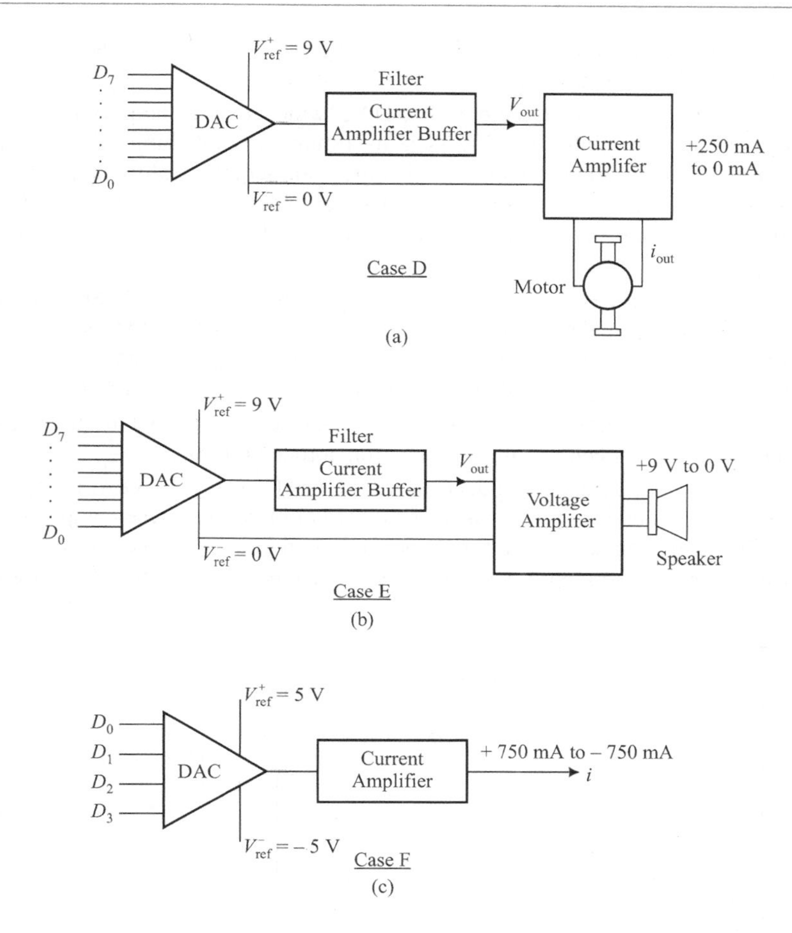

FIGURE 20.3 (a), (b) and (c) Three cases of digital input for the analog output to a motor, speaker and current-source using the DAC outputs.

TABLE 20.2 Digital to analog conversion and other characteristics for conversion in circuits of Figure 20.3(a)

Parameter	Case D	Case E	Case F
Number of DAC bits at the input	8	8	4
Exemplary Analog output from the DAC	1.96 V	500 mA	250 mA
Exemplary DAC input for an output	0110 0110 (102_d)	1010 1010 (170_d)	1011 (11_d)
Lowest Limit	0 mA	0 V	– 750 mA
Highest Limit	250 mA	9 V	+ 750 mA
Resolution *	0.98 mA/bit	35 mV/bit	100 mA/bit

* Resolution R_{dac} = (Highest analog output limit – Lowest analog output limit) / (2^n – 1), where n is the number of DAC bits at the input.

DAC Resolution R_{dac} = (Highest analog output limit – Lowest analog output limit)/ $(2^n - 1)$, where n is the number of DAC bits at the input.

Let V^+_{ref} is the analog reference to DAC, which gives an output equal to a proportionality constant β multiplied by V^+_{ref} as the maximum threshold limit, when all input bits are 1s. Let V^-_{ref} is the analog reference, which gives an output equal to the β multiplied by V^-_{ref}, which is the minimum threshold limit when all input bits are 0s. Each DAC has the lowest and highest limits of the analog outputs, which is defined by the reference voltage inputs and as per $\beta \times V^+_{ref}$ and $\beta \times V^-_{ref}$.

Most DACs have $V^-_{ref} = 0.00$V and V^+_{ref} between 2.5 V to 5V. Some DACs have $V^+_{ref}/2$ and analog ground pin and some V^+_{ref} and V^-_{ref} pins. A few DAC has V^-_{ref} settable to –ve values and few to –ve as well as positive value ($< V^+_{ref}$). V^+_{ref} and V^-_{ref} sets the values of lowest and highest measurable analog inputs as well as of DAC resolution.

Figure 20.4 shows the relationship between the input and out in a four-bit DAC. Sixteen dots show the sixteen possible outputs from a four-bit DAC.

The external analog outputs from a DAC can be to any source of voltages or currents or frequencies. For example, speaker can be feed through a frequency output through a voltage to frequency converter. The DAC converts the digital bits in the voltage or current form. Current and power amplifier setting is such that it gives at the DAC the currents, which pro-

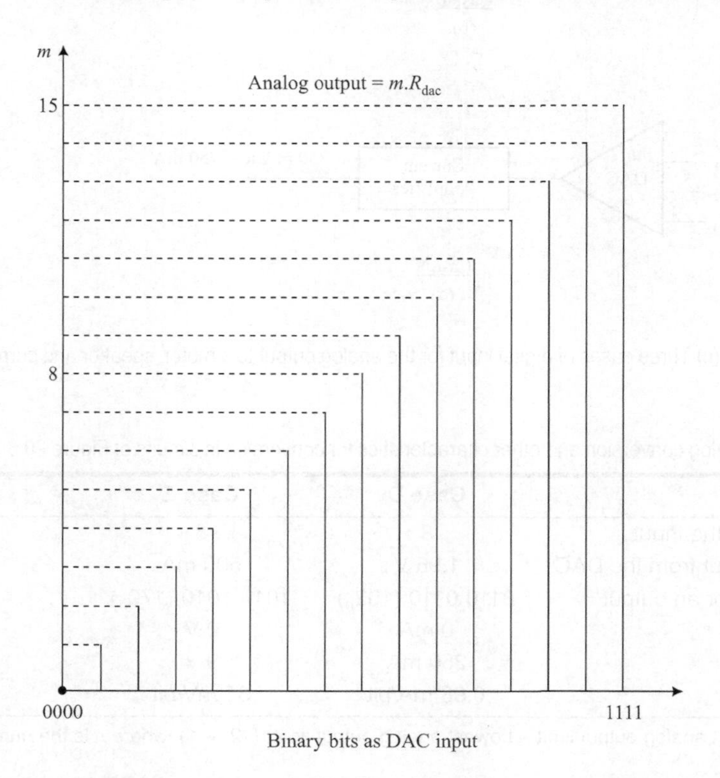

FIGURE 20.4 Relationship between the input and out in a four-bit single-reference DAC. Sixteen dots show the sixteen possible outputs from a four-bit DAC inputs.

vide outputs between the lowest and highest limits. The constant β is controlled by the amplifier design for feeding to the activator (motor, speaker, lamp).

20.2.1 Characteristics of *n*-bit DAC

Points to Remember

The following relationships are useful for understanding a DAC operation.

1. The DAC analog out resolution $R_{dac} = (V^{+}_{ref} - V^{-}_{ref})\beta / (2^n - 1)$
2. Range of output signal that is then obtained = $\beta.V^{+}_{ref}$ and $\beta.V^{-}_{ref}$. (This is why we need a suitable amplifier. This amplifier at the input is as an interface circuit. It also has a suitable filter. It has a scaling factor and/or offset. This sets the minimum threshold output = βV^{-}_{ref}. for all 0s at the DAC input. This also sets the maximum = βV^{-}_{ref} for all 1s at the DAC input).
3. If analog output of an *n* bit DAC is per the decimal *m*, input and *m* times the DAC resolution, when lowest limit is 0.00 V or 0.00 mA.
4. Input bits $d_{n-1}, d_{n-2} ..., d_1$ and d_0 give the analog output V^{-}_{out} by the equation:

$V_{out} = (d_{n-1} \times w_0^{n-1} + d_{n-2} \times w_0^{n-2} + d_{n-3} \times w_0^{n-3} + ... + d_2 \times w_0^2 + d_1 \times w_0^1 + d_0 \times w_0^0)R_{dac} = m\ R_{dac}$, where R_{dac} is DAC resolution part, where $w_0 = 2$ and $d_{n-1}, d_{n-2}, ..., d_1$ and d_0 are the input bits at the DAC when the output = V_{out}. ...(20.2)

Besides the resolution, bits *n* at the DAC input, R_{dac}, V^{+}_{ref} and V^{-}_{ref}, and β the following are other important points to remember:

1. A DAC conversion time is a time in which the expected digital input converts to the analog output result.
2. A DAC gain drift due to circuit temperature drift is defined in ppm/°C (part per million per °C).
3. A linearity error of DAC is defined as the maximum deviation from a straight-line relationship. The considered relationship is between the DAC input in decimal vs. the output voltage in Volt or in current. (Figure 20.4).
4. An accuracy of DAC output is defined by the DAC quantization error. A quantized level means a discrete level. Let us assume that we have a 10-bit DAC, which is providing the analog outputs. The peak voltage is say 2.0 V. The resolution is then equal to division of 2.0 V by 2^{10} = 1.95 mV/bit. We can then say that there are the 2^{10} = 1024 quantized levels. The levels are from 0 to 2000 mV. These are 0 mV, 1.95 mV, 3.90 mV, 5.85 mV, ..., 2000 mV. A quantization error is therefore within 0.5 times of resolution (± 0.5 DAC resolution). It will be ± 0.097 mV in the above example. The quantization error is the analog output range within which the converted output is the variable. It is due to the limited number of quantized levels between the minimum and maximum permitted analog measurable output.
5. A DAC settling time is the time in which the expected output voltage is within 0.5 times of resolution, i.e., within 1/2 lsb. Let us consider that the input to a DAC suddenly falls from 1023_d to 511_d. Let DAC is of high speed and is of resolution 5 mV/bit. Let in 8 μs, the expected output drop of 512 mV and the output settles between the (511 + 2.5) and (511 – 2.5) mV. The settling time is then 8 μs.

6. A DAC glitch is understood by the following example. Let us consider that the input to a DAC suddenly falls from 127_D to 63_D. Let resolution is 10 mV/bit. If DAC is of high speed and is of resolution 10 mV, then the output may first drop from 1.27 V to 0 V and then eventually settle at 630 mV. Appearance of false output signal is called a glitch.

There is a type of DAC called ladder network based DAC. Example 20.14 will explain *R*-2*R* based DAC.

20.3 ANALOG/DIGITAL SWITCH

Recall that an enhancement mode *n*-channel MOSFET shows the characteristics as follows: When voltage between gate and source, $V_{GS} = 0$, the current I_D between the drain and source is 0 and is independent of voltage between drain. Similarly, in the *p*-channel MOSFET when voltage between gate and source, V_{GS} threshold, the current I_D between the drain and source is 0 and is independent of voltage between drain. Therefore, a MOSFET analog switch can be designed to switch the currents *on* and *off* (Figure 20.5).

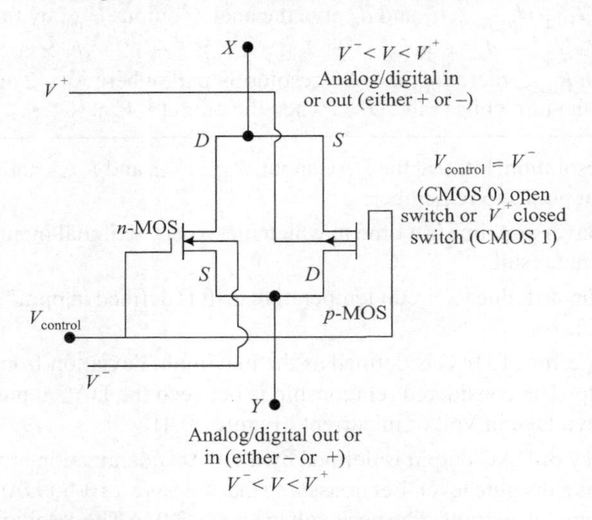

FIGURE 20.5 Bi-directional symmetrical analog switching circuit.

A CMOS analog bi-directional switch is made from a pair of *n*-channel and *p*-channel MOSFET. Figure 20.5 shows a circuit for it. It provides us two features (i) It gives a bi-directional current flow and its input and (ii) Its output is symmetrical. The basic design of Figure 20.5 has been improved in recent decades to protect the MOSFETs from the gate input form the over voltages due to a phenomenon known as SCR latch up.

The circuit of Figure 20.5 works as an analog switch. (i) When we connect the analog input at one end *X*, we shall get the same at the another end, *Y*, whenever the switch closes by applying a control voltage of about V^+ (CMOS logic state 1 voltage). (ii) No analog input appears at other end if the switch opens due to V^- (CMOS logic state is 0 voltage)

A digital circuit is also actually an analog circuit but is one where we consider only signals in terms of 1s and 0s, where the 1 and 0 represent two discrete states of voltages and currents. (Section 1.1).

Therefore, the circuit of Figure 20.5 also works like a digital switch. (i) When we connect either state 1 or 0 at one end X, we shall get state 1 or 0 another end, Y respectively, only, whenever the switch closes by applying a control voltage of about V^+. (ii) No logic output appears at other end if the switch opens by V^-.

IC chips 4016, 4051, 4052, 4053, 4066 have the circuit of Figure 20.5 as one of the basic element for analog switching and analog multiplexing.

20.4 ANALOG MULTIPLEXER

A set of N analog switches can be used to multiplex N analog channels like the channel between X and Y (Figure 20.5). Figure 20.6(a) shows how two analog switches can be used to multiplex the two analog inputs X_0 and $X1$ at an output Y. Multiplexing is a way of sending many, say, N signals at the inputs through a common channel at an output. Let A be an address bit or a channel select input. Figure 20.6(b) shows a four-channel analog multiplexer. Figure 20.6(c) shows how to connect the two enable (address) pins, A_0 and $A_{1,}$ to select four switches in Figure 20.6(b). A_0 and A_1 are called two address bits. These can also be called as two channels select bits (CS_0 and CS_1) or (CH_0 and CH_1).

20.5 ANALOG DEMULTIPLEXER

If X and Y path is a symmetrical and is unilateral, then the circuits of Figures 20.6(a) and (b) will also act as a two or four channel analog demultiplexer. Figure 20.7(a) shows an analog de-multiplexer. Figure 20.7(b) shows the circuit and table in the truth table like format for the four channel analog de-multiplexer. De-multiplexing is a way to direct one input signal to a selected channel among the many different output-signal channels.

20.6 INTERFACING LOGIC GATES WITH THE ANALOG CIRCUITS

When interfacing the following points one should remember:

1. Threshold input voltages are around 25% of the supply voltage in the TTL gates showing narrower noise margin at the logic input = 0 compared to the CMOS gates.
2. Never connect a higher voltage to input than the V_{supply}.

20.6.1 Interfacing Circuits

1. CMOS families of the gates show a sharp rise in the current from supply V at ≈ 50% intermediate V_{in} (Figure 20.8). From this, we conclude that CMOS gates should not be permitted to float in a circuit but should be tied to either V_{SS} or V_{DD} (i.e. V^+ or V^-) or to an output of a pervious stage. CMOS gates draw very high currents in its floating state compared to when there is the fixed 1 or 0 logic state at an input, while in TTL gates this do not happen. TTL gates inputs can float (can remain unconnected) as the current ratio are not rising at the intermediate V_{in} values for these gates, in contrast to the circuit made using CMOS gates.

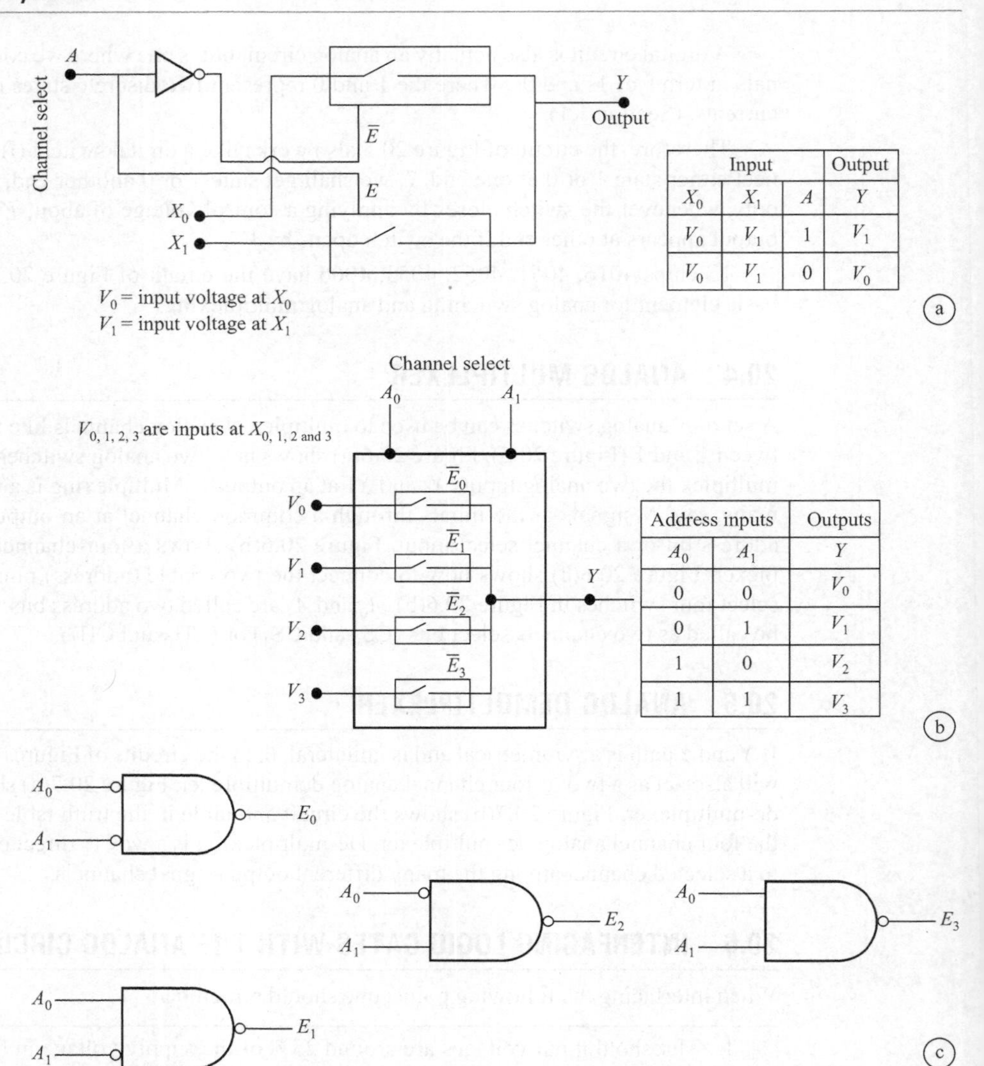

Input			Output
X_0	X_1	A	Y
V_0	V_1	1	V_1
V_0	V_1	0	V_0

Address inputs		Outputs
A_0	A_1	Y
0	0	V_0
0	1	V_1
1	0	V_2
1	1	V_3

FIGURE 20.6 (a) Two analog CMOS based switches used to multiplex the two analog inputs X_0 and X_1 at an output Y (b) Four-channel analog multiplexer (c) Connections of the two enable (address) pins, A_0 and A_1, to select four switches.

2. There is a sudden drop in currents in TTL families of the gates when the supply V is about ~1.25 volt (Figure 20.8). Both thresholds are not equal in a Schmitt Trigger circuit. A typical circuit exhibits two separate thresholds voltage values at 0.8 V and 1.6 V during V_{in} rising from 0 V to 5 V, and V_{in} falling from 0 V to 5 V in place of one threshold at about 1.25 volt here. A 0 to 1 transistor threshold is about twice the silicon *p-n* junction diode threshold and is about the band gap of silicon in electron volt.
3. There is a sudden change in the output voltage levels at ~ 1.25 V (Figure 20.8) for the TTL gates.

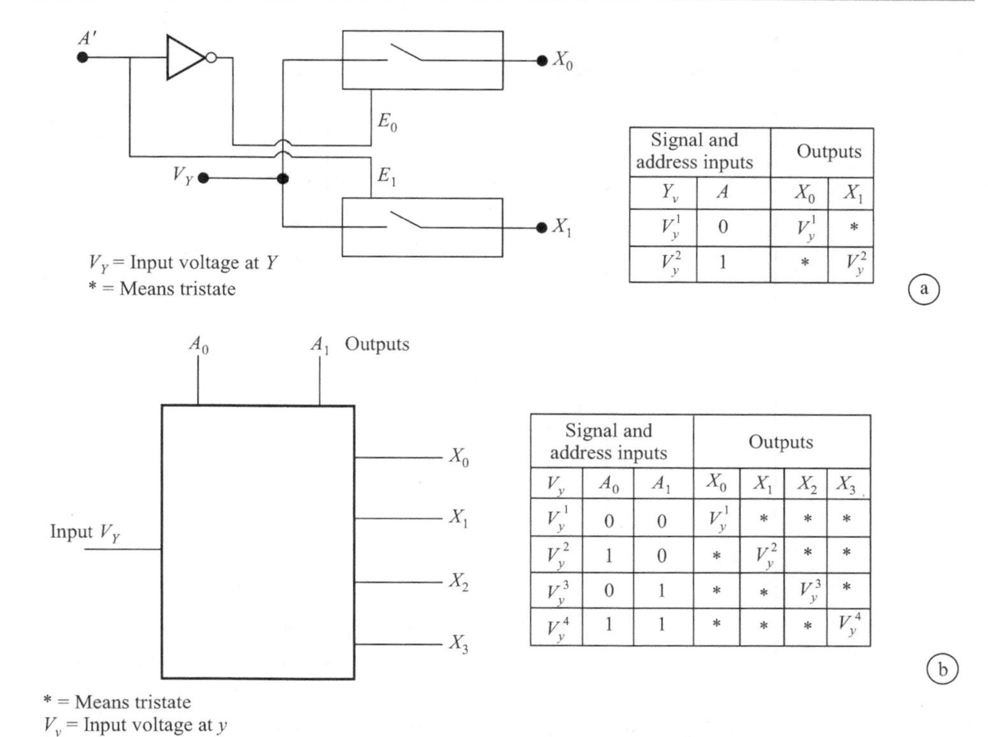

Signal and address inputs		Outputs	
Y_v	A	X_0	X_1
V_y^1	0	V_y^1	*
V_y^2	1	*	V_y^2

Signal and address inputs			Outputs			
V_y	A_0	A_1	X_0	X_1	X_2	X_3
V_y^1	0	0	V_y^1	*	*	*
V_y^2	1	0	*	V_y^2	*	*
V_y^3	0	1	*	*	V_y^3	*
V_y^4	1	1	*	*	*	V_y^4

FIGURE 20.7 (a) An analog de-multiplexer (b) The circuit and a table (in the truth table like format) for the four channel analog de-multiplexer.

4. CMOS gates have negligible current drains on the supply provided input is held steady at 0 or 1 i.e. at V_{SS} or at V_{DD} (Figure 20.8). In CMOS gates, the current ratio from supply is negligible as the currents are few mA only if the inputs are not floating but are kept at logic 0 or at logic 1. These circuit inputs can be given directly from V^- or to V^+ or to an output of a logic gate.
5. Currents at logic 1 inputs are much smaller compared to logic 0 inputs in the TTL families of the gates, while identical currents flows at logic 0 and 1 inputs for the CMOS family of the gates. Therefore, we should prefer a steady state at the output to be 1 in case of TTL families of the gates.
6. TTL gates drain the current, which is not negligible when the inputs are at logic 0s.

20.7 EXEMPLARY DIGITAL ANALOG INTERFACING CIRCUITS

Figures 20.9(a) to (d) show four interesting connections, which are permitted as per Table 7.3. Figure 20.9(a) shows how to connect a relay. Figure 20.9(b) shows how to connect an LED. Figure 20.9(c) shows how to connect a Darlington pair. Figure 20.9(d) shows a connection to a power transistor.

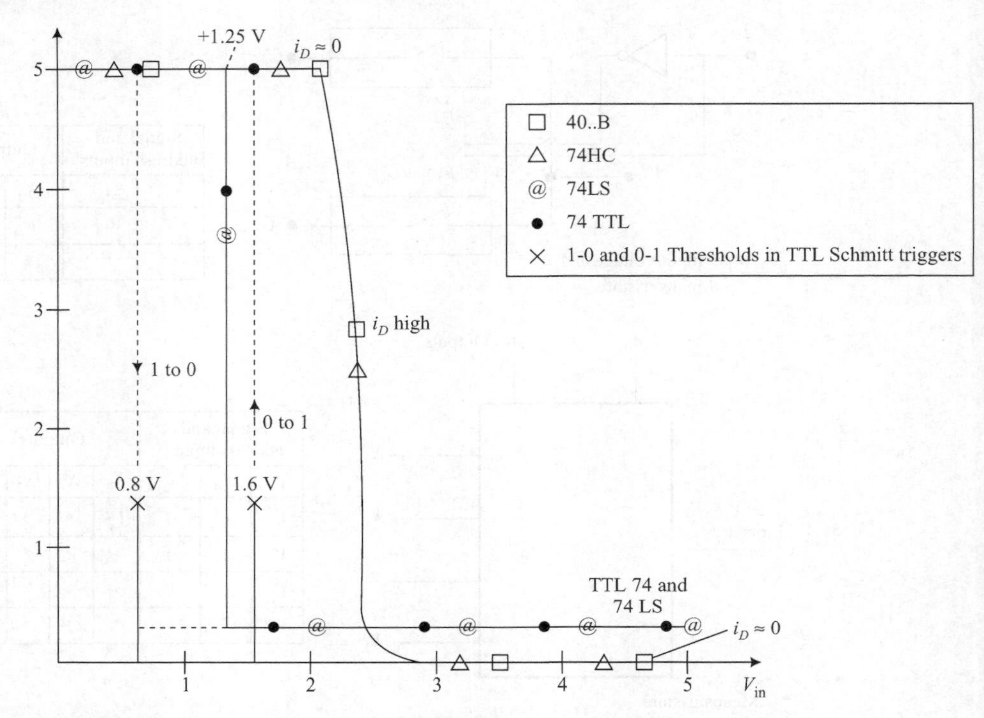

FIGURE 20.8 Voltage output as function of voltage input in CMOS and TTL gates.

EXAMPLES

Example 20.1

A doctor needs to measure the patient temperature within 37°C to 40°C with resolution of 1°C, how many bits ADC is needed.

Solution

Since only four readings are required, therefore output = 00, 01, 10 and 11 corresponding to 37, 38, 39 and 40°C patient temperature should suffice. A two bits ADC suffices.

Example 20.2

A furnace temperature is within 1000°C to 1600°C with resolution of 1°C, how many bits ADC is needed.

Solution

Since 601 readings are required between within 1000°C to 1600°C, therefore 10 bit ADC as needed, because $2^{10} = 1024$. Nine bit ADC will not suffice because $2^9 = 512$.

Example 20.3

If digital output of an 8 bit ADC is 1011 1111, find the ADC resolution and the input voltage at that instance when $V_{ref}/2 = 1$ V.

Solution

Since $V_{ref} = 2$ V, the $R_{adc} = V_{ref}/(2^8 - 1) = 2\text{ V}/ 255 = 7.8$ mV/bit

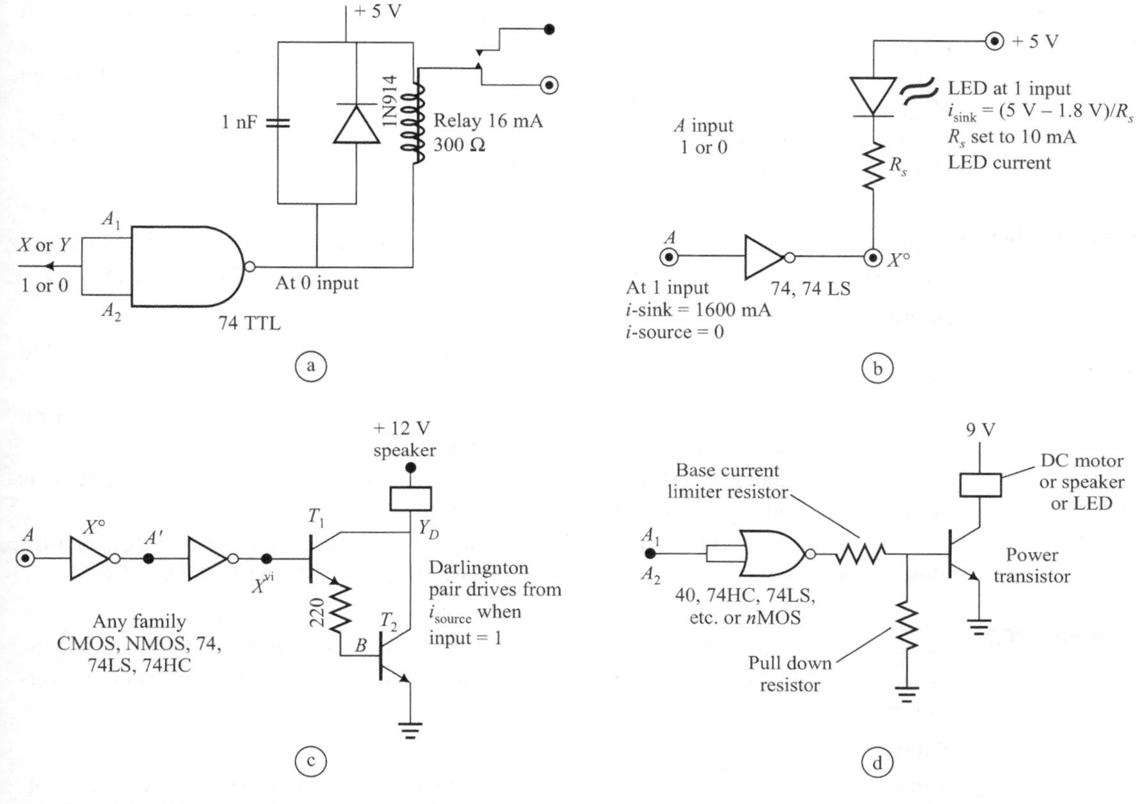

FIGURE 20.9 (a) Connection to a relay (b) Connection to an LED (c) Connection to a Darlington pair (d) Connection to a power transistor.

If digital output of an 8 bit ADC is 1011 1111 binary = decimal 191 then the input has to be 191 times the ADC resolution R_{adc}. Therefore, input voltage = 1.5 V.

Example 20.4

If digital output of a 12-bit ADC is 1000 0111 1111 then find the resolution and the input at that instance when $V_{ref}/2 = 0.75$ V

Solution

$V_{ref} = 1.5$ V. Therefore, $R = V_{ref}/(2^{12} - 1) = 1.5$ V/ 4095 = 0.37 mV/bit

If digital output of a 12-bit ADC is 1000 0111 1111 binary, which means 2175_d then the input is 2175 times the ADC resolution R_{adc}. Therefore, input voltage = 0.8 V.

Example 20.5

If an ADC of 8 –bits can measure inputs between 0 V and 1.023 V then find the resolution. If input voltage is 600 mV, what will be the ADC output?

Solution

Given $V^-_{ref} = 0$ V because the input can be minimum 0 V. Then for a digital 8-bit output to the ADC = 0000 0000, the input is then = 0 V.

Given V^+_{ref} = 1.023V because the input can be maximum 1.023 V. Then a digital 8 bit output of the ADC = 1111 1111 when the input is 1.023 V.

The ADC analog input resolution = $(V^+_{ref} - V^-_{ref}) / 2^8$ = 4 mV/bit and range is between V^+_{ref} and V^-_{ref} = between 1.023 V and 0 V.

Given input = 600 mV, therefore the ADC output decimal number m = Rounded-off closet integer of 600 mV/ 4 mV = 150. ADC output = 1001 0110.

Example 20.6

With a suitable amplifier that also has a filter and scaling factor, let the input to an 12-bit ADC be maximum 1.275 V and minimum 0 V then find the output when the input = 1 V.

Solution

Given V^-_{ref} = 0 V because the input can be minimum 0 V. Then for a digital 8-bit output to the ADC = 0000 0000, 0000 the input is then = 0 V.

Given V^+_{ref} = 1.275 V because the input can be maximum 1.275 V. Then a digital 8-bit output of the ADC = 1111 1111 1111 when the input is 1.275 V.

The ADC analog input resolution = $(V^+_{ref} - V^-_{ref}) / (2^{12} - 1)$ = 0.3 mV/bit and range is between V^+_{ref} and V^-_{ref} = between 1.275 V and 0 V.

Given input = 1000 mV, therefore the ADC output decimal number m = Rounded-off closet integer of 1000 mV/ 0.3 mV = 3333. ADC binary output = 1101 0000 0101.

Example 20.7

Let sample time for ADC input in is 0.25 μs. Let the conversion time of the circuit is 2 μs. What is the upper limit of frequency of input analog signal? Assume the continuous conversion mode.

Solution

Since ADC conversion time is 2 μs and the sample time = 0.25 μs, the input frequency cant' be more than (1/2.25 μs) = 444 kHz.

Example 20.8

An eight-bit DAC has (V^+_{ref}, V^-_{ref}) pins at 2, –2 V. The DAC circuit maximum output is 256 μA. What is the resolution?

Solution

The DAC analog output resolution $R_{dac} = \beta(V^+_{ref} - V^-_{ref}) / (2^n - 1)$ and range is between V^+_{ref} and V^-_{ref}.

β = output at the maximum/V^+_{ref} = 256 μA /2 V = 128 μA/V

R_{dac} = (128 μA/V)(2 V – (–2 V)) / $(2^n - 1)$ = 128 × 4/ 255 ≈ 2 μA/ bit.

Example 20.9

If digital input to an 8 bit DAC is 1011 1111 and resolution is 8 mV/bit, what will be the analog output?

Solution

If digital input to an n bit DAC is decimal m, the output is m times the DAC resolution. If digital input to an 8 bit DAC is 1011 1111 binary, i.e., decimal 191 then the output is 191 times the DAC resolution.

Therefore output = 191 × 8 mV= 1.528 V.

Example 20.10 Let V^-_{ref} = 0 V. Then for a digital 8 bit input to the DAC = 0000 0000, the output = 0 μA. With a suitable amplifier that also has a low pass filter and scaling factor, let the final output be 0 V (= analog ground). Let V^+_{ref} = 2.047 V. Then for a digital 8 bit input to the DAC = 1111 1111, the output = 100 μA. What is the DAC resolution?

Solution

$R_{dac} = \beta(V^+_{ref} - V^-_{ref})/(2^n - 1)$ = (Maximum threshold output – Minimum threshold output) / $(2^n - 1)$ = 100 μA/255 ≈ 0.39 μA/bit.

Example 20.11 Let V^-_{ref} = 0 V. Then for a digital 12 bit input to the DAC = 0000 0000, 0000 the output = 0 V. With a suitable amplifier that also has a low pass filter and scaling factor, let the final output be 1.0235 V. Let V^+_{ref} = 1.0235 V. Then for a digital 8 bit input to the DAC = 1111 1111 1111. What is the DAC resolution? If digital input is 0000 1011 1111, what will be analog output from the DAC?

Solution

Since final output and reference voltage is same β = 1,

$R_{dac} = \beta(V^+_{ref} - V^-_{ref}) / 2^n - 1)$ = (Maximum threshold output – Minimum threshold output) / 2^n = (1.0235 V – 0 V)/4095 = 1.0235/4095 V/bit = 0.25 mV/bit.

$V_{out} = \text{m}.R_{dac}$ = 191 × 0.25 mV = 47.75 mV.

Example 20.12 Give the circuit diagram and explain working of a successive approximation ADC employing a DAC circuit.

Solution

Figure 20.10(a) shows the successive approximating circuit. Principle of working is as follows. Analog input is compared with the DAC output to start with. Let it be 0. The digital inputs to it is from the counter = 00...00 at D_0 to D_{n-1} pins at the beginning. Now, since analog input is more than 0 V, the comparator keeps up on driving the counter. Counter successively increments till the analog input = DAC output. The DAC input at that instant is the ADC output also. Figure 20.10(b) shows the DAC output versus number of counts during successive built of the DAC output to match with the analog input.

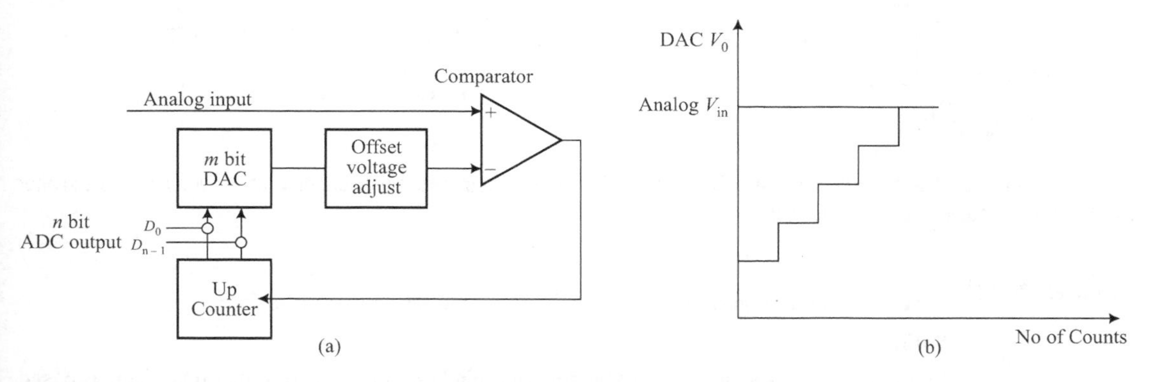

FIGURE 20.10 (a) Successive Approximation ADC (b) DAC output as counts increase.

Example 20.13 Give the circuit diagram and explain working of a dual slope ADC.

Solution

Figure 20.11(a) shows the dual slope circuit. Principle of working is as follows. Let n is the number of ADC bits, same that of a counter. Analog input is integrated with an integrator from a time when an analog switch closes at time $t = 0$. A comparator receives the output of the integrator as the input. It give output = 1 to an AND gate input. AND gate other inputs is to the clock pulses of a frequency. A counter counts from 0 to $2^n - 1$ counts of the pulses, when there is n bit counter. Now the counter last bit is given to a *T-FF*. On counter nth-bit becoming equal to 1, the *T-FF* toggles. If clock pulse period is T, then the counter lets this *T-FF* toggle. On toggling now the switch is cutoff from the integrator and an integrator is given the opposite polarity voltage through another switch, which now closes. The integrator circuit discharges, till the counter again reaches the output to re-toggle the *T-FF*. Integrator discharges till the integrator output equals the voltage reference input to ADC. The counting will stop at that point when comparator output = 0 to stop the counter. The output bits of the counter represent the converted ADC output bits. The integrator has a dual slope output. Figure 20.11(b) shows the waveform.

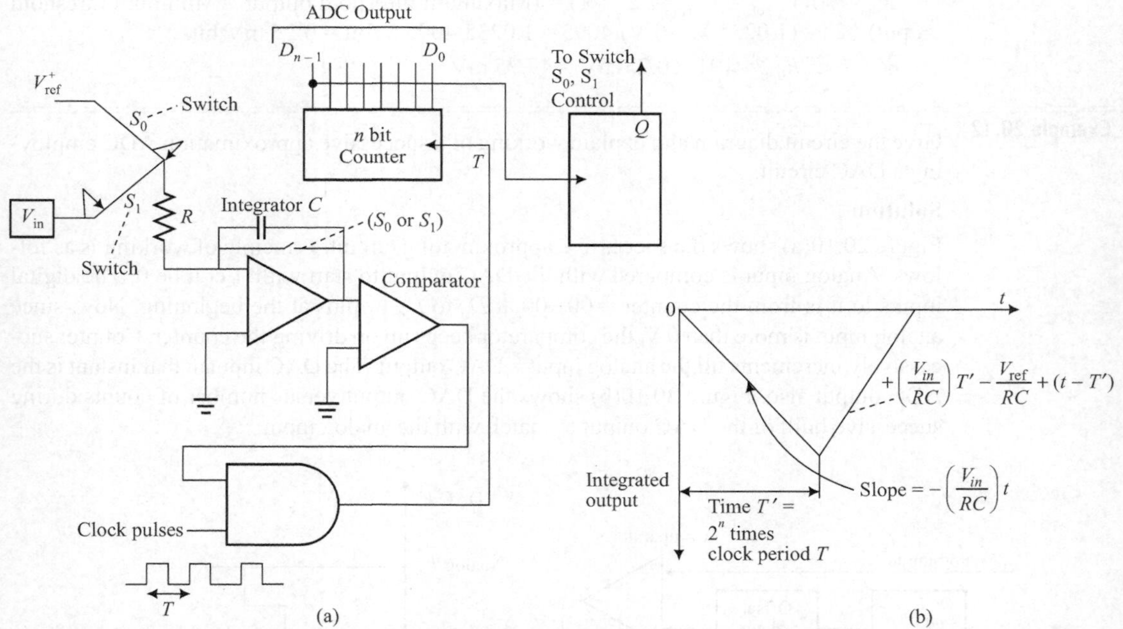

FIGURE 20.11 (a) Dual slope circuit for an ADC circuit (b) Integrator output versus time during which counter counts and after which the *T-FF* toggles to repeat the cycle.

Example 20.14 Give the circuit diagram and explain the *R*-2*R* ladder network of a DAC.

Solution

Figure 20.12 shows the *R*-2*R* ladder network of a DAC circuit. Principle of working is based on the operation amplifier as a summing amplifier. The lower bit of DAC faces higher resis-

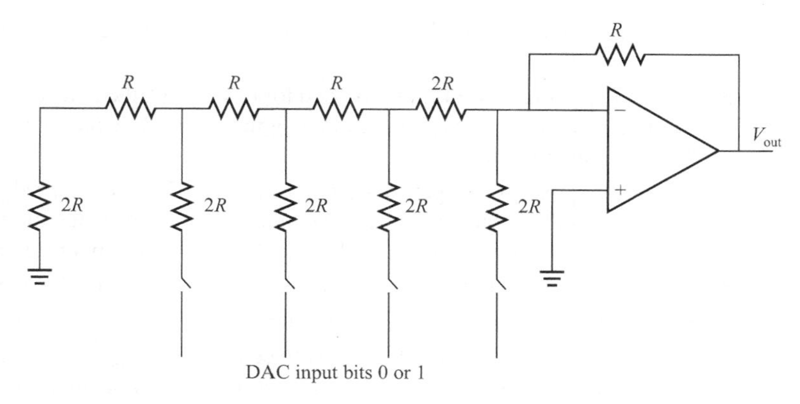

FIGURE 20.12 DAC using *R*-2*R* ladder network.

tance at the input. Therefore, its contribution is one half of the next stage bit. Using summing amplifier formula, the DAC

$$V_{out} = -\{(R_p)/R\}.[d_0.\{1/(2^{n-1})\}R_{dac} + d_1.\{1/(2^{n-2})\}(R_{dac}) + ... + d_{n-1}.\{1/(2^1)\}(R_{dac})^{n-1}].$$

R_{dac} is DAC resolution, voltage output per bit increment at the DAC input. The successive sum terms is double that of the preceding term. This is because, the resistance in the path of $(n-1)$ bit input $= R$, for $(n-2)^{th}$ bit, it is $2R$, and so on. It is 2^{n-2} time R in the path of lowest bit input.

EXERCISES

1. Let V^-_{ref} input = 0.000 V and a 12 bit ADC generates output 1111 1111 1111 when V^+_{ref} input = 1.5 V. What is the range of analog inputs? If same system generates output 0000 1111 1111, what is the range of analog inputs now?
2. Let V^-_{ref} input = 0.256 V and V^+_{ref} input = 1.280 V. An 8 bit ADC generates output 0101 1110. What is the analog input?
3. Let V^-_{ref} input 0.0 V and V^+_{ref} input = 2.5 V. Assume that maximum analog output is 5 V. An 8 bit DAC has the input 0111 1111. What are the analog output and its resolution?
4. Let V^-_{ref} input 0.512 V and V^+_{ref} input = 2.047 V and the maximum threshold and minimum threshold outputs are 100 μA and 25 μA. A 12 bit DAC input is 1111 0000 1111. What are the range of analog output and its resolution now?

QUESTIONS

1. Define the relationship between the analog input V_{in} and ADC output of n-bits.
2. What are the V^+_{ref} and V^-_{ref} inputs in an ADC? Most ADCs have only V^+_{ref} and analog ground inputs? Describe the advantage of providing V^+_{ref} and V^-_{ref} inputs? (Hint: Variable upper limit and lower limits of the input analog signals and hence.

variable resolution and also –ve inputs can be converted in case of dual polarity ADC).

3. In case V^-_{ref} input < 0 V is not provided for in an ADC, then the method that you will adopt to measure the –ve analog input signal. (Hint: Use an inverting OPAMP before the analog input).
4. In case V^+_{ref} input is not permissible to be above +2.5 V compared to V^-_{ref} then describe the method that we shall adopt to measure the +ve analog input signal of 12 V. (Hint: Use a non-inverting OPAMP that attenuates the signal by a known factor).
5. Define the relationship between the analog output V_{out} and DAC input of n-bits.
6. Suppose the V_{CC} supply to a TTL circuit is cutoff, is output logic state 0?
7. Assume that the potential difference between V_{DD} and V_{SS} is zero in a CMOS circuit. A logic probe or a voltmeter now measures an output of a circuit. It shows the LED off or no voltage between the output and the ground or V_{SS}. Is the output in logic state 0?
8. Describe an analog switch? Where are the analog switches used? How is an analog switch made from a CMOS pair?
9. A CMOS analog switch dissipates less power than a bipolar junction transistor based switch. It also shows no D.C. offset voltage. Inputs and outputs can be either up to V_{DD} or down to V_{SS}. When a channel, between the drain and source, is conducting, the resistance is very low less than 100 Ohm. Why are these properties helpful in making a switch? What will be disadvantages if a BJT is used to make an analog switch in place of MOSFETs?
10. CMOS switches without special protection circuits and latch proofing can be destroyed either due to noise or due to electrostatic charges or when the gate voltages exceed $V_{DD.}$ Why?
11. What do you mean by cross talk between two channels? (Hint: Noise in one effecting the working of a nearly channel due to capacitive effects).
12. Describe an analog multiplexer? Describe an analog demultiplexer?
13. Can you connect a LED to a standard TTL gate directly? (Hint: Open collector gates can be).
14. What can be the current rating of the D.C. motor of 9 V in circuit of Figure 20.9(d)?
15. What does the advantage a power MOSFET offers in place of a bipolar junction transistor (BJT) used for driving external loads from the TTL and CMOS families?
16. A relay of 500 Ω, 5 V can be driven by 74 standard TTL gate output. Why can't be by the 74LS or 74HC or 40 ... B family of the gates?
17. What are the changes in the circuit you should do for during the LED when using the CMOS gates?
18. Where are the circuits used which accept in their input the output from the logic gates and in their outputs drive analog loads?
19. Can you play a musical instrument using a logic gate output? If yes, how? If no, why not?
20. Can you drive a loud speaker from a logic gate output?
21. Why do you prefer 5 V supply when using the (MOS families of the gates)?

22. For power supplies of the 74, 7LS 40B TTL/CMOS gates, you wish its output impedance remains below 100 mΩ even up to MHz. Why?
23. Explain the working of a dual slope 8-bit ADC.
24. Describe the successive approximation ADC and show how a DAC is used to design an ADC.

CHAPTER 21

CPLDs and FPGAs

OBJECTIVE

We will learn in this chapter about CPLDs and FPGAs. These are advanced high-density field-programmable devices.

21.1 CPLDS

The PROM, PLA and PAL (PLDs) are programmable logic devices for combinational circuits. Following are the needs observed when designing the various application circuits:

1. The output from the OR gate(s) is sometimes needed in complement form.
2. Sometimes an output is needed through a tristate or tristate NOT form also. [Figure 21.1(a)]. Tristate output (s) enables the output availability on a common bus from several sources.
3. Sometimes an output with or without complementing (or both) is needed to be feedback to an input of the same stage to a previous stage or next stage. [Refer lower part of the circuit in Figure 21.1(a)].

PALs for these applications as per need are available in the various versions. CPLDs are the complex programmable logic devices of high logic gate-densities, that has many blocks made of PLAs, PALs or GALs (Sections 21.2 and 21.3). A block is a unit which repeats in rows or columns or in a hierarchial structure.

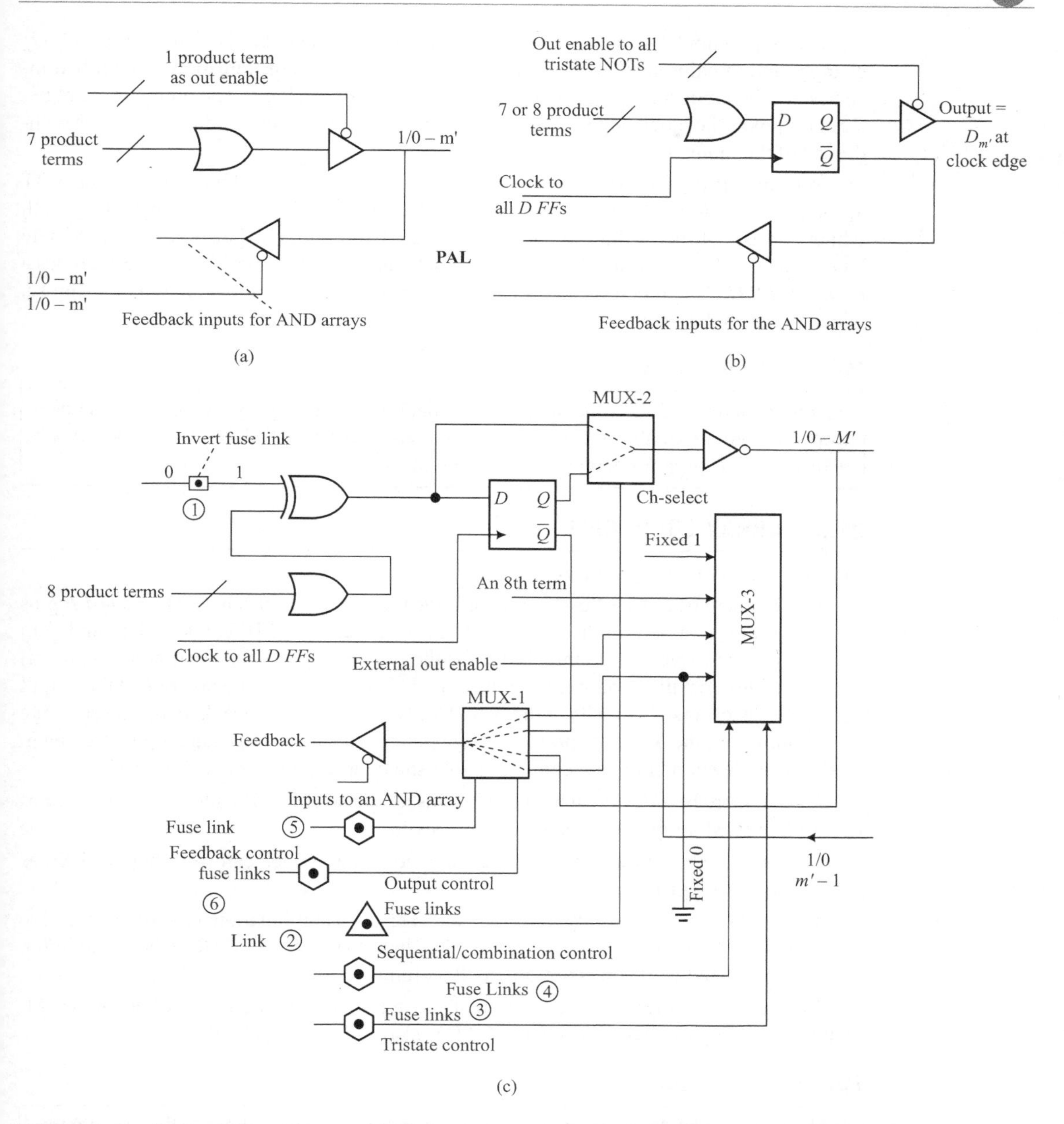

FIGURE 21.1 (a) Output stage cell of a PAL (b) Output stage cell of a registered PAL (c) Macrocell of a GAL.

21.2 REGISTERED PAL

Many a times, the output bit (s) needs to be stable like in a memory cell and is according to the inputs at the instance of a clock edge or clock pulse (master-slave) transition. For example, refer Figure 16.2(a) sequential logic circuit for a PIPO register. In that circuit an X_i

registers at Q_i output. Other example is a counter, on the count pulse the output $Q_D Q_C Q_B Q_A$ changes and remains stable till next clock edge. The process of transition at a defined instance is called registering the input (s) as per the in-between logic. The output is the registered output at the clock edge. [Registering means remembering the state even after withdrawal of the inputs.]

A form of PAL is *registered PAL*, for example, an IC 16R8. In addition to a tristate NOT at each of OR output [Figure 21.1(a) upper part] as in 16L8, there exists a output stage cell, which consists of the *D*-flip flop (*FF*) [Figure 21.1(b) upper part] also before the tristate NOT and each *D-FF* has an edge triggered clock input pin, then the PAL device becomes a ***registered PAL***. The input(s) can now be withdrawn after the application of a clock pulse, which registers the programmed output from the given set of inputs.

Point to Remember

> An unregistered PAL acts like a combinational circuit while a registered PAL also performs as a programmable general-purpose sequential circuit. (PAL has two planes–AND plane, which is programmable and fixed OR-plane.)

21.3 ARRAY LOGIC CELL

Consider the following questions:

1. Can there be a more flexible output stage then simply a tristate NOT or a *D-FF* plus tristate NOT like in the 16 L8 and 16R8, respectively? [Figures 21.1(a) and (b).] Flexible means programmability for the output stage either as (i) active 1, or (ii) active 0, or (iii) standby state either 0 or 1 or tristate or tristate not, or (iv) the output feedback as such or with complement, or (v) the output feedback to the present stage input (iv) the output of previous stage given as a succeeding stage input as given in a ripple counter or ring counter or Johnson counter. (Figures 16.7 to 16.10)
2. Can there be a device, which can be erased like an E^2PROM and so that it becomes E^2PROM array that is reconfigurable again?
3. Can there be a programmable logic device (PLD), which can work both as PAL as well as a registered PAL?
4. Can there be a flexibility to convert a PLD from a combinational circuit device, like PAL 16L8 to a sequential circuit device, like PAL 16R8 (or 21R8 where maximum inputs can be 20 corresponding to 40 columns)?

Lattice (a company) gave the solution of above questions in what is now known as GAL (generic array logic). Output stage of a GAL is also called a macrocell.

Points to Remember

> GAL is a PAL like device with an E^2PROM like erasing ability along with the programmable output stage and feedback stage for designing a combinational or sequential circuit.

Figure 21.1(c) shows, what a macrocell is. A popular GAL version is GAL20V8. S.G. Thomson, Lattice and other leading companies manufacture the GALs. Example 21.1 will explain the use of fuse links to control output and input stages. Due to use of the CMOS gates and flip-flops, the these PLDs consumes less power and current is $\approx$ 45 mA with access time for obtaining an output from a change at the inputs is $<$ 75 ns. These PLDs are called quarter power GALs. These new CMOS based GALs have made PALs and registered PALs less practical because of its excellent low power dissipation feature. Advanced Micro Devices made a 0.01 mA standby power dissipation version of GAL. A CPLD may contain approximately 5000-12000 gates and 50 GALs/PALs.

Example 21.2 give an example of a CMOS PLD. Table 21.1 gives the features of PAL, registered PAL, GAL and CMOS-array logic. PLDs. In the table, the abbreviations are as under:

n'_i = number of internal inputs to each OR from the AND arrays.

m_o = number of non-tristatable dedicated outputs.

m'_o = number of tristatable dedicated output.

n_i = number of external dedicated inputs to an AND array.

max. m_o = maximum number of outputs.

max. n_i = maximum number of inputs.

n_Q = number of Q outputs which are programmable as feedback inputs to an AND array.

TABLE 21.1 PLAs and array-logics

Type	n'_i, m'_o, n_i and n_Q	No. of ORs	Output stage of each OR	Max. m_o	Max. n_i	Designable circuits
16L8	7, 2, 10 and 6 from the tristate NOT	8	A tristate NOT	8	16	Combinational (C)
16R8	8, 8, 8 and 8 from $\overline{Q}$ of each *D-FF*	8	Synchronus clocked. An edge trigger *D-FF* before a tristate NOT	8	16	Sequintial (S)
20V8 (E^2PROM based)	8, - 12@ and 8*	8	Macrocell[1]	12*	20	Both C and S
85C224 (EPROM based)	8, -, 14! and 8	8	Macrocell[1]	8	22	Both C and S

* 8 of these through the macrocell. @ means a global clock and a global out-enable are in addition. ! means including clock and out-enable (non-global) – means not available.

Points to Remember

A PAL implements a combinational circuit of the multiple inputs, n, with multiple outputs, m, total $n + m$, say, 20. It essentially fuses the sum of products (SOPs) functions.

A GAL implements a combinational as well a sequential circuit of $(n + m)$ inputs and outputs. The GAL has a macrocell at each output stage of the SOPs. The macrocell makes it possible to select one or more options as follows:

(i) a feedback to input as addition input from a present stage or a neighbouring stage,
(ii) a complementary output,
(iii) a tristate output, and
(iv) a registered output.

A GAL provides a flexible output stage for each input stage implementing a SOP function. It does not provide for multiple inputs at its macrocell and the multiple outputs from its macrocell. Total number of input and output pins are 16, 20 or 22. Further, there is only one common clock input to all the output stage flip-flops.

21.4 FIELD PROGRAMMABLE GATE ARRAYS (FPGAS)

In a field programmable gate array (FPGA), there are no distinct input and output stages as in the AND-OR array and macrocell, of a PAL and GAL, respectively. Consider an exemplary FPGA with logic cell (for example, Figure 21.2 right side cell) with all the logic cells arranged in the array form (Figure 21.2 left side for an arrangement), 12×8 or 24×32 matrix (array) are examples of the array structures. A logic cell is also called CLB (configurable logic block). A 12×8 FPGA possess total 96 logic cells. A 24×32 FPGA possess 768 logic cells.

A typical logic cell consists of the followings:

(i) The gates to implement SOP function,

(ii) A *D-FF* with preset and clear, and

(iii) Data path selector multiplexer (MUX) at the input.

Xilinx has done pioneering work. There are 100 to 750 cells in a single EPLD IC for implementing FPGA circuit with features as follows:

(i) The number of IO (input and output) pins could be 50 to 200,

(ii) Number of clock inputs can be 8, and

(iii) Number of gates could be from 1000 to 10,000.

[Note: A latest FPGA XC2VP125 has 125136 logic cells.]

Point to Remember

The logic cell or a CLB can be thought of a combination circuit input stage for SOPs plus a macrocell designed in a much more flexible way with (i) Feedback possibilities not only from a neighbouring IO stage but from other stages as well, and (ii) provision of multiple outputs from a macrocell.

A typical logic cell shown at Figure 21.2 right side has (i) two number six inputs AND gates, four number two input AND gates, three number 2 to 1 MUXs and a clock input edge triggered *D FF* with the preset and clear inputs, and (ii) three number outputs from the ANDs, one output from the MUX and one output from the *D FF*. Per logic cell, the total number of inputs are 21 and outputs are 5.We can not only obtain SOP functions on the inputs but also the multiplexing and decoding functions at the inputs at the logic cell.

The above described logic cell design, with each cell arranged like an element in a matrix inside an FPGA, facilitates implementation of a multiple bits (16 or 32) complex combinational or sequential circuit. The examples are an adder or other arithmetic unit, a state machine, a shift register or an auto reloadable counter. The *D FF* in a cell of the FPGA is configurable as *JK FF*, *RS FF* or *T FF*, which facilitates an implementation of the 16 or 32 bit circuit for the various types of the shift registers and counters. Example 21.6 will give

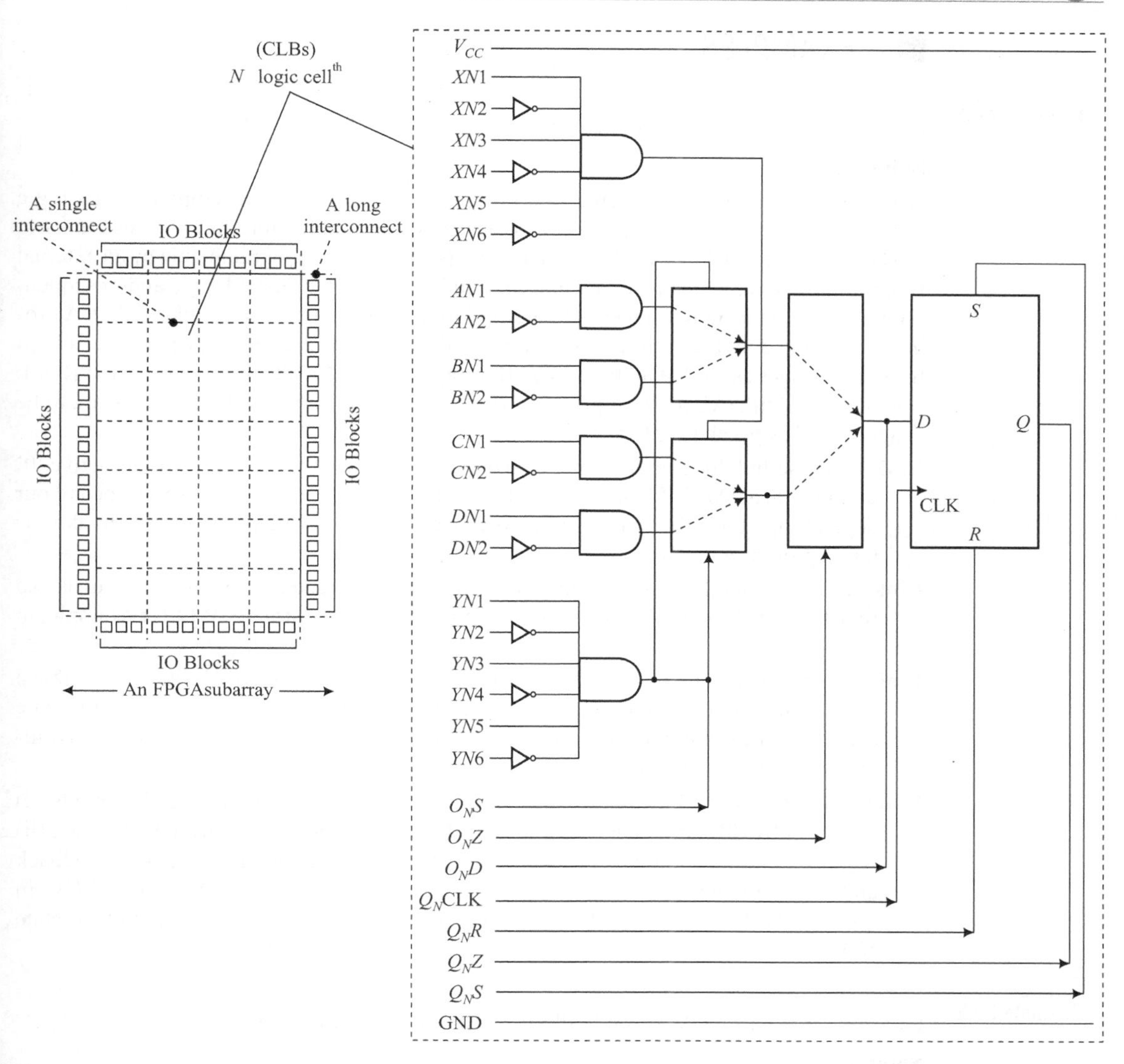

FIGURE 21.2 FPGA (Logic cell (right side) and the array structure of the logic cells (left side) (Dotted lines are single interconnects with programmable switches) (Dashed lines are long interconnects) logic cells from a CLB (Configurable logic block).

the forms of programmable links. The number of logic cells can be very large. For example, 125 to 136 in Xi link Xc 2 VP 125. It has four power PC processors also.

An FPGA finds the exemplary applications; data acquisition logic, plant automatic operation controller, graphic or image or voice processor, mouse interface, disk cache controller or parallel processor controller or encryption device, decryption device, pattern recognizer, DNA sequence storage, signature recorder. An FPGA can have 1000-50000 logic gates (NANDs or NORs).

EXAMPLES

Example 21.1 Give an example of fuse-links in an array logic for a CPLD.

Solution

A macrocell is at an output stage of a GAL. A macrocell in the exemplary array logic GAL20V8 has the additional fuse/links. These are programmable like the fuse/links in AND-OR arrays or PAL or PLA. The macrocell(s) of GAL are programmable by an external programming unit, which fuses an undesired link to obtain the desired logic and/or sequential functions at the output stage(s). Example 21.6 describe interconnect link-method A programming unit does the programming of the all the fuse/links present in the macrocells at all the output stages as well as at the AND- OR arrays. The unit is like the one for an E^2PROM. The macrocell in GAL20V8 (Figure 21.1(c)) has the following fuse/links, which are to be programmed as per the application circuit:

Link 1: A fuse/link for complementing an output. It controls the availability of an active '0' output as in a PAL 16L8 or the availability of active '1' output. The observed behaviour depends upon whether this fuse/link is linked or fused during programming by the external programming unit.

Link 2: There is another fuse/link to control the output as a sequential circuit output as in a registered PAL or as a combination circuit output as in a simple PAL, for example, in 16L8.

Links 3 and 4: There are two other fuse/links to decide one of the possibilities: (i) enable a permanent tristate at an output; (ii) no tristate at an output; (iii) the tristate NOT gate enable from an external out enable pin; (iv) tristate enabled by an 8th product term (output of the 8th AND array).

Links 5 and 6: There are two other fuse links to select one of the four possible outputs, (i) permanently '0' state and its complement '1' as the two feedbacks to an AND array; (ii) feedback and its complement feedback from the another output stage; (iii) feedback and its complement feedback from the *Q* output (registered output) of the *D-FF* as in 16R-8 or 20R8; (iv) feedback and its complement feedback from an output of a tristate NOT.

Example 21.2 Give a CMOS based UV-erasable PLD of fast 0.01 ns access time.

Solution

Intel UV-erasable CMOS PLD 85C224 is from 1990. Refer row 4 of Table 21.1. It has the two speed modes. One mode is a faster speed mode. It has 0.01ms access times (which is fast compared to 0.015ms in the GALs). In fast speed mode, it consumes 35 mA when in operation at 15 MHz inputs frequency, and 25 μA in its standby mode. In its slower speed mode, the access time is 0.02 μs. Its also has an automatic standby mode, which helps in obtaining a greatly reduced power dissipation when programmed like that.

One of the two speed modes is programmable by a fuse/link. This Intel PLD has another fuse/link to program the one of two possibilities: (i) I/O pin as an input for feedback and its complement feedback. (ii) *Q* output of *D-FF* as input for a feedback and its complement feedback in place of the two fuse links programming for four possibilities in a GAL macrocell shown in Figure 21.1(c). This Intel 85C224 PLD has a permanent additional external OE (output enable) dedicated line. It has all the 8 product terms connected to an OR input but not connected to another tristate control multiplexer (MUX) as in Figure 21.1(c). It uses EPROM cell which is at 1 when UV light erased (nonconducting) and at 0 programming the link between interconnects (Example 21.6). In other words, MUX-3 of Figure 21.1(c) is not present. MUX-1 has two channels only in place of four, and MUX-2 is available as per Figure 21.1(c). This PLD also emulates number of PLDs – for examples, 20V8, 20L8, 20L8, 20R8, 20R6, 20R4, 20P8, 20RP8, 20RP6 and 20RP4.

Example 21.3

Why do we need additional SOP terms when programming an array logic cell ?

Solution

While programming a GAL, one uses additional SOP terms for removing static hazards (Section 19.3). The static glitches occur due different propagation periods in different AND arrays. A dynamic glitch at an output can also be there. [Refer static and dynamic hazard described in Chapter 19.] This means that the output makes several transitions before a correct output is observed. The source of dynamic glitch is internal race problem, which should be eliminated by a careful planning by a user of the GAL.

Example 21.4

Give a CLB (configurable logic locu structure).

Solution

A CLB has look-up-table LUT and programmable *FF*-pair.

Example 21.5

What can be the possible structures of the CLBs?

Solution

The following can be possible structures of the CLBs (Figure 21.2).

(i) Matrix with single length interconnects and long interconnect.

(ii) Row-based Architecture with single length interconnects shown in Figure 21.2.

(iii) Hierarchial structures.

Example 21.6

What are possible ways of implementing programming links at the single length interconnects and long interconnects in the arrays?

Solution

Figure 21.3 shows the programmable switch designs using an SRAM cell (Figure 17.1(a)) or an E^2P ROM or EPROM transistor (Figure 17.1(b)) or an antifuse structure. The SRAM provides volatile links. The SRAM may connect a PROM (EPROM or E^2PROM) in the FPGA chip to make it non-volatile.

Example 21.7

How do we to use LUTs in the CLB or logic cell designs?

Solution

An LUT (look-up table) means there are an inputs (for a key or code or address) and for each input there is an one output. Suppose a look up table has eight rows. In each row, it can have output 1 or 0.

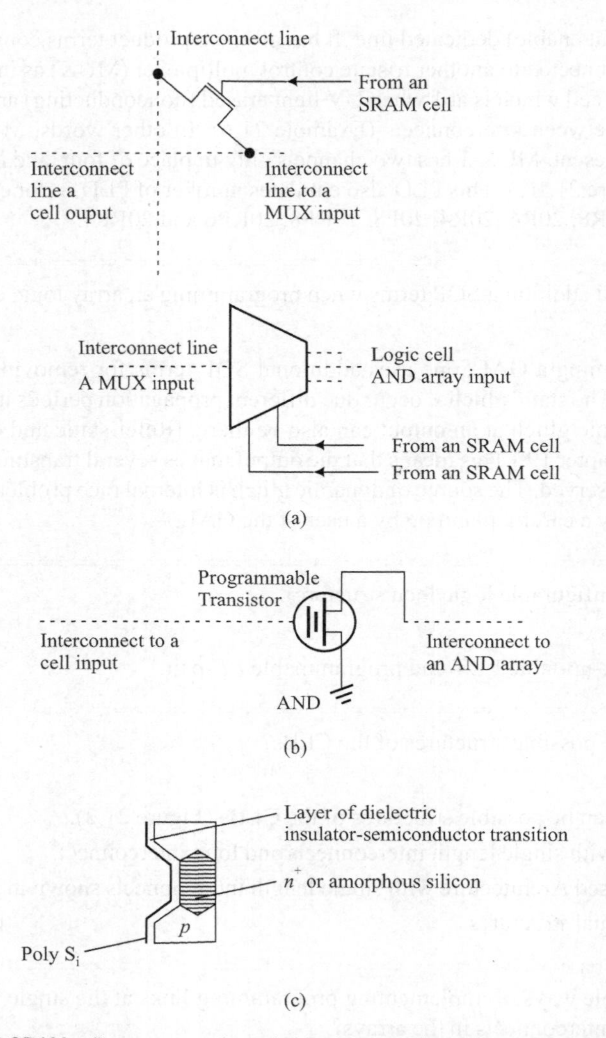

FIGURE 21.3 (a) SRAM cells to program the switching between two ends of an interconnect-line (b) E2PROM or EPROM cell (a transistor) to program the interconnect switch (on or off) interconnecting the input with AND array (c) An antifuse structure as an interconnect switch (An Actel Inc. technology) by insulator to semi conducting transition at a layer during programming.

Inputs	Output
000	0 or 1
001	0 or 1
010	0 or 1
.	.
.	.
.	.
111	0 or 1

An LUT in an FPGA logic cell is $2^m \times 1$ bit memory unit $m = 3$ for m-imput LUT. The LUT output is given to a MUX or MUXs at a cell (CLB) using the interconnect switch(es). An LUT output can also be an input to another LUT. An LUT may get input(s) from a selector bit(s) or one (or more) LUT output(s). The interconnect switches are programmed to connect the cell inputs through the LUTs or directly to the MUXs and other gates of the cell.

Example 21.8

How do the IO blocks connect in the FPGA using the interconnects the CLBs and IO blocks?

Solution

The IO blocks (Figure 21.2) connect the other IO blocks by long interconnects and connects the CLBs through single interconnects and also by long interconnects.

Example 21.9

Can we have on a single chip the CLBs as well as processors?

Solution

Yes, for example Micro Blaze for Xilinx has a 32-bit RISC soft processor core, memory and logic CLBs and peripherals on a single chip. XC2 VP125 has over 125 thousand logic cells plus four processors.

Example 21.10

List the characteristics that describe a CPLD or FPGA?

Solution

CPLD

(i) Constitution of the CPLB. Whether using PLA or PAL or registered PAL or array logic block.
(ii) Number of logic cells or macrocells used
(iii) Technology for programming the interconnects between inputs and output stages and between the CLBS.
(iv) Long or single or both interconnects programmable.
(v) Structure in the array row, matrix or heirarchial structure.
(vi) Delay time of interconnects formed by
(vii) Clock frequency limit-fused links.

FPGA

(i) Logic design of CLBs
(ii) Use of LUTs
(iii) Use of single interconnects
(iv) Use of long interconnects
(v) Use of SRAM interconnects.
(vi) Use of EPROM or E^2PROM cells with the SRAM interconnects.
(vii) Number of logic cells
(viii) Number of IO blocks horizontal and vertical
(ix) Delay between interconnects
(x) Clock frequency limit

Example 21.11 A Xilinx XC 4000 CLB has 2 *D-FF*s each with the *R* and *S* inputs (preset and set) and an enable input. How many CLBs shall be needed to design a (i) 32-bit PIPO and (ii) 32-bit ripple counter.

Solution

Sixteen CLBs the 32-bit-PIPO and counter will be needed.

EXCERCISES

1. Draw Xilinx XC 4000 logic block (CLB design).
2. Draw flash logic device from Altera which uses a 24 V 10 PAL.
3. Describe Xlinix XC·7000 and XC 9500 CPLDs.
4. Describe XC 4000 FPGA. CLB. Compare the XC 4000 and XC 801.00 features.
5. List the links to be programmed for a ripple counter of 4-bit using CLBs of XC 4000.
6. Show a design each of 32-bit ripple counter, 32-bit synchromous counter, 32-bit × 32 bit multiplier and a 8 × 8 bit multiply and then all 16-bit unit (colled MAC unit). Use XC 4000 FGGA CLBs.

QUESTIONS

1. Define PAL? Define registered PAL? Define GAL?
2. Define a Macrocell? How will you employ a GAL to implement a complicated combination circuit as well as a complicated sequential circuit?
3. Define an FPGA? What do you mean by a logic cell in a FPGA?
4. When do we use a FPGA, when a GAL, when a registered PAL, when a PAL, when an EPROM, and when an individual logic gates and *FF*s based logic design? Explain thoroughly.
5. Now FPGA have over 100000 logic cells and dissipates very low power. GALs and FPGAs have made many digital ICs a thing of past. How and why is it so?
6. Search the Web and find Xilinx Spartan 3 and XC2VP125 features.

Index